ARISTOTLE
XIII

LCL 366

ARISTOTLE

GENERATION OF ANIMALS

WITH AN ENGLISH TRANSLATION BY

A. L. PECK

HARVARD UNIVERSITY PRESS
CAMBRIDGE, MASSACHUSETTS
LONDON, ENGLAND

First published 1942
Reprinted 1953, 1963, 1979, 1990, 2000

LOEB CLASSICAL LIBRARY® is a registered trademark
of the President and Fellows of Harvard College

ISBN 0-674-99403-5

HARRIS RACKHAM
praeceptori socio amico
dedicat
A. L. P.

Printed in Great Britain by St Edmundsbury Press Ltd,
Bury St Edmunds, Suffolk, on acid-free paper.
Bound by Hunter & Foulis Ltd, Edinburgh, Scotland.

CONTENTS

PREFACE

In reviewing Karl Bitterauf's book[a] in 1914, H. Stadler[b] described the *Generation of Animals* as " this still inadequately appreciated work of Aristotle's," and it must be confessed that his description is not yet out of date. It has, perhaps, been more appreciated by men of science than by scholars and philosophers ; but it has a strong interest for both classes of students. Its neglect by scholars and philosophers[c] is the more surprising, since it may, I think, be justly claimed that in this treatise Aristotle's thought is to be seen integrated as it is nowhere else ; for in reproduction, as understood by Aristotle, not only the individual is concerned but the cosmos at large : it is a business in which the powers of the universe are concentrated and united ; and it is the means whereby that eternity, with which, if he could have done it, God would have filled the whole creation from one end to the other, is attained so far as is possible by the creatures that are subject to decay ; indeed, these very beings, animals and plants,[d] have in Aristotle's view the best claim to the title of " being " (οὐσία), a much better claim than the lifeless things out of which they are composed, or the objects made by human art ; and therefore they merit to an exceptional degree the attention of the student of reality.

[a] *Der Schlussteil der aristotelischen Biologie*; see below, p. xxv.

[b] In *Berliner Philologische Wochenschrift* (1914), p. 833.

[c] Among the less learned, however, the outstanding achievement of Aristotle in this branch of study has been for at least the last three centuries acknowledged by the title of the popular handbook known as *Aristotle's Masterpiece*.

[d] Aristotle's strong interest in plants is shown by the large number of references to them in *G.A.* ; see Index.

Perhaps philosophers, like the visitors who came to call on Heracleitus and found him in the kitchen, have felt embarrassed at finding Aristotle in his laboratory, and have thought it more dignified to wait until he came out ; failing to perceive that " there too gods are present." [a] And where the gods are, there too is beauty, however mean and however small the creature may be which is the subject of study—greater beauty than is to be found in the products of human skill ; for these are the workmanship of Nature, who does nothing idly or without purpose ; and in them too is to be found the activity of Soul, working through its instrument *pneuma*, which is the terrestrial counterpart of the celestial " quintessence," *aither*, the divine constituent of the heavenly spheres and of the stars ; in them, therefore, Form at its highest and Matter at its highest are seen operating in unison. For men of science, the *Generation of Animals* has a special interest, in that it is the first systematic treatise on animal reproduction and embryology, containing records of observations, marking out schemes of classification, and suggesting methods of dealing with problems, much of which has proved of permanent value ; indeed, Aristotle's work was not resumed until after the lapse of nearly two thousand years, and some of his observations were not repeated until comparatively recent times. Of this I shall have more to say presently.

ARISTOTLE'S EMBRYOLOGY

Aristotle's zoological works.

The *De generatione animalium* is the culminating

[a] See *P.A.* I. 645 a 20 ff.

portion of Aristotle's zoological works, of which the scheme may be exhibited as follows :

I. Record of observations.

  Historia animalium.

II. Theory based upon observations (including also many observational data).

(a)	*De partibus animalium* *De incessu animalium*	treating of the " matter " of animals and the way in which it is arranged to subserve their various purposes ; *i.e.*, their " parts," excluding those used in reproduction.
(b)	*De anima*	treating of the " form " of animals —*i.e.*, Soul, and its " parts " or functions.
(c)	*Parva naturalia* *De motu animalium*	treating of the functions " common to body and Soul," excluding reproduction.
(d)	*De generatione animalium*	treating of the " parts " used in reproduction, and of the reproductive functions (which are common to body and Soul).

The section (*b*) is necessary to the completeness of the scheme, but as it has given rise to a whole department of study, it is usually treated apart from the rest. Thus the main bulk of the zoological and biological works may be taken to consist of the three great treatises *H.A.*, *P.A.* and *G.A.*[a] It was these which, through Latin translations made from the Arabic, were restored to the West by those who revived scientific studies at the beginning of the 13th century.

The late D'Arcy W. Thompson, in the prefatory note to his translation of *H.A.*,[b] wrote : " I think it Date of composition.

[a] For abbreviations, see p. lxxvi.
[b] *The Works of Aristotle translated*, vol. iv., Oxford, 1910, p. vii.

can be shown that Aristotle's natural history studies were carried on, or mainly carried on, in his middle age, between his two periods of residence at Athens," *i.e.* in the Troad, in Lesbos and in Macedonia, between the years 347 and 335 ; and this view has recently received convincing support from Mr. H. D. P. Lee,[a] who bases his argument upon an examination of the place-names mentioned in *H.A.* This is opposed to the view which has been current for some years past,[b] that the zoological works belong to a late period in Aristotle's life, and has important consequences for the reconstruction of Aristotle's philosophical development, which cannot be dealt with here. It may, however, be remarked that, as Thompson said, it would follow that we might legitimately proceed to interpret Aristotle's more strictly philosophical work in the light of his work in natural history. But apart from these considerations, the great importance of the zoological works is that they represent the first attempt in Europe to observe and describe in a scientific way the individual living object.

Aristotle's method.   Aristotle's method may be described as substantially the same as that of modern scientific workers : it is inductive-deductive, as opposed on the one hand to earlier (and later) methods of pure deduction from *a priori* premisses, and on the other hand to the Baconian method of almost exclusive induction. Aristotle often complains that his predecessors' work was marred by insufficient observation, and the importance which he himself attached to careful and thorough observation is apparent throughout the zoological treatises. Of particular interest in this connexion are his observations of the viviparous dog-fish (*Mustelus laevis*), observations not repeated in

[a] *C.Q.* XLII (1948), 61 ff.
[b] See W. D. Ross, *Aristotle*, and W. W. Jaeger, *Aristotle*.

viii

modern times until the seventeenth century,[a] and his knowledge of the hectocotylization of one of the tentacles in the Octopus ; the problem involved in the latter case has not yet been solved. Other problems raised by him have found their solution only in very recent years ; among them may be mentioned the breeding of eels and the anatomy of the hyena.[b] His discussion of the reproduction of bees is a remarkable piece of analysis ; and here again the facts are not yet fully ascertained. It is in connexion with this problem that Aristotle makes his well-known dictum : " But the facts have not yet been sufficiently ascertained ; and if at any future time they are ascertained, then credence must be given to the direct evidence of the senses rather than to theories—and to theories too provided that the results which they show agree with what is observed." This, indeed, is the principle upon which his work is based ; and although he is often forced to rely upon bare theories, it is only because he was unable to obtain experimental data—most insects, he regretfully remarks, are too small to be observed—in other words, it is only because he lacked the necessary apparatus. For his magnificent *apologia* (if such it can be called—*protreptic* would be a better word) on the subject of the study of natural history, the reader should refer to the passage in the first Book of the *De partibus* (ch. 5). Nevertheless it is probable that his theories, though they sometimes led him astray, did in fact often help him to adopt a correct general outlook, even if the detailed working out of them is

[a] By Nicolaus Steno (1638–1686) ; although the facts were not widely known until the work of J. Müller in the 19th century (see 754 b 33, n.).　　　　　　[b] See p. 565.

erroneous. As examples of this we may quote his discussions and conclusions upon preformation and epigenesis and upon the time of sex-determination in the embryo.[a]

Aristotle's contributions to embryology.

The main contributions of Aristotle to embryology, as judged from the viewpoint of a modern embryologist,[b] may be stated as follows :

1. Following the lead of men like the author of the Hippocratic treatise π. γονῆς, Aristotle greatly extended the field of careful and accurate observation, and was thereby enabled to introduce for the first time the comparative method into embryology, and so to arrange the available data in an orderly way. This is expressed, e.g., by his classification of animals according to their methods of embryonic development.

2. He stated in the clearest terms the two rival theories of preformation and epigenesis, and decided in favour of the latter. He also laid down that the sex of the embryo was determined at the very beginning of its development.

3. He clearly stated that generic characteristics precede specific characteristics in embryonic development, and, by his theory that the various faculties of Soul developed successively in the embryo, foreshadowed the modern theory of " recapitulation." By his observation that the " upper " parts of the embryo develop more rapidly than the " lower " parts he foreshadowed

[a] For a useful general estimate of Aristotle's work, see E. S. Russell, *The Interpretation of Development and Heredity* (1930), pp. 11-24.

[b] See, *e.g.*, J. Needham, *A History of Embryology* (1934), pp. 36 ff.

x

the modern doctrine of " axial gradients " (see 741 b 28, n.).

4. He correctly understood the functions of the placenta and the umbilical cord ; and

5. He prefigured (see 772 b 13 ff.) with wonderful insight the cell-streams or morphogenetic movements which are fundamental in embryonic development during the period when the germ-layers are taking up their definitive positions.[a] His dynamic view of the origin both of normal structures and of monstrous deviations can be fully appreciated only in the light of modern knowledge of the great part played by movement, migration of cells, etc., in early embryonic development.

On the contrary side we must range such mistakes as these :

1. The insect larva, which Aristotle regarded as the earlier stage of an egg, " an egg laid too soon," has in fact passed the embryonic stage.

2. Observations of newly-castrated animals led him to regard the testes as of secondary importance.

With regard to his famous doctrine that the male supplies the Form and the female the Matter of the embryo (see 729 a 11), some misunderstanding may easily arise. And also, with regard to his insistence upon the importance of the Final Cause, we find that modern scientific opinion, following the lead of Francis Bacon, who led the attack upon Formal and Final Causes, often tends to consider Aristotle's talk

Theory of Form and Matter in reproduction.

[a] See J. Needham, *Biochemistry and Morphogenesis* (1942), where also the most modern views on the origin of monsters will be found. On this subject, C. Dareste's *Production artificielle des monstruosités* (1877) is still the classical work.

about these Causes as inferior to what he has to say on other matters. It is, however, open to question whether Aristotle would in fact have reached some of his most valuable conclusions apart from his insistence upon the pre-eminence of the Final Cause (any more than Harvey might have discovered the circulation of the blood unless he had tried to discover what was the Final Cause of the valves in the veins) ; and although Aristotle was of course ignorant of the existence of spermatozoa and of the mammalian ovum,[a] and although he considered that the menstrual fluid was the " matter " out of which the embryo was formed, it is not so certain that he was quite as wrong-headed as he is often said to be.

Before coming to a conclusion, we must consider what exactly Aristotle meant by Form and Matter in this connexion. In the first place, we must realize that the Form is not bare Form, nor is the Matter bare Matter : this, indeed, is a fundamental doctrine of Aristotle. Form is not found apart from Matter (that was a Platonic view), nor is Matter found which is not to some extent " informed " ; and Aristotle can say (end of *Met*. H) that Matter in its ultimate stage is identical with Form (see Introd. § 17). At any rate, the Matter with which we are concerned in the generation of animals is far from being " uninformed." Like the " residue " contributed by the male, the " residue " contributed by the female is " concocted blood " ; and, since blood is the " ultimate nourishment " which maintains the upkeep of

---

[a] Discovered by K. E. von Baer ; there is a complete facsimile of his fundamental memoir *De ovi mammalium et hominis genesi* (Leipzig, 1827) in Sarton's *Isis*, XVI (1931), 315 ff.

the body and its parts, both " residues " are *potentially* the body of a living creature of the same kind as that which produced them. Indeed, the only important difference between them is one of the degree of " concoction " which they have undergone, for the female, whose vital heat is weaker, cannot carry the " concoction " of blood as far as the male can. But the female's " residue " (viz., the menstrual fluid) is, potentially, all the parts of the body ; and hence, too, it is, or contains, Soul potentially (this is merely another way of saying the same thing, because just as any actual living body must possess Soul, which is its Form, actually, so a potential living body must possess Soul potentially). That the female's " residue " does in fact possess Soul potentially is shown, says Aristotle, by the occurrence of wind-eggs in birds : these possess nutritive Soul, and up to a point they grow and " are fertile." The Matter, therefore, is " informed " to a high degree ; and the only part of the Form which it lacks is sentient Soul. Hence, the meaning of the statement that " the male supplies the Form " can only be that the male supplies that part of the Form known as sentient Soul : everything else, including nutritive Soul, can be, and is, supplied by the female.

We may now go on to consider the " residue " contributed by the male. Aristotle, as we saw, held that Form is not normally found apart from Matter (*i.e.*, body) of some sort,[a] and besides that, according

[a] See Introd. § 42. An exception is *rational* Soul, which is not the Form of any body (§ 44), but this is a separate question, and in any case affects man only. We must also except the 55 immaterial unmoved movers, which Aristotle posits in the *Metaphysics* (1074 a) to account for the movements of the planets.

ARISTOTLE

to him, action can only be exerted, change can only
be brought about, by something that can come into
contact with another thing. Therefore in any case
something corporeal must be supplied by the male as
well as Form, and this is of course the substance which
carries the (potential) Form : it is the substance with
which the sort of Form known as Soul is specially and
regularly connected, and in which it resides, viz.,
connate *pneuma*. This *pneuma*, which is thus present
in the semen, is charged with the " movements "
proper to Soul, including (in the case of the male) the
" movements " proper to sentient Soul ; and these
" movements," when given the right material to
work upon (viz., material which is potentially an
animal of the right kind) and the right conditions, are
able to produce an animal of the same kind as that
which they would have produced or maintained in the
male parent even if the blood in which they were
originally present had not undergone the further
stage of being concocted into semen.

Hence it is clear that fundamentally the contributions of both parents in generation are identical ;
both are potentially a living animal of a certain kind,
and this involves that both possess the living animal's
Form, viz., its Soul, potentially ; and the only difference between them is that the male's contribution
possesses also *sentient* Soul potentially.

At the same time, this is an important difference,
and makes itself apparent in the difference of bulk
between the two : the female's is large in bulk, the
male's is small. And this difference of bulk is
accounted for by the fact that the female's is less
" concocted " than the male's—it is less concentrated.
Further, the only Matter that the semen need con-

tain is a sufficient amount to transmit the " movements " to the female's residue, and once this has been done—that is to say, once the embryo or rather its heart has been " constituted," once it has been given its " principle " and has the power to grow— then the " body " of the semen can " evaporate," for the Matter which provides the embryo with its wherewithal for *growth* is of course supplied by the female parent.

As a final word on the subject we may recall that, in addition to what we have already found Aristotle saying about the identity of Matter and Form in the long run, he finds no greater difficulty in identifying φύσις with Matter than he does in identifying it with Form or with the Motive Cause and the Final Cause (see Introd. § 14, end); and when all the attributes have been ascribed to Matter which Aristotle ascribes to it in spontaneous generation (see App. B § 17, additional note), there are few prerogatives still left to be bestowed upon it.

I have not thought it necessary to call attention to Scope of all Aristotle's mistakes, partly because of lack of this edition. space, but chiefly because it would serve no really useful purpose. Nor have I given an account of modern embryological theory. My main object has been to ensure that the reader shall be able to find out what Aristotle said, and to secure that Aristotle shall get neither credit nor discredit for things which he did not say. In a treatise such as *G.A.*, this means that fairly copious footnotes are necessary,[a] and as a further help to the reader I have provided not only a full account of Aristotle's technical terms (which gives an opportunity for explaining a good deal of the

---

[a] See also p. xxxiv.

ARISTOTLE

framework of his thought), but also, in the Appendix, accounts of his theory of the universe and movement (without which parts of Books II and IV cannot be understood) and of the functions of Σύμφυτον Πνεῦμα,[a] an essential factor in his doctrine of generation. On

Aristotle his own interpreter. the principle that, for the most part, Aristotle is his own best interpreter, these accounts are compiled almost entirely from passages taken direct from Aristotle's own treatises.

Aristotle's predecessors. In reading Aristotle's scientific works, it is important not only to recognize how great were the advances which he himself made in natural history, both in practical observation and in theory, but also to remember that his work was a continuation and an expansion of what had been begun by previous scientific workers.[b] Those to whom he most frequently refers by name are three : Anaxagoras, Empedocles, and Democritus, besides several references to theories which can be traced in the Hippocratic treatises [c] ; and the fact that he often quotes them in order to disagree with them should not lead us to underrate their achievement. It is not possible here to give any adequate account of these predecessors of his, and for details about them the reader must be referred to the standard works on

---

[a] The doctrine of ΣΠ was older than Aristotle (see Jaeger; references given Introd. § 46, n.), but in this volume I am concerned only with Aristotle's presentation of it.

[b] Aristotle calls them collectively φυσικοί or φυσιολόγοι, "physiologers," *i.e.*, writers on "Nature," "natural" scientists. See 741 b 10, n.

[c] There are also, of course, references to theories stated by Plato, to which attention is called in the notes ; but Plato is not mentioned by name. See also K. Prächter, *Platon Präformist ?* in *Philologus*, LXXXIII (1927), 18-30.

the early scientists and philosophers and to other works of reference.[a] Alcmeon, to whom also he refers, is an important figure, since it was he, apparently, who originated the famous doctrine of "passages" (or "pores," πόροι) in connexion with sensation, and held that the brain was the common sensorium, in which belief he was followed by Hippocrates and Plato, whereas Empedocles and Aristotle reverted to the older view that the heart is the central organ of sensation. Alcmeon also treated systematically of the special senses, in particular that of sight. Other theories of his mentioned by Aristotle may be found by reference to the Index.

Anaxagoras of Clazomenae, the last great name in the Ionian philosophic succession of Asia Minor, is well known for his theory that νοῦς is responsible for the order of the universe as a whole, just as it is for the order which is to be discerned in living creatures, and for his remarkable theory of matter, which he constructed specially with a view to accounting for generation and growth. I have treated fully of this elsewhere.[b]

Empedocles of Acragas, a striking figure, was a slightly younger contemporary of Anaxagoras, and was renowned as a politician, religious teacher, rhetorician, philosopher, and physician : he was the

---

[a] e.g., J. Burnet, Early Greek Philosophy ; see also for Hippocrates, W. H. S. Jones (Loeb ed.) ; for Alcmeon, J. Wachtler, De Alcmaeone Crotoniata (1896) ; and M. Wellmann, Die Schrift π. ἱρῆς νούσου, in Archiv f. Gesch. der Med. XXII (1929), 290-312. For a conspectus of ancient embryology, H. Balss, Die Zeugungslehre u. Embryologie in der Antike, in Quellen u. Studien zur Gesch. der Naturw. u. der Med. V (1936), 193-274.

[b] C.Q. XXV (1931), 27 ff., 112 ff. ; see also G.A. 723 a 7.

founder of the " Italian " school of medicine. Considerable portions of his poems on *Nature* and *Purifications* are extant. He adopted, perhaps formulated, the doctrine of the four Elements, which really means (see π. ἀρχαίης ἰητρικῆς, chh. 13 ff.) that he selected, as especially important, four out of the many substances already recognized as fundamental in traditional Greek medical theory (see Introd. § 24).

Democritus of Abdera, the follower of Leucippus, is best known for his advancement of the atomic theory originated by his master. Abdera is not far from Aristotle's birthplace, Stageira, and Aristotle seems to have been specially interested in Democritus.[a]

Dates. The following table will indicate roughly the dates of these early scientists :

Alcmeon of Crotona, probably a Pythagorean and a pupil of Pythagoras himself (he was " a young man in Pythagoras' old age "). Pythagoras is said to have gone to Italy in 529 B.C. and to have lived at Crotona for twenty years. Alcmeon was probably active, therefore, about 510–480.[b]

Anaxagoras of Clazomenae. Born about 500 B.C., died 428. Lived at Athens *c.* 480–450, and was a friend of Pericles. Mentioned by Socrates in a well-known passage in Plato's *Phaedo*.

Empedocles of Acragas (Agrigentum) in Sicily ; *c.* 494–434.

Democritus of Abdera in Thrace ; *c.* 460–370.

[a] For further details about Democritus, see C. Bailey, *The Greek Atomists and Epicurus.*

[b] According to W. A. Heidel, however, *Hippocratic Medicine* (1941), 43, and *American Journal of Philology*, LXI (1940), 3 ff., Alcmeon's *floruit* should be put considerably later, say at 450 B.C.

GENERATION OF ANIMALS

It is not possible to assign exact dates for all the treatises in the Hippocratic collection ; indeed they cannot all be ascribed to a single author, but one of the most important, the π. ἀρχαίης ἰητρικῆς, refers to Empedocles as having introduced new-fangled ideas into the long established science of medicine (ch. 20). Other treatises relevant to our subject are the π. ἀέρων ὑδάτων τόπων, the π. διαίτης, and the π. γονῆς καὶ π. φύσιος παιδίου. All of these are most interesting and will repay study. The last named in particular is the work of a most active and enterprising man, always ready to experiment and to record his results and to make use of them.

It should of course be remembered that although Aristotle introduced much new technical terminology and sometimes gave new content to what already existed, many of the terms which he uses were the common property of scientific writers, among them being such important ones as the following : δύναμις, κρᾶσις, σύντηγμα, συμμετρία, εἶδος, πνεῦμα and the like. I have attempted to trace the development of one such term in my account of δύναμις (Introd. §§ 23 ff.). Termino-
logy.

It is not possible here to say much about Aristotle's successors, but it is necessary to say enough to emphasize the important influence which they have had in the history of science. Hieronymus Fabricius ab Aquapendente (1537–1619) knew and admired Aristotle's work on embryology, and what is more, himself carried out further important observations on the same subject. His brilliant successor, William Harvey (1578–1657), was a student of Aristotle, and much of his inspiration came from Aristotle's works. Harvey was indeed the first to make any substantial advance in embryology since Aristotle him- Aristotle's
successors.

self. In other departments of study, however, during the 17th century, the authority of Aristotle and the scholastic doctrine with which he was identified were being combated in the name of " freedom," and so it came about that the zoological works too, which had been brought to light by the " dark " ages, were allowed to pass back into oblivion by the age of enlightenment. It was not until the end of the 18th century that they were rediscovered for the second time by Cuvier (1769–1832) and members of the Saint-Hilaire family.

EARLY TRANSLATORS

Early translators. Lack of space forbids reprinting here the account which I gave in the Introduction to *P.A.* of the fascinating history of the early translators of Aristotle's zoological works, and I must be allowed to refer the reader to that volume (pp. 39 ff.) for details and other references. A mere list of the four most important translations must here suffice :

(1) The physician Ibn al-Batriq translated the *H.A.*, *P.A.* and *G.A.* into Arabic at Bagdad during the time of the Caliphate of al-Mamun (813–833), son of Harun al-Rashid. There is a MS. of an Arabic translation, probably Ibn al-Batriq's, in the British Museum [a] ; and there

---

[a] B.M. Add. 7511, 13th–14th century (=Steinschneider B.M. 437). I have seen this MS. Judging from the passages which Dr. R. Levy kindly read for me in this MS., Scot's Latin version is a close translation from the Arabic. This is confirmed by the fact that the contents-preface which is found prefixed to Scot's translation corresponds exactly with the preface which precedes the Arabic version in this MS. (see B.M. *Catalogus codicum manuscriptorum orientalium*, p. 215).

can be little doubt that this is the translation from which Michael Scot made his Latin version.

(2) Michael Scot translated *H.A.*, *P.A.* and *G.A.* into Latin from the Arabic at Toledo. This translation was finished before 1217.

(3) William of Moerbeke translated the zoological works into Latin from the Greek, at Thebes, in or before 1260.

(4) Theodore of Gaza began at Rome in 1450 to make translations of Aristotle and other Greek authors. His translation of the zoological works of Aristotle is dedicated to Pope Sixtus IV, and this soon became the standard Latin version. It is printed in the Berlin edition of Aristotle.

The Text

It soon became clear that for the purpose of transla- Method. tion it was necessary to make a working version of the Greek text, and to this end I made my first draft with the Berlin edition, Aubert and Wimmer's edition, and Platt's translation and textual emendations before me. Next, I transcribed suspected passages with their contexts from the mss. of Scot's version, in order to give them fuller consideration. Then, having incorporated a large number of changes into the text, some of them my own, I took into consideration the work of Bitterauf and others. In some cases I found that the same emendation had been proposed by two or more scholars independently, and also that some of these emendations were confirmed by Scot. Finally, I found it necessary to transcribe further portions of Scot's version.

I do not wish to claim more for the text here offered than that it is a better text than any hitherto available. I have done my best with the data at my disposal, but I am well aware that many passages yet remain to which I have not been able to offer any satisfactory solution.

Apparatus criticus. When I have accepted the reading of Bekker's edition, I have not normally given the MSS. variants. These will be found in Bekker's *apparatus*. All citations of Z are from my own collation ; corrected reports of other MSS. readings as given by Bitterauf I have distinguished by an asterisk ; the other readings are as reported by Bekker [a] (sometimes confirmed by Bitterauf). Every departure from Bekker's text is recorded.

Arrangement. The text has been reparagraphed throughout, and in many places the punctuation has been corrected.

Manuscripts :
(a) Greek text. The following manuscripts [b] are cited for the Greek text :

Z Oxoniensis Collegii Corporis Christi W.A. 2. 7 (= Coxe 108). About A.D. ·1000. Presented to the College by Henry Parry, Fellow, in 1623. It contains *P.A., I.A., G.A.*, some of the *Parva Naturalia*, and *De spiritu*. *G.A.* begins f. 74[r], and ends f. 161[r], but this page is identical with 62[r]. Two folios of Book II have been lost, and a passage in Book III omitted.[c]

S Laurentianus Mediceus 81, 1. Written in different hands, some of the 12th, some of the 13th

---

[a] A few (for m and E) are as reported by Bussemaker.

[b] For further details, see Bitterauf (below, p. xxv), Dittmeyer, *H.A.* (Introd.), Jaeger, *M.A. and I.A.* (Introd.), etc.

[c] 738 b 1 βελτ[ίονος . . . 740 b 6 τὸ] γενόμενον ; and 760 a 12 πως [ἡ γένεσις . . . 760 b 26 μὲν] ἐλάττω, the latter passage having been supplied by a later hand.

century. *G.A.* is written in a 12th century hand.

P Vaticanus graecus 1339. Great variation of opinion upon the date of this manuscript has been expressed by various scholars. Some date it 12th century, others 15th.

Y Vaticanus graecus 261. 14th century (Btf.).

The following are cited for a few places only :

m Parisinus 1921. 14th century. In this ms. *G.A.* is accompanied by the commentary of Michael of Ephesus.

O^b Riccardianus 13. Late 14th century.

E Parisinus regius 1853. Written in various hands, from 10th to 15th century. *G.A.* is in a 15th century hand.

The following manuscripts of Michael Scot's translation are to be found in this country :

(b) Michael Scot's translation.

Cambridge, University Library Ii. 3. 16.
Cambridge, University Library Dd. 4. 30.
Cambridge, Gonville and Caius College 109.
Oxford, Merton College 278.
Oxford, Balliol College 252.
London, British Museum Royal 12. C. XV.
London, British Museum Harl. 4970.

All these are of the 13th or 14th century. I have seen all these mss. of Scot's translation, but chiefly owing to present conditions I have worked with the two first mentioned only.[a]

The chief mss. cited by Bekker for *G.A.* (namely, PSYZ) are identical with four of the six cited by him

The text:
(a) Bekker's
edition;

[a] Lists of mss. of William of Moerbeke's translation will be found in G. Rudberg, *Textstudien zur Tiergeschichte des Aristoteles* (1908), and L. Dittmeyer (see below, p. xxix).

for *P.A.*[a] Some years ago, when working on *P.A.* for the Loeb edition, my examination of the ms. Z at several places led me to state (*P.A.* Introd. p. 46) that a more reliable collation of the chief mss. than Bekker's *apparatus criticus* afforded was clearly needed. This view is amply confirmed by K. E. Bitterauf, who has in fact undertaken such a collation for *G.A.* (see below), and he shows that there are several errors and misleading reports on every page in Bekker's *apparatus*. I have myself collated the ms. Z throughout from photostats.

(b) the manuscript tradition.
A comparison of the text of *P.A.* exhibited by our Greek mss. with the translation of Michael Scot showed me that the former had all suffered identical corruptions or losses (or both) in certain passages (*e.g.*, *P.A.* 684 b 22 ff.), by which the Greek ms. from which Scot's Arabic original was translated had not been affected ; and I found exactly the same when I came to work on *G.A.* (see, *e.g.*, 722 a 20, 766 b 35). My conclusion about the common origin of our Greek mss. is also supported by Bitterauf, who comes independently to the conclusion, based exclusively upon a study of the Greek mss., that they are all derived from a single archetype, which, in his opinion, contained a number of variant readings.

Modern work on the mss.
This brings us to a consideration of the ms. tradition of *G.A.* After the publication of Bekker's Berlin edition in 1831, very little work was done on the mss. of *G.A.* for about eighty years. Bussemaker, who edited *G.A.* in the Didot edition (Paris, 1854), cites many readings from the two Paris mss. E and m, and several times quotes the authority of William of Moerbeke, less frequently that of Michael Scot, and

---

[a] Of the other two, U does not contain *G.A.*, and in E *G.A.* is written in a later hand.

in a few cases quotes their Latin versions. Aubert and Wimmer, in their Leipzig edition published in 1860, took into account throughout the commentary of Michael of Ephesus and Gaza's Latin translation, but they too relied upon Bekker for the MSS. readings. The first to go back again direct to the MSS. was F. Susemihl,[a] at whose request in 1885 Bywater and Vitelli inspected a number of selected places in Aristotle's zoological works in the MSS. Z and S respectively, and of these fourteen are places in *G.A.* The majority of these, however, are of minor importance. A really serious attempt to revise the text throughout on the basis of a new collation of the MSS. was made about 1913 by K. E. Bitterauf in preparation for a new Teubner edition, which however was never published.[b] In all, Bitterauf enumerates 31 MSS. containing *G.A.*, and of these he collated three in full himself from photographs (Z, Y and E), and a single selected Book (not the same Book in each case) in eight more (of which m was one). He also had at his disposal collations of seven others, of which five were apparently collated direct by Hugo Tschierschky

---

[a] *Kritische Studien zu den zoologischen Schriften des Aristoteles*, in *Rhein. Mus.* XL (1885), 563 ff., and a very convenient summary of his proposals there made in *Bursian*, XLII, 245 f.

[b] But he published some of his results in two preliminary pamphlets : (1) *Der Schlussteil der aristotelischen Biologie : Beiträge zur Textgeschichte und Textkritik der Schrift " De generatione animalium."* (Wissenschaftliche Beilage zum Jahresbericht des kgl. humanistischen Gymnasiums Kempten für das Schuljahr 1912/13). Kempten im Allgäu, 1913. (2) *Neue Textstudien zum Schlussteil der aristotelischen Biologie.* (Ibid., 1913/14.) Kempten im Allgäu, 1914. These are the source of the readings (except of Z) recorded throughout the text where they differ from Bekker's *apparatus.*

(these include S and O^b, and another ms. called β which contains only a very small part of the beginning of *G.A.*) and the remaining two (one of which is P) were collated by L. Dittmeyer from photographs. Five others were collated (apparently from photographs) by Bitterauf sufficiently to establish their character ; of the remaining eight he gives no report on the character of their text. The upshot of Bitterauf's work is to show that Bekker was right in basing the text upon PSYZ, and that although the most faithful witness to the original text is Z, with P a good second, no ms. has a monopoly of the truth, since their common descent gives them all a fair chance of preserving a good reading, just as it has undoubtedly ensured, as I mentioned above, that they have all failed to preserve the text in certain passages.

With regard to the defective nature of Bekker's *apparatus*, the corrections which Bitterauf gives are of value primarily in determining the comparative trustworthiness of the mss. rather than in yielding substantial improvements of the text [a] ; but there are a good many places where they do make an improvement possible, and all the suggestions which Bitterauf makes for so doing I have carefully considered, and many I have adopted.[b] When the changes indicated are of a minor character, for

[a] Examples are: 718 a 36, Bekker's *app.* αὐταῖς Z, actually αὐταῖς SZ ; 719 a 31, Bekker's *app.* ἐντός, τὰ δ' ἐκτός Y, but actually PZ. Bitterauf had access to Bekker's own copy of the Basel Aristotle (1550), and shows that some of Bekker's errors are due to his having used one set of symbols for the mss. in his collation and another set in his *apparatus*.

[b] It should be remembered that Bitterauf's pamphlets are merely " foretastes " of his projected edition, and therefore the list of passages dealt with by him cannot be treated as exhaustive.

instance those affecting merely the order of words, I have not always felt it necessary to alter Bekker's text, though it might be held that *ceteris paribus* Z's reading should be preferred.

Bitterauf does not appear, at any rate from what he has published, to have envisaged the existence of deep-seated corruptions or serious losses from the text. The furthest he ventures along this path is to suggest that αἷμα and σάρξ should be written twice instead of once at 722 b 34, and that καὶ θώων has dropped out at 746 a 34 ; but the latter suggestion, which is certainly right, is taken over from Bussemaker. However, that loss of phrases and corruption of the text have occurred is sometimes clear from intrinsic evidence, and loss can sometimes be proved by the survival of the original words in M. Scot's translation.

Apart from re-examination of the MSS., proposals to improve the text by conjectural emendation have been made by the following :

Conjectural emendation.

(1) Wimmer, who was responsible for the textual work in Aubert and Wimmer's edition of 1860, made a number of conjectures, some of which he incorporated in the text and others he printed in the footnotes. Many of them are undoubtedly correct, and some I have found are supported by Scot (though I have no reason to think that Wimmer himself was aware of this).

(2) F. Susemihl,[a] beside the work which he did on the MSS., dealt with the question of duplicate recensions in the text, and also that of interpolations

---

[a] *Rhein. Mus.* XL (1885), 563 ff.

by commentators, and made a number of conjectural emendations.

(3) Arthur Platt, in his translation of *G.A.* in the Oxford Translations of Aristotle, published 1910, suggests a number of emendations, many of which have been adopted in the present text ; and some of these, again, I have found to be confirmed by Scot's translation, though Platt himself was unaware of this. Platt also detected many corrupt places and misplaced passages or interpolations.

(4) Bitterauf himself puts forward about ten conjectural emendations in addition to his other suggestions for improving the text, but few of them are of major importance.

A few suggestions for emendation were made by :

(5) H. Bonitz,[a] *en passant*, as asides to his treatment of passages in other works of Aristotle, and by

(6) H. Richards [b] ; some of these will be found recorded in their proper places.

Single small emendations are proposed by M. Hayduck [c] and E. Zeller.[d] A few are proposed by H. Diels and one by W. Kranz.[e] J. G. Schneider, too, in his edition of *H.A.* (1811) made some suggestions for improving the text of *G.A.* based partly on the Latin versions, but most of his work is superseded by Bekker's edition. Some passages are also discussed by J. Zahlfleisch.[f]

[a] *Aristot. Studien* (1866), IV. 363, 378, 413.
[b] *J. of Philology*, XXXIV (1918), 254.
[c] *Emendationes Aristoteleae*, in *Neue Jahrbücher f. Philol. u. Pädagog.* CXIX (1879), 111.
[d] *Phil. der Gr.* II. 2³, 569-570¹.
[e] *Die Fragmente der Vorsokratiker* (5th edn., ed. Kranz, 1934–1937).
[f] *Philologus*, LIII (1894), 39-44.

GENERATION OF ANIMALS

Platt seems to have known nothing of Bonitz' or Susemihl's work, and Bitterauf seems to have known nothing of Platt's work. Bitterauf refers to and quotes Susemihl's article, but puts forward as an original conjecture one which Susemihl had already made (756 a 24).

Several emendations have been put forward by various scholars, beginning with Schneider, on the strength of Gaza's Latin version, others on that of William of Moerbeke. As a contribution to the projected Teubner edition of *G.A.*, Dittmeyer [a] published in 1915 the first part of William's version (up to 737 b 5). Although this version gives support to two small emendations already adopted in my present text (at 733 b 34 and 734 b 18), and at 775 a 11 ff. (*teste* Schneider) preserves a passage which our Greek mss. have lost, in general it does not yield anything that is independent of our existing Greek mss. and is, as Dittmeyer himself agreed, of little value for the restoration of the text.[b] _{William of Moerbeke's translation.}

The case is far different with Michael Scot's version. This was made about 1217, not from a Greek text, but from an Arabic translation, itself made at the beginning of the 9th century, and hence the Greek text involved must have been considerably older than any of our present mss. and *a priori* may have represented an independent tradition of the text; indeed, my examination of Scot's version has proved this to be so. Dittmeyer quotes Schneider's opinion (IV. xxxvii) that Scot's version is of little value for restor- _{Michael Scot's translation.}

---

[a] *Guilelmi Moerbekensis translatio commentationis Aristotelicae De generatione animalium.* Edidit Leonardus Dittmeyer. Programm des K. humanistischen Gymnasiums Dillingen a. D. für das Schuljahr 1914/15.

[b] See also *P.A.* (Loeb ed.), p. 47.

ing the text, but it is obvious that neither he nor Bitterauf [a] had troubled to read Scot's version of *G.A.* beyond the tiny fragments (*frustula*, Dittmeyer's own word) quoted by previous scholars. Against this we have the opinion of G. Rudberg,[b] who had made a considerable study of it in connexion with *H.A.* and published its version of *H.A.* X *in extenso*, that there is no doubt of its critical value for rectifying the text; and this judgement I can confirm from my own experience. Naturally, the circumstances dictate that proper safeguards must be adopted in using it for correcting the Greek text; and what these are can be learnt only by fairly wide experience of the version itself; any judgement given,[c] either for or against, without this experience as a foundation is worse than useless. My own method has involved the transcription of a large number of continuous passages from the MSS. of Scot's version, containing places which some previous editor or I myself had already felt for some reason to be doubtful; and the pertinent parts of these, where they have anything to contribute, I have given in the *apparatus*. Scot's version sometimes confirms conjectures previously made, sometimes it confirms the suspected corruption of the text either through glosses or otherwise, and in these cases may suggest means for remedying the trouble. Often it clearly confirms the existing text; sometimes it gives no clear indication, and sometimes

[a] Bitterauf quotes Scot only once, and that quotation is taken from Bussemaker.

[b] *Kleinere Aristotelesfragen*, in *Eranos*, IX (1909), 92 ff.; see also *Zum so-genannten 10. Buch der Tiergeschichte*, Upsala, 1911.

[c] *e.g.*, D. W. Thompson, *C.R.* LII (1938), 15 " the dubious aid of an Arabic version "; see also *ibid.*, p. 89.

it simply omits the passage. I consider the time and trouble spent upon Scot's version as well spent.

The Greek commentary of Michael of Ephesus (formerly attributed to Johannes Philoponus), 11–12th century A.D., has been edited by Michael Hayduck (Berlin, 1903), but it is of little use for textual criticism. Commentary of Michael of Ephesus.

Apart from manuscript errors of the usual kind, and losses of words or phrases due to *homoioteleuton*, etc., which will be found noted in their places where they can be detected, the chief points of note in the text of *G.A.* may be classed as follows : Interpolations, etc.

A. Paragraphs, occasionally sentences only, which obviously interrupt the line of argument or are superfluous to it. Of these, some seem to be

  (a) genuine Aristotelian material, but misplaced, perhaps incorporated at the wrong place, or perhaps originally supplementary notes never intended to stand in the text ;

  (b) alternative versions of matter already in the text ;

  (c) extraneous matter, derived from commentators' remarks and wrongly incorporated in the text (*e.g.*, 724 b 12-23, 726 b 25-30).

These are often found at the beginning or end of a section, which suggests that they were originally appended in the margin. There is no need to give a full list of these passages, but a list of (a) and (b) may be useful. They are :

715 b 26-30 ; 718 a 27-34 ; 726 a 16-25 ; 732 a 12-23 ; 737 a 35-b 7 ; 760 a 26-27 ; 760 b 2-8 ; 760 b 33—761 a 2 ; 781 a 21-b 6.

B. Short passages, often only a few words, derived

from glosses which have either (*a*) supplanted the text or (*b*) been incorporated into it.

There are a great many short interpolations, and I have frequently omitted them from the translation.

Modern Editions

1. The Berlin edition of Aristotle, by Immanuel Bekker. Vol. I includes *G.A.*, pages 715-789 (by the columns and lines of which the work is normally cited). Berlin, 1831.
1a. The Oxford edition (a reprint of the preceding). Vol. V includes *G.A.* Oxford, 1837.
2. One-volume edition of Aristotle's works, by C. H. Weise.[a] Leipzig, 1843.
3. The Didot edition. Edited by Bussemaker. Vol. III includes *G.A.* Paris, 1854.
4. The Leipzig edition. Vol. III contains *G.A.*, edited and translated into German by H. Aubert and Fr. Wimmer. Leipzig, 1860. Contains a useful introduction, table of animals, and Greek index.

Translations Only

5. Thomas Taylor. English translation of Aristotle in ten volumes. Vol. IV contains *G.A.* (pp. 243 ff.). London, 1808.
6. J. Barthélemy-Saint-Hilaire.[b] Introduction, French translation of *G.A.* and notes. In two volumes. Paris, 1887.

[a] The text of this edition is the pre-Bekker *vulgata*, founded on Sylburg and Casaubon.
[b] Saint-Hilaire argues (I. cclix ff.) that Book V of *G.A.* does not belong with the rest of the treatise, but goes rather

7. Arthur Platt. In the Oxford series of translations of Aristotle. Vol. V contains Platt's translation of *G.A.*, with notes. Oxford, 1910.
8. Francisco Gallach Palés. Aristóteles : Obras completas. Vol. XII contains *G.A.* Spanish translation, without notes. Vol. lxxi of *Nueva Biblioteca Filosófica*. Madrid, 1933.

THE TRANSLATION

In translating *G.A.* I have followed two main principles, with the aim of presenting Aristotle as faithfully as possible to the English reader : Principles of translation ;

(1) I have attempted to translate *G.A. into English*, and therefore I have not felt obliged to write in Aristotelian, or even in Greek, idiom. Hence, for example, I have not uniformly translated γάρ by " for," καί by " and," or δέ by " but " : unfortunately, it is still necessary to point out, even to learned reviewers, that there is a better way than that of " stock " translations ; and a translator is not automatically a traitor if he sometimes *omits* γάρ—as the most idiomatic way of translating it.

(2) Technical terms, on the other hand, must whenever possible be uniformly represented by an invariable term in the English. Sometimes this rule must be broken, either (*a*) because the original term has a variety of meanings (*e.g.*, δύναμις), sometimes (*b*) because there is no English word which will do (*e.g.*, συνιστάναι). I have avoided modernizing Aristotle's terms, so as to avoid misleading any modern

with *P.A.* The same suggestion, unknown to him, had been made by Weise (p. xxix) in 1843. Saint-Hilaire thinks that its inclusion with *G.A.* dates from the time of Andronicus of Rhodes, head of the Peripatetic School at Rome, who edited Aristotle's works from the MSS. belonging to Apellicon's library brought to Rome by Sulla in 84 B.C.

readers who may have but little Greek ; and on the positive side I have given a full account of many of these terms in the Introduction. In my opinion, it is essential that the Introduction be read before undertaking any study of the treatise itself.

The purpose I have had in mind, therefore, is to ensure so far as possible, that the reader shall not have the unnecessary difficulty of " translators' English " to overcome, but shall be able to give his full attention to Aristotle's thought and argument : this is especially necessary in the present case, where we are dealing primarily with a scientific treatise. My aim has not been to paraphrase Aristotle or to " improve " upon him, but to represent what he says as closely and as faithfully as possible in English.

of annotation ; Since, however, *G.A.* is not intelligible, even to a Greek scholar, without some familiarity with Aristotle's general thought and some of his main doctrines, I have provided an outline of these in the Introduction and in the Appendix ; and in the footnotes I have given many cross-references to other passages in *G.A.* and other treatises ; attention is also called to points of special interest. One of these, which I think has not hitherto been noticed, may be mentioned here : the possibility that there is an allusion at 735 b 17 to an early process of oil-flotation in ore-dressing.

of the Index. The Index is not intended to be exhaustive, but forms a supplement to the Contents-Summary (p. lxxi) and the Introduction. Particular attention is given to certain key-phrases and ideas. It covers the Preface, Introduction, footnotes and Appendix as well as the translation.

A glance through the Index may help a reader with special interests to find the passages most relevant to his subject : *e.g.*, the entry " causation, mechanical " gives a reference to the passage, specially interesting to modern readers, which compares the development of the embryo to the action of automatic puppets.

A number of books which the student of Aristotle's zoological works will find useful are mentioned in the footnotes throughout the volume ; to them may be added the following : Additional bibliography.

F. J. Cole, *Early Theories of Sexual Generation*, Oxford, 1930.

C. H. Haskins, *Studies in the History of Medieval Science*, ed. 2, Cambridge, Mass., 1927.

T. E. Lones, *Aristotle's Researches in Natural Science*, London, 1912.

A. W. Meyer, *The Rise of Embryology*, Stanford, Calif., 1939.

C. Singer, *Studies in the History and Method of Science*, Oxford, 1921.

C. Singer, *Greek Biology and Greek Medicine*, Oxford, 1922.

H. B. Torrey and F. Felin, *Was Aristotle an Evolutionist ?* in *Qu. Rev. of Biology* (Baltimore), XII (1937), 1-18.

D'Arcy W. Thompson, Essay on " Natural Science " in *The Legacy of Greece*, Oxford, 1924.

S. D. Wingate, *The Medieval Latin Versions of the Aristotelian Scientific Corpus*, London, 1931.[a]

In addition to Ross's *Aristotle* and Jaeger's *Aristotle* (English translation by R. Robinson) and *Diokles von Karystos*, which are of special importance, the fol-

---

[a] For other works on the early translations, see my edition of *P.A.* (Loeb Library).

lowing bear upon certain subjects dealt with in
G.A. :

P. Bochert, *Aristoteles' Erdkunde von Asien und Libyen*,
1908, and

H. Diller, *Wanderarzt und Aitiologe*, 1934 (for the
effects of climate, etc.).

H. Meyer, *Das Vererbungsproblem bei Aristoteles*, in
Philologus, LXXV (1919), 323 ff.

M. Wellmann, *Fragmentsammlung der sikelischen
Ärzte*, 1901.

The following more general technical works may
also be mentioned :

J. S. Huxley and G. R. de Beer, *The Elements of
Experimental Embryology*, Cambridge, 1934.

H. G. Müller-Hess, *Die Lehre von der Menstruation
vom Beginn der Neuzeit bis zur Begründung der
Zellenlehre, Abhandl. z. Gesch. d. Naturw. u.
Med.*, 1938, no. 27.

Aute Richards, *Outline of Comparative Embryology*,
New York, 1931.

D'Arcy W. Thompson, *Growth and Form*, Cambridge,
1917 (new ed., 1942).

P. Weiss, *Principles of Development*, New York, 1939.

The standard work on its subject is *Geschlecht und
Geschlechter im Tierreiche*, by Johannes Meisenheimer
(1921).

ACKNOWLEDGEMENTS

It is a great pleasure to acknowledge here the help
which I have received from many friends in many
ways, and above all to thank them for their continuous
interest and encouragement. First I thank Dr.
W. H. D. Rouse, my old teacher and present colleague,

to whom I owe, among many other things, the opportunity of undertaking this translation. The whole of the translation has been read through by my colleagues Mr. H. Rackham and Dr. F. H. A. Marshall, F.R.S., and also by Dr. Sydney Smith ; for valuable help with some difficult passages in the Greek I am indebted to Professor R. Hackforth, and to Mr. Hugh Tredennick, who has also read part of the translation ; for much assistance in biological matters I am indebted to Dr. Marshall, to Dr. Joseph Needham, F.R.S., to Dr. Smith, and to Miss M. E. Brown. Professor A. S. Pease of Harvard University has placed me under a great obligation by most courteously securing for me microfilms of Bitterauf's two pamphlets and of Dittmeyer's edition of Moerbeke's translation, none of which I could find in this country. It is a special pleasure to acknowledge this help from America. I am indebted to the staff of Cambridge University Library for excellent arrangements made for me to read the microfilms and also the Scot manuscripts. Professor Sir Ellis Minns kindly dated the MS. Z for me. Dr. P. J. Durrant suggested to me that the mention of oil in connexion with lead-ore (see Bk. II. 735 b 17) might indicate an early process of flotation. Finally but not least I should like to express my appreciation of the kindness of Mr. R. Elmhirst, Director of the Marine Biological Station at Millport, Great Cumbrae, who gave me a room in which to work at my translation during a visit to Millport in the summer of 1938 and also included me in an expedition to Loch Goil for collecting marine animals closely allied to those often mentioned by Aristotle.

The present edition has been revised.

A. L. P.

Easter Eve, 1949

INTRODUCTION

The " Causes "

The four
Causes.
(1) Aristotle begins and ends the *G.A.* with a paragraph about Causes [a] ; and indeed Causes are at the foundation of all his thought, especially of his theories about animal reproduction and development.

To know, says Aristotle, is to know by means of Causes (see *Anal. Post.* 94 a 20). A thing is explained when you know its Causes. And a Cause is that which is responsible, in any of four modes, for a thing's existence. The four Causes are :

(1) The Final Cause, the End or Object towards which a formative process advances, and *for the sake of which* it advances—the *logos*,[b] the rational purpose.

(2) The Motive (or Efficient) Cause, the agent which is responsible for having set the process going ; it is that *by which* the thing is made.

(3) The Formal Cause, or Form, which is responsible for the *character* of the course which the process follows (this also is described as the *logos*,[b] as expressing *what* the thing is, or is to be).

(4) The Material Cause, or Matter, *out of which* the thing is made.

(2) As an illustration of the theory of Causes the following will serve. Suppose the thing to be explained is a dog. The chronological order of the Causes is different from their logical order.

(1) The Motive Cause : the male parent which supplied the " movement " that set the process of development going.

(2) The Material Cause : the menstrual fluid and the nourishment supplied by the female parent and other nourishment taken after birth.

(3) The Formal Cause. The embryo and the puppy as it grew into a dog followed a process of development which had the special character proper to dogs.

---

[a] In the translation I have retained the traditional rendering "cause" for αἰτία, although perhaps in some contexts " reason " or " explanation " might have been a closer rendering ; but a variation in the English term might well produce more obscurity than clarity.

[b] See § 10 below.

(4) The Final Cause : the end towards which the process was directed, the perfect and full-grown dog.

A similar set of examples could be constructed to suit the case of artificial objects, though some adjustments would have to be made. In both cases the Formal Cause comes from the same source as the Motive Cause, but with a difference : in the case of natural objects, the parent already possesses the Form fully realized in himself ; in the case of artificial objects, the craftsman possesses the Form " in his Soul." Both the parent and the craftsman normally employ " instruments " to deal with the " material " ; these are not mentioned in the table given above, but will be dealt with in Appendix B §§ 9 ff.

(3) Very often the Final and Motive Causes tend naturally to coalesce with the Formal Cause, in opposition to the Material Cause ; and this opposition is found in *G.A.* (*e.g.*, Book II, *init.*), where Aristotle regards the male (which possesses the Form and which supplies the " movement " and therefore acts as a Motive Cause) as superior and more " divine " than the female (which supplies the raw material for the embryo, *i.e.*, supplies the Material Cause). At the same time, we shall find (below, §§ 6, 7) that the Motive and Material Causes are often together contrasted with the Final Cause, just as Necessity is contrasted with the Good.

(4) In modern parlance the term Cause has become generally limited to Motive (Efficient) Causes, as is shown by the common phrase " cause and effect " ; and hence when Aristotle is concerned especially with the operation of Motive Causes (as *e.g.* at 734 b 9 ff.) his words have a more modern and familiar ring than when he is speaking of Final Causes.

(5) For Aristotle himself, however, it is the Final Cause, the End, which is of paramount importance and which dominates every process. This is abundantly clear in the *P.A.*, where Aristotle endeavours throughout to provide a Final Cause which will explain the existence and structure of the various parts ; and it is no less clear in the *G.A.*, where the whole process of development of the embryo from start to finish is subservient to the Final Cause : the course of the process is deter-

" Teleo-logy."

mined by the nature of the product which is to result from it, not the other way round : things γίγνεται as they do because they *are* what they are.[a] We are therefore justified in describing Aristotle's outlook as " teleological " ; but we must not read too much into this description. " Nature does nothing without a purpose " ; but if we ask what that purpose is, we may find that the answer is not quite what we had expected, that the purpose is not so grand as we had hoped. Aristotle seems to be satisfied when the τέλος has been realized in each individual's full development ; and this is because for him Form is not normally independent of Matter (as it is for Plato) ; Form must be embodied in matter, that is, in individuals. Each complete and perfect embodiment and realization of Form in Matter is therefore for him the crowning achievement of the efforts of the four Causes—it is the End towards which they were directed. We might, then, describe this " teleology " in Bergson's phrase as a doctrine of " internal finality " : each individual is " complete " in itself.[b] Aristotle does, however, maintain that the " most natural " thing for an animal to do is to produce another one like itself, and hence the species is implicated in so far as it is the individual's business to perpetuate it (see App. A §§ 15 ff.). We must also remember that the continuity of γένεσις, one department of which is the continuous succession of generations of animals, is, for Aristotle, " necessary " (App. A § 14) ; and it is also part of the general purpose of " God," who always aims at " the better," and who, because he was unable to fill the whole universe from circumference to centre with eternal " being," filled up the central region of it with the next best available, viz., continuous γένεσις.[c] In another connexion, too, in the *Ethics*, we find that Aristotle looks further than the individual, at any rate so far as man is concerned, for there he tells us that man cannot attain his τέλος in the fullest sense—the " good life "—except in association (τὸ συζῆν) with other men

" Internal finality " ; but modified in various ways : (*a*) perpetuation of species ;

(*b*) human societies ;

---

[a] *Cf.* quotation from Dante, *Paradiso* xx. 78, on p. 1.

[b] *Cf.* § 16.

[c] For further details see App. A § 12.

in a πόλις. But this seems to be due exclusively to the fact that man possesses Reason; and so far as other animals are concerned, Aristotle does not appear to have envisaged any such widening of the τέλος.[a] From yet another point of view, however, when discuss- *(c) sub-*ing the subject of property and household management *or lination* at the beginning of the *Politics* (1256 b 15), Aristotle *of animals* says that just as Nature provides sustenance for animals *to man.* from the very beginning of their existence in the larva, in the egg, or in the uterus, so we must hold that after birth as well Nature provides plants *for the sake of* animals, and also that she provides animals other than man *for the sake of* men, for food and service. And if we are right in holding that Nature makes nothing without a purpose (ἀτελές) or pointlessly, we must of necessity say that " Nature has made all the animals *for the sake of* men."

(6) As Aristotle says at the beginning of *G.A.* I, the two *Grouping of* Causes with which he is chiefly concerned in this treatise *the Causes.* are (1) the Motive (or Efficient) Cause, with which he had not dealt in *P.A.*, and (2) the Material Cause. In zoology, of course, the Material Cause is represented by the " parts " of the body of an animal, and all of these except the generative " parts "[b] he had dealt with in *P.A.*; hence in *G.A.* the Material Cause is represented chiefly by the parts concerned in generation—those, in fact, through which and upon which the Motive Cause operates. At the beginning of Books II and V and at the end of Book V we have further discussions about Causes, and here we find these two Causes identified with " that which is *of necessity* " (ἐξ ἀνάγκης) ; while *Necessity* on the other side and contrasted with them is the Final *versus the* Cause (the Cause " for the sake of which "), which is *Better.* equated with τὸ βέλτιον or τἀγαθόν (*cf. Met.* A 983 a 33, etc.). Indeed, this contrast of Necessity and the Better is continually confronting us throughout the *G.A.* For instance (717 a 15 ff.), whatever Nature does or makes is done or made either διὰ τὸ ἀναγκαῖον or διὰ τὸ βέλτιον ; one or other of these will account for every

---

[a] Perhaps Aristotle would have been willing to include Bees, which possess some " divine " ingredient (see 761 a 5).
[b] It should be remembered that " parts " includes semen, milk, etc. See §§ 18 ff.

phenomenon in the realm of Nature. The whole of Book V is devoted to those features—" conditions " (πάθη) as Aristotle calls them—in animals which are in no way due to a Final Cause but are due purely to Necessity, *i.e.*, to Material and Motive Causes.

Necessity: (7) We must, however, distinguish two sorts of Necessity (the second of which will be the one just described):

(i) "conditional"; (1) The first is that which elsewhere (*e.g.*, *P.A.* 642 a 7 ff., a 32 ff.; *cf.* 639 b 25 ff., *Phys.* 199 b 33 ff.) Aristotle describes as " conditional " (ἐξ ὑποθέσεως) Necessity; that is to say, assuming that some end or purpose is to be achieved, certain means are necessary in order to achieve it. In other words, this is the sort of Necessity which is implied by the Final Cause being what it is. Thus, if a piece of wood is to be split, an axe or some such instrument is necessary, and the axe must, *owing to the nature of the circumstances*, be hard and sharp, hence of necessity bronze or iron must be used to make it. The same sort of Necessity is obviously involved in the construction by Nature of the living body and its various parts : certain materials must of necessity be used and certain processes gone through if this or that living body is to be produced.[a]

(ii) "absolute," (2) The other sort of Necessity is that which Aristotle (*Phys.* 199 b 33 *et al.*) calls " simple " or " absolute " Necessity (ἁπλῶς). This applies in cases (*a*) where the presence of a material object or set of objects (*i.e.*, a Material Cause), and the fact that their nature is what it is, entails as a necessary consequence a certain result or set of results; (*b*) where the nature of the " movement " set up by an activating agent (a Motive Cause) similarly entails certain results. This " simple " or " absolute " Necessity may therefore be regarded as the sort of Necessity involved in the Material and Motive Causes—as a reassertion of themselves by these Causes against the Final Cause (*G.A.* 778 b 1) and against Nature as she advances towards her achievement of it. " In the field of natural objects, Necessity is what

---

[a] Thus even this Necessity can be said to be located " in the matter " (*Phys.* 200 a 15).

we call matter and the κινήσεις of matter " (*Phys.* 202 a 32).[a]

(8) Aristotle, however, is continually drawing our attention to the adroitness of Nature in employing the results of this latter sort of Necessity in order to serve her *purpose*, in order to achieve her *end*. For example (738 a 33 ff.), the production of " residue " by females is ἐξ ἀνάγκης, simply because the female is not hot enough to effect complete concoction ; but Nature makes use of this residue to provide the material out of which the embryo is to be formed. Other instances of things which, though occurring ἐξ ἀνάγκης, are nevertheless employed by Nature ἕνεκά τινος, will be found at, *e.g.*, 739 b 28, 743 a 36 ff., 755 a 22, 776 a 15 ff., b 33. See also *P.A.* 642 a 31, 663 b 13, b 20 ff. On the other hand, Nature cannot always manage to do this, and what results then is a *useless* residue (*e.g.*, excrements), or a " colliquescence " (*P.A.* 677 a 12 ff.). These by-products, however, may still be regarded as " natural," [b] because they are of general occurrence (that is one definition of what is "natural"; see *G.A.* 727 b 29, 770 b 10 ff., 777 a 20 ff., *P.A.* 663 b 28). When, however, Nature is more seriously thwarted by the indeterminateness or the unevenness of matter (*G.A.* 778 a 7 ; *cf.* App. A § 11), we find unnatural results occurring, such as monstrosities and deformities (see *G.A.* IV. 766 a 18 *et passim*).[c]

used by Nature to serve a purpose ;

(9) The " simple " or " absolute " Necessity described in the preceding paragraphs refers only to the limited field of some particular γιγνόμενον, *i.e.*, to the process by means of which some particular natural object is produced and to the Causes therein concerned. But there is a wider and more universal meaning of " simple " or " absolute " Necessity (which we may, if we like, consider as being an extension by Aristotle of the narrower meaning of Necessity as applied to the γένεσις of individual things, though it is really on a different

(iii) "absolute" Necessity in the universe as a whole.

---

[a] The verb συμβαίνειν (sometimes in the phrase συμβαίνει ἐξ ἀνάγκης) is frequently used with reference to the results of this sort of Necessity, as being facts which merely " occur " and are not designed to forward any purpose.

[b] A " colliquescence " may be an unnatural by-product ; see *G.A.* 724 b 26-29 and § 67 below.

[c] For further notes on " Nature," see §§ 12 ff.

ARISTOTLE

plane)—a Necessity which embraces the whole field of γένεσις in the universe at large, *i.e.*, the whole process of the seasonal and cyclic transformations of the " elements," and the whole process of the cyclical generation of animals and plants (see App. A §§ 12 ff.) ; and which even further still (*ibid.* ; and see *P.A.* 639 b 24) includes those things which do not pass through a process of formation (γένεσις) at all, but persist eternal and immutable. In this context Aristotle lays down (*G. & C.* 337 b 35) that ἐξ ἀνάγκης and ἀεί coincide ; thus " eternity "—whether it be individual eternity, as of the stars, or specific eternity, as of plants and animals—and Necessity are mutually interconnected (see App. A § 14) ; thus, that which *always* is or *always* γίγνεται, is, or γίγνεται, *of necessity* ; that which is, or γίγνεται, *of necessity*, is, or γίγνεται, *always*. This meaning of " absolute " Necessity, however, does not enter directly into the *G.A.*, though it is once touched upon in passing (at 770 b 12 ; *cf.* 742 b 26 ff.), and it is incidentally implied to some extent in the passages of Books II and IV referred to and supplemented in the Appendix, A and B.[a]

Λόγος

(10) Frequently in the translation, rather than represent λόγος by an inadequate or misleading word, I have transliterated it by *logos*.[b] This serves the useful purpose of reminding the reader that we have here a term of wide and varied application, with which a number of correlated conceptions are associated, one or other of which may be uppermost in a particular case. The fundamental idea of λόγος, as its connexion with λέγειν shows, is that of *something spoken or uttered*, more especially a *rational utterance* or *rational explanation*, expressing a thing's *nature* and the *plan* of it ; hence λόγος can denote the *defining formula*, the *definition* of a thing's *essence*, of its *essential being* (as often in the phrase λόγος τῆς οὐσίας), expressing the structure or character of the object to be defined. See also § 1 above.

---

[a] Other modes of Necessity not relevant to *G.A.* are here omitted.
[b] The less technical meanings are translated in the normal way.

GENERATION OF ANIMALS

Ἀρχή

(11) For want of a better term, and in order to preserve the line of Aristotle's thought, I have usually translated ἀρχή by " principle," or " first principle." There is, however, really but little difficulty about this term, for the context will usually indicate what its connotation is. A few examples of its use may be given. (1) Often, as at 715 a 6, it is a principle or source of " movement " (ἀρχὴ τῆς κινήσεως). Hence, obviously, (2) the Motive Cause may be described as an ἀρχή, and so too may the other Causes (e.g., 716 a 5 ff., 778 a 7), including Matter; and for the same reasons the sexes also are ἀρχαί; so is semen. (3) An ἀρχή is something which though small in itself is of great importance and influence as being the source or starting-point upon which other things depend, and which causes great changes (κινήσεις) in them (cf. 716 b 3, 763 b 23 ff., 766 a 14 ff.). An ἀρχή may, of course, be of greater or less fundamental importance; and the ultimate ἀρχή of an animal is its heart (e.g., 766 a 35), though there are also ἀρχαί that are external to the animal, e.g., the sun and moon (777 b 24).

Φύσις, πήρωμα

(12) Πήρωμα, ἀναπηρία, and cognate words occur several times in G.A., and for convenience I have translated them " deformation " or " deformity." Other possible renderings, none of which fully brings out the meaning of the Greek word, are given in the note on 737 a 25. The underlying notion is that φύσις has not succeeded in achieving her proper τέλος; and this close connexion of πήρωμα with a falling short of natural completeness is clearly brought out by the reasons given at 724 b 33 why semen cannot be a πήρωμα, viz., because it is found in all individuals (for that which is " general " is " natural," see § 8), and because ἡ φύσις γίγνεται out of semen. " Deformation."

(13) Perhaps the most striking instance of Aristotle's application of this idea is his statement (775 a 15) that female-ness (θηλύτης) is " as it were a natural ἀναπηρία." Here we have two conceptions of Nature asserting themselves The female a " deformity."

ARISTOTLE

in Aristotle's mind : (1) that the male represents the full development of which Nature is capable ; it is hotter than the female, and more " able " to effect concoction, etc. ; but at the same time (2) the female is so universal and regular an occurrence that it cannot be dismissed out of hand as " unnatural " ; besides, the female is essential for generation, which is a typically " natural " process (see § 5).

Nature (14) **versus Nature.** This opposition of " Nature " to " Nature " is, however, not unique, for it is found elsewhere in Aristotle ; *e.g.*, at *G.A.* 770 b 20 he can say that τὸ παρὰ φύσιν is in a way κατὰ φύσιν, viz., when ἡ κατὰ τὸ εἶδος φύσις has not mastered ἡ κατὰ τὴν ὕλην φύσις ; and at *P.A.* 663 b 22 he speaks of ἡ κατὰ τὸν λόγον φύσις making use of the products of ἡ ἀναγκαία φύσις in order to serve a purpose (*cf.* also *P.A.* 641 a 26, 642 a 17 ; at *Phys.* 199 a 31 Aristotle distinguishes φύσις ὡς ὕλη and φύσις ὡς μορφή, the latter being a τέλος and ἡ αἰτία ἡ οὗ ἕνεκα. *Cf.* 729 a 34, n.).

Nature: (15) **as purpose ;** It is impossible and unnecessary to provide here a full account of what Aristotle intended by the term φύσις, since a proper understanding of it can best be obtained by reading Aristotle's works themselves, and for this *G.A.* is one of the most useful, because it is pervaded by references to φύσις. A few remarks may however be made here about φύσις in its highest manifestation.

comparable (16) **to an artist.** By Aristotle, φύσις and the products of φύσις are constantly compared with τέχνη and the products of τέχνη : φύσις works to produce a finished product, a τέλος, just as the artist or craftsman does [a] ; and φύσις, again like the artist, uses " instruments," charged with a specific " movement," in order to bring these products to fulfilment.

Nature as Soul. And the most typical of the products of φύσις are, of course, living creatures ; indeed, Aristotle can speak of the φύσις of *each* living thing as being identical with nutritive Soul (741 a 1, where see note, and *cf.* *P.A.* 641 b 9), the Soul which generates and fashions it and promotes its growth ; and again (*De caelo* 301 b 17), φύσις is to be regarded as a principle of movement *in the*

---

[a] φύσις is also compared (744 b 16) to a careful housekeeper, who throws away nothing that is useful ; or to a cook (743 a 31 ; *cf.* 767 a 17 ff.), tempering the heat of her stove so that the food she is preparing may be done to a turn. See also Συμμετρία, § 39.

thing itself. An artist, then, at work—yes, but *in* each several thing ; and it is doubtful whether Aristotle had, or intended to have, any idea of Nature over and above, outside, the individual things *a* which he described as her " works." In fact, he goes so far as to say (*P.A.* 641 b 11) that no abstraction can be the object of study for Natural science, because Nature makes all that she makes *to serve some purpose* (ἕνεκά του). Nature aims always at producing a τέλος in the sense of a completely formed *individual*, and that is the Final Cause in each case, for that is what has the best claim to be called a " being " (οὐσία).*b* There is, says Aristotle, more beauty and purpose (τὸ οὗ ἕνεκα καὶ τὸ καλόν) to be found in the works of Nature than in those of art (*P.A.* 639 b 20).

(17) Nevertheless, we must remember that Nature is not, in Aristotle's view, a term to be exclusively reserved for the Final Cause, with which are associated the Formal and often the Motive Causes ; it may be applied also, as we saw just now (§ 14), to the Material Cause ; and in this connexion we may recall that, for Aristotle, Matter and Form themselves pervade all the strata of existence, for even the simplest sort of Matter is to some extent " informed," and Matter in its highest phase is identical with Form (see 729 a 34, n.). Nature as matter.

Μόριον, μέρος, *" part "*

(18) The term " part," which occurs in the title of the treatise Meaning. *De partibus animalium,* περὶ ζώων μορίων (or, as Aristotle himself calls it at *G.A.* 782 a 21, " the treatise *Of the Causes of the Parts of Animals* "), includes considerably more than is normally included by the English " part of the body." For instance, we should not normally call blood a " part," but Aristotle applies the term μόριον to all the constituent substances of the body as well as to the limbs and organs. For him, blood is one of the ζώων μόρια (see *P.A.* 648 a 2 ; and note on *G.A.* 720 b 31). Since, however, all the " parts " are either " uniform " or " non-uniform," a detailed description of them will be more appropriate in the following paragraphs.

---

a See however § 5 above.
b See App. A § 18.

ARISTOTLE

Τὰ ὁμοιομερῆ μόρια, *the " uniform parts "*
Τὰ ἀνομοιομερῆ μόρια, *the " non-uniform parts "*

Two sorts of " parts." (19) At *G.A.* 724 b 23 ff., Aristotle classifies the substances found in the body into five divisions, one of which is " natural parts," [a] and this division he subdivides into " uniform parts " and " non-uniform parts." As examples of " uniform parts " he cites (*P.A.* 647 b 10 ff.) blood, serum, lard, suet, marrow, semen, bile, milk, flesh [b] (these are soft and " fluid " [c] ones) ; also bone, fish-spine, sinew, blood-vessel (these are hard and " solid " ones). And although in some cases the same name is applied to the substance out of which the whole is made and to the whole that is made out of it,[d] this is not true in all cases. Examples of " non-uniform " parts are (*P.A.* 640 b 20) face, hand, foot.

Relation between them. (20) The relation of the " uniform " to the " non-uniform " parts Aristotle describes as follows (*P.A.* 647 b 22 ff.) :

(a) some of the uniform parts are the material out of which the non-uniform are made (*i.e.*, each instrumental part is made out of bones, sinews, flesh, etc.) ;

(b) some, viz., " fluid " ones, serve as nourishment for those in class (a), since all growth is derived from fluid matter ;

(c) some are " residues " [e] from those in class (b), *e.g.*, faeces, urine.

Thus it is not possible to equate this division into uniform and non-uniform parts with the more modern division into tissues and organs ; for instance, blood, though a uniform part, is not a tissue. The term " organs," on the other hand, corresponds closely with Aristotle's own description of the non-uniform parts (*P.A.* 647 b 23) as τὰ ὀργανικὰ μέρη, " the instrumental parts."

(21) The fundamental difference between the two sorts of " parts " is that each of the uniform parts has its own definite character as a *substance* (in the modern sense),

---

[a] This must not be taken to imply the existence of *unnatural* " parts."
[b] Some of these are also " residues " ; see below, § 65.
[c] For the meaning of " fluid " and " solid," see below, § 38.
[d] *e.g.*, we speak of " bone " and " a bone " ; Aristotle's own example is " blood-vessel."
[e] See § 65.

xlviii

while each of the non-uniform parts has its own definite character as a *conformation* or organ. The heart is the only " part " which belongs to both classes (*P.A.* 647 a 25 ff.) : it is made out of one uniform part only, but at the same time it has essentially a definite configuration or shape, and thus it is a non-uniform part.

(22) The four stages or " degrees of composition," so far as biology is concerned, are thus enumerated by Aristotle (*G.A.* 715 a 10 ff. ; *cf. P.A.* 646 a 13 ff.) :

(1) The four " Elements," Fire, Air, Water, Earth [a] ;
(2) the uniform parts ;
(3) the non-uniform parts ;
(4) the animal organism as a whole.

We thus begin from the simplest sorts of matter (Aristotle calls the four Elements " simple bodies ") and proceed upwards by stages until the most organized or most " informed " sort of matter is reached : each stage is the " material " for the stage next above it (*G.A.* 715 a 9 ff.).

[marginal note: The stages o composition.]

Δύναμις

(23) This term has a number of different, though related, meanings, and it is not always easy to determine precisely which one Aristotle has uppermost in mind. Unlike some other terms, therefore, this one cannot always be represented by the same term in English, and sometimes it is best left untranslated.

(24) (A) To begin with, we will examine the pre-Aristotelian meaning of δύναμις, as found for instance in the Hippocratic corpus and in Plato's *Timaeus*. Δύναμις was the old technical term for the simplest sorts of matter, *i.e.*, for what came later to be called στοιχεῖα (" elements "). Δύναμις was however applied exclusively to substances of a particular class, viz., τὸ ὑγρόν, τὸ ξηρόν, τὸ θερμόν, τὸ ψυχρόν, τὸ πικρόν, τὸ γλυκύ, τὸ δριμύ, etc., etc. In the Hippocratic treatise περὶ ἀρχαίης ἰητρικῆς (*The Ancient and Genuine Art of Medicine*) these substances are regarded as being the constituents both of the body and

[marginal note: Dynameis as element-ary kinds of matter.]

---

[a] In the *P.A.* passage Aristotle says it might be better to substitute for these " the δυνάμεις," or rather four of them ; see below, § 24. Fire, Air, Water, Earth are of course the constituents of non-living things as well ; see App. A § 2.

of its foods. The δυνάμεις are referred to by Aristotle at the beginning of *P.A.* II (see § 22, note), where he speaks of "the 'elements' as they are called, viz., Earth, Air, Water, Fire, or perhaps it is better to say the δυνάμεις— not all the δυνάμεις, of course, but these four, ὑγρόν, ξηρόν, θερμόν, ψυχρόν." The explanation of this is that although Aristotle held that in a sense Earth, Air, Water, and Fire were "elements," *i.e.*, that they were the simplest states of matter actually found in the world as we know it, yet theoretically each of them could be resolved into a pair of δυνάμεις : thus Fire is θερμόν and ξηρόν, Air θερμόν and ὑγρόν, Water ψυχρόν and ὑγρόν, Earth ψυχρόν and ξηρόν (*G. & C.* 330 a 30 ff.), each of them being characterized by one constituent *par excellence*, Fire by θερμόν, Air by ὑγρόν, Water by ψυχρόν, Earth by ξηρόν. According to Aristotle, all other physical " differences " are consequent upon these four fundamental ones.

Origin of (25) The meaning implied in this use of δύναμις seems to have *this usage.* been " substance of a specific character " (perhaps the adjective " strong " should be prefixed : this would of course be very appropriate to δυνάμεις such as τὸ δριμύ, τὸ πικρόν, etc.). But originally, no doubt, the term was an item in the Pythagorean political metaphor terminology, as would appear for instance from the theory held by Alcmeon [a] that bodily health was maintained by the ἰσονομία τῶν δυνάμεων, and that the " monarchy " of any one of them produced disease. It is important to notice that there is no notion here of the substance *having* power in the sense of power *to produce a specific effect* [b] upon a body, though this was a meaning

Earth, Air, Water, Fire, resolvable into dynameis.

---

[a] See Aëtius v. 30. 1 (Diels, *Doxographi Graeci* 442).

[b] *e.g.*, causing stomach-ache. In Plato's *Timaeus* we find this extended meaning of δ ναμις (*i.e.*, power to produce a specific effect) side by side with the old meaning of specific substance ; and it is frequent in π διαίτης. Clearly, this marks a change over from the medical theory originally associated with the political metaphor terminology ; and we find that, as δύναμις takes on the meaning of " power to produce a specific effect," the term " humour " comes in to denote the specific substances to which οὐναμις was originally applied. Thus Diocles (*apud* Galen vi. 455) can argue against doctors who hold that all things which possess similar " humours " also possess the same δυνάμεις (powers of producing specific effects on the body), *e.g.*, are laxative, diuretic, etc. There is no space to say more here on this development, which I dealt with fully in my thesis *Pseudo-Hippocrates Philosophus* (1928). Studies

which developed later. A δύναμις is rather a substance which *is* a power, which can assert itself, and by the simple act of asserting itself, by being too strong, stronger than the others, can cause trouble. The remedy in such a case is to deprive it of some of its strength, until it again takes its proper place among its peers, or, in the language of medicine, to " concoct " it or otherwise bring it into a harmless condition by " blending " [a] it with the other substances.

(26) (B) As each of the substances known as δυνάμεις had its own specific and peculiar character, sharply marked off from the others, it was easy for the meaning " peculiar and distinctive character " to become closely associated with the term δύναμις, quite apart from any reference to these particular substances. In fact, it almost comes to mean any " substance of a distinctive quality " ; and in this sense it is found in *G.A.*, for instance at 720 b 32 (ἄλλη τις δύναμις) and 736 a 21 (Aphrodite was called after " this δύναμις," *sc.* ἀφρός, foam). From this it is an easy step to " distinctive physical quality," or simply " distinctive character " (as, *e.g.*, at 731 b 19, where it is joined with λόγος τῆς οὐσίας ; at 751 a 33, where it refers to the distinctive character of the yolk and white of an egg respectively [b] ; and *cf.* 733 b 15 ἔχει δύναμιν ᾠοῦ— it has the distinctive character of an egg, it is equivalent to an egg ; and 780 b 8, 784 b 15) ; or " characteristic " (applied to the sexes at 756 a 1, 763 b 23 ; *cf.* 760 a 19). *D nameis as substances of distinctive character.*

(27) In the sense of " (substance of) distinctive character " it can be used practically as an alternative to φύσις, or in conjunction with φύσις (as indeed it often is in Hippocrates and Plato), and this seems to be the use of it in *P.A.* 655 b 12 ἐξ ἀνάγκης δὲ ταῦτα πάντα γεώδη καὶ στερεὰν ἔχει τὴν φύσιν· ὅπλου γὰρ αὕτη δύναμις (*cf. P.A.* 651 b 21).

(28) (C) From this usage it is not far to the idiomatic, pleonastic usage, *e.g.*, ἡ τῶν ἐντέρων δύναμις (almost=τὰ ἔντερα *P.A.* 678 a 13) ; ἡ τῶν πτερῶν δύναμις (=τὰ πτερά, 682 b 15) ; and this is paralleled by the similar usage *Idiomatic usage.*

on some of the uses of δύναμις have been made by J. Souilhé, *Étude sur le terme* δύναμις *dans les dialogues de Platon*, Paris, 1919, and A. Keus, *Über philosophische Begriffe u. Theorien in den hippokratischen Schriften*, Cöln, 1914, pp. 46 ff.

[a] See § 40.   [b] φύσις is used in a similar context at 753 a 35.

of φύσις, e.g., ἡ τῶν ὀστρακοδέρμων φύσις (G.A. 761 b 24), ἡ τοῦ αἰδοίου φύσις (717 b 18 ; cf. also 755 a 20), ἡ τῶν πτερῶν φύσις (749 b 7, a striking instance, because φύσις is used in an entirely different sense, " Nature," in the very next line) ; and even σύστασις is sometimes used in a similarly weakened sense, e.g., ἡ τῶν ὄρχεων σύστασις (G.A. 717 a 15), ἡ τῶν καταμηνίων σύστασις (G.A. 727 b 33) ; and σύστασις appears in two manuscripts as a variant for φύσις at G.A. 717 b 20.

Dynamis in generation.

(29) (D) In the passages dealing with the rôle of the male parent in generation we find phrases such as " the δύναμις in the semen," " the δύναμις in the male " (e.g., 726 b 19,[a] 727 b 14, 729 b 27, 730 a 3, a 14, 736 a 27, etc.). The meaning of δύναμις here would appear to be fundamentally the same as that dealt with in § 26 above, i.e., δύναμις here is the physical substance by means of which impregnation is effected ; and the distinctive physical characteristic with which we find this δύναμις closely associated by Aristotle is " vital heat " or " Soul-heat."[b] The most distinctive characteristic, however, of this substance is that it is charged with a specific

Associated with "movement."

" movement," capable of constituting and developing an embryo out of the matter supplied by the female ; and hence we also find a close association of δύναμις with κίνησις.[c] This is the most important extension of δύναμις in its ancient sense made by Aristotle, for it links up the old sense of the term with the typically and peculiarly Aristotelian sense of δύναμις = " potentiality " (see §§ 34 ff. below).

(30) (E) Under the same category comes the use of δύναμις and ἀδυναμία as applied to male and female respectively (G.A. 765 b 9 ff., 766 a 32 ff.), for these are explained by Aristotle as the ability and inability respectively to effect " concoction " of the ultimate nourishment (blood) into semen, and this is directly dependent upon the possession of sufficient " natural *heat*."

[a] An interesting example, because δυνάμει (= *potentially*) occurs in the previous line.
[b] Not to be confused with the *ordinary* δύναμις " θερμόν " ; see App. B §§ 13, 18.
[c] References for δύναμις associated with " vital heat " and κίνησις, e.g., 726 b 19 ff., 729 b 6 ff., 738 b 12, 739 b 24, 740 b 30 ff., 767 b 17 ff. (cf. 755 a 20 " the φύσις of the Soul-heat "). See also κίνησις, § 50.

(31) (F) Under the same category too must be placed the use of the term δύναμις in the remarkable discussion on heredity in Book IV. This is admittedly a particularized use of the term,[a] and Aristotle carefully explains its meaning when he first introduces it (767 b 23 ff., *q.v.*). But here too it is applied to special and distinctive characteristics, be it those of genus, species, or individual, and therefore this use of it stands in the same line of succession as the meaning already described in §§ 24 ff. As for the way in which Aristotle conceived these δυνάμεις to operate, it is clear that, as they were present both in the semen and in the menstrual fluid (see *loc. cit.*) and gave rise to κινήσεις (767 b 36), they must have been closely associated with Soul and inherent in its instrument *pneuma*.

(32) It may be noted here that the physical substance concerned throughout the theory of generation is *pneuma* (a substance " analogous to *aither*," the " fifth element," the " element of the stars "), with which Soul is " associated " ; and it is this *pneuma* which Soul charges with a specific " movement " and uses as its " instrument " in generation just as it does in locomotion, and as an artist uses his instruments, to which he imparts " movement," in order to create his works of art. (For fuller details about *pneuma*, see App. B, and *cf.* § 45.) *Pneuma.*

(33) Thus δύναμις, even at its most glorified, still retains the marks of its descent from the historic δύναμις of the early medicine, for, although Soul-heat is something different from the old θερμόν and superior to it, nevertheless it is still θερμόν. And there is another respect in which its descent is still to be seen, though this time it may be fortuitous and perhaps no more than a verbal coincidence. This physical substance is the vehicle for the activity of Form (εἶδος) ; and in the Hippocratic treatise π. ἀρχαίης ἰητρικῆς each of the innumerable physical substances known as δυνάμεις had also been called an εἶδος. Continuity in meaning of *dynamis.*

(34) (G) We now come to the last and most typically Aristotelian of the meanings of δύναμις ; and although it is Dynamis as " potentiality."

---

[a] And therefore I have felt justified in translating it "faculty" in this sense, to avoid repeated recurrence of the Greek word transliterated. It may perhaps be simply an extension of the meaning dealt with in the last section but one.

ARISTOTLE

usually considered independently of the ones we have already described, it is clear from Aristotle's own words that he did not so regard it himself, for he associates it very closely with κίνησις. In *Met.* Δ 1019 a 15 ff. and Θ 1046 a 10 f., he defines the primary and fundamental sense of δύναμις in this connexion in the following words : δύναμις is ἀρχὴ κινήσεως ἢ μεταβολῆς ἐν ἑτέρῳ ἢ ᾗ ἕτερον : δύναμις is a principle (or source) of κίνησις or of change—a principle either (a) subsisting in some other thing than that which is to be affected by the κίνησις or change, or (b) subsisting in the thing itself *qua* other than changeable in that respect. An example of (a) is building ; an example of (b) is the science of medicine in the case of a person who is being healed but not *qua* being healed (a man doctoring himself). That is the fundamental sense of this δύναμις ; but Aristotle goes on at once to mention the complementary sense of it, which in fact is the sense in which he commonly uses it, viz., the δύναμις of being acted upon (παθεῖν), which he describes as the ἀρχή in the thing acted upon of a passive change caused either by some other thing or by itself *qua* other (ἡ ἐν αὐτῷ τῷ πάσχοντι ἀρχὴ μεταβολῆς παθητικῆς ὑπ' ἄλλου ἢ ᾗ ἄλλο).

(35) It is therefore clear that there is the closest possible connexion between this notion also of δύναμις and κίνησις : δύναμις is in fact the capacity to set up "movement" or (more commonly) to be set in "movement" : it is a "dynamic" conception. To say that A is B δύναμει (*potentially*) means that A is a Material Cause capable of being set moving with a certain κίνησις by a Motive Cause, which κίνησις will result in A acquiring the Form of B, thus attaining the Final Cause (becoming a B itself). It is thus a conception which integrates the four Causes through the process of κίνησις.

(36) The correlative of δυνάμει (*potentially*) is ἐνεργείᾳ (*in actuality*) ; " X ἐνεργείᾳ " means something in which the Form X has been realized—something which already possesses the Form X, and further, in the case of animals, something which can reproduce the Form X in other matter which is so far only " δυνάμει X."

(37) Of all the possible translations or mistranslations of δύναμις, " force " is one of the most misleading ; for

liv

there is nothing more fundamental in Aristotle's—and in his predecessors'—idea of δύναμις than that it is something *natural*; and the associations of the term "force" run counter to this. Aristotle himself contrasts "natural" and "enforced" movement (see App. B § 22, and *cf.* 739 a 4, 788 b 27, *Politics* 1253 b 22). It is also important that any notion of a vague and indefinite urge, even (and perhaps especially) where Soul is involved, should be excluded; for, as we have seen, δύναμις is associated primarily with some material substance of a specific character or with some κίνησις (carried in a definite substance) of a specific character. From every point of view it is best to avoid "force" altogether as a translation of δύναμις.

Τὸ ὑγρὸν καὶ τὸ ξηρόν, "*fluid substance and solid substance*"

(38) These are two of the original δυνάμεις (§ 24); and following Ogle in his translation of *P.A.* I use the above renderings as being more in conformity with the definitions given by Aristotle himself than "moist" and "dry" which have sometimes been used. Actually neither pair of English words quite expresses the Greek. Aristotle's definition of them (at *G. & C.* 329 b 30) is this: "ὑγρόν is that which is not bounded by any boundary of its own but can readily be bounded; ξηρόν is that which is readily bounded by a boundary of its own but can with difficulty be bounded"; at the end of each definition there should of course be understood "by a boundary imposed from without." (ὑγρόν is τὸ ἀόριστον οἰκείῳ ὅρῳ εὐόριστον ὄν, ξηρόν is τὸ εὐόριστον μὲν οἰκείῳ ὅρῳ δυσόριστον δέ.)

Translation.

Definition.

Συμμετρία, κρᾶσις

(39) An idea which recurs a number of times in *G.A.* is that of συμμετρία. In this treatise the majority of the references[a] to συμμετρία are concerned with the relative amounts of residue contributed in generation by the two

Correct proportion: (*a*) in generation;

[a] See list of passages in the Index.

parents, or to the heat or " movement " contributed by the male or otherwise provided (*e.g.*, by the Sun). Σύμμετρος κίνησις is also mentioned in connexion with the amount of fluid in the pupil of the eye (779 b 25 ; *cf.* 780 b 24). The meaning throughout is that the amount of substance, or of heat, must be adjusted in the correct proportion ; and this, as the context at 786 b 5 indicates, means suitably adjusted between the two extremes of too much and too little. This at once recalls to mind the famous doctrine of the " mean " in the *Ethics*, where goodness (or " virtue," ἀρετή) is held to be a mean between the two extremes of excess and deficiency ; indeed, at *E.N.* 1104 a 12 ff. Aristotle says that whereas the moral ἀρεταί are destroyed by excess and deficiency, they are produced and preserved by the mean, just as excessive food and drink destroy health, whereas τὰ σύμμετρα produce and preserve it.[a] Similarly, at *Phys.* 246 b 4 he says " we posit that the ἀρεταί of the body, viz., health and fitness, lie in the κρᾶσις (blend) and συμμετρία of hot things and cold,[b] either as regards each other internally, or as regards the surrounding environment ; and the same applies to the other ἀρεταί and κακίαι." This reference to κρᾶσις and to the environment is closely parallel to the most important passage on συμμετρία in *G.A.*, 767 a 14 ff., where Aristotle says that the male and female need συμμετρία as towards each other, because all things formed by Nature or by Art λόγῳ τινί ἐστιν—depend upon a certain proportional relationship, or ratio. Just as in cooking, the heat must strike the due proportion,.the mean, or your meat will be either overdone or underdone.[c] So too in the mixture of male and female, συμμετρία is required. He then goes on to speak of the dependence of our bodily condition upon the κρᾶσις of the environing air (*cf.* 777 b 7) and of the foods we take, and especially the water.

(40) This is not the place to discuss the origin of the doctrine

(b) in ethics and politics;

(c) in bodily health.

Blend.

---

[a] The importance of συμμετρία in the growth of a State is also emphasized by comparing it with the growth of the body (*Pol.* 1302 b 35 ff.).

[b] *Cf.* the phrase ψυχρότερα τῆς συμμέτρου κράσεως used of the parts around the brain (*P.A.* 652 b 36).

[c] *Cf.* § 16 above.

of the mean, nor of the closely allied doctrine of κρᾶσις, except that it should be noted that great importance is attached in the Hippocratic treatise π. ἀρχαίης ἰητρικῆς to securing proper κρᾶσις for the ingredients of the food we take and of the constituents of our bodies (the two sets of substances being identical); and that in π. διαίτης the κρᾶσις of Fire and Water in the Soul is responsible for its health and sensitivity (*cf. G.A.* 744 a 30). References to the pertinent passages of the Hippocratic treatises will be found in the notes; see also *P.A.* (Loeb ed.), pp. 37 f. It should also be noted that Alcmeon of Crotona (Aëtius v. 30; see Diels, *Doxographi* 442) held that health was the σύμμετρος τῶν ποιῶν κρᾶσις (*cf.* § 25). It is important to realize that some, at any rate, of Aristotle's terminology was the common property of scientific writers.

Ψυχή, " Soul "

(41) The English word Soul, as will be seen, owing to its associations is not entirely satisfactory as a rendering of ψυχή, but it is by far the most convenient one, and I have used it in preference to " life " or " vital principle " (for which Aristotle employs other terms).

(42) Animate bodies, bodies " with Soul in them " (ἔμψυχα), are " concrete entities " made up of Form and Matter, Soul being the Form and body the Matter; indeed, Soul is the Form of the body. (*Cf. G.A.* 738 b 27, n., 741 a 1.) Aristotle also describes this relationship by saying that Soul is the " realization " (ἐντελέχεια, " actuality ") of the animal body. Strictly speaking, Soul is the " first realization " of an animal body, for an animal can " have Soul in it " and yet be asleep; its active, waking life will be its " second realization." Further, Aristotle tells us that Soul is the first realization of a body *furnished with organs*. The importance of this is clear: the body is *for the sake of* the Soul (because the Soul is the Final Cause as well); and hence (*P.A.* 687 a 8 ff.) Aristotle maintains that man has hands because he is the most intelligent animal, not, as some had said, the most intelligent animal because he has hands. Soul is " prior " to body, and the body is such as it is because that is the sort of body the Soul

Soul the Form of Body.

lvii

requires in order to function. Indeed, the Soul cannot function without a body; it cannot, we may say, exist (*De anima* 414 a 19).

The (43) faculties of Soul.

This will be clear if we distinguish the different parts or "faculties" of Soul. They can be arranged in a definite order, so that the possession of any one of them implies the possession of all those which precede it in the list; and it will be seen that all except the last of them obviously require a body for their functioning.

(1) Nutritive and generative Soul,[a] in all plants;
(2) sentient Soul ⎫
(3) appetitive Soul ⎰   in all animals;
(4) locomotive Soul,   in some animals;
(5) rational Soul,   in man only.

Rational (44) Soul.

It is the last faculty of Soul which stands out by itself. Aristotle feels that he cannot admit that Soul is *wholly* dependent upon body for its functioning; there may, he says, be some "part" of Soul which is not the "realization" of any body, a "part" whose activities have nothing whatever to do with any physical activities (*G.A.* 736 b 28). This part, which is "rational Soul," comes in over and above, from without (*G.A.* 736 b 25 ff.), and continues to exist after the death of the body (*De anima* 413 a 6, b 24 ff., 430 a 22, etc., *Met.* Λ 1070 a 26). The problems raised by this belief are, however, not fully dealt with by Aristotle even in *G.A.*, where he has much to say about the development of Soul in the embryo; indeed, he nowhere offers any solution of them.

Soul (45) subsists in *pneuma.*

So much then for the theoretical relationship of Soul and body. What is their practical relationship? How precisely does Soul function through the body? The answer to these questions is one of the most striking parts of all Aristotle's philosophical work. Soul, says Aristotle, is not, as some have wrongly supposed, Fire or any such stuff (δύναμις); it is better to say that it "subsists in *some such substance*" as Fire (ἐν τοιούτῳ τινὶ σώματι συνεστάναι), viz., in "hot substance" (τὸ θερμόν), which is the most serviceable of all substances for the activities of Soul (*P.A.* 652 b 8); and elsewhere (*G.A.* 736 b 30 ff.; see App. B § 13) he is more explicit.

---

[a] See also 744 b 33, n.

This θερμόν is no ordinary θερμόν, but it is *pneuma*, a substance " more divine " than Fire, Air, Water, or Earth, and " analogous to " the fifth element, *aither*, the element of the Upper Cosmos. It is this *pneuma*, and the substance (φύσις) in the *pneuma*, which is the vehicle of Soul, and it is *pneuma* which Soul uses as its " instrument," through which it brings about κίνησις, both in moving the full-grown body and in " moving " *i.e.*, developing the embryo. Here, then, we have reached the heart of the business : *pneuma* is the last physical term of the series ; *pneuma* is the immediate instrument of Soul, and it is through *pneuma* first of all that Soul expresses itself.

(46) It must not be supposed that this *pneuma* is the breath breathed in by the animal from outside ; Aristotle is most explicit on this point, and he often describes this *pneuma* as " connate " (σύμφυτον). Owing to the important place of Σύμφυτον Πνεῦμα in Aristotle's biology, I have provided a full account of its nature and functions in Appendix B.[a]

This *pneuma* is " connate."

Κίνησις

(47) Κίνησις is a term of wider range than the English " movement," though it is useful to retain " movement " as a translation in order to preserve the line of Aristotle's thought. Κίνησις is one department of μεταβολή (Change), of which there are three divisions :

Meaning.

Two, which are changes affecting οὐσία :

(1) γένεσις, change from the non-existent to the existent ;
(2) φθορά, change from the existent to the non-existent.

And one, which comprises changes affecting categories other than οὐσία :

(3) κίνησις, change in existing things.

(48) Κίνησις has three subdivisions :

(a) as regards Quantity : Growth and diminution ;
(b) as regards Quality : " Alteration " (ἀλλοίωσις) ;

Varieties of " movement."

---

[a] See also G. L. Duprat, *La théorie du* πνεῦμα *chez Aristote, Archiv f. Gesch. d. Phil.* XII (1899), 305 ff., and W. W. Jaeger, *Das Pneuma im Lykeion, Hermes,* XLVIII (1913), 29 ff.; the latter also gives a history of the *pneuma*-doctrine. See also W. W. Jaeger, *Diokles von Karystos* (1938) and J. I. Beare, *Greek Theories of Elementary Cognition from Alcmeon to Aristotle* (1906).

(c) as regards Place : Locomotion (φορά), either (i) in a circle, or (ii) in a straight line.

Sometimes Aristotle includes γένεσις and φθορά as a fourth subdivision of κίνησις, thus making κίνησις embrace every variety of change. (See also *Met.* Λ 1069 b 8 ff.)

Definition. (49) The definition of κίνησις which he gives at *Phys.* 201 a 11 ff. is this : ἡ τοῦ δυνάμει ὄντος ἐντελέχεια, ᾗ τοιοῦτον, κίνησίς ἐστιν : " Movement " is the realizing of that which is potentially X, *qua* potentially X. For example, to take the case of ἀλλοίωσις, κίνησις is the altering of a thing which is alterable, *qua* alterable ; and so with the other modes of potentiality.

"Move-ment" and Form. (50) It will be seen at once that, in order to set going the κινήσεις by which the various potentialities are to be realized, Motive Causes are required. And the thing which causes the "movement," says Aristotle (*Phys.* 202 a 10), will always bring with it some Form (maybe some οὐσία, or some quality, or some quantity), which will be a " principle " and a cause of " movement." In other words, the " movement " will be informed, determined, characterized, in such a way that it will produce a thing which has a certain οὐσία, or quality, or quantity. The agent (or Motive Cause), then, will set up in the material a " movement " which will result in the material which is *potentially* A becoming A *in actuality*, that is to say, in its acquiring the same Form as that which the agent possessed. And this result is brought about, generally, by the use of an intermediary, an " instru-ment " (see App. B §§ 6, 15), to which the agent imparts the " movement " for transmission.

"Move-ment" derived from Soul, (51) All these sorts of κίνησις, Aristotle points out (*De anima* 415 b 22 ff.), are derived from Soul ; they are not found apart from Soul. This is because Soul is the Cause (αἰτία) and principle (ἀρχή) of the living body : it is alike its Motive Cause, its Final Cause and its Formal Cause (*ibid.* 415 b 8 ff.), and it is situated in the heart. We must not forget, however, that in the long run κίνησις, at any rate κίνησις of inorganic things, is due to the

and from the Unmoved Mover. Unmoved Mover, from whom " movement " is mediated by the heavenly bodies to the Lower Cosmos (App. A §§ 3 ff.) ; and even in the case of living things (" things

GENERATION OF ANIMALS

with Soul in them "), the heavenly bodies act as a Motive Cause, for " man is begotten by man and by the Sun " (see App. A §§ 6, 9).

Γένεσις, γίγνεσθαι

(52) Γένεσις, as we have already seen (§ 47), is a process Meaning. of change ; in fact, it is the most fundamental sort of change, viz., " coming into being " ; hence, the product resulting from a process of γένεσις is some οὐσία, for although some sorts of οὐσία persist eternally, there are others which are " perishable," *i.e.*, which are subject to γένεσις and φθορά (see App. A §§ 1, 12, 16). Indeed, the sort of οὐσία produced by the γένεσις with which our present treatise is concerned—animals and plants— is the οὐσία which Aristotle considers to have the best claim to the name (App. A § 18).

(53) Γένεσις, and its verb γίγνεσθαι, are terms of frequent Translation. occurrence in Aristotle, and especially in *G.A.* In the title of the treatise, γένεσις is commonly translated " generation," and this is a convenient rendering of it there ; but we must not forget that γένεσις also refers to the whole process of an animal's development until it has reached its completion ; that is to say, γένεσις includes the whole subject of reproduction and embryo- logy. In the body of the treatise " generation " is often not satisfactory as a translation ; nor is " coming-to-be" particularly neat or indeed appropriate in a biological work. I have therefore commonly used " formation," " process of formation " and the like to render γένεσις, and for γίγνεσθαι " to be formed," " to come to be formed," etc.

Συνιστάναι, συνίστασθαι

(54) Another verb closely connected with γίγνεσθαι is the Meaning, verb συνιστάναι, which might almost be regarded as the active voice of γίγνεσθαι, though συνιστάναι tends rather to refer to the beginning of the process. It is specially frequent in passages describing the initial action of the semen in constituting a " fetation " out of the menstrual fluid of the female, and it is also used by Aristotle to describe the action of rennet upon milk, a parallel

instance which he cites by way of illustration (739 b 23). Συνιστάναι therefore denotes the first impact of Form upon Matter, the first step in the process of actualizing the potentiality of Matter. The meaning of συνιστάναι therefore is plain enough, but there is no really convenient English word to translate it; and in consequence makeshift devices have to be adopted. Sometimes I have used "constitute," sometimes "set," sometimes "cause to take shape"; and for συνίστασθαι, which is also very frequent, "set" (intransitive), "take shape," "arise," etc. I decided against "composit," chiefly because I found it essential to introduce the term "fetation" for κύημα (q.v.), and as the two so often occur together, the outlandish phrase "composits the fetation" would have been frequently occurring. Nevertheless, it would have represented Aristotle's thought much more precisely, and for that reason alone I am convinced that it would have been amply justified.

"Organ- (55) Another possible rendering would have been "organ-
izers." ize": and indeed "organizers" is a term which has recently been introduced into embryology to denote substances which are responsible for bringing about the differentiation of the parts of the embryo. It is interesting in this connexion to note that Aristotle seems to be working on a similar theory in *G.A.* IV, viz., that there is a κίνησις (*i.e.*, a specific "movement," implying a δύναμις or specific substance) for each part of the body, which brings about its development in the embryo. We should, however, note that the "organizers" are not found until after impregnation is effected, whereas the distinctive "movements" proper to sensitive Soul are *ex hypothesi* already in the semen.

Κύημα

"Feta- (56) This is a term which occurs very frequently in *G.A.* At
tion": 728 b 34 we read that by κύημα is meant "the first (or
(a) in sexual primary) mixture of male and female"; and although
generation; the term is very often so used, it is also used by Aristotle to include more than this. Actually it covers all stages of the living creature's development from the time when the "matter" is first "informed" (a common phrase is

lxii

κύημα συνίσταται; see § 54) to the time when the creature is born or hatched. Hence we find κύημα applied to the embryo or fetus of Vivipara; to the " perfect " eggs of birds and to the " imperfect " eggs [a] of Cephalopods, etc. (733 a 24; they are still so called after deposition), to the roe of fishes (741 a 37), and to larvae (758 a 12); indeed, the larva is compared with the earliest stage of the κύημα in viviparous animals (758 a 33).

(57) In all the foregoing cases, the " matter " for the κύημα is supplied by the female parent; but in the case of spontaneous generation there is of course no female parent, and the κύημα is formed, e.g., out of the sea-water by the *pneuma* acting upon it (762 b 17). *(b)* in spontaneous generation;

(58) There are, however, some κυήματα which never reach the point of hatching (e.g., " wind-eggs "); thus a κύημα is not necessarily fertilized. Such a κύημα is, however, to some extent " informed " and can develop up to a point because it possesses nutritive Soul potentially. *(c)* infertile fetations;

(59) There is no English word which covers the wide range of the term κύημα, and I have therefore introduced the term " fetation," by which I invariably translate it.

(60) Aristotle holds that the seeds of plants are " as it were a κύημα," because in them male and female are not separated; hence the seed of a plant begins with the male factor and the female factor already mixed in it; and that is why only one stalk or plant can be formed from one seed: there is no such opportunity available, as there is in the case of animals, for the male *dynamis* to " set " numerous fetations out of the material supplied. *(d)* fetations of plants.

Nourishment, Residues, etc.

(61) Several important terms in Aristotle's technical vocabulary are connected with the processes through which the food passes in the living body, and therefore an account of these processes will most conveniently explain the meaning of the terms.

(62) After mastication, the food passes into the stomach, where it is " concocted " [b] by means of the " natural (or Concoction.

[a] See also § 77 below.
[b] The Greek word for concoction is the same as that employed to denote the process of ripening or maturing of fruit, corn and the like by means of heat—also that of baking and cooking (see

ARISTOTLE

vital) heat " resident there. Any living thing (anything " with Soul in it ") possesses " natural heat," and the chief seat of the Soul and the source of the vital heat is the heart (or its analogue). But also, every part of the body as well has its own natural heat (*cf.* 784 b 26 ff.), derived from the heart through the blood : thus, the stomach concocts the nourishment before passing it on to the heart, and other parts may concoct it still further when the heart has sent it on to them. Beside the stomach, the liver and the spleen assist in the concoction of the nourishment (*P.A.* 670 a 20 ff.).

Blood. (63) Having received its first stage of concoction in the stomach, the nourishment passes on to the heart, where as we should expect it undergoes the most important stage of its concoction, and is thereby turned into blood, the " ultimate nourishment " for the whole body (*P.A.* 647 b 5, *cf.* 666 a 8). It is probable that, in Aristotle's view, an important part of this process was the " pneumatization " of the blood (see App. B §§ 31, 32), *i.e.*, the charging of it with Σύμφυτον Πνεῦμα and with the special " movement " requisite to enable it (*a*) to maintain the " being " of the animal and (*b*) to supply its growth.

"Pneumatization" of blood.

Two grades (64) These two functions of nourishment, and the consequent of nourishment. distinction of two grades of nourishment, which is made by Aristotle at 744 b 33 ff. (where see note ; and *cf.* list of passages in the Index), enable us also to distinguish the different classes of residues. The first-grade nourishment (*a*), which is described as " nutritive " and " seminal," provides the whole animal and its parts with " being " ; the second-grade (*b*) is described as " growth-promoting," and causes increase of bulk. In the development of the embryo, it is the leavings of the first-grade nourishment, or " nutritive residue," left over after the " supreme parts "—flesh and the other sense-organs—have been provided for, which are used to form the bones and sinews ; the second-grade, inferior, nourishment (which is taken in by way of supplement from the mother or from outside) is used to form nails, hair, horns, etc. The latter is more " earthy "

715 b 24, n.). Indeed, the processes are regarded by Aristotle as being fundamentally identical. (*Cf.* 743 a 31 ff.) It is also applied by him to the " maturing " of the embryo (719 a 34).

than the former; indeed, with such residue in mind, Aristotle can say (745 b 19) that " residue is unconcocted substance, and the most unconcocted substance in the body is earthy substance " ; see also § 66 below.

(65) Generally, then, more blood is produced than is required for the purposes mentioned at the end of § 63, and the surplus may then undergo a further stage of concoction, and Nature is often able to turn it to some useful purpose (*cf.* § 8 above). These are the *useful* " residues," and Nature has provided each with its proper place (*G.A.* 725 b 1) ; indeed, it is only in its proper place that each " residue " is formed (739 a 2). Examples of useful residues are semen, menstrual fluid, milk. Marrow, which gives the backbone coherence and elasticity, is produced when " the surplus of bloodlike nourishment is shut up in the bones " and concocted by their heat (*P.A.* 652 a 5, a 20). Sometimes, when the nourishment is particularly abundant, the surplus blood is concocted into fat, such as lard and suet (651 a 20). Also, some of the blood, reaching the extremities of the vessels in which it is carried, makes its way out in the form of nails, claws or hair.[a] — Residues: (a) useful;

(66) Residues may appear at various stages (725 a 13) ; they may appear before, as well as after, the nourishment has been turned into blood ; and then they are residues of " nourishment at its first stage " ; thus (653 a 2, *cf.* 458 a 1 ff.), after a meal, the nourishment rises as vapour through the vessels to the brain, where it is cooled, and then condenses into *phlegma* and *ichor* (serum). But both of these, it seems, may also be *useless* residues, for at 677 b 8 *phlegma* is mentioned in company with " the sediment from the stomach," though perhaps it is most often a residue of the *useful* nourishment (725 a 14). *Ichor*, too, the " watery part of the blood," is sometimes unconcocted blood, sometimes corrupted blood (653 a 2 ; *cf.* 458 a 1 ff., 651 a 15 ; no doubt εἴ τι ἄλλο τοιοῦτον at *G.A.* 725 a 15 refers to *ichor*). — (b) ambiguous;

(67) Residues, then, are " the surplus of the nourishment " — (c) useless;

---

[a] The Aristotelian doctrine of " residues " came down to Shakespeare, as is shown by the passage in *Hamlet* (III. IV), where the Queen says to Hamlet :

" Your bedded haire, like life in excrements,
    Start up, and stand an end."

(724 b 26); but there are *useless* as well as *useful* residues, for residue may come either from the useful or the useless nourishment (725 a 4). Useless nourishment is " that which can contribute nothing further to the natural organism, and if too much of it is consumed it causes very great injury to it " (725 a 5 ff.). Among the *useless* residues are the excrements; these are natural useless residues; but there are also some un-natural ones, as has already been hinted. Among them perhaps should be included bile, which serves no useful purpose whatever. It is a residue produced by the liver (677 b 1), it is the residue of blood in those animals which are made out of less pure blood; it is merely a "necessary" product, an "offscouring," a "colliques-cence." Colliquescence (σύντηγμα, σύντηξις) is defined at 724 b 26 ff. as that which is produced as an ἀπόκρισις from the material that supplies growth, as the result of decomposition proceeding contrary to Nature " (τὸ ἀποκριθὲν ἐκ τοῦ αὐξήματος ὑπὸ τῆς παρὰ φύσιν ἀναλύσεως). Colliquescence, then, is an unnatural residue,[a] and therefore there is no proper place set apart for it by Nature (725 a 1); it just runs about wherever it can in the body. (See also 726 a 11 ff.) Colliquescence is a very common term in the Hippocratic treatise περὶ διαίτης, where its effect is said to be the production of an unhealthy ἀπόκρισις (abscession), and both there and in Aristotle ἀπόκρισις is specially associated with re-sidues, useful, or useless, or even harmful ones. A great deal of π. διαίτης is taken up with suggestions for getting rid of harmful ἀποκρίσεις.

(d) un-natural: Colliques-cence.

Generative residues. (68) The most important residues so far as *G.A.* is concerned are of course semen and menstrual fluid; natural and useful residues, for which Nature has set apart special places in the body. The difference between them is one of degree of concoction: semen is a residue of the final stage of useful nourishment (726 a 26); so is menstrual fluid (738 a 36), but the female has not sufficient natural heat to carry the concoction far enough to produce semen. Hence, the difference between male and female

Source of

---

[a] It seems however that a " colliquescence " may sometimes be a *natural* residue, for at *P.A.* 677 a 13 bile is said to be " a residue or a colliquescence," and it is classed with the sediment in the stomach and intestines. See also *P.A.* (Loeb ed.), pp. 38 f.

is to be traced back to the innermost source of the organism, viz., the heart; the sexual organs may serve as an outward expression of the difference, but the difference is not due to them. Like the blood, of which it is a more fully concocted form, semen derives its character primarily from the heart, where the blood is pneumatized and charged with the requisite specific " movements " (see § 63 and *G.A.* 737 a 19). Semen, therefore, like blood, is the vehicle of " Soul," and especially so in virtue of the Σύμφυτον Πνεῦμα which it contains, for Σύμφυτον Πνεῦμα is the physical substance with which Soul is most intimately " associated." In terms of Soul, the difference between semen and menstrual fluid is that semen possesses the principle of sentient Soul, menstrual fluid possesses only nutritive Soul (potentially): the fluid has not been charged with the " movement " proper to sentient Soul owing to deficiency of heat in the female. The other " movements " in these generative residues are a most important factor in the determination of generic, specific, sexual, and even individual characteristics : see the discussion in *G.A.* IV. 766 a 13 ff., 767 b 15 ff.

sex-differ-ence is the heart.

(69) It should be noted that the heat both of blood and of semen (the concocted residue of blood) is not inherent, but is acquired from a source other than themselves. The *logos* of blood, it is true, includes the term " hot," but only in the same sense that the *logos* of " boiling water " (if we had one word for that as we have for blood) would include the term " hot." In other words, the permanent substratum of blood is not hot; and thus, although in one way blood is " essentially " hot, in another way it is not " essentially " hot (*P.A.* 649 b 21 ff.). Similarly, the " matter " of semen is " watery " (*i.e.*, the substratum of it is the Element Water; *cf.* 736 a 1 and preceding passage) ; and its heat is a supplementary acquisition (ἐπίκτητος : *G.A.* 747 a 18, *cf.* 750 a 9, 10). The explanation of these statements, as will be obvious from the preceding sections, is that blood is produced by the heat of the heart out of the fluid matter supplied by the stomach from the food (§ 63), and semen of course has to undergo still further concoction by the vital heat in the appropriate parts (§ 62).

Heat of blood and of semen not inherent but " acquired."

ARISTOTLE

Two modes of difference; Blood; Classification of Animals

(1) "The
more and
less." (70) Differences " by the more and less," or " of excess and deficiency "—differences of degree, as we should say, are minor differences such as are found as between different species of one and the same genus or of any larger group. Thus (*P.A.* 644 a 19, 692 b 24) the parts of birds differ in this way, some having long legs, or feathers, others short ones ; some a broad tongue, others a narrow one. Again, the male will have the same defensive or offensive organ as the female, but " to a greater degree," and this sometimes holds good of organs essential for food and nutrition [a] (661 b 28 ff.). Difference " by the more and less " can also be applied to skin, blood-vessel, membrane, sinew : these are substances which differ among themselves in this way (*G.A.* 737 b 4 ; *cf.* 739 b 32).

(2) "Counter-
parts." (71) Where the divergence is wider, as for instance between different groups of animals such as birds and fishes, the difference is no longer τῷ μᾶλλον καὶ ἧττον, but τῷ ἀνάλογον (*P.A.* 644 a 21) : the corresponding parts, *e.g.*, the feathers of birds, the scales of fishes, and the scales of reptiles, differ " by analogy," *i.e.*, they are merely the " counterparts," the " opposite numbers " of each other, as indeed the large groups of animals themselves may be (see *G.A.* 761 a 27 and context ; *cf.* also 784 b 16 ff., and 737 b 4, n.).

(72) Many examples of this usage occur in *G.A.* ; we find mention of τὸ ἀνάλογον of the heart ; of the blood, and of the menstrual fluid, in bloodless creatures ; of teeth ; of flesh ; of fat ; of hair ; of sinew. Menstrual fluid in females is ἀνάλογον to semen in males (727 a 3) ; we might have expected this difference to be only a difference " by the more and less," but no doubt the reason why it is a wider divergence is that menstrual fluid lacks sentient Soul (see § 68). The most frequent references to τὸ ἀνάλογον in *G.A.* are the counterparts of the heart and of the blood. And the most important of all the counterparts is of course " the substance in the *pneuma*,"

[a] *Cf.* the view that the female is a " deformity," § 13.

which is ἀνάλογον to the element of the stars, *aither* (736 b 37).

(73) It should be noted that by " blood " Aristotle means red blood only, and he makes a division of animals into " blooded " (ἔναιμα) and " bloodless " (ἄναιμα). These two classes do not quite coincide with vertebrates and invertebrates, for there are some invertebrates which have red blood, *e.g.*, molluscs (*Planorbis*), insect larvae (*Chironomus*), and worms (*Arenicola*). In other invertebrates the blood may be blue (Crustacea and most molluscs) or green (Sabellid worms), or there may be no respiratory pigment at all (most Insects). Blood.

(74) The following table shows how Aristotle's division works out : Blooded and Bloodless animals.

Blooded animals	*Bloodless animals*
Man.	Crustacea.
Viviparous quadrupeds.	Cephalopods.
Oviparous quadrupeds and footless animals (=reptiles and amphibians).	Insects.
	Testacea.
Birds.	
Fishes.	

It may be convenient to give here the Greek names used by Aristotle for the four classes of Bloodless animals, together with their literal translation and the terms which I have used to translate them :

τὰ μαλακόστρακα	soft-shelled animals	Crustacea.
τὰ μαλάκια	softies	Cephalopods.
τὰ ἔντομα	insected animals	Insects.
τὰ ὀστρακόδερμα	shell-skinned animals	Testacea.[a]

(75) The Testacea were a source of considerable embarrassment to Aristotle, who considered them to be intermediate between animals proper and plants. Nor, according to him, did they reproduce sexually, but arose from spontaneous generation. In his treatise on the *Progression of Animals*, he defers mention of them to The Testacea.

[a] In using " Testacea " to translate τὰ ὀστρακόδερμα (" the animals with earthenware skins ") I use it in the old-fashioned sense, so as to include a number of shelled invertebrates, comprising Gasteropods, Lamellibranchs, and some Echinoderms. Modern zoologists apply the term Testacea to the Foraminifera, which are shelled Protozoa. The term Ostracoderms (a transliteration of Aristotle's word) is now given by zoologists to a group of primitive fossil fishes.

the very end and then says that strictly speaking they ought not to move about at all, yet in fact we see them moving : anyway, their movement is " contrary to nature," because they " have no right and left ." (The mechanism of their movement can be detected only by the microscope, and is known as ciliary.)

Classification according to method of reproduction. (76) In *G.A.*, however, although Aristotle adheres to his classification into Blooded and Bloodless animals, perhaps a more important classification is that which is based upon their method of reproduction. This classification will be found in the Contents-Summary, pp. lxxii ff. And in this connexion we must notice that the list is headed by the Viviparous animals, of which "Perfect" animals. the first is Man : these are the " most perfect animals," and therefore they produce their offspring in the most perfected condition. And by " most perfect " (732 b 29) Aristotle means the animals which are " in their nature hotter and more fluid (ὑγρότερα), and are not earthy " ; and, as the test of natural heat is the presence of the lung, and further, a lung well supplied with blood, no animal can be internally viviparous unless it respires. (See the whole passage 732 a 26—733 b 16.)

Distinction of " perfect " and " imperfect " eggs. (77) It should be noted that Aristotle clearly distinguishes between what he calls " perfect " and " imperfect " eggs, that is to say between eggs which do not and those which do increase in size after deposition. This is the basis of the modern distinction between cleidoic and non-cleidoic eggs (see 718 b 7, n.). He also clearly Distinction of egg and larva. distinguishes between an egg and a larva : an egg is that from *part* of which the young creature is formed, the remainder serving as nourishment for it ; a larva is something of which the *whole* is used to form the young animal. (See 732 a 29 and note, and 758 b 10 ff.) The fact that Aristotle drew these distinctions so clearly is particularly noteworthy. He was, of course, unaware of the existence of the mammalian ovum, which cannot be detected without the aid of the microscope. It should also be noted that Aristotle compares the growth of a non-cleidoic egg with the action of yeast in fermentation ; see 755 a 18.

CONTENTS-SUMMARY

Introductory

Classification of the various methods of Generation

3 (resumed)—Theory of Sexual Generation

---

[a] The larva represents a stage previous to that of the egg, for, according to Aristotle, the larva develops into an egg-like object.

GENERATION OF ANIMALS

Generation in Blooded Animals—I. Vivipara

Generation in Blooded Animals—II. Ovipara (laying perfect eggs)

Generation in Blooded Animals—III. Ovovivipara (laying perfect eggs)

ARISTOTLE

GENERATION OF ANIMALS

ABBREVIATIONS USED IN
THIS VOLUME

Works of Aristotle

H.A.	*Historia animalium*	*Phys.*	*Physica*
P.A.	*De partibus anima-*	*Met.*	*Metaphysica*
	lium	*Meteor.*	*Meteorologica*
G.A.	*De generatione ani-*	*Pol.*	*Politica*
	malium	*E.N.*	*Ethica Nicomachea*
I.A.	*De incessu anima-*	*Cat.*	*Categoriae*
	lium	*De an.*	*De anima*
M.A.	*De motu animalium*	*De resp.*	*De respiratione*

G. & C. De generatione et corruptione

Other Works

L. & S.	Liddell and Scott's *Greek-English Lexicon* (1925–1940)
Diels, *Vorsokr.*	*Die Fragmente der Vorsokratiker*, by Hermann Diels, fifth ed., edited by W. Kranz, 1934–1937
C.Q.	*Classical Quarterly*
C.R.	*Classical Review*

Other abbreviations are self-explanatory.

SIGLA

GENERATION OF ANIMALS

ΑΡΙΣΤΟΤΕΛΟΥΣ

ΠΕΡΙ ΖΩΙΩΝ ΓΕΝΕΣΕΩΣ

Α

I Ἐπεὶ δὲ περὶ τῶν ἄλλων μορίων εἴρηται τῶν ἐν
τοῖς ζῴοις καὶ κοινῇ καὶ καθ' ἕκαστον γένος περὶ
τῶν ἰδίων χωρίς, τίνα τρόπον διὰ τὴν τοιαύτην
αἰτίαν ἐστὶν ἕκαστον, λέγω δὲ ταύτην τὴν ἕνεκά
του· ὑπόκεινται γὰρ αἰτίαι τέτταρες, τό τε οὗ
ἕνεκα ὡς τέλος, καὶ ὁ λόγος τῆς οὐσίας (ταῦτα
5 μὲν οὖν ὡς ἕν τι σχεδὸν ὑπολαβεῖν δεῖ), τρίτον
δὲ καὶ τέταρτον ἡ ὕλη καὶ ὅθεν ἡ ἀρχὴ τῆς κινήσεως—περὶ μὲν οὖν τῶν ἄλλων εἴρηται (ὅ τε γὰρ
λόγος καὶ τὸ οὗ ἕνεκα ὡς τέλος ταὐτόν, καὶ ὕλη
10 τοῖς ζῴοις τὰ μέρη, παντὶ μὲν τῷ ὅλῳ τὰ ἀνομοιομερῆ, τοῖς δ' ἀνομοιομερέσι τὰ ὁμοιομερῆ,

---

[a] i.e., in the De partibus animalium.
[b] See Introd. § 18.
[c] i.e., the Final Cause appropriate to each part, either qua
part belonging to all animals, or qua part belonging to some
special group of animals.　　　　　[d] See Introd. §§ 1 ff.
[e] See Introd. § 10.　　　　　　　[f] See Introd. § 11.
[g] See Introd. §§ 19 ff.

ARISTOTLE

GENERATION OF ANIMALS

BOOK I

WITH one exception we have now *a* spoken about
all the parts *b* that are present in animals, both generally concerning them, and also taking them group by
group and dealing separately with the parts peculiar
to each, and have shown in what way each part exists
on account of the Cause which is of a corresponding
kind : I refer to the Cause which is " that for the
sake of which " a thing exists.*c* As we know, there
are four basic Causes *d* : (1) " that for the sake of
which " the thing exists, considered as its " End " ;
(2) the *logos*[e] of the thing's essence (really these
first two should be taken as being almost one and the
same) ; (3) the matter of the thing, and (4) that from
which comes the principle *f* of the thing's movement.
And with one exception I have already spoken about
all of these Causes, since the *logos* of a thing and
" that for the sake of which " it exists, considered as
its End, are the same ; and, for animals, the matter
of them is their parts (the non-uniform *g* parts are
the matter for the animal as a whole in each case ;
the uniform parts are the matter for the non-uniform

3

τούτοις δὲ τὰ καλούμενα στοιχεῖα τῶν σωμάτων),
λοιπὸν δὲ τῶν μὲν μορίων τὰ πρὸς τὴν γένεσιν
συντελοῦντα τοῖς ζῴοις, περὶ ὧν οὐθὲν διώρισται
πρότερον, περὶ αἰτίας δὲ τῆς κινούσης, τίς αὕτη.
τὸ δὲ περὶ ταύτης σκοπεῖν καὶ τὸ περὶ τῆς
15 γενέσεως τῆς ἑκάστου τρόπον τινὰ ταὐτόν ἐστιν·
διόπερ ὁ λόγος εἰς ἓν συνήγαγε, τῶν μὲν περὶ τὰ
μόρια τελευταῖα ταῦτα, τῶν δὲ περὶ γενέσεως τὴν
ἀρχὴν ἐχομένην τούτων τάξας.

Τῶν δὴ ζῴων τὰ μὲν ἐκ συνδυασμοῦ γίνεται
θήλεος καὶ ἄρρενος, ἐν ὅσοις γένεσι τῶν ζῴων ἐστὶ
20 τὸ θῆλυ καὶ τὸ ἄρρεν· οὐ γὰρ ἐν πᾶσίν ἐστιν, ἀλλ'
ἐν μὲν τοῖς ἐναίμοις ἔξω ὀλίγων ἅπασι τὸ μὲν
ἄρρεν τὸ δὲ θῆλυ τελειωθέν ἐστι, τῶν δ' ἀναίμων
τὰ μὲν ἔχει τὸ θῆλυ καὶ τὸ ἄρρεν, ὥστε τὰ ὁμογενῆ
γεννᾶν, τὰ δὲ γεννᾷ μέν, οὐ μέντοι τά γε ὁμογενῆ·
τοιαῦτα δ' ἐστὶν ὅσα γίνεται μὴ ἐκ ζῴων συνδυαζο-
25 μένων, ἀλλ' ἐκ γῆς σηπομένης καὶ περιττωμάτων.[1]
ὡς δὲ κατὰ παντὸς εἰπεῖν, ὅσα μὲν κατὰ τόπον
μεταβλητικὰ τῶν ζῴων ἐστὶ[2] τὰ μὲν νευστικὰ τὰ

---

[1] huc procul dubio transferenda vv. 715 b 25-30 ἔστι δὲ . . .
ἰξός, quae ibi aliena, hic congrua.
[2] ἐστὶ Peck : ὄντα vulg. : locus hic corruptus.

---

[a] Elements : στοιχεῖα. The term is a metaphor taken
from " letters of the alphabet," the original meaning of the
term. In the physical sense, "element" may be defined as
ἐξ οὗ σύγκειται πρώτ\υ ἐνυπάρχοντος ἀδιαιρέτου τῷ εἴδει εἰς
ἕτερον εἶδος (Met. 1014 a 26). See Introd. § 24.
[b] i.e., after the De partibus and the De incessu animalium.
[c] See Introd. § 74.
[d] The exceptions are the erythrinus and the channa: see
741 a 35, 755 b 21 ; cf. 760 a 8.
[e] See Introd. § 67. Here probably=excrements : cf.
H.A. 551 a 6. See however 737 a 4, 762 a 3 ff.

4

parts ; and the corporeal " elements," [a] as they are called, are the matter for the uniform parts). Consequently, of the parts it remains to describe those which subserve animals for the purpose of generation, about which I have so far said nothing definite, and of Causes we still have the Motive Cause to deal with, and to explain what it is. And, in a way, consideration of this Cause and consideration of the generation of each animal comes to the same thing : and that is why our treatise has brought the two together, by placing these parts at the end of our account of the parts,[b] and by putting the beginning of the account of generation immediately after them.

Now of course some animals are formed as a result of the copulation of male and female, namely, animals belonging to those groups in which there exist both male and female, for we must remember that not all groups have both male and female. Among the blooded [c] animals, with a few exceptions,[d] the individual when completely formed is either male or female ; but among the bloodless animals, while some groups have both male and female and hence generate offspring which are identical in kind with their parents, there are other groups which, although they generate, do not generate offspring identical with their parents. Such are the creatures which come into being not as the result of the copulation of living animals, but out of putrescent soil and out of residues.[e] [f] Speaking generally, however, we may say that (a) in the case of all those animals which have the power of locomotion, whether they are adapted

Distinction of sexes not universal.

[f] The passage 715 b 25-30 should be inserted here, if anywhere.

715 a

δὲ πτηνὰ τὰ δὲ πεζευτικὰ τοῖς σώμασιν, ἐν πᾶσι
τούτοις ἐστὶ[1] τὸ θῆλυ καὶ τὸ ἄρρεν, οὐ μόνον[2]
30 τοῖς ἐναίμοις, ἀλλὰ ἐνίοις καὶ ἀναίμοις[3]· καὶ

715 b

τούτων τοῖς μὲν καθ' ὅλον τὸ γένος, οἷον τοῖς
μαλακίοις καὶ τοῖς μαλακοστράκοις· ἐν δὲ τῷ τῶν
ἐντόμων γένει τὰ πλεῖστα. τούτων δ' αὐτῶν ὅσα
μὲν ἐκ συνδυασμοῦ γίνεται τῶν συγγενῶν ζώων,
καὶ αὐτὰ γεννᾷ κατὰ τὴν συγγένειαν· ὅσα δὲ μὴ
5 ἐκ ζώων ἀλλ' ἐκ σηπομένης τῆς ὕλης, ταῦτα δὲ
γεννᾷ μὲν ἕτερον δὲ γένος, καὶ τὸ γιγνόμενον
οὔτε θῆλύ ἐστιν οὔτε ἄρρεν. τοιαῦτα δ' ἐστὶν ἔνια
τῶν ἐντόμων. καὶ τοῦτο συμβέβηκεν εὐλόγως·
εἰ γὰρ ὅσα μὴ γίγνεται ἐκ ζώων, ἐκ τούτων
ἐγίνετο ζῷα συνδυαζομένων, εἰ μὲν ὁμοιογενῆ,[4] καὶ
10 τὴν ἐξ ἀρχῆς τοιαύτην ἔδει τῶν τεκνωσάντων
εἶναι γένεσιν (τοῦτο δ' εὐλόγως ἀξιοῦμεν· φαίνεται
γὰρ συμβαῖνον οὕτως ἐπὶ τῶν ἄλλων ζώων)· εἰ
δ' ἀνόμοια μὲν δυνάμενα δὲ συνδυάζεσθαι, πάλιν
ἐκ τούτων ἑτέρα τις ἂν ἐγίνετο φύσις, καὶ πάλιν
ἄλλη τις ἐκ τούτων, καὶ τοῦτ' ἐπορεύετ' ἂν εἰς
15 ἄπειρον. ἡ δὲ φύσις φεύγει τὸ ἄπειρον· τὸ μὲν
γὰρ ἄπειρον ἀτελές, ἡ δὲ φύσις ἀεὶ ζητεῖ τέλος.
ὅσα δὲ μὴ πορευτικά, καθάπερ τὰ ὀστρακόδερμα
τῶν ζώων καὶ τὰ ζῶντα τῷ προσπεφυκέναι, διὰ
τὸ παραπλησίαν αὐτῶν εἶναι τὴν οὐσίαν τοῖς
φυτοῖς, ὥσπερ οὐδ' ἐν ἐκείνοις, οὐδ' ἐν τούτοις

---

[1] ἐν πᾶσι τούτοις ἔστιν Z[1] : ἐν ἐνίοις μὲν τούτων ἅπαν τὸ γένος
ἔχει vulg., Z[2].   [2] μόνον SZ : μόνον ἐν vulg.
[3] sic PZ[2] (ἀνέμοις Z[1]) : ἀλλὰ καὶ τῶν ἀναίμων ἔν τισιν vulg.
[4] ὁμογενῆ P*Z[1] : ὁμοιοειδῆ Z[2].

---

[a] See Introd. § 74.

to be swimmers, or fliers, or walkers, male and female
are found ; and this applies not only to the blooded
animals but to some of the bloodless ones as well.
And among the latter, in some cases it holds good of
a whole group, as for instance the Cephalopods and
the Crustacea [a] ; and it holds good of most of the
Insects. Among animals of this class, those which
are formed as the result of the copulation of animals
of the same kind, themselves generate in turn after
their own kind ; those, however, which arise not
from living animals but from putrescent matter,
although they generate, produce something that is
different in kind, and the product is neither male nor
female. Some of the Insects are like this.[b] And
this is what we should expect; for supposing that
creatures which are produced otherwise than from
living animals copulated and produced living animals :
if these products were similar in kind to their parents,
then the manner of their parents' original generation
should have been like theirs. This we may reasonably claim, because it is evident that this is so with
all other animals. If, on the other hand, the products were dissimilar from their parents, and yet able
to copulate, we should then get arising from them
yet another different manner of creature, and out of
their progeny yet another, and so it would go on *ad
infinitum.* Nature, however, avoids what is infinite,
because the infinite lacks completion and finality,
whereas this is what Nature always seeks. (*b*) The
creatures which cannot move about, like the Testacea and those which live by being attached to some
surface, are in their essence similar to plants, and
therefore, as in plants, so also in them, male and

[b] See 732 a 25 ff., 758 b 6 ff.

20 ἐστὶ τὸ θῆλυ καὶ τὸ ἄρρεν, ἀλλ' ἤδη καθ' ὁμοιότητα καὶ κατ' ἀναλογίαν λέγεται· μικρὰν γάρ
τινα τοιαύτην ἔχει διαφοράν.   καὶ γὰρ ἐν τοῖς
φυτοῖς ὑπάρχει τὰ μὲν καρποφόρα δένδρα τοῦ
αὐτοῦ γένους, τὰ δ' αὐτὰ μὲν οὐ φέρει καρπόν,
συμβάλλεται δὲ τοῖς φέρουσι πρὸς τὸ πέττειν, οἷον
25 συμβαίνει περὶ τὴν συκῆν καὶ τὸν ἐρινεόν.

¹[Ἔστι δὲ καὶ ἐπὶ τῶν φυτῶν τὸν αὐτὸν τρόπον·
τὰ μὲν γὰρ ἐκ σπέρματος γίνεται, τὰ δ' ὥσπερ
αὐτοματιζούσης τῆς φύσεως· γίνεται γὰρ ἢ τῆς
γῆς σηπομένης ἢ μορίων τινῶν ἐν τοῖς φυτοῖς·
ἔνια γὰρ αὐτὰ μὲν οὐ συνίσταται καθ' αὐτὰ χωρίς,²
30 ἐν ἑτέροις δ' ἐγγίνεται δένδρεσιν, οἷον ὁ ἰξός.]

---

¹ quae sequuntur vv. 25-30 plane huc aliunde tralata, hic
enim iamdudum de plantis sermo.   transferenda censeo ad
715 a 25 post περιττωμάτων.
² χωρὶς ἐκ γῆς ΖΣ.

ᵃ The concoction referred to here is that which produces
the ripening of fruit.   See Introd. § 62.   The use of the
same word πέττειν both for the fruit of plants and for the
semen of animals is appropriate, in that both, according to
Aristotle, are produced out of "nourishment" by a process
of "concoction."

ᵇ See 755 b 10, and H.A. 557 b 31.   The fig tree commonly cultivated in S. Europe is *Ficus carica.*   This species
includes two kinds of individual trees : (1) those whose inflorescences contain fully-developed female flowers only ;
(2) those whose inflorescences contain male flowers near
the opening, and lower down aborted female flowers known
as " gall-flowers " owing to their being specially prepared
to receive the eggs of the fig-wasp (*Blastophaga grossorum*),
which turns the ovary of the flower into a " gall."   The
latter trees are known as *Caprificus.*   The female wasps,
after impregnation by the male wasps within the gall,
emerge from it and get dusted with pollen from the male

female are not found, although they are called male
and female just by way of similarity and analogy,
since they exhibit a slight difference of this sort.
Thus among plants also we find that in one and the
same kind some individual trees bear fruit, while
some, although they do not bear any themselves,
assist in the concocting[a] of that which is borne by
the others. An instance of this is the fig and the
caprifig.[b]

[c][The same sort of thing is found in plants too :
some are formed out of seed, others as it might be
by some spontaneous activity of Nature—they are
formed when either the soil or certain parts[d] in
plants become putrescent, since some of them do
not take shape[e] independently on their own, but
grow upon other trees, as for instance the mistletoe
does.]

<hr />

flowers as they leave the inflorescence, and then pollinate
female flowers elsewhere. Caprification is the name given
to the artificial assistance of this process by hanging inflorescences of the caprifig on to trees of class (1). The
growers believe that the fruit of the *Ficus* is improved by the
wasps ; but in fact excellent fruit is produced by these trees
without pollination, though of course *no fertile seeds*. Hence
caprification must be a traditional usage dating from the
time when fertile seeds were required for propagation, which
is now done by means of cuttings. See Kerner and Oliver,
Natural History of Plants, ii. 160-162 ; H. Müller, *Fertilization of Flowers*, tr. p. 521 and bibliography. *Cf. H.A.*
557 b 26 ff., where the wasp is mentioned.

[c] The following sentence is obviously out of place here, as
is shown (*a*) by the opening words, which must mark the
beginning of a reference to plants, whereas here plants are
already being discussed ; and (*b*) by its inappropriateness to
the particular point under discussion. It would be relevant if
transferred to 715 a 25. *Cf. H.A.* 539 a 16 ff.

[d] *Cf.* 762 b 19.

[e] See Introd. § 54.

716 a

Περὶ μὲν οὖν φυτῶν, αὐτὰ καθ᾽ αὑτὰ χωρὶς
ἐπισκεπτέον.

II Περὶ δὲ τῶν ἄλλων ζῴων τῆς γενέσεως λεκτέον
κατὰ τὸν ἐπιβάλλοντα λόγον καθ᾽ ἕκαστον αὐτῶν,
ἀπὸ τῶν εἰρημένων συνείροντας. καθάπερ γὰρ
5 εἴπομεν, τῆς γενέσεως ἀρχὰς ἄν τις οὐχ ἥκιστα
θείη τὸ θῆλυ καὶ τὸ ἄρρεν, τὸ μὲν ἄρρεν ὡς τῆς
κινήσεως καὶ τῆς γενέσεως ἔχον τὴν ἀρχήν, τὸ δὲ
θῆλυ ὡς ὕλης. τοῦτο δὲ μάλιστ᾽ ἄν τις πιστεύσειε
θεωρῶν πῶς γίνεται τὸ σπέρμα, καὶ πόθεν· ἐκ
τούτου μὲν γὰρ τὰ φύσει γινόμενα συνίσταται, τοῦτο
10 δὲ πῶς ἀπὸ τοῦ θήλεος καὶ τοῦ ἄρρενος συμβαίνει
γίγνεσθαι, δεῖ μὴ λανθάνειν· τῷ γὰρ ἀποκρίνεσθαι
τὸ τοιοῦτον μόριον ἀπὸ τοῦ θήλεος καὶ τοῦ ἄρρενος,
καὶ ἐν τούτοις τὴν ἀπόκρισιν εἶναι καὶ ἐκ τούτων,
διὰ τοῦτο τὸ θῆλυ καὶ τὸ ἄρρεν ἀρχαὶ τῆς γενέσεώς
15 εἰσιν. ἄρρεν μὲν γὰρ λέγομεν ζῷον τὸ εἰς ἄλλο
γεννῶν, θῆλυ δὲ τὸ εἰς αὑτό· διὸ καὶ ἐν τῷ ὅλῳ τὴν
τῆς γῆς φύσιν ὡς θῆλυ καὶ μητέρα ὀνομάζουσιν,[1]
οὐρανὸν δὲ καὶ ἥλιον ἤ τι τῶν ἄλλων τῶν τοιούτων
ὡς γεννῶντας καὶ πατέρας προσαγορεύουσιν.

Τὸ δ᾽ ἄρρεν καὶ τὸ θῆλυ διαφέρει κατὰ μὲν τὸν
λόγον τῷ δύνασθαι ἕτερον ἑκάτερον, κατὰ δὲ τὴν

---

[1] ὀνομάζουσιν Z : νομίζουσιν vulg.

---

[a] It is impossible to represent the force of the Greek neuter
in English.
[b] See note on *Causes*, Introd. §§ 1 ff. This statement,
here unexplained and unjustified, will be fully dealt with
later on.
[c] See Introd. § 54.

Still, plants will have to be considered independently all by themselves.

As far as animals are concerned, we must describe their generation just as we find the theme requires for each several kind as we go along, linking our account on to what has already been said. As we mentioned, we may safely set down as the chief principles of generation the male ⟨factor⟩ [a] and the female ⟨factor⟩ ; the male as possessing the principle of movement and of generation, the female as possessing that of matter.[b] One is most likely to be convinced of this by considering how the semen is formed and whence it comes ; for although the things that are formed in the course of Nature no doubt take their rise [c] out of semen,[d] we must not fail to notice how the semen itself is formed from the male and the female, since it is because this part [e] is secreted from the male and the female, and because its secretion takes place in them and out of them, that the male and the female are the principles of generation. By a " male " animal we mean one which generates in another, by " female " one which generates in itself. This is why in cosmology too they speak of the nature of the Earth as something female and call it " mother," while they give to the heaven and the sun and anything else of that kind the title of " generator," and " father."

Now male and female differ in respect of their *logos*,[f] in that the power or faculty possessed by the one differs from that possessed by the other ; but they differ also to bodily sense, in respect of certain

II
Definition of male and female.

The sexual parts :

---

[d] *Cf.* the definition given at 724 a 17 ff., and also 721 b 6.
[e] See Introd. § 18.
[f] See Introd. § 10. With this passage *cf.* 766 a 18 ff.

11

716 a

20 αἴσθησιν μορίοις τισίν, κατὰ μὲν τὸν λόγον τῷ
ἄρρεν μὲν εἶναι τὸ δυνάμενον γεννᾶν εἰς ἕτερον,
καθάπερ ἐλέχθη πρότερον, τὸ δὲ θῆλυ τὸ εἰς αὑτό,
καὶ ἐξ οὗ γίνεται ἐνυπάρχον ἐν τῷ γεννῶντι τὸ
γεννώμενον. ἐπεὶ δὲ δυνάμει διώρισται καὶ ἔργῳ
τινί, δεῖται δὲ πρὸς πᾶσαν ἐργασίαν ὀργάνων,
25 ὄργανα δὲ ταῖς δυνάμεσι τὰ μέρη τοῦ σώματος,
ἀναγκαῖον εἶναι καὶ πρὸς τὴν τέκνωσιν καὶ τὸν
συνδυασμὸν μόρια, καὶ ταῦτα διαφέροντ᾽ ἀλλήλων,
καθὸ τὸ ἄρρεν διοίσει τοῦ θήλεος. εἰ γὰρ καὶ καθ᾽
ὅλου λέγεται τοῦ ζῴου τοῦ μὲν τὸ θῆλυ τοῦ δὲ τὸ
ἄρρεν, ἀλλ᾽ οὐ κατὰ πᾶν γε [τὸ]¹ αὐτὸ θῆλυ καὶ
30 ἄρρεν ἐστίν, ἀλλὰ κατά τινα δύναμιν καὶ κατά τι
μόριον, ὥσπερ καὶ² ὁρατικὸν καὶ πορευτικόν, ὅπερ
καὶ φαίνεται κατὰ τὴν αἴσθησιν. τοιαῦτα δὲ τυγχάνει μόρια ὄντα τοῦ μὲν θήλεος αἱ καλούμεναι
ὑστέραι, τοῦ δ᾽ ἄρρενος τὰ περὶ τοὺς ὄρχεις καὶ τοὺς
περινέους ἐν πᾶσι τοῖς ἐναίμοις· τὰ μὲν γὰρ ὄρχεις
35 ἔχει αὐτῶν, τὰ δὲ τοὺς τοιούτους πόρους. εἰσὶ δὲ

716 b
διαφοραὶ τοῦ θήλεος καὶ ἄρρενος καὶ ἐν τοῖς ἀναίμοις, ὅσα αὐτῶν ἔχει ταύτην τὴν ἐναντίωσιν. διαφέρει δ᾽ ἐν τοῖς ἐναίμοις τὰ μέρη τὰ πρὸς τὴν μίξιν
τοῖς σχήμασιν. δεῖ δὲ νοεῖν ὅτι μικρᾶς ἀρχῆς μετα-
κινουμένης πολλὰ συμμεταβάλλειν εἴωθε τῶν μετὰ

---

¹ seclusit Rackham, om. Z¹.　　　² καὶ PZ¹ : καὶ τὸ vulg.

---

[a] The force of this important remark will be explained
later.　Cf. 734 b 35.
　[b] Cf. 766 b 2 ff.; also 729 b 12 ff.
　[c] This introduces what is to some extent a modification of

12

physical parts. They differ in their *logos*, because the male is that which has the power to generate in another (as was stated above), while the female is that which can generate in itself, *i.e.*, it is that out of which the generated offspring, which is present in the generator,[a] comes into being. Very well, then : they are distinguished in respect of their faculty, and this entails a certain function. Now for the exercise of every function instruments are needed, and the instruments for physical faculties are the parts of the body. Hence it is necessary that, for the purpose of copulation and procreation, certain parts should exist, parts that are different from each other, in respect of which the male will differ from the female ; for although male and female are indeed used as epithets of the whole of the animal, it is not male or female in respect of the whole of itself, but only in respect of a particular faculty and a particular part [b] —just as it is " seeing " and ' walking " in respect of certain parts—and this part is one which is evident to the senses. Now in the female this special part is what is called the uterus, and in the male the regions about the testes and the penis, so far as all the blooded animals are concerned : some of them have actual testes, some testicular passages. There are also differences between male and female in those of the bloodless creatures which have this opposition of the sexes. In the blooded animals the parts which serve for copulation differ in their shapes. We must note, however,[c] that when a small principle [d] changes, usually many of the things which depend upon it

the statement just made (716 a 27 ff.). And *cf.* the passage *H.A.* 583 b 31 ff. *Cf.* also 764 b 28, 766 a 24 ff.

[d] See Introd. § 11.

13

5 τὴν ἀρχήν. δῆλον δὲ τοῦτο ἐπὶ τῶν ἐκτεμνομένων·
τοῦ γεννητικοῦ γὰρ μορίου διαφθειρομένου μόνον
ὅλη σχεδὸν ἡ μορφὴ συμμεταβάλλει τοσοῦτον ὥστε
ἢ θῆλυ δοκεῖν εἶναι ἢ μικρὸν ἀπολείπειν, ὡς οὐ
κατὰ τὸ τυχὸν μόριον οὐδὲ κατὰ τὴν τυχοῦσαν
10 δύναμιν θῆλυ ὂν καὶ ἄρρεν τὸ ζῷον. φανερὸν οὖν
ὅτι ἀρχή τις οὖσα φαίνεται τὸ θῆλυ καὶ τὸ ἄρρεν·
πολλὰ γοῦν συμμεταβάλλει μεταβαλλόντων ᾗ θῆλυ
καὶ ἄρρεν, ὡς ἀρχῆς μεταπιπτούσης.

III Ἔχει δὲ τὰ περὶ τοὺς ὄρχεις καὶ τὰς ὑστέρας
οὐχ ὁμοίως πᾶσι τοῖς ἐναίμοις ζῴοις, καὶ πρῶτον
15 τὰ περὶ τοὺς ὄρχεις τοῖς ἄρρεσιν. τὰ μὲν γὰρ
ὅλως ὄρχεις οὐκ ἔχει τῶν τοιούτων ζῴων, οἷον τό
τε τῶν ἰχθύων γένος καὶ τὸ τῶν ὄφεων, ἀλλὰ
πόρους μόνον δύο σπερματικούς· τὰ δ' ἔχει μὲν
ὄρχεις, ἐντὸς δ' ἔχει τούτους πρὸς τῇ ὀσφύι κατὰ
20 τὴν τῶν νεφρῶν χώραν, ἀπὸ δὲ τούτων ἑκατέρου
πόρον, ὥσπερ ἐν τοῖς μὴ ἔχουσιν ὄρχεις, συνάπ-
τοντας εἰς ἕν, καθάπερ καὶ ἐπ' ἐκείνων, οἷον οἵ τε
ὄρνιθες πάντες καὶ τὰ ᾠοτοκοῦντα τετράποδα τῶν
δεχομένων τὸν ἀέρα καὶ πλεύμονα ἐχόντων· καὶ
γὰρ ταῦτα πάντα ἐντὸς ἔχει πρὸς τῇ ὀσφύι τοὺς
25 ὄρχεις, καὶ δύο πόρους ἀπὸ τούτων ὁμοίως τοῖς
ὄφεσιν, οἷον σαῦροι καὶ χελῶναι καὶ τὰ φολιδωτὰ

---

[a] In this passage Aristotle prefigures the distinction made
to-day between primary sex-characters, *i.e.*, the genital organs
themselves including testis or ovary ; and the secondary sex-
characters, *e.g.*, the cock's comb or the hen's special feather-
ing, which, as is now known, depend on the secretion of the

14

undergo an accompanying change.[a] This is clear
with castrated animals, where, although the generative part alone is destroyed, almost the whole form
of the animal thereupon changes so much that it
appears to be female or very nearly so, which
suggests that it is not merely in respect of some
casual part or some casual faculty that an animal is
male or female. It is clear, then, that " the male "
and " the female " are a principle. At any rate,
when animals undergo a change in respect of that
wherein they are male and female, many other things
about them undergo an accompanying change, which
suggests that a principle undergoes some alteration.

The testicles and the uterus are not of similar III
arrangement in all the blooded animals. Consider
first the males, and their testicles. Some blooded
animals (as the groups of Fishes and Serpents) have
no testicles at all, only two seminal passages.[b] Others
have testicles, but they are inside, by the loin, near
the place where the kidneys are ; from each of them
runs a passage (as in those animals which have no
testicles), and these two passages join up together
(again like those other animals) : among the class
of animals which breathe air and have a lung, this
occurs in all the Birds and in the oviparous quadrupeds, for all these as well have their testicles inside,
by the loin, and two passages leading from them, just
as the Serpents have : examples are the lizards, the
tortoises, and all the animals with horny scales. All

sex hormones from the interstitial cells of the testis and ovary
respectively.
   [b] These are in fact the testes, but Aristotle reserves this
name for the firm, oval-shaped testes. This negative statement does not of course include the *cartilaginous* fishes, the
Selachia, many of which are viviparous.

716 b

πάντα. τὰ δὲ ζῳοτόκα πάντα μὲν ἐν τῷ ἔμπροσθεν
ἔχει τοὺς ὄρχεις, ἀλλ' ἔνια αὐτῶν ἔσω πρὸς τῷ
τέλει τῆς γαστρός, οἷον ὁ δελφίς, καὶ οὐ πόρους
ἀλλ' αἰδοῖον ἀπὸ τούτων περαῖνον εἰς τὸ ἔξω,
καθάπερ οἱ βόες,[1] τὰ δ' ἔξω, καὶ τούτων τὰ μὲν
30 ἀπηρτημένους, ὥσπερ ἄνθρωπος, τὰ δὲ πρὸς τῇ
ἕδρᾳ, καθάπερ οἱ ὗες. διώρισται δὲ περὶ αὐτῶν
ἀκριβέστερον ἐν ταῖς ἱστορίαις ταῖς περὶ τῶν ζῴων.

Αἱ δ' ὑστέραι πᾶσι[2] μέν εἰσι διμερεῖς, καθάπερ
καὶ οἱ ὄρχεις τοῖς ἄρρεσι δύο πᾶσιν· ταύτας δ'
35 ἔχουσι τὰ μὲν πρὸς τοῖς ἄρθροις, καθάπερ αἵ τε
γυναῖκες καὶ πάντα τὰ ζῳοτοκοῦντα μὴ μόνον

717 a θύραζε ἀλλὰ καὶ ἐν αὑτοῖς, καὶ οἱ ἰχθύες ὅσοι
ᾠοτοκοῦσιν εἰς τοὐμφανές, τὰ δὲ πρὸς τῷ ὑποζώματι, καθάπερ οἵ τ' ὄρνιθες πάντες καὶ τῶν
ἰχθύων οἱ ζῳοτοκοῦντες. ἔχουσι δὲ δικρόας καὶ
τὰ μαλακόστρακα τὰς ὑστέρας καὶ τὰ μαλάκια· τὰ
5 γὰρ καλούμενα τούτων ᾠὰ τοὺς περιέχοντας
ὑμένας ὑστερικοὺς ἔχει.

Μάλιστα δὲ ἀδιόριστον ἐπὶ τῶν πολυπόδων ἐστίν,
ὥστε δοκεῖν μίαν εἶναι· τούτου δ' αἴτιον ὁ τοῦ
σώματος ὄγκος πάντῃ ὅμοιος ὤν. δικρόαι δὲ καὶ

---

[1] καθάπερ οἱ βόες delet Platt, qui tauros credit significari.
[2] πᾶσι(ν) PSY*Z : πᾶσαι Bekker.

---

[a] In front : that is, with reference to the ideal posture of
an animal, viz., that of man.

[b] The term αἰδοῖον seems to be used inclusively by Aristotle for any genital organs ; often it means " penis," but
obviously it cannot mean this here. *Cf. H.A.* 509 b 27-29.

[c] For the *bōs*, one of the Selachia or cartilaginous fishes,
cf. H.A. 540 b 17 ff., 566 b 4. It is probably either *Notidanus griseus*, which has very large eyes, or *Cephaloptera
giorna* (= *Dicerobatis g.*), the " ox-ray."

This reference to βόες is excised from the text by Platt, who

16

the Vivipara, however, have their testicles in front,[a] though some of them have them inside by the end of the abdomen—*e.g.*, the dolphin—and have no passages, but a sexual duct[b] which leads from them to the outside, as the ox-fish[c] have ; while some have the testicles outside, and of these some are pendent (as in man), others fastened by the fundament (as in swine). I have given a more accurate account of these in the *Researches upon Animals.*[d]

The uterus[e] is always double without exception, just as in males there are always two testes without exception. In some animals the uterus is by the pudenda (as it is in women and in all animals that are viviparous internally as well as externally, and such of the fishes as lay their eggs visibly) ; in other animals the uterus is up towards the diaphragm[f] (as it is in all birds and in the viviparous fishes). The Crustacea, too, and the Cephalopods have a double uterus, since the membranes which surround their " eggs "[g] as they are called are uterine in nature.

The uterus is particularly indistinct in the Octopuses, so that it appears to be single.[h] The reason for this is that the whole bulk of the creature's body is of similar consistency throughout. In the large

supposes βόες here to be " oxen." A.-W. translate " wie die Stiere."

[d] See *H.A.* Bk. III, ch. 1.

[e] It should be noted, once for all, that this term includes what are now known as oviducts.

[f] Aristotle does not confine his use of this term to mammals, which alone have a diaphragm in the usual sense of that term, and hence it must be understood to refer also to the corresponding position in lower animals, as in the present passage; *cf.* also *De respiratione* 475 a 8, where the ὑπόζωμα of wasps, crickets, etc., is mentioned.

[g] See *H.A.* Bk. V, ch. 18.   [h] *Cf.* 758 a 8.

αἱ τῶν ἐντόμων εἰσὶν ἐν τοῖς μέγεθος ἔχουσιν·
ἐν δὲ τοῖς ἐλάττοσιν ἄδηλοι διὰ τὴν μικρότητα
10 τοῦ σώματος.

Τὰ μὲν οὖν εἰρημένα μόρια τοῖς ζῴοις τοῦτον
ἔχει τὸν τρόπον·

IV   Περὶ δὲ τῆς ἐν τοῖς ἄρρεσι διαφορᾶς τῶν σπερματικῶν ὀργάνων, εἴ τις μέλλει θεωρήσειν τὰς
αἰτίας δι’ ἃς εἰσιν, ἀνάγκη λαβεῖν πρῶτον τίνος
15 ἕνεκέν ἐστι ἡ τῶν ὄρχεών ἐστι σύστασις. εἰ δὴ πᾶν ἡ
φύσις ἢ διὰ τὸ ἀναγκαῖον ποιεῖ ἢ διὰ τὸ βέλτιον,
κἂν τοῦτο τὸ μόριον εἴη διὰ τούτων θάτερον. ὅτι
μὲν τοίνυν οὐκ ἀναγκαῖον πρὸς τὴν γένεσιν, φα-
νερόν· πᾶσι γὰρ ἂν ὑπῆρχε τοῖς γεννῶσι, νῦν δ’
οὔθ’ οἱ ὄφεις ἔχουσιν ὄρχεις οὔθ’ οἱ ἰχθύες· ὠμμένοι
20 γάρ εἰσι συνδυαζόμενοι καὶ πλήρεις ἔχοντες θοροῦ
τοὺς πόρους.   λείπεται τοίνυν βελτίονός τινος
χάριν.   ἔστι δὲ τῶν μὲν πλείστων ζῴων ἔργον
σχεδὸν οὐθὲν ἄλλο πλὴν ὥσπερ τῶν φυτῶν σπέρμα
καὶ καρπός.   ὥσπερ δ’ ἐν τοῖς περὶ τὴν τροφὴν
τὰ εὐθυέντερα λαβρότερα πρὸς τὴν ἐπιθυμίαν τὴν
25 τῆς τροφῆς, οὕτω καὶ τὰ μὴ ἔχοντα ὄρχεις πόρους
δὲ μόνον, ἢ ἔχοντα μὲν ἐντὸς δ’ ἔχοντα, πάντα
ταχύτερα πρὸς τὴν ἐνέργειαν τῶν συνδυασμῶν. ἃ
δὲ δεῖ σωφρονέστερα εἶναι, ὥσπερ ἐκεῖ οὐκ εὐθυέντερα, καὶ ἐνταῦθ’ ἕλικας ἔχουσιν οἱ πόροι πρὸς
τὸ μὴ λάβρον μηδὲ ταχεῖαν εἶναι τὴν ἐπιθυμίαν.
30 οἱ δ’ ὄρχεις εἰσὶ πρὸς τοῦτο μεμηχανημένοι· τοῦ

---

^a The Final Cause.
^b See Introd. § 6.
^c Cf. the reason given in Plato, Timaeus 73 A, for the coil-
ing of the intestines. See also P.A. 675 a 19 ff., 675 b 23 ff.
^d See below, 718 a 15.

Insects too the uterus is double, whereas in the smaller ones it is indistinct on account of the smallness of the creatures' body.

This describes the arrangement of those parts of animals which I have mentioned.

Returning to the subject of the difference of the seminal organs in various groups of male animals : If we are to consider the causes to which this is due, we must first of all understand the purpose for the sake of which [a] testes exist. If we agree that everything which Nature does is done either because it is *necessary* or else because it is *better*,[b] we should expect to find that this part, like the rest, exists for one or the other of these two reasons. Now it is evident that it is not *necessary* for generation, otherwise all animals that generate would have it, whereas actually neither Serpents nor Fishes have testes, and these do in fact generate, because they have been observed copulating, with their passages full of milt. The other reason then remains : testes exist for some purpose—because it is *better* that they should exist. Now the business of most animals may be summed up pretty much as that of plants is—viz., seed and fruit; and, just as (to take a parallel case) animals which have straight intestines are more violent in their desire for food,[c] so here also, animals which have no testes but passages only, or which have testes but not external ones, are all quicker with the business of copulation. Those, however, which have to be more sober (*a*) in the case of feeding, have not straight intestines, and (*b*) in the case of copulation, have passages which are twisted,[d] so that their desire shall not be violent or speedy. This then is the object for which the testes have been contrived : they make

IV
(i.) In blooded animals : (*a*) sexual parts in male;

19

717 a

γὰρ σπερματικοῦ περιττώματος στασιμωτέραν
ποιοῦσι τὴν κίνησιν, ἐν μὲν τοῖς ζῳοτόκοις, οἷον
ἵπποις τε καὶ τοῖς ἄλλοις τοῖς τοιούτοις καὶ ἐν
ἀνθρώποις, σώζοντες τὴν ἐπαναδίπλωσιν (ὃν δὲ
τρόπον ἔχει αὕτη, ἐκ τῶν ἱστοριῶν τῶν περὶ τὰ
35 ζῷα δεῖ θεωρεῖν)· οὐθὲν γάρ εἰσι μόριον τῶν πόρων
οἱ ὄρχεις, ἀλλὰ πρόσκεινται, καθάπερ τὰς λαιὰς
προσάπτουσιν αἱ ὑφαίνουσαι τοῖς ἱστοῖς· ἀφαιρου-
717 b
μένων γὰρ αὐτῶν ἀνασπῶνται οἱ πόροι ἐντός, ὥστ'
οὐ δύνανται γεννᾶν τὰ ἐκτεμνόμενα, ἐπεὶ εἰ μὴ
ἀνεσπῶντο, ἐδύναντο ἄν, καὶ ἤδη ταῦρός τις μετὰ
τὴν ἐκτομὴν εὐθέως ὀχεύσας ἐπλήρωσε διὰ τὸ
5 μήπω τοὺς πόρους ἀνεσπάσθαι. τοῖς δ' ὄρνισι καὶ
τοῖς ᾠοτόκοις τῶν τετραπόδων δέχονται τὴν σπερματικὴν περίττωσιν, ὥστε βραδυτέραν εἶναι τὴν
ἔξοδον[1] ἢ τοῖς ἰχθύσιν. φανερὸν δ' ἐπὶ τῶν ὀρνίθων·
περὶ γὰρ τὰς ὀχείας πολὺ μείζους ἴσχουσι[2] τοὺς
ὄρχεις, καὶ ὅσα γε τῶν ὀρνέων καθ' ὥραν μίαν
10 ὀχεύει, ὅταν ὁ χρόνος οὗτος παρέλθῃ, οὕτω μικροὺς
ἔχουσιν ὥστε σχεδὸν ἀδήλους εἶναι, περὶ δὲ τὴν
ὀχείαν σφόδρα μεγάλους. θᾶττον μὲν οὖν ὀχεύουσι
τὰ ἐντὸς ἔχοντα· καὶ γὰρ τὰ ἐκτὸς ἔχοντα οὐ
πρότερον τὸ σπέρμα ἀφίησι πρὶν ἀνασπάσαι τοὺς
ὄρχεις.

V   Ἔτι δὲ τὸ ὄργανον τὸ πρὸς τὸν συνδυασμὸν τὰ
15 μὲν τετράποδα ἔχει· ἐνδέχεται γὰρ αὐτοῖς ἔχειν·
τοῖς δ' ὄρνισι καὶ τοῖς ἄποσιν οὐκ ἐνδέχεται διὰ τὸ

---

[1] διέξοδον PZ.   [2] ἔχουσι PSY.

20

the movement of the seminal residue more steady. (1) In the Vivipara, as for instance in horses and other such animals, and also in man, they do this by maintaining in position the doubling-back of the passages (for a description of this reference must be made to the *Researches upon Animals*),[a] since the testes are no integral part of the passages : they are merely attached thereto, just like the stone weights which women hang on their looms when they are weaving.[b] When the testes are removed, the passages are drawn up within ; this is why castrated animals cannot generate, whereas if the passages were not so drawn up they would be able to do so. A bull immediately after castration has been known to mount a cow and effect impregnation,[c] because the passages had not yet been drawn up. (2) In Birds and in the oviparous quadrupeds the testes receive the seminal residue, so that its emission is slower than it is in the case of Fishes.[d] This is clearly to be seen in Birds : their testes are much larger at the time of copulation.[e] Those birds which copulate at one season only of the year have such tiny testes when this period is over that they are almost indistinguishable, whereas during the breeding season they are very big. So then the animals whose testes are internal accomplish their copulation more quickly, since in fact those with external testes do not emit the semen until the testes have been drawn up.

Another point. The organ for copulation is present V in the quadrupeds because it is possible for them to have it, whereas it is not possible for birds and foot-

---

[a] *H.A.* 510 a 20 ff., and 718 a 15 below.
[b] *Cf.* 787 b 26.   [c] *Cf. H.A.* 510 b 3.
[d] Which have no " testes " in Aristotle's sense.
[e] *Cf. H.A.* 509 b 35 ff.

τῶν μὲν τὰ σκέλη ὑπὸ μέσην εἶναι τὴν γαστέρα,
τὰ δ' ὅλως ἀσκελῆ εἶναι, τὴν δὲ τοῦ αἰδοίου φύσιν
ἠρτῆσθαι ἐντεῦθεν καὶ τῇ θέσει κεῖσθαι ἐνταῦθα.
διὸ καὶ ἐν τῇ ὁμιλίᾳ ἡ σύντασις γίνεται τῶν σκε-
20 λῶν· τό τε γὰρ ὄργανον νευρῶδες καὶ ἡ φύσις[1]
τῶν σκελῶν νευρώδης. ὥστ' ἐπεὶ τοῦτ' οὐκ ἐνδέχεται ἔχειν, ἀνάγκη καὶ ὄρχεις ἢ μὴ ἔχειν ἢ μὴ
ἐνταῦθ' ἔχειν· τοῖς γὰρ ἔχουσιν ἡ αὐτὴ θέσις
ἀμφοτέρων αὐτῶν.

Ἔτι δὲ τοῖς γε τοὺς ὄρχεις ἔχουσιν ἔξω διὰ τῆς
κινήσεως θερμαινομένου τοῦ αἰδοίου προέρχεται τὸ
25 σπέρμα συναθροισθέν, ἀλλ' οὐχ ὡς ἕτοιμον ὂν
εὐθὺς θιγοῦσιν, ὥσπερ τοῖς ἰχθύσιν.

Πάντα δ' ἔχει τὰ ζῳοτόκα τοὺς ὄρχεις ἐν τῷ
πρόσθεν [ἢ ἔξω],[2] πλὴν ἐχίνου· οὗτος δὲ πρὸς τῇ
ὀσφύι μόνος, διὰ τὴν αὐτὴν αἰτίαν δι' ἥνπερ καὶ οἱ
ὄρνιθες, ταχὺν γὰρ ἀναγκαῖον γίνεσθαι τὸν συν-
30 δυασμὸν αὐτῶν[3]· οὐ γὰρ ὥσπερ τὰ ἄλλα[4] τετράποδα ἐπὶ τὰ πρανῆ ἐπιβαίνει, ἀλλ' ὀρθοὶ μίγνυνται
διὰ τὰς ἀκάνθας.

Δι' ἣν μὲν οὖν αἰτίαν ἔχουσι τὰ ἔχοντα ὄρχεις,
εἴρηται, καὶ δι' ἣν αἰτίαν τὰ μὲν ἔξω τὰ δ' ἐντός.

---

[1] φύσις Z, vulg. : σύστασις S.
[2] aut ἢ ἔξω secludenda (om. Σ), aut ⟨ἢ ἐντός⟩ addenda
(Platt).
[3] διὰ . . . αὐτῶν fortasse secludenda ; sed cf. 769 b 34 seqq.
[4] ἄλλα Z : om. vulg.

---

[a] But the goose has a penis, H.A. 509 b 30.
[b] Cf. 718 a 5, 739 a 10.
[c] Omit, or read " either outside or inside."
[d] Inside, of course.

less animals. It is impossible for birds [a] because their legs are under the middle of the abdomen. It is impossible for the other creatures because they have no legs at all, and that is the place where the penis is always suspended and that is the position for it. (This also is the reason why there is strain on the legs during sexual intercourse : both the organ itself and the legs are by their nature sinewy.) And so, since it is impossible for them to have this organ, they must of necessity have no testes either, or else not have them in that place, since in those animals which possess both penis and testes the situation of both is one and the same.

Another point. As far as the animals with external testes are concerned, as the penis is set in movement and gets heated, the semen first collects itself together, and then advances : it is not ready immediately contact is established, as it is in fishes.[b]

All the Vivipara have their testes in front, [or outside,[c]] except the hedgehog. This is the only one that has them by the loin,[d] and the reason is the same as for the birds,[e] since they must of necessity accomplish their copulation quickly, for they do not mount on the back as the other quadrupeds do, but on account of their spines stand upright for intercourse.

We have now said why those animals which have testes have them, and why some have them outside

[e] This remark, if it remains in the text, obviously cannot refer to the only reason so far given for birds at 717 b 15-17 ; if taken as referring to the reason which immediately follows, this will roughly correspond to the statement in *H.A.* 539 b 34 that some birds copulate quickly. But no doubt the reason Aristotle has in mind is the one mentioned below at 719 b 11 ff., viz., that the skin is too hard.

717 b

VI ὅσα δὲ μὴ ἔχει, καθάπερ εἴρηται, διά τε τὸ μὴ εὖ ἀλλὰ
35 τὸ ἀναγκαῖον μόνον οὐκ ἔχει τοῦτο τὸ μόριον, καὶ
διὰ τὸ ἀναγκαῖον εἶναι ταχεῖαν γίνεσθαι τὴν ὀχείαν·
τοιαύτη δ' ἐστὶν ἡ τῶν ἰχθύων φύσις καὶ ἡ τῶν
ὄφεων. οἱ μὲν γὰρ ἰχθύες ὀχεύουσι παραπίπτοντες
καὶ ἀπολύονται ταχέως. ὥσπερ γὰρ ἐπὶ τῶν ἀνθρώπων καὶ πάντων τῶν τοιούτων ἀνάγκη κατα-
σχόντας τὸ πνεῦμα προΐεσθαι τὴν γονήν, τοῦτο δ'
5 ἐκείνοις συμβαίνει μὴ δεχομένοις τὴν θάλατταν,
εἰσὶ δὲ εὔφθαρτοι τοῦτο μὴ ποιοῦντες, οὔκουν δεῖ
ἐν τῷ συνδυασμῷ τὸ σπέρμα πέττειν αὐτούς,
ὥσπερ τὰ πεζὰ καὶ ζῳοτόκα, ἀλλ' ὑπὸ τὴν ὥραν¹
τὸ σπέρμα πεπεμμένον ἀθρόον ἔχουσιν, ὥστε μὴ
ἐν τῷ θιγγάνειν ἀλλήλων πέττειν,² ἀλλὰ προΐεσθαι
10 πεπεμμένον. διὸ ὄρχεις οὐκ ἔχουσιν, ἀλλ' εὐθεῖς
καὶ ἁπλοῦς τοὺς πόρους, οἷον μικρὸν μόριον τοῖς
τετράποσιν ὑπάρχει περὶ τοὺς ὄρχεις· τῆς γὰρ
ἐπαναδιπλώσεως τοῦ πόρου τὸ μὲν ἔναιμον μέρος
ἐστὶ τὸ δ' ἄναιμον, ὃ δέχεται καὶ δι' οὗ ἤδη σπέρμα
ὂν πορεύεται, ὥσθ' ὅταν ἐνταῦθα ἔλθῃ ἡ γονή,
15 ταχεῖα καὶ τούτοις γίνεται ἡ ἀπόλυσις. τοῖς δ'
ἰχθύσι τοιοῦτος ὁ πόρος πᾶς ἐστιν οἷος ἐπὶ τῶν

---

¹ ὑπὸ τὴν ὥραν A.-W., cf. H.A. 509 b 20, 35 : πρὸ τῆς ὥρας
coniecerat Platt : ὑπὸ τῆς ὥρας vulg.
² πέττειν A.-W., digestio Σ : ποιεῖν vulg.

ᵃ See ch. 4, *init.* For *necessity,* see Introd. § 6.
ᵇ This appears to be the meaning ; Michael Scot renders
eiciunt sperma velociter : *cf.* the English phrase " relieve
themselves." Also at 718 a 14.
ᶜ Viz., all that breathe.
ᵈ This, according to Aristotle, corresponds to breathing ;
it is their method of self-refrigeration : see *De respiratione*
476 a 1 ff.

and others inside. And as for those which have no VI
testes, they lack this part, as we have said, because
such absence is not *good*, but *necessary* merely[a];
and also because it is necessary that their copulation should be accomplished quickly. Fishes and
serpents come under this class. Fishes copulate
by placing themselves alongside each other and
quickly ejaculate.[b] Just as men and all such animals[c]
in order to emit the semen must of necessity hold
their breath, so fishes must refrain from taking in the
seawater,[d] and when they omit to do this they easily
come to grief. On this account they are bound to
avoid concocting[e] the semen during the act of copulation (which is what the viviparous land-animals do);
instead, they have their semen ready concocted and
collected at the proper time, so that they do not concoct it while in contact with each other, but emit it
already concocted. For this reason they have no
testes, but passages which are straight and simple.
In the testes of quadrupeds there is a small portion
of a similar character : I refer to the latter portion of
that length of the passage which is doubled back.[f]
One portion of this length has blood in it and one has
not, and by the time the fluid enters this latter portion and passes through it, it is already semen ; so
that when it arrives there, ejaculation quickly takes
place[g] in these animals too. In Fishes the whole of
the passage is of the same character as this latter

[e] *Cf.* 717 b 25 above.
[f] The *vas deferens*; *cf.* above 717 a 33 ; and *H.A.*
510 a 23 ff.
[g] *Cf.* above, 718 a 1 : Scot's Arabic original seems to
have been extremely cautious and to have given both possible
meanings of ἀπόλυσις; for Scot has *eius exitus est velox, et
cum exit sperma separantur mas et femina.*

ἀνθρώπων καὶ τῶν τοιούτων ζῴων κατὰ τὸ ἕτερον
μέρος τῆς ἐπαναδιπλώσεως.

VII Οἱ δὲ ὄφεις ὀχεύονται περιελιττόμενοι ἀλλήλοις,
οὐκ ἔχουσι δ᾽ ὄρχεις οὐδ᾽ αἰδοῖον, ὥσπερ εἴρηται
πρότερον, αἰδοῖον μὲν ὅτι οὐδὲ σκέλη, ὄρχεις δὲ διὰ
20 τὸ μῆκος, ἀλλὰ πόρους, ὥσπερ οἱ ἰχθύες· διὰ γὰρ
τὸ εἶναι αὐτῶν προμήκη τὴν φύσιν, εἰ ἔτι ἐπί-
στασις ἐγίγνετο περὶ τοὺς ὄρχεις, ἐψύχετ᾽ ἂν ἡ
γονὴ διὰ τὴν βραδυτῆτα. ὅπερ συμβαίνει καὶ ἐπὶ
τῶν μέγα τὸ αἰδοῖον ἐχόντων· ἀγονώτεροι γάρ εἰσι
τῶν μετριαζόντων διὰ τὸ μὴ γόνιμον εἶναι τὸ
25 σπέρμα τὸ ψυχρόν, ψύχεσθαι δὲ τὸ φερόμενον λίαν
μακράν. δι᾽ ἣν μὲν οὖν αἰτίαν τὰ μὲν ὄρχεις ἔχει
τὰ δ᾽ οὐκ ἔχει τῶν ζῴων, εἴρηται.

[1][Περιπλέκονται δ᾽ ἀλλήλοις οἱ ὄφεις διὰ τὴν
ἀφυΐαν τῆς παραπτώσεως. μικρῷ γὰρ προσαρ-
μόττοντες μορίῳ λίαν μακροὶ ὄντες οὐκ εὐσυν-
30 άρμοστοί εἰσιν· ἐπεὶ οὖν οὐκ ἔχουσι μόρια οἷς
περιλήψονται, ἀντὶ τούτων[2] τῇ ὑγρότητι χρῶνται
τοῦ σώματος, περιελιττόμενοι ἀλλήλοις. διὸ καὶ
δοκοῦσι βραδύτερον ἀπολύεσθαι τῶν ἰχθύων, οὐ
μόνον διὰ τὸ μῆκος τῶν πόρων ἀλλὰ καὶ διὰ τὴν
περὶ ταῦτα σκευωρίαν.]

VIII 35 Τοῖς δὲ θήλεσι τὰ περὶ τὰς ὑστέρας ἀπορήσειεν
ἄν τις ὃν τρόπον ἔχει· πολλαὶ γὰρ ὑπεναντιώσεις

---

[1] quae sequuntur non proprio loco posita videntur.
[2] τούτου PZ.

---

[a] Which is the place where it would have to be : 717 b 17,
18.
[b] As the preceding sentence would normally mark the

portion of it in man and other such animals (*i.e.*, the latter portion of that length of it which is doubled back).

Serpents copulate by twisting round each other, VII but they have no testes and not even a penis, as I said earlier : no penis, because they have no legs either,[a] and no testes because of their length—instead, they have passages just as fish do—since as their bodies are so very long, if there were to be yet further delay in the region of the testes, the semen would be cooled off owing to its slow rate of progress. This does in fact happen with men who have a large penis : they are less fertile than those who have a moderately large one, because the semen gets cooled off by being transported too great a distance, and cold semen is not generative. I have now stated why some animals have testes and others not.

[b] [Serpents intertwine because they are not naturally fitted for placing themselves alongside each other ; their bodies are so long, and the part by which they unite is so small, that they find difficulty in achieving union ; and so, as they have no parts by which they can take hold of each other, they make use of the suppleness of their bodies instead, and twist around each other. On this account, they seem, too, to take longer to ejaculate than fish do, not only because of the length of the passages but also because of the intricacy of the manœuvre.]

One may well be puzzled concerning the arrange- VIII ment of the uterus in the various female animals ; (b) sexual many instances of quite contrary arrangements parts in female.

conclusion of the chapter, the remarks which follow are probably a supplementary note, or an alternative version, incorporated in the text.

ὑπάρχουσιν αὐτοῖς. οὔτε γὰρ τὰ ζῳοτοκοῦντα
ὁμοίως ἔχει πάντα, ἀλλ' ἄνθρωποι μὲν καὶ τὰ πεζὰ

πάντα κάτω πρὸς τοῖς ἄρθροις, τὰ δὲ σελάχη ⟨τὰ⟩[1]
ζῳοτοκοῦντα ἄνω πρὸς τῷ ὑποζώματι, οὔτε τὰ
ᾠοτοκοῦντα, ἀλλ' οἱ μὲν ἰχθύες κάτω καθάπερ
ἄνθρωπος καὶ τὰ ζῳοτοκοῦντα τῶν τετραπόδων,
οἱ δ' ὄρνιθες ἄνω, καὶ ὅσα ᾠοτοκεῖ τῶν τετρα-
5 πόδων. οὐ μὴν ἀλλ' ἔχουσι καὶ αὗται αἱ ὑπεναντιώσεις κατὰ λόγον. πρῶτον μὲν γὰρ τὰ ᾠοτοκοῦντα ᾠοτοκεῖ διαφερόντως· τὰ μὲν γὰρ ἀτελῆ
προΐεται τὰ ᾠά, οἷον οἱ ἰχθύες· ἔξω γὰρ ἐπιτελεῖται καὶ λαμβάνει αὔξησιν τὰ τῶν ἰχθύων.
αἴτιον δ' ὅτι πολύγονα ταῦτα, καὶ τοῦτ' ἔργον
10 αὐτῶν ὥσπερ τῶν φυτῶν· εἰ οὖν ἐν αὑτοῖς ἐτελεσι-
ούργουν, ἀναγκαῖον ὀλίγα τῷ πλήθει εἶναι· νῦν δὲ
τοσαῦτα ἴσχουσιν ὥστε δοκεῖν ⟨ἓν⟩[2] ᾠὸν εἶναι τὴν
ὑστέραν ἑκατέραν ἔν γε τοῖς μικροῖς ἰχθυδίοις·
ταῦτα γὰρ πολυγονώτατά ἐστιν, ὥσπερ καὶ ἐπὶ
τῶν ἄλλων τῶν ἀνάλογον τούτοις ἐχόντων τὴν
φύσιν, καὶ ἐν φυτοῖς καὶ ἐν ζῴοις· ἡ γὰρ τοῦ

---

[1] ⟨τὰ⟩ Peck, vel fortasse ζῳοτόκα ⟨ὄντα⟩.
[2] ⟨ἓν⟩ Peck, unum ovum Σ. ἓν supplendum esse suspicati
erant A.-W. (collato H.A. 510 b 24), Schneider.

---

[a] Selachia : the cartilaginous fishes, including the Sharks.
The " fishing-frog " is not viviparous (see 754 a 26, n.).

[b] The observation of Aristotle that the eggs of many
organisms swell during their development, though unap-
preciated for many centuries, is the basis of the modern
distinction between cleidoic and non-cleidoic eggs. The
walls of a cleidoic egg are permeable only to matter in the
gaseous state (e.g., the hen's egg). Most aquatic animals,
however, lay non-cleidoic eggs, i.e., eggs which, though they
have a sufficiency of organic material (such as proteins, fats,

28

occur. To begin with, not all the Vivipara have the same arrangement. All that are land-animals, including human beings, have the uterus placed low down by the pudenda, whereas the viviparous Selachia [a] have it higher up by the diaphragm. And then again, the Ovipara show the same variations. Fishes have the uterus low down like human beings and the viviparous quadrupeds, whereas birds have it higher up, and so do the oviparous quadrupeds. Nevertheless, there is rhyme and reason even in these contradictory phenomena. First of all, the egg-laying animals have different ways of laying their eggs. (a) Some creatures' eggs are imperfect when laid—e.g., those of fishes, which become perfected, i.e., grow, outside the creature which produces them.[b] The reason is that these animals are very prolific and this is their function,[c] as it is that of plants ; so that if they brought the eggs to a state of perfection inside their bodies, the eggs would of necessity be few in number, whereas in actual fact they produce so many that each uterus seems to be just one mass of egg, at any rate in the very small fishes, which are the most prolific of all. The same is true both of those plants and of those animals which are of a corresponding nature [d] in their own classes ; what

carbohydrates, etc.) to make each an embryo, are insufficiently supplied with water and inorganic materials; these they have to absorb from their environment. Hence their swelling. Though the main bulk of this is due to water-intake, it is interesting that the greater part of the copper, for example, which is present in the respiratory blood-pigment of the octopus at the time of hatching is derived, not from the egg as laid, but from the surrounding sea-water. See also 732 b 5, etc.

[c] Cf. 717 a 22.   [d] i.e., small.

718 b

15 μεγέθους αὔξησις τρέπεται εἰς τὸ σπέρμα τούτοις.
οἱ δ᾽ ὄρνιθες καὶ τὰ τετράποδα τῶν ᾠοτόκων
τέλεια ᾠὰ τίκτουσιν, ἃ δεῖ πρὸς τὸ σῴζεσθαι
σκληρόδερμα εἶναι (μαλακόδερμα γὰρ ἕως ἂν αὔξησιν ἔχῃ ἐστίν), τὸ δ᾽ ὄστρακον γίνεται ὑπὸ
θερμότητος ἐξικμαζούσης τὸ ὑγρὸν ἐκ τοῦ γεώδους.
20 ἀναγκαῖον οὖν θερμὸν εἶναι τὸν τόπον ἐν ᾧ τοῦτο
συμβήσεται. τοιοῦτος δ᾽ ὁ περὶ τὸ ὑπόζωμα· καὶ
γὰρ τὴν τροφὴν πέττει οὗτος. εἰ οὖν τὰ ᾠὰ
ἀνάγκη ἐν τῇ ὑστέρᾳ εἶναι, καὶ τὴν ὑστέραν ἀνάγκη
πρὸς τῷ ὑποζώματι εἶναι τοῖς τέλεια τὰ ᾠὰ τίκτουσι, τοῖς δ᾽ ἀτελῆ κάτω· πρὸ ὁδοῦ γὰρ οὕτως
25 ἔσται. καὶ πέφυκε δὲ μᾶλλον ἡ ὑστέρα κάτω εἶναι
ἢ ἄνω, ὅπου μή τι ἐμποδίζει ἕτερον ἔργον τῆς
φύσεως· κάτω γὰρ αὐτῆς καὶ τὸ πέρας ἐστίν· ὅπου
δὲ τὸ πέρας, καὶ¹ τὸ ἔργον· αὕτη² δ᾽ οὗ τὸ ἔργον.

IX Ἔχει δὲ καὶ τὰ ζῳοτοκοῦντα πρὸς ἄλληλα διαφοράν. τὰ μὲν γὰρ οὐ μόνον θύραζε ζῳοτοκεῖ
30 ἀλλὰ καὶ ἐν αὑτοῖς, οἷον ἄνθρωποί τε καὶ ἵπποι
καὶ κύνες καὶ πάντα τὰ τρίχας ἔχοντα, καὶ τῶν
ἐνύδρων δελφῖνές τε καὶ φάλαιναι καὶ τὰ τοιαῦτα
κήτη.

X Τὰ δὲ σελάχη καὶ οἱ ἔχεις θύραζε μὲν ζῳοτοκοῦσιν, ἐν αὑτοῖς δ᾽ ᾠοτοκοῦσι πρῶτον. ᾠοτοκοῦσι δὲ τέλειον ᾠόν· οὕτως γὰρ γεννᾶται ἐκ

---

¹ καὶ Z : om. vulg.   ² αὕτη P*SYZ : αὐτὴ vulg.

ᵃ i.e., do not increase in size after being laid.
ᵇ i.e., without first producing an egg internally. Aristotle
knew nothing of the existence of the mammalian egg, which
is a single cell of microscopic size.

30

would have produced increase of size is in them diverted to form seed. (b) Birds, however, and quadrupedal Ovipara lay eggs that are perfect,[a] and these eggs for safety's sake are bound to have a hard skin (while they are still growing, they have a soft skin), and the shell is formed by heat, which evaporates the fluid from the earthy substance ; hence the place where this is to be done must of necessity be hot—a condition which is fulfilled by the region round the diaphragm, as the fact that it concocts the food shows. So, if the eggs must of necessity be within the uterus, the uterus must of necessity be alongside the diaphragm in those animals whose eggs are in a perfected condition when laid, while it must be low down in those whose eggs are imperfect when laid ; it will be advantageous so. Further, it is more natural that the uterus should be low down than high up (unless there is some other business of Nature's which prevents it), since its conclusion is down below too ; and where the conclusion is, there also the function is ; thus the uterus is where the function is.

Similarly, the Vivipara differ from one another. IX Some of them bring forth their young alive not externally only but also within themselves,[b] as for instance, human beings, horses, dogs and all haired animals, also such water-animals as dolphins, whales and such cetacea.[c]

Selachia and vipers, though they bring forth their X young alive externally, first of all produce eggs internally. And the egg they produce is a perfected one, for thus only is an animal generated from the

[c] *Cf. H.A.* 566 b 2, where Aristotle explains this to mean those creatures which have no gills, but a blowhole.

35 τοῦ ᾠοῦ τὸ ζῷον, ἐξ ἀτελοῦς δὲ οὐθέν. θύραζε δὲ
οὐκ ᾠοτοκοῦσι διὰ τὸ ψυχρὰ τὴν φύσιν εἶναι καὶ
XI οὐχ ὡς τινές φασι θερμά. μαλακόδερμα γοῦν τὰ
ᾠὰ γεννῶσιν· διὰ γὰρ τὸ εἶναι ὀλιγόθερμα οὐ ξηραίνει αὐτῶν ἡ φύσις τὸ ἔσχατον. διὰ μὲν οὖν
719 a τὸ ψυχρὰ εἶναι μαλακόδερμα γεννῶσι, διὰ δὲ τὸ
μαλακόδερμα οὐ θύραζε· διεφθείρετο γὰρ ἄν.

῞Οταν δὲ τὸ[1] ζῷον ἐκ τοῦ ᾠοῦ γίγνηται, τὸν αὐτὸν
τρόπον τὰ πλεῖστα γίγνεται ὅνπερ ἐν τοῖς ὄρνισιν·
καταβαίνει γὰρ κάτω,[2] καὶ γίγνεται ζῷα πρὸς τοῖς
5 ἄρθροις, καθάπερ καὶ ἐν[3] τοῖς ἐξ ἀρχῆς εὐθὺς ζῳοτοκοῦσιν. διὸ καὶ τὴν ὑστέραν τὰ τοιαῦτα ἔχει
ἀνομοίαν καὶ τοῖς ζῳοτόκοις καὶ τοῖς ᾠοτόκοις,
διὰ τὸ ἀμφοτέρων μετέχειν τῶν εἰδῶν· καὶ γὰρ
πρὸς τῷ ὑποζώματι ἔχουσι καὶ κάτω πάρηκουσαν
πάντα τὰ σελαχώδη. δεῖ δὲ καὶ περὶ ταύτης καὶ
10 περὶ τῶν ἄλλων ὑστερῶν, ὃν τρόπον ἔχουσιν, ἔκ
τε τῶν ἀνατομῶν τεθεωρηκέναι καὶ τῶν ἱστοριῶν.
ὥστε διὰ μὲν τὸ ᾠοτόκα εἶναι τελείων ᾠῶν ἄνω
ἔχει, διὰ δὲ τὸ ζῳοτοκεῖν κάτω, καὶ ἀμφοτέρων
μετειλήφασιν.

Τὰ δ᾽ εὐθὺς ζῳοτοκοῦντα πάντα κάτω· οὐ γὰρ

[1] τὸ SY : om. vulg.   [2] sed vid. p. 562.
[3] καὶ ἐν Z : ἐν Y : καὶ vulg.

---

[a] According to Aristotle, Empedocles had said that those
animals which are hottest live in the water to counteract the
excess of heat in their constitution (*De respir.* 477 b 1 ff.).

[b] The *Dissections*, in seven Books, is no longer extant.
Aristotle several times refers to the " diagrams in the *Dissections* " and the like (*e.g.* 746 a 14), and it was no doubt a
collection of material with anatomical diagrams prepared
for use in the lecture-room. Jaeger (*Aristotle*, Eng. trans.,
336), following V. Rose, describes it as an anatomical atlas.
See also Jaeger, *Diokles von Karystos*, 165-167.

egg : nothing is generated from an imperfect egg. The reason why they do not lay their eggs externally is because they are by nature cold creatures, not hot, as some persons allege.[a]   Anyway, the eggs they produce are soft-skinned—because the creatures have so little heat in them that their natural constitution does not dry off the outermost part of the eggs.   Thus the coldness of the creatures is the reason why the eggs they produce are soft-skinned, and the fact that the eggs are soft-skinned is the reason they are not produced externally : if they were, they would come to grief.

XI

When the animal is formed out of the egg, the process of formation is for the most part the same as for birds : ⟨the eggs⟩ descend, and the young animals are formed close by the pudenda, as occurs also in creatures which are viviparous right from the outset.   Another result of this is that in animals such as we are now discussing the uterus differs both from that of the Vivipara and from that of the Ovipara, since they have a share in both these groups ; that is to say, in all the Selachians the uterus is at the same time close by the diaphragm and also extends along downwards. (However, to ascertain the arrangement of the uterus of the Selachians and other kinds as well, the *Dissections* [b] should be inspected and also the *Researches* [c]).   Thus the Selachians have their uterus high up because they are oviparous and lay perfected eggs, while they have it low down because they are viviparous ; thus they have a share in both.

Animals which are viviparous from the outset [d] all have the uterus low down, since they have no natural

[c] *H.A.* 510 b 5 ff.
[d] See above, 718 b 30.

15

20

25

ἐμποδίζει τῆς φύσεως οὐδὲν ἔργον, οὐδὲ διττο-
γονεῖ. πρὸς δὲ τούτοις ἀδύνατον ζῷα γίγνεσθαι
πρὸς τοῖς ὑποζώμασιν· τὰ μὲν γὰρ ἔμβρυα βάρος
ἔχειν ἀναγκαῖον καὶ κίνησιν, ὁ δὲ τόπος ἐπίκαιρος
ὢν τοῦ ζῆν οὐκ ἂν δύναιτο ταῦθ᾽ ὑπενεγκεῖν. ἔτι
δ᾽ ἀνάγκη δυστοκίαν εἶναι διὰ τὸ μῆκος τῆς φορᾶς,
ἐπεὶ καὶ νῦν ἐπὶ τῶν γυναικῶν, ἐὰν περὶ τὸν τόκον
ἀνασπάσωσι χασμησάμεναι ἤ τι τοιοῦτον ποιή-
σασαι, δυστοκοῦσιν. καὶ κεναὶ δ᾽ οὖσαι αἱ ὑστέραι
ἄνω προσιστάμεναι πνίγουσιν· καὶ γὰρ ἀνάγκη τὰς
μελλούσας ζῷον ἕξειν ἰσχυροτέρας εἶναι, διὸ σαρ-
κώδεις εἰσὶν αἱ τοιαῦται πᾶσαι, αἱ δὲ πρὸς τῷ
ὑποζώματι[1] ὑμενώδεις. καὶ ἐπ᾽ αὐτῶν δὲ τῶν
διγονίαν ποιουμένων ζῴων φανερὸν τοῦτο συμ-
βαῖνον· τὰ μὲν γὰρ ᾠὰ ἄνω καὶ ἐν τῷ πλαγίῳ
ἴσχουσι, τὰ δὲ ζῷα ἐν τῷ κάτω μέρει τῆς ὑστέρας.

Δι᾽ ἣν μὲν οὖν αἰτίαν ὑπεναντίως ἔχουσι τὰ περὶ
τὰς ὑστέρας ἔνια τῶν ζῴων, καὶ ὅλως διὰ τί τοῖς
μὲν κάτω τοῖς δὲ ἄνω πρὸς τῷ ὑποζώματί εἰσιν,
εἴρηται.

XII 30 Διότι δὲ τὰς μὲν ὑστέρας ἔχουσι πάντα ἐντός,
τοὺς δ᾽ ὄρχεις τὰ μὲν ἐκτὸς τὰ δ᾽ ἐντός, αἴτιον τοῦ
μὲν τὰς ὑστέρας ἐντὸς εἶναι πᾶσιν, ὅτι ἐν ταύταις
ἐστὶ τὸ γινόμενον, ὃ δεῖται φυλακῆς καὶ σκέπης[2]
καὶ πέψεως, ὁ δ᾽ ἐκτὸς τοῦ σώματος τόπος εὔβλα-

---

[1] sic PZ : τοῖς ὑποζώμασιν vulg.
[2] καὶ σκέπης om. PΖΣ, A.-W.

---

[a] They omit the internally oviparous stage.
[b] See Introd. § 62, and n.

function that prevents this, nor do they produce their young by the two-stage process.[a] Besides, it is impossible for young animals to be formed near the diaphragm ; embryos are bound to be heavy and to move about, and that part of the body is a vital spot and would not be able to put up with such things. Further, ⟨if the uterus were placed high,⟩ parturition would of necessity be difficult on account of the distance to be covered, since even as it is, in the case of women, if they draw up the uterus at the time of parturition by yawning or by doing something of the sort, difficulty in delivery is the result. Even when empty the uterus produces a stifling sensation if pushed upwards. Besides, a uterus which is destined to contain ⟨not an egg but⟩ an actual animal must of necessity be a stronger thing ; that is why the uterus of all viviparous animals is fleshy, whereas in those cases where it is near the diaphragm the uterus is membranous. This is clearly to be seen in the case of those animals which produce their young by the two-stage process : the eggs are carried high up and towards one side, whereas the young creatures are carried in the lower part of the uterus.

We have now explained the reason why contrary arrangements of the uterus are found in certain animals, and in general why in some the uterus is placed low down and in others high up by the diaphragm.

We have seen too that while all animals have their XII uterus inside, some have their testes inside and others (c) General remarks. outside. The reason why the uterus is always inside is that it is the container for the young creature while it is being formed, and this needs protection, shelter, and concoction,[b] which the outer part of the body

35

35 πτος καὶ ψυχρός. οἱ δ᾽ ὄρχεις τοῖς μὲν ἐντὸς τοῖς
δ᾽ ἐκτός[1]· διὰ ⟨δὲ⟩[2] τὸ δεῖσθαι καὶ τούτους σκέπης
καὶ καλύμματος πρός τε σωτηρίαν καὶ πρὸς τὴν τοῦ
σπέρματος πέψιν (ἀδύνατον γὰρ ἐψυγμένους καὶ
πεπηγότας ἀνασπᾶσθαι καὶ προΐεσθαι τὴν γονήν),
[διόπερ][3] ὅσοις ἐν φανερῷ εἰσὶν οἱ ὄρχεις, ἔχουσι
5 σκέπην δερματικὴν τὴν καλουμένην ὀσχέαν· ὅσοις
δ᾽ ἡ τοῦ δέρματος φύσις ἐναντιοῦται διὰ σκληρότητα πρὸς τὸ μὴ περιληπτικὴν εἶναι μηδὲ μαλ-
θακὴν καὶ δερματικήν,[4] οἷον τοῖς τ᾽ ἰχθυῶδες ἔχουσι
τὸ δέρμα καὶ τοῖς φολιδωτόν, τούτοις δ᾽ ἀναγκαῖον
10 ἐντὸς ἔχειν. διόπερ οἵ τε δελφῖνες καὶ ὅσα τῶν
κητωδῶν ὄρχεις ἔχουσιν, ἐντὸς ἔχουσι, καὶ τὰ
ᾠοτόκα καὶ τετράποδα τῶν φολιδωτῶν. καὶ τὸ
τῶν ὀρνίθων δὲ δέρμα σκληρόν, ὥστε κατὰ μέγεθος
ἀσύμμετρον εἶναι περιλαβεῖν, καὶ ταύτην αἰτίαν
εἶναι πᾶσι τούτοις πρὸς ταῖς εἰρημέναις πρότερον
15 ἐκ τῶν περὶ τὰς ὀχείας συμβαινόντων ἀναγκαίων.
διὰ τὴν αὐτὴν δ᾽ αἰτίαν καὶ ὁ ἐλέφας καὶ ὁ ἐχῖνος
ἔχουσιν ἐντὸς τοὺς ὄρχεις· οὐδὲ γὰρ τούτοις εὐφυὲς
τὸ δέρμα πρὸς τὸ χωριστὸν ἔχειν τὸ σκεπαστικὸν
μόριον.

[Κεῖνται δὲ καὶ τῇ θέσει ὑπεναντίως αἱ ὑστέραι
τοῖς τε ζῳοτοκοῦσιν ἐν αὑτοῖς καὶ τοῖς ᾠοτοκοῦσι
20 θύραζε, καὶ τούτων τοῖς τε τὰς ὑστέρας ἔχουσι
κάτω καὶ τοῖς πρὸς τῷ ὑποζώματι, οἷον τοῖς

---

[1] ἐκτὸς τοῖς δ᾽ ἐντός SZ.
[2] sic interpungunt A.-W., qui et ⟨δὲ⟩ addunt.
[3] διόπερ seclusi.
[4] μηδὲ . . . δερματικήν secludunt A.-W.

---

[a] Not in the *Generation of Animals*; but see 717 b 29.

cannot provide, being easily injured and cold. The testicles, however, are inside in some animals but outside in others : since, however, they also need shelter and covering to keep them safe and to secure concoction for the semen (for if they have been exposed to cold and rendered stiff they cannot be drawn up and emit the semen), those animals whose testes are in the open have a covering of skin over them known as the scrotum; while those animals the nature of whose skin is so hard that it is not amenable to this arrangement, and cannot be used for a wrapping and is not soft or like ordinary skin (*e.g.*, animals whose skin is like that of fish, and those whose skin is made of horny scales)—they must of necessity have their testes inside. On this account the dolphins and those cetacea which possess testes have them inside ; so do those horny-scaled animals which are oviparous and four-footed. Birds, too, have hard skin, which will not accommodate itself to the size of the testes and make a wrapping for them, and this makes another reason why in all these cases the testes are inside in addition to the reasons (previously mentioned [a]) due to the necessary exigencies of copulation. And for this selfsame reason the testes are also inside in the elephant and in the hedgehog ; the skin of these two animals, as of the others, is not well adapted for having the protective part separate.

[b][Contrary positions of the uterus are found in those animals which are internally viviparous and in those which are externally oviparous ; and again in some of the latter class it is placed low down, in others by the diaphragm, as for instance in fishes on the one

[b] The following paragraph is simply a hash-up of parts of the preceding chapters.

719 b

ἰχθύσι πρός τε τοὺς ὄρνιθας καὶ τὰ ᾠοτόκα τῶν
τετραπόδων, καὶ τοῖς κατ' ἀμφοτέρους τοὺς τρόπους γεννῶσιν, ἐν ἑαυτοῖς μὲν ᾠοτοκοῦσιν, εἰς
δὲ τὸ φανερὸν ζῳοτοκοῦσιν. τὰ μὲν γὰρ ζῳο-
25 τοκοῦντα καὶ ἐν αὑτοῖς καὶ ἐκτὸς ἐπὶ τῆς γαστρὸς
ἔχει τὰς ὑστέρας, οἷον ἄνθρωπος καὶ βοῦς καὶ
κύων καὶ τἆλλα τὰ τοιαῦτα· πρὸς γὰρ τὴν τῶν
ἐμβρύων σωτηρίαν καὶ αὔξησιν συμφέρει μηθὲν
ἐπεῖναι βάρος ἐπὶ ταῖς ὑστέραις.][1]

XIII   Ἔστι δὲ καὶ ἕτερος ὁ πόρος δι' οὗ ἥ τε ξηρὰ
30 περίττωσις ἐξέρχεται καὶ δι' οὗ ἡ ὑγρὰ τούτοις
πᾶσιν. διὸ ἔχουσιν αἰδοῖα τὰ τοιαῦτα πάντα καὶ τὰ
ἄρρενα καὶ τὰ θήλεα, καθ' ἃ[2] ἐκκρίνεται τὸ περίττωμα τὸ ὑγρὸν καὶ τοῖς μὲν ἄρρεσι τὸ σπέρμα, τοῖς
δὲ θήλεσι τὸ κύημα.[3] οὗτος δ' ἐπάνω καὶ ἐν τοῖς
προσθίοις ὑπάρχει ὁ πόρος ⟨τοῦ⟩[4] τῆς ξηρᾶς τροφῆς.
35 [ὅσα δ' ᾠοτοκεῖ μὲν ἀτελὲς δ' ᾠόν, οἷον ὅσοι τῶν

720 a   ἰχθύων ᾠοτοκοῦσιν, οὗτοι δ' οὐχ ὑπὸ τῇ γαστρὶ
ἀλλὰ πρὸς τῇ ὀσφύι ἔχουσι τὰς ὑστέρας· οὔτε γὰρ
ἐμποδίζει ἡ τοῦ ᾠοῦ αὔξησις, διὰ τὸ ἔξω τελειοῦσθαι καὶ προϊέναι τὸ αὐξανόμενον.][5] ὅ τε πόρος
ὁ αὐτός ἐστι [καὶ][6] ἐν τοῖς μὴ ἔχουσι γεννητικὸν
5 αἰδοῖον τῷ[7] τῆς ξηρᾶς τροφῆς, πᾶσι τοῖς ᾠοτόκοις
καὶ τοῖς ἔχουσιν αὐτῶν κύστιν, οἷον ταῖς χελώναις·
τῆς γενέσεως γὰρ ἕνεκεν, οὐ τῆς τοῦ ὑγροῦ περιττώματος ἐκκρίσεως, εἰσὶ διττοὶ οἱ πόροι· διὰ δὲ
τὸ ὑγρὰν εἶναι τὴν φύσιν τοῦ σπέρματος καὶ ἡ τῆς

---

[1] κεῖνται . . . ὑστέραις secludit Platt.
[2] ἃ EmZ[2] : ὃ vulg.   [3] τὰ καταμήνια A.-W.
[4] ⟨τοῦ⟩ Aldus, A.-W.
[5] ὅσα δ' . . . αὐξανόμενον secludit Platt.
[6] καὶ secl. A.-W.   [7] τῷ Z[1] : ὁ vulg.

hand as against birds and oviparous quadrupeds on the other ; and then again it is different in those animals which produce their young by both of the two methods, viz., which are internally oviparous and outwardly viviparous. Those animals which are both internally and externally viviparous have their uterus placed against the abdomen, as for instance man, ox, dog, and other such animals, since it is expedient for the safety and growth of the embryo that no weight should be put upon the uterus.]

In all these animals the passage through which the XIII solid residue issues is other than that through which the fluid issues. On this account all such animals, both male and female, have pudenda by which the fluid residue is voided, and thereby too in males the semen passes out and in the females the fetation.[a] This passage is situated higher up than the passage for the solid nourishment and in front of it. [b] [Those animals which lay eggs, but lay imperfect ones, e.g., the oviparous fishes, have their uterus not under the abdomen but by the loin, since the growth of the egg causes no obstruction, because the growing object comes to its perfection and makes its advance outside the animal.] In all those animals which have no pudendum which serves for generation, this passage is the same as that for the solid nourishment, viz., in all the Ovipara, including those Ovipara which have a bladder, e.g., the tortoises. The existence of *two* passages, it must be remembered, is for the sake of generation, not for the sake of voiding the fluid residue, and it is only because the semen is fluid in

[a] See Introd. § 56.
[b] This sentence is a continuation of the previous interpolation.

ὑγρᾶς τροφῆς περίττωσις κεκοινώνηκε τοῦ αὐτοῦ
10 πόρου. δῆλον δὲ τοῦτο ἐκ τοῦ σπέρμα μὲν πάντα
φέρειν τὰ ζῷα, περίττωμα δὲ μὴ πᾶσι γίνεσθαι
ὑγρόν.

Ἐπεὶ οὖν δεῖ καὶ τοὺς τῶν ἀρρένων πόρους
τοὺς σπερματικοὺς ἐρηρεῖσθαι καὶ μὴ πλανᾶσθαι,
καὶ τοῖς θήλεσι τὰς ὑστέρας, τοῦτο δ' ἀναγκαῖον
ἢ πρὸς τὰ πρόσθια τοῦ σώματος ἢ πρὸς τὰ πρανῆ
15 συμβαίνειν, τοῖς μὲν ζῳοτόκοις διὰ τὰ ἔμβρυα ἐν
τοῖς προσθίοις αἱ ὑστέραι, τοῖς δ' ᾠοτόκοις πρὸς
τῇ ὀσφύι καὶ τοῖς πρανέσιν· ὅσα δ' ᾠοτοκήσαντα
ἐν αὑτοῖς ζῳοτοκεῖ ἐκτός, ταῦτα δ' ἀμφοτέρως
ἔχει διὰ τὸ μετειληφέναι ἀμφοτέρων καὶ εἶναι καὶ
ζῳοτόκα καὶ ᾠοτόκα· τὰ μὲν γὰρ ἄνω τῆς
20 ὑστέρας, καὶ ᾗ γίγνεται τὰ ᾠά, ὑπὸ τὸ ὑπόζωμα
πρὸς τῇ ὀσφύι ἐστὶ καὶ τοῖς πρανέσι, προϊοῦσα[1] δὲ
κάτω ἐπὶ τῇ γαστρί· ταύτῃ γὰρ ζῳοτοκεῖ ἤδη. ὁ
δὲ πόρος εἷς καὶ τούτοις τῆς τε ξηρᾶς περιττώσεως
καὶ τῆς ὀχείας· οὐδὲν γὰρ ἔχει τούτων αἰδοῖον,
25 καθάπερ εἴρηται πρότερον, ἀπηρτημένον. ὁμοίως
δ' ἔχουσι καὶ οἱ τῶν ἀρρένων πόροι, καὶ τῶν ἐχόντων καὶ τῶν μὴ ἐχόντων ὄρχεις, ταῖς τῶν ᾠοτόκων
ὑστέραις· πᾶσι γὰρ πρὸς τοῖς πρανέσι προσπεφύ-
κασι καὶ κατὰ τὸν τόπον τὸν[2] τῆς ῥάχεως· δεῖ μὲν
γὰρ μὴ πλανᾶσθαι ἀλλ' ἑδραίους εἶναι, τοιοῦτος δ'
30 ὁ ὄπισθεν τόπος· οὗτος γὰρ τὸ συνεχὲς παρέχει καὶ
τὴν στάσιν. τοῖς μὲν οὖν ἐντὸς ἔχουσι τοὺς ὄρχεις
εὐθὺς ἐρηρεισμένοι εἰσὶν [ἅμα τοῖς πόροις],[3] καὶ τοῖς

---

[1] προϊοῦσα Platt, προϊούσης vulg. : cf. 719 a 7, H.A. 511 a 7
seqq. : προϊούσης δὲ ⟨τὰ⟩ κάτω Sus.
[2] τὸν P : om. vulg.
[3] ἅμα τοῖς πόροις secl. Platt.

nature that the residue from the fluid nourishment shares the use of the same passage. This is clear from the fact that although all animals produce semen, fluid residue is not formed in all of them.

Now in males the seminal passages must have a fixed position and not stray about, and the same is true of the uterus in females ; and this fixed position must of necessity be either towards the front or the back of the body. Hence, (a) in the Vivipara the uterus is in front, on account of the embryo ; (b) in the Ovipara it is by the loin and at the back ; (c) in those animals which begin by producing eggs within themselves and later bring their young forth exter-nally, both positions are found combined, because the animals share the characteristics of both classes ; they are viviparous and oviparous alike ; thus, the upper portion of the uterus, in which the eggs are formed, is below the diaphragm by the loin, and towards the back ; but its continuation is lower down, by the abdomen, for from this point onwards the production of live young begins. In these animals also there is one passage only for the solid residue and for copulation ; none of them has a pudendum projecting from the body, as has been said before. What is true of the uterus in Ovipara is true also of the passages in the males, both those which have testes and those which have not. In all of them the passages are fastened towards the back near the region of the spine ; fastened, because they may not stray about, but must have a settled position, which is just what the back part of the body provides ; it gives continuity and stability. Indeed, in those animals which have their testes inside, the passages acquire their fixed position at

ἐκτὸς δ' ὁμοίως· εἶτ' ἀπαντῶσιν εἰς ἓν πρὸς τὸν
τοῦ αἰδοίου τόπον. ὁμοίως δὲ καὶ τοῖς δελφῖσιν
οἱ πόροι ἔχουσιν· ἀλλὰ τοὺς ὄρχεις ἔχουσι κεκρυμ-
35 μένους ὑπὸ τὸ περὶ τὴν γαστέρα κύτος.

Πῶς μὲν οὖν ἔχουσι τῇ θέσει περὶ τὰ μόρια τὰ
720 b συντελοῦντα πρὸς τὴν γένεσιν, καὶ διὰ τίνας αἰτίας,
εἴρηται.

XIV   Τῶν δ' ἄλλων ζῴων τῶν ἀναίμων οὐχ ὁ αὐτὸς
τρόπος τῶν μορίων τῶν πρὸς τὴν γένεσιν συντελούντων οὔτε τοῖς ἐναίμοις οὔθ' ἑαυτοῖς.   ἔστι δὲ
5 γένη τέτταρα τὰ λοιπά, ἓν μὲν τὸ τῶν μαλακοστράκων, δεύτερον δὲ τὸ τῶν μαλακίων, τρίτον δὲ τὸ
τῶν ἐντόμων, καὶ τέταρτον τὸ τῶν ὀστρακοδέρμων
(τούτων δὲ περὶ μὲν πάντων ἄδηλον, τὰ δὲ πλεῖστα
ὅτι οὐ¹ συνδυάζεται φανερόν· τίνα δὲ συνίσταται
τρόπον, ὕστερον λεκτέον).   τὰ δὲ μαλακόστρακα
10 συνδυάζεται μὲν ὥσπερ τὰ ὀπισθουρητικά, ὅταν τὸ
μὲν ὕπτιον τὸ δὲ πρανὲς ἐπαλλάξῃ τὰ οὐραῖα· τοῖς
γὰρ ὑπτίοις πρὸς τὰ πρανῆ ἐπιβαίνειν ἐμποδίζει τὰ
οὐραῖα μακρὰν ἔχοντα τὴν ἀπάρτησιν τῶν πτε-
ρυγίων. ἔχουσι δ' οἱ μὲν ἄρρενες λεπτοὺς πόρους
θορικούς, αἱ δὲ θήλειαι ὑστέρας ὑμενώδεις παρὰ
15 τὸ ἔντερον, ἔνθεν καὶ ἔνθεν ἐσχισμένας, ἐν αἷς ἐγ-
XV γίνεται τὸ ᾠόν.   τὰ δὲ μαλάκια συμπλέκεται μὲν
κατὰ τὸ στόμα, ἀντερείδοντα καὶ διαπύττοντα
τὰς πλεκτάνας, συμπλέκεται δὲ τὸν τρόπον τοῦτον
ἐξ ἀνάγκης· ἡ γὰρ φύσις παρὰ τὸ στόμα τὴν τελευτὴν τοῦ περιττώματος συνήγαγε κάμψασα, καθ-
20 άπερ εἴρηται πρότερον [ἐν τοῖς περὶ τῶν μορίων

---

¹ οὐ Z² in ras., vulg., Σ : om. PY, Platt.

ᵃ Snails are the exception (762 a 33).

the very outset [at the same time as the passages];
and similarly in those animals whose testes are external. Afterwards they meet and unite towards
the region of the pudendum. The arrangement of
the passages is the same as this in dolphins, although
their testes are hidden below the abdominal cavity.

We have now described the situation of the parts
which are concerned with generation in the blooded
animals and have stated the causes.

In the other class of animals, viz., the bloodless **XIV**
ones, the manner of the parts concerned with genera- (ii.) In
tion is quite different from what it is in the blooded bloodless
ones ; and what is more they differ among them- animals.
selves. We have here four groups still left to deal
with : (1) Crustacea, (2) Cephalopods, (3) Insects,
(4) Testacea (with regard to all of these the facts
are obscure, but it is plain that most of them do
not copulate *a* ; as for the manner in which they
arise, we must describe this later on).*b* (1) The (1) Crus-
Crustacea copulate as the retromingent animals do : taeca.
one lies prone and the other supine and they fit their
tail-parts one to the other. The males are prevented
from mounting the females belly to back by their
tail-parts which have long flaps attached to them.
The males have narrow seminal passages, and the
females have a membranous uterus by the side of the
gut, divided on either side, and in this the egg is
formed. (2) The Cephalopods copulate by the **XV**
mouth, pushing against each other and intertwining (2) Cepha-
their tentacles. This manner of copulation is due to lopods.
necessity, because nature has bent the end of the
residual passage so as to bring it round by the side of
the mouth, as I have previously said [in the treatise

b Book III, ch. 11.

λόγοις].[1] ἔχει δ' ἡ θήλεια μὲν ὑστερικὸν μόριον
φανερῶς ἐν ἑκάστῳ τούτων τῶν ζῴων· ᾠὸν γὰρ
ἴσχει τὸ μὲν πρῶτον ἀδιόριστον, ἔπειτα διακρι-
νόμενον γίνεται πολλά, καὶ ἀποτίκτει ἕκαστον
τούτων ἀτελές, καθάπερ καὶ οἱ ᾠοτοκοῦντες τῶν
25 ἰχθύων. ὁ δὲ πόρος ὁ αὐτὸς τοῦ περιττώματος
καὶ τοῦ ὑστερικοῦ μορίου καὶ τοῖς μαλακοστράκοις
καὶ τούτοις· [ἐστὶ γὰρ ᾗ τὸν θορὸν ἀφίησι διὰ τοῦ[2]
πόρου·][3] τοῦτο[4] δ' ἐστὶν ἐν τοῖς ὑπτίοις τοῦ σώματος, ᾗ τὸ κέλυφος ἀφέστηκε καὶ ἡ θάλαττα
εἰσέρχεται. διὸ ὁ συνδυασμὸς κατὰ τοῦτο γίνεται
30 τῷ ἄρρενι πρὸς τὴν θήλειαν· ἀναγκαῖον γάρ, εἴπερ
ἀφίησί τι[5] ὁ ἄρρην εἴτε σπέρμα εἴτε μόριον εἴτε
ἄλλην τινὰ δύναμιν, κατὰ τὸν ὑστερικὸν πόρον
πλησιάζειν. ἡ δὲ τῆς πλεκτάνης τοῦ ἄρρενος διὰ
τοῦ αὐλοῦ δίεσις ἐπὶ τῶν πολυπόδων, ᾗ φασὶν
ὀχεύειν πλεκτάνῃ οἱ ἁλιεῖς, συμπλοκῆς χάριν ἐστίν,
35 ἀλλ' οὐχ ὡς ὀργάνου χρησίμου πρὸς τὴν γένεσιν·
ἔξω γάρ ἐστι τοῦ πόρου καὶ τοῦ σώματος.

---

[1] om. PZ (Z et πρότερον om.), secl. A.-W.
[2] τὸ Bekker per typothetae errorem.
[3] seclusi.   [4] τοῦτο Peck, ταῦτα vulg.
[5] τι PZ : om. vulg.

---

[a] 684 a 15, 685 a 1.   [b] Cf. 718 b 11.
[c] Cf. P.A. 684 a 17 ff.

[d] See Introd. § 18.   " Part " does not necessarily imply
a limb, and the fact that it is mentioned here between
semen and *dynamis* suggests that " limb " is not the meaning here (cf. P.A. 648 a 2, where blood is described as a
" part "). All the same, Aristotle *may* here be intending to
use " part " in the sense of limb, for in three genera of the
Octopoda the hectocotylized arm (see note below, on l. 32)
becomes detached from the male and remains within the

on *The Parts of Animals*].[a] The female of each of these animals has a part like a uterus, which is plain to be seen ; it contains an egg which at first is indistinct,[b] but later divides up and is formed into a number of eggs, each of which the creature deposits in an imperfect state, just as the oviparous fishes do. In these animals as well as in the Crustacea the passage which serves for the residue and connects with the uterus-like part, is one and the same (it is on the under surface of the body, where the " mantle " lies open and the sea water enters in [c]). Hence it is through this that the male effects copulation with the female, since if the male discharges anything, be it semen, or some part,[d] or some other substance,[e] he must of necessity unite with the female through the passage which leads to the uterus. In the case of the Octopuses, the male inserts his tentacle through the funnel of the female, and the fishermen allege that copulation is effected by means of this tentacle,[f] but its purpose is really to link the two creatures together ; it has no instrumental use so far as generation is concerned, because it is outside the passage ⟨of the male⟩ and outside his body.[g]

mantle of the female. Aristotle however does not explicitly mention this detaching of the arm.

  [e] *Dynamis*; see Introd. §§ 23 ff.

  [f] This refers to the remarkable phenomenon in the Dibranchiata of the " hectocotylization " of one of the arms of the male, by means of which copulation is effected, as is stated in *II.A.* 524 a 5 ff., 541 b 9, 544 a 8 ff. Here, however, Aristotle denies that the arm is so used, and his argument is not unreasonable, for it is not yet known how the arm becomes charged with the spermatophores. For details and diagrams see P. Pelseneer, *Mollusca* (tr. G. Bourne), 323 ff.

  [g] *i.e.*, not a part of the main bulk of the body and not directly connected with the seminal passage.

Ἐνίοτε δὲ συνδυάζονται καὶ ἐπὶ τὰ πρανῆ τὰ
μαλάκια· πότερον δὲ γενέσεως χάριν ἢ δι' ἄλλην
αἰτίαν, οὐθὲν ὦπταί πω.

XVI Τῶν δ' ἐντόμων τὰ μὲν συνδυάζεται, καὶ ἡ γένεσις αὐτῶν ἐστὶν ἐκ ζῴων συνωνύμων, καθάπερ
ἐπὶ τῶν ἐναίμων, οἷον αἵ τε ἀκρίδες καὶ οἱ τέτ-
5 τιγες καὶ τὰ φαλάγγια καὶ οἱ σφῆκες καὶ οἱ μύρ-
μηκες, τὰ δὲ συνδυάζεται μὲν καὶ γεννῶσιν, οὐχ
ὁμογενῆ δ' αὑτοῖς ἀλλὰ σκώληκας μόνον, οὐδὲ
γίγνονται ἐκ ζῴων ἀλλ' ἐκ σηπομένων ὑγρῶν, τὰ
δὲ ξηρῶν, οἷον αἵ τε ψύλλαι καὶ αἱ μυῖαι καὶ αἱ
κανθαρίδες· τὰ δ' οὔτ' ἐκ ζῴων γίνονται οὔτε
10 συνδυάζονται, καθάπερ ἐμπίδες τε καὶ κώνωπες
καὶ πολλὰ τοιαῦτα γένη. τῶν δὲ συνδυαζομένων
ἐν τοῖς πλείστοις τὰ θήλεα μείζω τῶν ἀρρένων
ἐστίν. πόρους δὲ τὰ ἄρρενα θορικοὺς οὐ φαίνεται
ἔχοντα. ἀφίησι δὲ ὡς ἐπὶ τὸ πλεῖστον εἰπεῖν τὸ
ἄρρεν εἰς τὸ θῆλυ οὐδὲν μόριον, ἀλλὰ τὸ θῆλυ εἰς
15 τὸ ἄρρεν κάτωθεν ἄνω. τεθεώρηται δὲ τοῦτο ἐπὶ
πολλῶν, [καὶ περὶ τοῦ ἀναβαίνειν ὡσαύτως,][1] τοὐναντίον δ' ἐπ' ὀλίγων· ὥστε δὲ γένει διελεῖν, οὔπω
συνεώραται. σχεδὸν δὲ τοῦτο καὶ ἐπὶ τῶν ᾠοτόκων ἰχθύων τῶν πλείστων ἐστί, καὶ ἐπὶ τῶν
τετραπόδων καὶ ᾠοτόκων· τὰ γὰρ θήλεα μείζω
20 τῶν ἀρρένων ἐστὶ διὰ τὸ συμφέρειν[2] πρὸς τὸν γινό-

---

[1] seclusi. [2] συμφέρειν PZ : συμφέρον vulg.

---

[a] See *Categ.* 1 a 5 " Things are called ' synonymous ' when
their name is common and the *logos* of the essence corresponding to the name is the same." For ὁμώνυμον, see
note on 726 b 24. A useful mnemonic is : συνώνυμον is same

Sometimes too Cephalopods copulate while both creatures are lying prone, but it has not yet been observed whether this is done for the purpose of generation or for some other cause.

(3) As regards Insects, some of them copulate, and XVI in those cases the young are generated from animals (3) Insects. which are of the same name [a] and nature as themselves, just as happens in the blooded creatures ; instances of this are locusts, cicadas, spiders, wasps, ants. Others, although they copulate and generate, generate not creatures of the same kind as themselves but only larvae [b] ; and these insects moreover are not produced out of animals at all but out of putrefying fluids (in some cases, solids) ; instances of this are fleas, flies, cantharides. Others neither are produced out of animals nor do they copulate ; such are gnats, mosquitoes [c] and many similar kinds of insects. In most of the sorts which copulate the females are larger than the males ; and the males do not seem to have any seminal passages. Speaking generally, the male does not insert any part into the female ; but the female does so into the male upwards from below : this has been observed in many instances, [and similarly as concerns mounting,] the opposite in a few ; but we have not yet enough observations to enable us to classify them distinctly. We find that the females are larger than the males not only in Insects but also in most of the oviparous fishes, and likewise in those quadrupeds which are oviparous ; the reason being that the size is an advantage to them when a great bulk is produced inside

in name and same in nature ; ὁμώνυμον is same in name but not in nature. [b] See Introd. § 77.
  [c] It is not possible to say exactly what insects are meant.

721 a

μενον αὐτοῖς ὑπὸ τῶν ῳῶν ὄγκον ἐν τῇ κυήσει.
τοῖς δὲ θήλεσιν αὐτῶν τὸ ταῖς ὑστέραις ἀνάλογον
μόριον ἐσχισμένον ἐστὶ παρὰ τὸ ἔντερον, ὥσπερ
καὶ τοῖς ἄλλοις, ἐν ᾧ ἐγγίγνεται τὰ κυήματα.
δῆλον δὲ τοῦτο ἐπί τε τῶν ἀκρίδων, καὶ ὅσα μέ-
25 γεθος αὐτῶν ἔχει, συνδυάζεσθαι πεφυκότων· τὰ
γὰρ πλεῖστα μικρὰ λίαν τῶν ἐντόμων ἐστίν.

Τὰ μὲν οὖν περὶ τὴν γένεσιν ὄργανα τοῖς ζῴοις,
περὶ ὧν οὐκ ἐλέχθη πρότερον, τοῦτον ἔχει τὸν
τρόπον· τῶν δ' ὁμοιομερῶν ἀπελείφθη περὶ γονῆς
καὶ γάλακτος, περὶ ὧν καιρός ἐστιν εἰπεῖν, περὶ
30 μὲν γονῆς ἤδη, περὶ δὲ γάλακτος ἐν τοῖς ἐχομένοις.

XVII Τὰ μὲν γὰρ προΐεται φανερῶς σπέρμα[1] τῶν
ζῴων, οἷον ὅσα αὐτῶν ἔναιμα τὴν φύσιν ἐστί, τὰ
δ' ἔντομα καὶ τὰ μαλάκια ποτέρως, ἄδηλον. ὥστε
τοῦτο θεωρητέον, πότερον πάντα προΐεται σπέρμα
τὰ ἄρρενα ἢ οὐ πάντα, καὶ εἰ μὴ πάντα, διὰ τίν'
35 αἰτίαν τὰ μὲν τὰ δ' οὔ· καὶ τὰ θήλεα δὲ πότερον

721 b

συμβάλλεται σπέρμα τι ἢ οὔ, καὶ εἰ μὴ σπέρμα,
πότερον οὐδ' ἄλλο οὐθέν, ἢ συμβάλλεται μέν τι,
οὐ σπέρμα δέ. ἔτι δὲ καὶ τὰ προϊέμενα σπέρμα τί
συμβάλλεται διὰ τοῦ σπέρματος πρὸς τὴν γένεσιν
σκεπτέον, καὶ ὅλως τίς ἐστιν ἡ τοῦ σπέρματος φύσις
5 καὶ ἡ τῶν καλουμένων καταμηνίων, ὅσα ταύτην
τὴν ὑγρότητα προΐεται τῶν ζῴων.

Δοκεῖ δὲ πάντα γίνεσθαι ἐκ σπέρματος, τὸ δὲ

---

[1] σπέρμα om. SY*.

---

[a] It will be noticed that Aristotle omits to describe the
Testacea, which would naturally follow at this point. The

them by the eggs at the time of breeding. In the females the part that answers to the uterus is divided and extends alongside the gut, as in other animals ; this is where the fetations are formed. This can be clearly seen in locusts and in any insect whose nature it is to copulate, provided it is large enough ; most insects however are too small.[a]

Such is the manner of animals' instrumental parts connected with generation, which I had not dealt with in my previous treatise.[b] Of the " uniform " [c] parts, semen and milk were there left undescribed, and the time has now come to speak of these. We will deal with semen without delay, and with milk in the chapters which are to follow.[d]

Some animals discharge semen plainly, for instance those which are by nature blooded animals ; but it is not clear in which way Insects and Cephalopods do so. Here then is a point we must consider : Do all male animals discharge semen, or not all of them ? and if not all, why is it that some do and some do not ? and further, Do females contribute any semen, or not ? and if they contribute no semen, is there no other substance at all which they contribute, or is there something else which is not semen ? And there is a further question which we must consider : What is it which those animals that discharge semen contribute towards generation by means of it ? and generally, what is the nature of semen, and (in the case of those animals which discharge this fluid) what is the nature of the menstrual discharge ?

It is generally held that all things are formed and

XVII
Semen.

Theory of

reason is that, according to him, they do not copulate: see
731 b 8 ff.   [b] *De partibus.*
    [c] See Introd. § 19.   [d] Book IV, ch. 8.

49

σπέρμα ἐκ τῶν γεννώντων. διὸ τοῦ αὐτοῦ λόγου
ἐστί, πότερον καὶ τὸ θῆλυ καὶ τὸ ἄρρεν προΐενται¹
ἄμφω ἢ θάτερον μόνον, καὶ πότερον ἀπὸ παντὸς
10 ἀπέρχεται τοῦ σώματος ἢ οὐκ ἀπὸ παντός· εὔλογον γάρ, εἰ μὴ ἀπὸ παντός, μηδ' ἀπ' ἀμφοτέρων
τῶν γεννώντων. διόπερ ἐπισκεπτέον, ἐπειδὴ φασί
τινες ἀπὸ παντὸς ἀπιέναι τοῦ σώματος, περὶ τούτου
πῶς ἔχει πρῶτον. ἔστι δὲ σχεδόν, οἷς ἄν τις χρή-
15 σαιτο τεκμηρίοις ὡς ἀφ' ἑκάστου τῶν μορίων
ἀπιόντος τοῦ σπέρματος,² τέτταρα, πρῶτον μὲν ἡ
σφοδρότης τῆς ἡδονῆς· μᾶλλον γὰρ ἡδὺ πλέον
ταὐτὸ γινόμενον πάθος, πλέον δὲ τὸ πᾶσι τοῖς
μορίοις ἢ τὸ ἑνὶ ἢ ὀλίγοις συμβαῖνον αὐτῶν. ἔτι
τὸ ἐκ κολοβῶν κολοβὰ γίνεσθαι· διὰ μὲν γὰρ τὸ
20 τοῦ μορίου ἐνδεὲς εἶναι οὐ βαδίζειν σπέρμα ἐν-
τεῦθέν φασιν, ὅθεν δ' ἂν μὴ ἔλθῃ, τοῦτο συμβαίνειν
μὴ γίνεσθαι. πρὸς δὲ τούτοις αἱ ὁμοιότητες πρὸς
τοὺς γεννήσαντας· γίνονται γὰρ ἐοικότες, ὥσπερ³
καὶ ὅλον τὸ σῶμα, καὶ μόρια μορίοις· εἴπερ οὖν
καὶ τῷ ὅλῳ⁴ αἴτιον τῆς ὁμοιότητος τὸ ἀφ' ὅλου
ἐλθεῖν τὸ σπέρμα, καὶ τοῖς μορίοις αἴτιον ἂν εἴη τὸ

---

¹ προΐενται PSZ : προΐεται vulg.
² ὡς . . . σπέρματος om. PZ¹.
³ ὥσπερ om. P.
⁴ τῷ ὅλῳ Z : τοῦ ὅλου vulg.

ᵃ This is a view which is found in the remarkable Hippocratic treatise π. γονῆς κτλ. 3 and 8 (vii. 474 and 480 Littré),
and seems also to have been held by Democritus (see Diels,
Vorsokr.⁵ 68 A 141 and 68 B 32). It closely resembles
the hypothesis (" pangenesis ") which was put forward by
Darwin, that every unit of an organism contributes its share
to the germ of the future offspring ; in other words, that the

come to be out of semen, and semen comes from
the parents. And so one and the same inquiry will
include the two questions : (1) Do both the male and
the female discharge semen, or only one of them ?
and (2) Is the semen drawn from the whole of the
parent's body [a] or not ?—since it is reasonable to hold
that if it is not drawn from the whole of the body it
is not drawn from both the parents either.[b] There
are some who assert that the semen is drawn from
the whole of the body, and so we must consider the
facts about this first of all. There are really four
lines of argument which may be used to prove that
the semen is drawn from each of the parts of the body.
The first is, the intensity of the pleasure involved ;
it is argued that any emotion, when its scope is
widened, is more pleasant than the same emotion
when its scope is less wide ; and obviously an emotion
which affects all the parts of the body has a wider
scope than one which affects a single part of a few
parts only. The second argument is thát mutilated
parents produce mutilated offspring, and it is alleged
that because the parent is deficient in some one part
no semen comes from that part, and that the part
from which no semen comes does not get formed in
the offspring. The third argument is the resemblances shown by the young to their parents : the
offspring which are produced are like their parents
not only in respect of their body as a whole, but part
for part too ; hence, if the reason for the resemblance
of the whole is that the semen is drawn from the

carriers of heredity move centripetally from all the parts of
the body to the germ, thus involving the inheritance of acquired characteristics (for which inheritance, however, there
is no evidence).—See also Hippocrates, περὶ ἀέρων ὑδάτων
τόπων 16.   [b] Cf. 724 a 9-10.

721 b

ἀφ' ἑκάστου τι τῶν μορίων ἐλθεῖν. ἔτι δὲ καὶ
25 εὔλογον ἂν εἶναι δόξειεν, ὥσπερ καὶ τοῦ ὅλου ἐστί
τι ἐξ οὗ γίνεται πρῶτον, οὕτω καὶ τῶν μορίων
ἑκάστου, ὥστ' εἰ ἐκείνου σπέρμα, καὶ τῶν μορίων
ἑκάστου εἴη ἄν τι σπέρμα ἴδιον. πιθανὰ δὲ καὶ τὰ
τοιαῦτα μαρτύρια ταύταις ταῖς δόξαις· οὐ γὰρ
μόνον τὰ σύμφυτα προσεοικότες γίνονται τοῖς γο-
30 νεῦσιν οἱ παῖδες, ἀλλὰ καὶ τὰ ἐπίκτητα· οὐλάς τε
γὰρ ἐχόντων τῶν γεννησάντων ἤδη τινὲς ἔσχον ἐν
τοῖς αὐτοῖς τόποις τῶν ἐκγόνων τὸν τύπον τῆς
οὐλῆς, καὶ στίγμα ἔχοντος ἐν τῷ βραχίονι τοῦ
πατρὸς ἐπεσήμηνεν ἐν Χαλκηδόνι τῷ τέκνῳ συγ-
κεχυμένον μέντοι καὶ οὐ διηρθρωμένον τὸ γράμμα.
35 ὅτι μὲν οὖν ἀπὸ παντὸς ἔρχεται τὸ σπέρμα, σχεδὸν

722 a   ἐκ τούτων μάλιστα πιστεύουσί τινες.

XVIII   Φαίνεται δ' ἐξετάζουσι τὸν λόγον τοὐναντίον
μᾶλλον· τά τε γὰρ εἰρημένα λύειν οὐ χαλεπόν, καὶ
πρὸς τούτοις ἄλλα συμβαίνει λέγειν ἀδύνατα. πρῶτον μὲν οὖν ὅτι οὐθὲν σημεῖον ἡ ὁμοιότης τοῦ
5 ἀπιέναι ἀπὸ παντός, ὅτι καὶ φωνὴν καὶ ὄνυχας καὶ
τρίχας ὅμοιοι γίγνονται καὶ τὴν κίνησιν, ἀφ' ὧν
οὐθὲν ἀπέρχεται. ἔνια δ' οὐκ ἔχουσί πω ὅταν γεννῶσιν, οἷον τρίχωσιν πολιῶν ἢ γενείου. ἔτι τοῖς
ἄνωθεν γονεῦσιν ἐοίκασιν, ἀφ' ὧν οὐθὲν ἀπῆλθεν·

---

whole, then the reason for the resemblance of the parts is surely that something is drawn from each of the parts. Fourthly, it would seem reasonable to hold that just as there is some original thing out of which the whole creature is formed, so also it is with each of the parts; and hence if there is a semen which gives rise to the whole, there must be a special semen which gives rise to each of the parts. And these opinions derive plausibility from such evidence as the following : Children are born which resemble their parents in respect not only of congenital characteristics but also of acquired ones [a] ; for instance, there have been cases of children which have had the outline of a scar in the same places where their parents had scars, and there was a case at Chalcedon of a man who was branded on his arm, and the same letter, though somewhat confused and indistinct, appeared marked on his child. These are the main pieces of evidence which give some people ground for believing that the semen is drawn from the whole of the body.

Upon examination of the subject, however, the opposite seems more likely to be true ; indeed, it is not difficult to refute these arguments, and besides that, they involve making further assertions which are impossible. First of all, then, resemblance is no proof that the semen is drawn from the whole of the body, because children resemble their parents in voice, nails, and hair and even in the way they move ; but nothing whatever is drawn from these things ; and there are some characteristics which a parent does not yet possess at the time when the child is generated, such as grey hair or beard. Further, children resemble their remoter ancestors, from whom nothing has been drawn for the semen. Resemblances

XVIII

ἀποδιδόασι γὰρ διὰ πολλῶν γενεῶν αἱ ὁμοιότητες,
10 οἷον καὶ ἐν Ἤλιδι ἡ τῷ Αἰθίοπι συγγενομένη· οὐ
γὰρ ἡ θυγάτηρ ἐγένετο, ἀλλ' ὁ ἐκ ταύτης Αἰθίοψ.
καὶ ἐπὶ τῶν φυτῶν δὲ ὁ αὐτὸς λόγος· δῆλον γὰρ ὅτι
καὶ τούτοις ἀπὸ πάντων ἂν τῶν μερῶν τὸ σπέρμα
γίγνοιτο. πολλὰ δὲ τὰ μὲν οὐκ ἔχει, τὰ δὲ καὶ
ἀφέλοι τις ἄν, τὰ δὲ προσφύεται. ἔτι οὐδ' ἀπὸ
15 τῶν περικαρπίων ἀπέρχεται· καίτοι καὶ ταῦτα γίνεται τὴν αὐτὴν ἔχοντα μορφήν.

Ἔτι πότερον ἀπὸ τῶν ὁμοιομερῶν μόνον ἀπέρχεται ἀφ' ἑκάστου, οἷον ἀπὸ σαρκὸς καὶ ὀστοῦ
καὶ νεύρου, ἢ καὶ ἀπὸ τῶν ἀνομοιομερῶν, οἷον
προσώπου καὶ χειρός; εἰ μὲν γὰρ ἀπ' ἐκείνων
20 μόνον, ⟨ἐοικέναι ἔδει ἐκεῖνα μόνον·⟩[1] ἐοίκασι δὲ
μᾶλλον ταῦτα τοῖς γονεῦσι [τὰ ἀνομοιομερῆ],[2]
οἷον πρόσωπον καὶ χεῖρας καὶ πόδας· εἴπερ οὖν
μηδὲ ταῦτα τῷ ἀπὸ παντὸς ἀπελθεῖν, τί κωλύει
μηδ' ἐκεῖνα τῷ ἀπὸ παντὸς ἀπελθεῖν ὅμοια εἶναι,
ἀλλὰ δι' ἄλλην αἰτίαν; εἰ δ' ἀπὸ τῶν ἀνομοιομερῶν μόνον, οὐκ ἄρα[3] ἀπὸ πάντων. προσήκει
25 δὲ μᾶλλον ἀπ' ἐκείνων· πρότερα γὰρ ἐκεῖνα, καὶ
σύγκειται τὰ ἀνομοιομερῆ ἐξ ἐκείνων, καὶ ὥσπερ
πρόσωπον καὶ χεῖρας γίγνονται ἐοικότες, οὕτω καὶ

---

[1] ⟨ἐοικέναι ἔδει ἐκεῖνα μόνον·⟩ Peck ; monuerant A.-W.
intellegi debere, e.g., ἔδει ἐκεῖνα μόνον ἐοικέναι.
[2] τὰ ἀνομοιομερῆ secludenda, nam ταῦτα hoc significat.
[3] ἄν PYZ.

of this sort recur after many generations, as the following instance shows. There was at Elis a woman who had intercourse with a blackamoor; her daughter was not a black, but that daughter's son was. And the same argument will hold for plants. We should have to say that the seed was drawn from the whole of the plant, just as in animals. But many plants lack certain parts; you can if you wish pull some of the parts off, and some parts grow on afterwards. Further, nothing is drawn from the pericarp to contribute to the seed, yet pericarp is formed in the new plant and it has the same fashion as that in the old one.

Here is a further question. Is the semen drawn only from each of the " uniform " parts of the body, such as flesh, bone, sinew, or is it drawn from the " non-uniform " parts as well, such as face and hand? Consider the possibilities : (1) The semen may be drawn from the uniform parts only. If so, ⟨then children ought to resemble their parents in respect of these only,⟩ but the resemblance occurs rather in the non-uniform parts such as face, hands, and feet. Therefore if even these resemblances in the non-uniform parts are not due to the semen being drawn from the whole body, why must the resemblances in the uniform parts be due to that and not to some other cause? (2) The semen may be drawn from the non-uniform parts only. This means that it is not drawn from all the parts. Yet it is more in keeping that it should be drawn from the uniform parts, because they are prior to the non-uniform, and the non-uniform are constructed out of them ; and just as children are born resembling their parents in face and hands, so they resemble them in flesh and

σάρκας καὶ ὄνυχας. εἰ δ' ἀπ' ἀμφοτέρων, τίς ὁ
τρόπος ἂν εἴη τῆς γενέσεως; σύγκειται γὰρ ἐκ
τῶν ὁμοιομερῶν τὰ ἀνομοιομερῆ, ὥστε τὸ ἀπὸ
30 τούτων ἀπιέναι τὸ ἀπ' ἐκείνων ἂν εἴη ἀπιέναι καὶ
τῆς συνθέσεως· ὥσπερ κἂν εἰ ἀπὸ τοῦ γεγραμ-
μένου ὀνόματος ἀπῄει τι,[1] εἰ μὲν ἀπὸ παντός, κἂν
ἀπὸ τῶν συλλαβῶν ἑκάστης, εἰ δ' ἀπὸ τούτων, ἀπὸ
τῶν στοιχείων καὶ τῆς συνθέσεως. ὥστ' εἴπερ ἐκ
πυρὸς καὶ τῶν τοιούτων σάρκες καὶ ὀστᾶ συνεστᾶ-
35 σιν, ἀπὸ τῶν στοιχείων ἂν εἴη μόνον[2]· ἀπὸ γὰρ
722 b τῆς συνθέσεως πῶς ἐνδέχεται; ἀλλὰ μὴν ἄνευ γε
ταύτης οὐκ ἂν εἴη ὅμοια. ταύτην δ' εἴ τι δημιουργεῖ ὕστερον, τοῦτ' ἂν εἴη τὸ τῆς ὁμοιότητος
αἴτιον, ἀλλ' οὐ τὸ ἀπελθεῖν ἀπὸ παντός.

Ἔτι εἰ μὲν διεσπασμένα τὰ μέρη ἐν τῷ σπέρματι,

---

[1] ἀπῄει τι PZ², ἀπίῃ τι Z¹ : om. vulg.
[2] μόνων Z, ex μόνον ut vid. : μᾶλλον vulg.

---

[a] The point of the argument is this. There is no additional *material* in the non-uniform parts beyond what there
was in the uniform ones ; the only additional factor is the
assemblage (composition, combination, arrangement) of the
uniform parts so as to make the non-uniform ones (*e.g.*, of
flesh, bone, blood, sinew, etc., so as to make a face or an
arm). And as the assemblage, the fact that the uniform
parts are arranged in a particular manner, is not a material
thing, obviously nothing can be drawn from it as an ingredient
for the semen. The argument can be carried a stage further
still, as Aristotle points out, for the uniform parts themselves
are merely assemblages of the elementary forms of matter,
Earth, Air, Fire, Water. (See Introd. § 24, and 715 a 10 ff.)

nails. (3) The semen may be drawn from both uniform and non-uniform parts. The question then
arises : What can be the manner in which generation
takes place ? The non-uniform parts are constructed
out of uniform ones assembled together ; so that
being drawn from the non-uniform parts would come
to the same thing as being drawn from the uniform
parts *plus* the assemblage of them.[a] (It is just like
the case of a word written down on paper : if there
were anything drawn from the whole of the word, it
would be drawn from each of the syllables also,[b] and
this of course means that it would be drawn from the
letters *plus* the assemblage of them together.) Now
flesh and bones, we should agree, are constructed
out of fire and the like substances [c] ; which means
that the semen would be drawn from the elements
only, for how can it possibly be drawn from the
assemblage of them ? And yet without this assemblage the parts would not have the resemblance ; so
if there is something which sets to work later on to
bring this assemblage about, then surely this something, and not the drawing of the semen from the
whole of the body, will be the cause of the
resemblance.

Further, if the parts of the body are scattered about

Hence, the theory boils down to an assertion that the semen
is drawn from the simplest forms of matter, and as this
excludes any distinctive characteristics, the theory loses all
meaning.

[b] Contrast the interesting theory examined in Plato, *Theae-
tetus* 201 D ff., that " elements " (στοιχεῖα), whether physical
elements or " letters " of the alphabet, are " ἄλογα " and
cannot be known, until they are assembled into a " syllable,"
which is an entity over and above its components, and " has
a λόγος," and so can be known.—See also 715 a 12, n.

[c] The " elements " ; see Introd. § 24.

5 πῶς ζῇ; εἰ δὲ συνεχῆ, ζῷον ἂν εἴη μικρόν. καὶ τὰ
τῶν αἰδοίων πῶς; οὐ γὰρ ὅμοιον τὸ ἀπιὸν ἀπὸ
τοῦ ἄρρενος καὶ τοῦ θήλεος.

Ἔτι εἰ ἀμφοτέρων ὁμοίως ἀπὸ πάντων ἀπέρχεται, δύο γίγνεται ζῷα· ἑκατέρων γὰρ ἅπαντα
ἕξει. διὸ καὶ Ἐμπεδοκλῆς ἔοικεν, εἴπερ οὕτω
λεκτέον, μάλιστα λέγειν ὁμολογούμενα τούτῳ τῷ
10 λόγῳ [τό γε τοσοῦτον, ἀλλ’ εἴπερ ἑτέρᾳ πη, οὐ
καλῶς]¹· φησὶ γὰρ ἐν τῷ ἄρρενι καὶ τῷ θήλει οἷον
σύμβολον ἐνεῖναι, ὅλον δ’ ἀπ’ οὐδετέρου ἀπιέναι,

ἀλλὰ διέσπασται μελέων φύσις, ἡ μὲν ἐν ἀνδρός . . .

διὰ τί γὰρ τὰ θήλεα οὐ γεννᾷ ἐξ αὑτῶν, εἴπερ ἀπὸ
παντός τε ἀπέρχεται καὶ ἔχει ὑποδοχήν; ἀλλ’
15 ὡς ἔοικεν ἢ οὐκ ἀπέρχεται ἀπὸ παντός, ἢ οὕτως
ὥσπερ ἐκεῖνος λέγει, οὐ ταὐτὰ ἀφ’ ἑκατέρου, διὸ
καὶ δέονται τῆς ἀλλήλων συνουσίας. ἀλλὰ καὶ
τοῦτ’ ἀδύνατον. ὥσπερ γὰρ καὶ μεγάλα ὄντ’ ἀδύνατον διεσπασμένα σώζεσθαι καὶ ἔμψυχα εἶναι,

¹ seclusi.

---

[a] *i.e.*, which generative organs is the offspring to have—
male or female ones ?

[b] *Sc.*, in this respect, though it may be identical in respect
of hand, nose, eyes, etc.

[c] " Nature " seems to mean here, as often, " natural
substance," or " substance."

[d] Emped. fr. 63 (Diels, *Vorsokr.*⁵) ; it probably continued,
e.g., " Seed, and the other portion is in woman's."

within the semen, how do they live ? If on the other
hand they are connected with each other, then surely
they would be a tiny animal. And what about the
generative organs ? [a] because that which comes from
the male will be different from that which comes
from the female.[b]

Further, if the semen is drawn from all the parts
of both parents alike, we shall have two animals
formed, for the semen will contain all the parts of
each of them. If this sort of view is to be adopted,
the statement most closely in accord with it appears to be that of Empedocles [at any rate up to
a point ; if we take any other view, he appears
wrong]. Empedocles says that in the male and
in the female there is as it might be a tally—a
half of something—and that the whole is not drawn
from either of the parents. " But " (I quote his
words)

torn asunder stands
The substance [c] of the limbs ; part is in man's . . .[d]

Otherwise the question arises, why is it that female
animals do not generate out of themselves, if so be
that the semen is drawn from the whole body and a
receptacle for it is at hand ? No ; so far as we can
see, either the semen is not drawn from the whole
body, or if it is, it happens in the way described by
Empedocles—the two parents do not both supply the
same portions, and that is why they need intercourse
with each other. But even Empedocles' explanation is impossible. The parts cannot remain sound
and living if " torn asunder " from each other when
small, any more than they can when they are fully
grown. Empedocles, however, implies that they

59

καθάπερ Ἐμπεδοκλῆς γεννᾷ ἐπὶ τῆς φιλότητος,
20 λέγων

ᾗ πολλαὶ μὲν κόρσαι ἀναύχενες ἐβλάστησαν.

εἶθ᾽ οὕτως συμφύεσθαί φησιν. τοῦτο δὲ φανερὸν
ὅτι ἀδύνατον· οὔτε γὰρ μὴ ψυχὴν ἔχοντα οὔτε
μὴ ζωήν τινα δύναιτ᾽ ἂν σώζεσθαι, οὔτε ὥσπερ
ζῷα ὄντα πλείω συμφύεσθαι ὥστ᾽ εἶναι πάλιν ἕν.
25 ἀλλὰ μὴν τοῦτον τὸν τρόπον συμβαίνει λέγειν τοῖς
ἀπὸ παντὸς ἀπιέναι φάσκουσιν, ὥσπερ τότε ἐν
τῇ γῇ ἐπὶ τῆς φιλότητος, οὕτω τούτοις ἐν τῷ
σώματι. ἀδύνατον γὰρ συνεχῆ τὰ μόρια γίγνεσθαι, καὶ ἀπιέναι εἰς ἕνα τόπον συνιόντα. εἶτα
πῶς καὶ " διέσπασται " τὰ ἄνω καὶ κάτω καὶ
δεξιὰ καὶ ἀριστερὰ καὶ πρόσθια καὶ ὀπίσθια;
30 πάντα γὰρ ταῦτα ἄλογά ἐστιν.

Ἔτι τὰ μέρη τὰ μὲν δυνάμει τὰ δὲ πάθεσι διώρισται, τὰ μὲν ἀνομοιομερῆ τῷ δύνασθαί τι ποιεῖν,
οἷον γλῶττα καὶ χείρ, τὰ δ᾽ ὁμοιομερῆ σκληρότητι
καὶ μαλακότητι καὶ τοῖς ἄλλοις τοῖς τοιούτοις
πάθεσιν. οὐ πάντως οὖν ἔχον αἷμα οὐδὲ σάρξ.[1]
35 δῆλον τοίνυν ὅτι ἀδύνατον τὸ ἀπελθὸν εἶναι συν-

---

[1] οὐ . . . σάρξ om. Σ: ⟨αἷμα⟩ αἷμα οὐδὲ ⟨σὰρξ⟩ σάρξ Btf.

[a] According to Empedocles, there were alternating periods
during which Love and Strife respectively gained the mastery; for details see Burnet, *Early Greek Philosophy*[4],
pp. 231 ff.
[b] Emped. fr. 57 (Diels).
[c] See Introd. §§ 41 ff.
[d] Viz., in the formation of the embryo.
[e] *Cf.* below, 723 b 14 ff., 729 a 7 ff.

can when he says in his account of their generation
during the " Reign of Love," [a]

> There many neckless heads sprang up and grew [b];

later on, he says, they grew on to each other. This
is clearly impossible : on the one hand, if they had
not Soul [c] or life of some sort in them they could
not remain safe and sound ; and on the other hand,
if they were a number of separate living animals,
as one might say, they could not grow on to each
other so as to become one animal again. Yet this
is actually the kind of thing which those people
have to say who allege that the semen is drawn
from the whole of the body : just as it was in the
beginning in the earth in the Reign of Love, so
it is, according to them, in the living body. [d] Of
course it is impossible that the parts should become
connected, i.e., come off from the parents so that they
go together into one place. [e] Besides, in any case,
how were the upper and lower parts, the right and
left, the front and the back, " sundered " ? All
these ideas are fantastic.

Further, among the parts, some are distinguished
by some faculty they possess, others by having certain physical qualities [f] : thus, the non-uniform parts
(such as the tongue or the hand) are distinguished by
possessing the faculty to perform certain actions, the
uniform parts by hardness or softness or other such
qualities. Unless, therefore, it possesses certain
special qualities, a substance is not blood or flesh ;
and hence it is plain that the substance which is

---

[f] One of the definitions of πάθος given at *Met.* 1022 b 15
is " a quality (ποιότης) in virtue of which a thing may be
altered, *e.g.*, whiteness, blackness, heaviness, lightness, etc."

723 a

ὤνυμον τοῖς μέρεσιν, οἷον αἷμα ἀπὸ αἵματος ἢ
σάρκα ἀπὸ σαρκός. ἀλλὰ μὴν εἴ γ’ ἐξ ἑτέρου τινὸς
ὄντος αἷμα γίνεται, οὐδ’ ἂν τῆς ὁμοιότητος αἴτιον
εἴη, ὡς λέγουσιν οἱ φάσκοντες οὕτω, τὸ ἀπελθεῖν
ἀπὸ πάντων τῶν μορίων· ἱκανὸν γὰρ ἀφ’ ἑνὸς
5 ἀπιέναι μόνον, εἴπερ μὴ ἐξ αἵματος αἷμα γίγνεται.
διὰ τί γὰρ οὐκ ἂν καὶ ἅπαντα ἐξ ἑνὸς γίγνοιτο;
ὁ αὐτὸς γὰρ λόγος ἔοικεν εἶναι οὗτος τῷ Ἀναξ-
αγόρου, τῷ μηθὲν γίγνεσθαι τῶν ὁμοιομερῶν·
πλὴν ἐκεῖνος μὲν ἐπὶ πάντων, οὗτοι δ’ ἐπὶ τῆς
γενέσεως τῶν ζώων τοῦτο ποιοῦσιν. ἔπειτα τίνα
10 τρόπον αὐξηθήσεται ταῦτα τὰ ἀπελθόντα ἀπὸ παντός; Ἀναξαγόρας μὲν γὰρ εὐλόγως φησὶ σάρκας
ἐκ τῆς τροφῆς προσιέναι ταῖς σαρξίν· τοῖς δὲ ταῦτα
μὲν μὴ λέγουσιν, ἀπὸ παντὸς δ’ ἀπιέναι φάσκουσι,
πῶς ἑτέρου προσγιγνομένου ἔσται μεῖζον, εἰ μὴ

---

[a] Cf. note on 721 a 3. It has no right to be called by the
same name (συνώνυμον, implying the same λόγος of its
essence) because it has not the same qualities, which clearly
shows that it has not the same essence.

[b] This phrase, which at once calls to mind the question
asked by Anaxagoras (Diels 59 B 10) πῶς γὰρ ἂν ἐκ μὴ τριχὸς
γένοιτο θρὶξ καὶ σὰρξ ἐκ μὴ σαρκός; leads so naturally to the
reference to Anaxagoras which immediately follows.

[c] According to Anaxagoras, the " uniform " substances,
such as flesh, bone, blood, etc., were to be ranked as elements,
i.e., as ultimate forms of matter, and therefore ex hypothesi did
not come into being or pass out of being; and there was a
portion of every one of them in every thing. Hence, there
was a portion of flesh, bone, blood, etc., in all nourishment
taken by the embryo, and so Anaxagoras could easily
account for the growth in bulk of the flesh, bone, and blood
in the embryo. The theory now being examined, says
Aristotle, seems to make a similar assertion about the *semen*
only—this, it holds, contains a portion of flesh, bone, blood,
etc.—but it does not go on to assert that the nourishment

drawn from the various parts of the parent has no
right to the same name [a] as those parts—we may not
call that " blood " which is drawn from the parents'
blood, and the same with flesh. This means that the
offspring's blood is formed out of something which is
other than blood, and if so, then the cause of its
resemblance will not be due to the semen's being
drawn from all the parts of the parent's body, as the
supporters of this theory assert—because if blood is
formed from something that is not blood,[b] the semen
need only be drawn from one part, there being no
reason why all the other constituents as well as blood
should not be formed out of the one substance. This
theory seems to be identical with that of Anaxagoras,[c]
in asserting that none of the uniform substances
comes into being ; the only difference is that whereas
he applied the theory universally, these people apply
it to the generation of animals. Again, how are
these parts which were drawn from the whole of the
parent's body going to grow ? Anaxagoras gives a
reasonable answer ; he says that the flesh already
present is joined by flesh that comes from the nourishment. Those people however, who do not follow
Anaxagoras in the statement just quoted, yet hold
that the semen is drawn from the whole body, are
faced with this question : how is the embryo to grow
bigger by the addition of different substance to it

which the embryo takes in afterwards also contains these
substances. Hence the theory gets into a difficulty when the
question arises of how the *growth* of the embryo is effected.
This difficulty is avoided by Anaxagoras, because he makes
his principle " a portion of every element in every thing "
apply *universally*, and does not limit its application to the
semen only. (For Anax., see A. L. Peck, *C.Q.* XXV (1931),
27 ff., 112 ff.)

μεταβάλλει[1] τὸ προσελθόν; ἀλλὰ μὴν εἴ γε
15 δύναται μεταβάλλειν τὸ προσελθόν, διὰ τί οὐκ
εὐθὺς ἐξ ἀρχῆς τὸ σπέρμα τοιοῦτόν ἐστιν ὥστ'
ἐξ αὐτοῦ δυνατὸν εἶναι γίνεσθαι αἷμα καὶ σάρκας,
ἀλλὰ μὴ αὐτὸ εἶναι ἐκεῖνο καὶ αἷμα καὶ σάρκας;
οὐ γὰρ δὴ οὐδὲ τοῦτο ἐνδέχεται λέγειν, ὡς τῇ
κατακεράσει αὐξάνεται ὕστερον οἷον οἶνος ὕδατος
προσεγχυθέντος· αὐτὸ γὰρ ἂν πρῶτον μάλιστα
20 ἦν ἕκαστον ἄκρατον ὄν· νῦν δὲ ὕστερον μᾶλλον
καὶ σὰρξ καὶ ὀστοῦν καὶ τῶν ἄλλων ἕκαστόν ἐστι
μορίων. τοῦ δὲ σπέρματος φάναι τι νεῦρον εἶναι
καὶ ὀστοῦν λίαν ἐστὶν ὑπὲρ ἡμᾶς τὸ λεγόμενον.

Πρὸς δὲ τούτοις εἰ τὸ θῆλυ καὶ τὸ ἄρρεν ἐν τῇ
κυήσει διαφέρει, καθάπερ Ἐμπεδοκλῆς λέγει

25 ἐν δ' ἐχύθη καθαροῖσι· τὰ μὲν τελέθουσι γυναῖκες
ψύχεος ἀντιάσαντα. . . .

φαίνονται δ' οὖν μεταβάλλουσαι καὶ[2] γυναῖκες καὶ
ἄνδρες, ὥσπερ ἐξ ἀγόνων γόνιμοι, οὕτω καὶ ἐκ
θηλυτόκων ἀρρενοτόκοι, ὡς οὐκ ἐν τῷ ἀπελθεῖν
ἀπὸ παντὸς ἢ μὴ τῆς αἰτίας οὔσης, ἀλλ' ἐν τῷ
30 σύμμετρον ἢ ἀσύμμετρον εἶναι τὸ ἀπὸ τῆς γυ-
ναικὸς καὶ τοῦ ἀνδρὸς ἀπιόν, ἢ καὶ δι' ἄλλην τινὰ
τοιαύτην αἰτίαν. δῆλον τοίνυν, εἰ τοῦτο θήσομεν
οὕτως, ὅτι οὐ τῷ ἀπελθεῖν ἀπό τινος τὸ θῆλυ, ὥστ'

---

[1] μὴ μεταβάλλει Z : μένει vulg.
[2] καὶ PYZ : om. vulg.

---

unless the substance that is added changes ? If however it is admitted that this added substance can
change, why not admit straight away that the semen
at the outset is such that out of it blood and flesh can
be formed, instead of maintaining that the semen is
itself both blood and flesh ? They might try to argue
that it grows at a later stage by admixture, just as
wine is increased in bulk by pouring in water ; but
even this line of argument proves impossible, because
if that were so, then it would surely be at the outset
that each of the parts was its own proper self, before
it was mixed, whereas in actual fact it is at a later
stage that this occurs (I refer of course to flesh and
bone and every one of the rest of them). And as for
asserting that some of the semen is sinew and bone—
well, the statement is quite over our heads.

Here is another objection. Suppose it is true that
the differentiation between male and female takes
place during conception, as Empedocles says [a] :

> Into clean vessels were they pourèd forth ;
> Some spring up to be women, if so be
> They meet with cold. . . .

Anyway, both men and women are observed to
change : not only do the infertile become fertile,
but also those who have borne females bear males ;
which suggests that the cause is not that the semen
is or is not drawn from the whole of the parents, but
depends upon whether or not that which is drawn
from the man and from the woman stand in the right
proportional relation to each other.[b] Or else it is
due to some other cause of this sort. Thus, if we are
to assume this as true, viz., that the same semen is

---

[b] *Cf.* 767 a 16, 772 a 17, and Introd. § 39.

οὐδὲ τὸ μέρος ὃ ἔχει ἴδιον τό τε ἄρρεν καὶ τὸ θῆλυ,
εἴπερ τὸ αὐτὸ σπέρμα καὶ θῆλυ καὶ ἄρρεν δύναται
35 γίγνεσθαι ὡς οὐκ ὄντος τοῦ μορίου ἐν τῷ σπέρματι.

τί οὖν διαφέρει ἐπὶ τούτου λέγειν ἢ ἐπὶ τῶν ἄλλων
μορίων; εἰ γὰρ μηδ' ἀπὸ τῆς ὑστέρας σπέρμα
γίνεται, ὁ αὐτὸς λόγος καὶ ἐπὶ τῶν ἄλλων ἂν εἴη
μορίων.

Ἔτι ἔνια γίνεται τῶν ζῴων οὔτ' ἐξ ὁμογενῶν
οὔτε τῷ γένει διαφόρων, οἷον αἱ μυῖαι καὶ τὰ γένη
5 τῶν καλουμένων ψυλλῶν.[1] ἐκ δὲ τούτων γίνεται
μὲν ζῷα, οὐκέτι δ' ὅμοια τὴν φύσιν, ἀλλὰ γένος τι
σκωλήκων. δῆλον οὖν ὅτι οὐκ ἀπὸ παντὸς μέρους
ἀπιόντος γίγνονται ὅσα ἑτερογενῆ· ὅμοια γὰρ ἂν
ἦν, εἴπερ τοῦ ἀπὸ παντὸς ἀπιέναι σημεῖόν ἐστιν ἡ
ὁμοιότης.

Ἔτι ἀπὸ μιᾶς συνουσίας καὶ τῶν ζῴων ἔνια
10 γεννᾷ πολλά, τὰ δὲ φυτὰ καὶ παντάπασιν· δῆλον
γὰρ ὅτι ἀπὸ μιᾶς κινήσεως τὸν ἐπέτειον πάντα
φέρει καρπόν. καίτοι πῶς δυνατόν, εἰ ἀπὸ παντὸς
ἀπεκρίνετο τὸ σπέρμα; μίαν γὰρ ἀπόκρισιν ἀπὸ
μιᾶς ἀναγκαῖον γίνεσθαι συνουσίας καὶ μιᾶς διακρίσεως. ἐν δὲ ταῖς ὑστέραις χωρίζεσθαι ἀδύνατον·

---

[1] ψυλλῶν F.m, Aldus, Buss., A.-W., Platt : ψυλῶν SZ : ψυχῶν
vulg.; cf. 721 a 8 supra.

---

[a] And that the differentiation takes place in the uterus.

[b] This does not of course imply a belief in plant fertilization ; but the precise meaning of the remark is not clear.
On comparison with 728 b 35 ff., it appears that the product
of the " one act of coition " in animals corresponds to the
" seed " of plants, which also is a " fetation," in which male
and female are not separate, just as male and female are

able to be formed into either male or female [a] (implying that the sexual part is not present in the semen), it is clear that it is not the semen's being drawn from some one part which causes the offspring to be female, nor, in consequence, is it responsible for the special physical part which is peculiar to the two sexes. And what can be asserted about the sexual part can equally well be asserted about the other parts ; since if no semen comes even from the uterus, the same will surely hold good of the other parts as well.

Further, some animals are formed neither from creatures of the same kind as themselves nor from creatures of a different kind ; examples are : flies and the various kinds of fleas as they are called. Animals are formed from these, it is true, but in these cases they are not similar in character to their parents ; instead we get a class of larvae. Thus in these creatures which differ in kind from their parents we clearly have animals which are *not* formed out of semen drawn from every part of the body, for if resemblance is held to be a sure sign that this has occurred, then they would resemble their parents.

Further, even among the animals there are some which generate numerous offspring from one act of coition, a phenomenon which is, indeed, universal with plants ; these, as is manifest, produce a whole season's fruit as the result of one single movement.[b] Now how is this possible on the supposition that the semen is secreted from the whole body ? One act of coition, and one effort of segregation, ought necessarily to give rise to one secretion and no more. That it should get divided up in the uterus is impossible,

combined in the " fetation " of an animal. See also 728 a 27, 731 a 1.

723 b

15 ἤδη γὰρ ὥσπερ ἀπὸ νέου φυτοῦ ἢ ζῴου, οὐ σπέρματος εἴη[1] ἡ διαχώρισις.

Ἔτι τὰ ἀποφυτευόμενα ἀπ' αὐτοῦ φέρει σπέρμα·
δῆλον οὖν ὅτι καὶ πρὶν ἀποφυτευθῆναι ἀπὸ τοῦ
αὐτοῦ μεγέθους[2] ἔφερε τὸν καρπόν, καὶ οὐκ ἀπὸ
παντὸς τοῦ φυτοῦ ἀπῄει τὸ σπέρμα.

Μέγιστον δὲ τούτων τεκμήριον τεθεωρήκαμεν
20 ἱκανῶς ἐπὶ τῶν ἐντόμων. καὶ γὰρ εἰ μὴ ἐν πᾶσιν,
ἀλλ' ἐπὶ τῶν πλείστων ἐν τῇ ὀχείᾳ τὸ θῆλυ εἰς τὸ
ἄρρεν μέρος τι αὐτοῦ ἀποτείνει [διὸ καὶ τὴν ὀχείαν,
καθάπερ εἴπομεν πρότερον, οὕτω ποιοῦνται][3]· τὰ
γὰρ κάτωθεν εἰς τὰ ἄνω φαίνεται ἐναφιέντα, οὐκ
ἐν πᾶσιν, ἀλλ' ἐν τοῖς πλείστοις τῶν τεθεωρη-
25 μένων. ὥστε φανερὸν ἂν εἴη ὅτι οὐδ' ὅσα προΐεται
γονὴν τῶν ἀρρένων, οὐ τὸ ἀπὸ παντὸς ἀπιέναι τῆς
γενέσεως αἴτιόν ἐστιν, ἀλλ' ἄλλον τινὰ τρόπον,
περὶ οὗ σκεπτέον ὕστερον. καὶ γὰρ εἴπερ τὸ ἀπὸ
παντὸς ἀπιέναι συνέβαινεν, ὥσπερ φασίν, οὐθὲν
ἔδει ἀπὸ πάντων ἀξιοῦν ἀπιέναι, ἀλλὰ μόνον ἀπὸ
30 τοῦ δημιουργοῦντος, οἷον ἀπὸ τοῦ τέκτονος ἀλλὰ
μὴ ἀπὸ τῆς ὕλης. νῦν δ' ὅμοιον λέγουσιν ὥσπερ
κἂν εἰ ἀπὸ τῶν ὑποδημάτων· σχεδὸν γὰρ ὁ ὅμοιος[4]
υἱὸς τῷ πατρὶ ὅμοια φορεῖ.

Ὅτι δ' ἡδονὴ σφοδρὰ γίνεται ἐν τῇ ὁμιλίᾳ τῇ

---

[1] ἀπὸ . . . εἴη] ἀπὸ ζῴου σπέρμα ποιεῖ Z, sim. Σ.
[2] μέρους coni. Bonitz.   [3] seclusi : om. Σ.
[4] ὁ ὅμοιος P : ὅμοιός τις vulg. : ὅμοιος Z.

---

[a] The text is probably corrupt: for the sense cf. 729 a 6 ff.
[b] Ch. 16.

for by that time the division would be made as it were from a new plant or animal, not of semen.[a]

Further, transplanted cuttings bear seed—derived, of course, from themselves; which is proof positive that the fruit they bore before they were transplanted was derived from that identical amount of the plant which is now the cutting, and that the seed was not drawn from the whole of the plant.

The weightiest proof of all, however, we have sufficiently established by our observations of Insects. Perhaps not in all Insects, but certainly in most, during copulation the female extends a part of itself into the male [so, as we said earlier,[b] this is actually the way in which they effect copulation]: the females can be seen inserting something into the males upwards from below. This does not apply to all Insects, but to most of those which have been observed. Hence surely it is clear that even in the case of those males which discharge semen generation is not caused by the semen's being drawn from the whole of the body, but it is brought about in some other way, which we must consider later on. And indeed, if it were really true that the semen is drawn from the whole body, as these people say, still they ought not to assert that it is drawn from all the parts; they ought to say it is drawn only from the creative part which does the fashioning—from the artificer, in other words, not from the material which he fashions. As it is, they talk as though even the shoes which the parent wears were included among the sources from which the semen is drawn, for on the whole a son who resembles his father wears shoes that resemble his.

It is true that there is intense pleasure in sexual

69

τῶν ἀφροδισίων, οὐ τὸ ἀπὸ παντὸς ἀπιέναι αἴτιον,
35 ἀλλ' ὅτι κνησμός ἐστιν ἰσχυρός· διὸ καὶ εἰ πολλάκις

συμβαίνει ἡ ὁμιλία αὕτη, ἧττον γίνεται τὸ χαίρειν
τοῖς πλησιάζουσιν. ἔτι πρὸς τῷ τέλει ἡ χαρά· ἔδει
δὲ ἐν ἑκάστῳ τῶν μορίων, καὶ μὴ ἅμα, ἀλλ' ἐν
μὲν τοῖς πρότερον ἐν δὲ τοῖς ὕστερον.

Τοῦ δ' ἐκ κολοβῶν γίνεσθαι κολοβὰ ἡ αὐτὴ αἰτία
5 καὶ διὰ τί ὅμοια τοῖς γονεῦσιν. γίνεται δὲ καὶ οὐ
κολοβὰ ἐκ κολοβῶν, ὥσπερ καὶ ἀνόμοια τοῖς τε-
κνώσασιν· περὶ ὧν ὕστερον τὴν αἰτίαν θεωρητέον·
τὸ γὰρ πρόβλημα τοῦτ' ἐκείνοις ταὐτόν ἐστιν.

Ἔτι εἰ τὸ θῆλυ μὴ προΐεται σπέρμα, τοῦ αὐτοῦ
λόγου μηδ' ἀπὸ παντὸς ἀπιέναι. κἂν εἰ μὴ ἀπὸ
10 παντὸς ἀπέρχεται, οὐθὲν ἄλογον τὸ μηδ' ἀπὸ τοῦ
θήλεος, ἀλλ' ἄλλον τινὰ τρόπον αἴτιον εἶναι τὸ
θῆλυ τῆς γενέσεως. περὶ οὗ δὴ ἐχόμενόν ἐστιν
ἐπισκέψασθαι, ἐπειδὴ φανερὸν ὅτι οὐκ ἀπὸ πάντων
ἀποκρίνεται τὸ σπέρμα τῶν μορίων.

Ἀρχὴ δὲ καὶ ταύτης τῆς σκέψεως καὶ τῶν ἑπο-
15 μένων πρῶτον λαβεῖν περὶ σπέρματος τί ἐστιν·
οὕτω γὰρ καὶ περὶ τῶν ἔργων αὐτοῦ καὶ τῶν περὶ
αὐτὸ συμβαινόντων ἔσται μᾶλλον εὐθεώρητον. βούλεται δὲ τοιοῦτον τὴν φύσιν εἶναι τὸ σπέρμα, ἐξ

intercourse. The cause of this however is not that the semen is drawn from the whole body, but that there is violent stimulation; and that of course is why those who indulge often in such intercourse derive less pleasure from it. Moreover, the pleasure in fact comes at the end, but according to the theory it should occur (a) in every one of the parts, and (b) not simultaneously, but earlier in some and later in others.

As for mutilated offspring being produced by mutilated parents, the cause is the same as that which makes offspring resemble their parents. And anyway, not all offspring of mutilated parents are mutilated, any more than all offspring resemble their parents. The cause of these things we must consider later [a]; the problem in both cases is the same.

Moreover, if the female does not discharge any semen, then it is consistent to say that the semen is not drawn from the whole body either; or again, if it is not drawn from the whole body, there is nothing inconsistent in saying that it is not drawn from the female either,[b] but that the female is responsible for generation in some other way than this. This, in fact, will be the next subject for us to investigate, now that it is clear that the semen is not secreted from all the parts of the body.

We must begin this investigation and those which are to follow by discovering first of all what semen is; this will enable us to consider more easily its functions and everything connected with it. Now the aim of semen is to be, in its nature, the sort of stuff from which the things that take their rise in the realm

οὗ τὰ κατὰ φύσιν συνιστάμενα γίνεται πρώτου [οὗ
τῷ ἐξ ἐκείνου τι εἶναι τὸ ποιοῦν, οἷον τοῦ ἀνθρώπου·
20 γίγνεται γὰρ ἐκ τούτου, ὅτι τοῦτό ἐστι τὸ σπέρμα]¹.
ἐπεὶ δὲ πολλαχῶς γίγνεται ἄλλο ἐξ ἄλλου—ἕτερον
γὰρ τρόπον, ὡς ἐξ ἡμέρας φαμὲν νὺξ γίγνεται καὶ
ἐκ παιδὸς ἀνήρ, ὅτι τόδε μετὰ τόδε· ἄλλον δὲ τρόπον, ὡς ἐκ χαλκοῦ ἀνδριὰς καὶ ἐκ ξύλου κλίνη,
25 καὶ τἆλλα ὅσα ὡς ἐξ ὕλης γίγνεσθαι τὰ γιγνόμενα
λέγομεν, ἔκ τινος ἐνυπάρχοντος καὶ σχηματι-
σθέντος τὸ ὅλον ἐστίν. ἕτερον δὲ τρόπον ὡς ἐκ
μουσικοῦ ἄμουσος καὶ ὡς ἐξ ὑγιοῦς κάμνων, καὶ
ὅλως ὡς τὸ ἐναντίον ἐκ τοῦ ἐναντίου. ἔτι δὲ
παρὰ ταῦτα, ὡς Ἐπίχαρμος ποιεῖ τὴν ἐποικοδό-
μησιν, ἐκ τῆς διαβολῆς ἡ λοιδορία, ἐκ δὲ ταύτης
30 ἡ μάχη· ταῦτα δὲ πάντα ἔκ τινος ἦ² ἀρχὴ τῆς
κινήσεως, τῶν δὲ τοιούτων ἐνίων μὲν ἐν αὐτοῖς

---

¹ seclusi : οὐ τῷ ἐξ ἐκείνου τινός, οἷον ἐκ τοῦ ἀνθρώπου, ὅτι
τούτου τί ἐστι τὸ σπέρμα· ἐπειδὴ δὲ etc. Aldus : alia alii edd.
εἶναι . . . σπέρ- Z² in ras. in loco pauciorum. vide et not.
Anglice scriptam. ² ἦ Platt : ἡ vulg.

ᵃ With this definition, cf. 716 a 7 ff., 721 b 6, and *Phys.*
190 b 3-5.—At this point in the Greek text there follow
some unintelligible phrases which I have omitted from the
translation. The version of them given in the *ed. princeps*
differs considerably from that in the Berlin edition, and they
may be fragments of some annotation upon the definition
(founded perhaps on some such passage as 765 b 12, 13 (*q.v.*)
or, more probably, on ll. 724 b 2-4, where *cf.* note and reference to *Physics*; ἄνθρωπος is there used as an illustration and
there may have been a similar illustration here, which has
been corrupted). Actually any addition to the definition,
as apart from an illustration of it, at this point is inappropriate, as Aristotle is here giving the simplest and basic
definition, from which he builds up his final definition; this

72

of Nature are originally formed.[a] There are, however, numerous senses in which one thing is formed or comes into being " from " another [b] : (1) as we say " from day comes night," and " from boy comes man," meaning that the one comes *after* the other ; (2) as a statue is formed from bronze, or a bedstead from wood, and all those cases where we describe things as being formed from some material ; here the finished whole has been fashioned into a certain shape from something which was there to begin with ; (3) as a person may become uncultured from being cultured or ailing from healthy, *i.e.*, all cases of a contrary coming from its contrary ; (4) as in a " cumulative " passage in Epicharmus [c] : *e.g.*, from slander comes abuse, from abuse a fight ; in all these cases " from so-and-so " means that so-and-so is the source of the movement,[d] and in some instances

is also abundantly clear from the argument which immediately follows.

  [b] *Cf.* the similar discussion, with some of the same examples, on the meaning of " from " in *Met.* 1023 a 26 ff. ; also *Phys.* 190 a 22 ff.

  [c] Epicharmus of Sicily (Aristot. *Poet.* 1448 a 33) was the chief Dorian comic poet. Aristotle may have in mind a passage of his similar to that quoted by Athenaeus (ii. 36 c, d), and Suidas, which G. Kaibel (*Comicorum Graecorum Fragmenta*, I. i. p. 118) prints as follows, with the Doric vowels restored and with the emendations of various scholars :

A. ἐκ μὲν θυσίας θοῖνα,
    ἐκ δὲ θοίνας πόσις, ἐγένετο. B. χαρίεν, ὥς γ᾽ ἐμὶν ⟨δοκεῖ⟩.
A. ἐκ δὲ πόσιος μῶκος, ἐκ μώκου δ᾽ ἐγένεθ᾽ ὑανία,
    ἐκ δ᾽ ὑανίας ⟨δίκα . . ., ἐκ δίκας δὲ κατα⟩δίκα,
    ἐκ δὲ καταδίκας πέδαι τε καὶ σφαλὸς καὶ ζαμία.

See also A. Lorenz (*Leben u. Schriften des Koers Epicharmos*, p. 271). *Cf.* Aristot. *Met.* 1023 a 30, 1013 a 10, *Rhet.* 1365 a 16.

  [d] *i.e.*, the " Efficient " or " Motive " Cause.

724 a

ἡ ἀρχὴ τῆς κινήσεώς ἐστιν, οἷον καὶ ἐν τοῖς νῦν
εἰρημένοις (μέρος γάρ τι ἡ διαβολὴ τῆς πάσης
ταραχῆς ἐστίν), ἐνίων δ' ἔξω, οἷον αἱ τέχναι τῶν
δημιουργουμένων καὶ ὁ λύχνος τῆς καιομένης
35 οἰκίας.

Τὸ δὲ σπέρμα φανερὸν ὅτι δυοῖν τούτοιν ἐν θατέρῳ ἐστίν· ἢ γὰρ ὡς ἐξ ὕλης αὐτοῦ ἢ ὡς ἐκ πρώ-

724 b
του κινήσαντός ἐστι τὸ γινόμενον. οὐ γὰρ δὴ ὡς
τόδε μετὰ τόδε, οἷον ἐκ τῶν Παναθηναίων ὁ πλοῦς,
οὐδ' ὡς ἐξ ἐναντίου· φθειρομένου τε γὰρ γίγνεται
τὸ ἐναντίον ἐκ τοῦ ἐναντίου, καὶ ἕτερόν τι δεῖ
ὑποκεῖσθαι ἐξ οὗ ἔσται πρώτου ἐνυπάρχοντος. τοῖν
5 δυοῖν δὴ ληπτέον ἐν ποτέρῳ θετέον τὸ σπέρμα,
πότερον ὡς ὕλην καὶ πάσχον ἢ ὡς εἶδός τι καὶ
ποιοῦν, ἢ καὶ ἄμφω. ἅμα γὰρ ἴσως δῆλον ἔσται
καὶ πῶς ἡ[1] ἐξ ἐναντίων γένεσις ὑπάρχει πᾶσι τοῖς
ἐκ τοῦ σπέρματος· φυσικὴ γὰρ καὶ ἡ ἐκ τῶν
ἐναντίων γένεσις· τὰ μὲν γὰρ ἐξ ἐναντίων γίγνεται,
10 ἄρρενος καὶ θήλεος, τὰ δ' ἐξ ἑνὸς μόνου, οἷον τά
τε φυτὰ καὶ τῶν ζῴων ἔνια, ἐν ὅσοις μή ἐστι
διωρισμένον τὸ ἄρρεν καὶ τὸ θῆλυ χωρίς.

---

[1] ἡ Z : om. vulg.

---

[a] *i.e.*, either (2) or (4) above.
[b] *Cf.* the discussion on the meaning of γίγνεσθαι and
γίγνεσθαι ἔκ τινος in *Phys.* 190 a 5 ff. These contraries are
merely attributes of something else, something which has
being (οὐσία), is a concrete existing thing, and is the " substrate ": καὶ γὰρ ποσὸν καὶ ποιὸν . . . γίνεται ὑποκειμένου
τινός (190 a 35). If we say that a man " becomes " cultured
" from " being uncultured, it is " man " that persists through-

of this sort the source of the movement is within the things themselves, as in the ones just quoted (where slander is actually one part of the whole to-do) ; in others it is external to them ; *e.g.*, craftsmanship of every kind is external to the works which the craftsman produces, and the torch is external to the house which is set on fire.

Now it is clear that the case of semen falls under one or other of these two senses [a] : the offspring is formed " from " it either (*a*) as " from " material, or (*b*) as " from " a proximate motive cause. It is definitely not an instance of (1) above, where " from " means " after," *e.g.*, " from the Panathenaean festival comes the sea-voyage " ; nor of (3), *i.e.*, of coming into being " from " a contrary ; for the one contrary is destroyed as the other comes into being from it, and so there must be present besides them some primary substrate, from which the new contrary is to come into being.[b] Thus we now have to discover in which of the two classes semen is to be placed : Is it to be regarded as matter, *i.e.*, as something which is acted upon, or as a form, *i.e.*, as something which acts of itself—or even as both ? for perhaps at the same time it will also be clear in what way formation from contraries has its place in all things that arise from semen. (After all, formation from contraries as well as the other methods of formation is found in nature ; some animals are formed from contraries—male and female, though some are formed from one parent only, as are plants and certain of the animals in which there is no definite separation of male and female.)

out the change. Clearly, says Aristotle, this is not the meaning of γίγνεσθαι required here.

¹[Γονὴ μὲν οὖν τὸ ἀπὸ τοῦ γεννῶντος καλεῖται
ἀπιόν,² ὅσα συνδυάζεσθαι πέφυκε, τὸ πρῶτον ἔχον
ἀρχὴν γενέσεως, σπέρμα δὲ τὸ ἐξ ἀμφοτέρων τὰς
15 ἰρχὰς ἔχον τῶν συνδυασθέντων, οἷον τά τε τῶν
φυτῶν καὶ ἐνίων ζώων, ἐν οἷς μὴ κεχώρισται τὸ
θῆλυ καὶ τὸ ἄρρεν,³ ὥσπερ τὸ γιγνόμενον ἐκ θήλεος
καὶ ἄρρενος πρῶτον μίγμα, οἷον κύημά τι ὂν ἢ
ᾠόν⁴· καὶ γὰρ ταῦτα ἤδη ἔχει τὸ ἐξ ἀμφοῖν.

Σπέρμα δὲ καὶ καρπὸς διαφέρει τῷ ὕστερον καὶ
20 πρότερον· καρπὸς μὲν γὰρ τῷ ἐξ ἄλλου εἶναι,
σπέρμα δὲ τῷ ἐκ τούτου ἄλλο, ἐπεὶ ἄμφω γε
ταὐτόν⁵ ἐστιν.

Ἡ δὲ τοῦ λεγομένου σπέρματος⁶ φύσις, ἡ πρώτη,
πάλιν λεκτέα τίς ἐστιν.]

Ἀνάγκη δὴ πᾶν, ὃ ἂν λαμβάνωμεν ἐν τῷ σώματι, ἢ μέρος εἶναι τῶν κατὰ φύσιν, καὶ τοῦτο ἢ
25 τῶν ἀνομοιομερῶν ἢ τῶν ὁμοιομερῶν, ἢ τῶν παρὰ
φύσιν, οἷον φῦμα, ἢ περίττωμα ἢ σύντηγμα ἢ
τροφήν. λέγω δὲ περίττωμα μὲν τὸ τῆς τροφῆς
ὑπόλειμμα, σύντηγμα δὲ τὸ ἀποκριθὲν ἐκ τοῦ⁷

---

¹ vv. 12-22 inepta seclusi.
² ἀπιόν P : exit Σ : αἴτιον vulg.
³ οἷον . . . ἄρρεν secluserat Platt.
⁴ ᾠόν Wimmer, ωιον Z¹, sicut ovum Σ : ζῷον vulg., Z².
⁵ fortasse τοῦ αὐτοῦ scribendum.
⁶ ὡς γονῆς addit Z.
⁷ ἐκ τοῦ] ἑκάστου PSZ¹.

ᵃ The following paragraphs seem to be an interpolation.
They interrupt the argument ; further definitions are here
inappropriate, and one of those here given is incorrect.
Besides, Aristotle does not in the *Generation of Animals*
make the distinction between γονή and σπέρμα. These
definitions seem to have been put in here because the following passage contains some definitions.

[a] [Seminal fluid is the name given to that which comes from the generating parent, in the case of those animals whose nature it is to copulate, and it is that in which a generative principle is first found. Semen (seed) is the name given to that which contains the principles derived from *both* the parents which have copulated, as in the case of the plants and certain animals in which male and female are not separate, like the first mixture which is formed from the male and female, being as it were a sort of fetation or egg—for these objects too already contain that which comes from both parents.

Semen (seed) and fruit differ by the " prior and posterior [b] ": fruit ⟨is posterior⟩ in that it is derived from something else, whereas seed ⟨is prior⟩ in that something else is derived from it, since in fact they are both one and the same thing.

We must now resume and state what is the primary nature of semen, as it is called.]

Now every substance, whatever it may be, that we find in the body, must of necessity be one of the following : (1) one of the parts which are there in accordance with nature, in which case it will be one of the uniform or non-uniform parts ; (2) one which is there contrary to nature, *e.g.*, a tumour ; (3) residue [c] ; (4) colliquescence [d] ; (5) nourishment. By residue I mean that which is left over as surplus from the nourishment ; by colliquescence that which is given off as an abscession [e] from the material that

[b] The meaning of these terms is discussed in *Met.* 1018 b 9 ff.

[c] See Introd. §§ 65 ff.

[d] See Introd. § 67 : also 725 a 27 ff. and *De somno et vig.* 456 b 34 ff.

[e] See Introd. § 67.

724 b

αὐξήματος ὑπὸ τῆς παρὰ φύσιν ἀναλύσεως. ὅτι
μὲν οὖν οὐκ ἂν εἴη μέρος, φανερόν· ὁμοιομερὲς
30 μὲν γάρ ἐστιν, ἐκ δὲ τούτου οὐθὲν σύγκειται,
ὥσπερ ἐκ νεύρου καὶ σαρκός. ἔτι δὲ οὐδὲ κεχωρισμένον, τὰ δ' ἄλλα πάντα μέρη. ἀλλὰ μὴν
οὐδὲ τῶν¹ παρὰ φύσιν, οὐδὲ πήρωμα· ἐν ἅπασί τε
γὰρ ὑπάρχει, καὶ ἡ φύσις ἐκ τούτου γίγνεται. ἡ
δὲ τροφὴ φανερῶς ἐπείσακτον. ὥστ' ἀνάγκη ἢ
35 σύντηγμα ἢ περίττωμα εἶναι. οἱ μὲν οὖν ἀρχαῖοι
ἐοίκασιν οἰομένοις εἶναι σύντηγμα· τὸ γὰρ ἀπὸ
παντὸς ἀπιέναι φάναι διὰ τὴν θερμότητα τὴν ἀπὸ
725 a τῆς κινήσεως συντήγματος ἔχει δύναμιν. τὸ δὲ
σύντηγμα² τῶν παρὰ φύσιν τι, ἐκ δὲ τῶν παρὰ
φύσιν οὐθὲν γίνεται τῶν κατὰ φύσιν. ἀνάγκη ἄρα
περίττωμα εἶναι. ἀλλὰ μὴν περίττωμά γε πᾶν ἢ
5 ἀχρήστου τροφῆς ἐστιν ἢ χρησίμης. ἄχρηστον
μὲν οὖν λέγω ἀφ' ἧς μηθὲν ἔτι συντελεῖται εἰς τὴν
φύσιν, ἀλλ' ἀναλισκομένου πλέονος μάλιστα κα-
κοῦται, τὴν δὲ χρησίμην τὴν ἐναντίαν. ὅτι μὲν
δὴ τοιοῦτον περίττωμα οὐκ ἂν εἴη, φανερόν· τοῖς

¹ οὐ τῶν Z : οὐδὲ vulg. : correxerunt A.-W.
² τὸ δὲ σύντηγμα SZ¹ : τὰ δὲ συντήγματα vulg.

---

[a] And therefore would have to be reckoned as one of the
uniform parts.

[b] Viz., the non-uniform parts, for the construction of which
the uniform parts act as the " material."

[c] This may mean that it is not present continuously as
such, but has to be " concocted " and " collected " on each
occasion for which it is required : see 717 b 25.

[d] See Introd. § 12 and 737 a 26, n.

[e] e.g., Hippocrates, π. γονῆς 1 (vii. 470 Littré), where this
statement occurs. Aristotle's equation of this view with the
belief that semen is a σύντηγμα is hardly fair, in face of the
context, q.v. Compare, e.g., the statement ἡ δὲ γονὴ . . .

supplies growth, as the result of decomposition proceeding contrary to nature. Now it is clear that
semen cannot possibly be (1) one of the parts ; since
although it is uniform,[a] it does not serve as the
material out of which any other parts [b] are composed,
as sinew and flesh do ; nor again is it separate and
distinct,[c] whereas all the other parts are. Nor (2) is
it something contrary to nature, or a deformation,[d]
(a) because it is present in every single individual, and
(b) because the natural organism develops out of it.
As for (5) nourishment, obviously this is introduced
into the body from without. It must therefore be
either (4) a colliquescence or (3) a residue. The
early thinkers appear to have supposed it was a
colliquescence, because to say that it is drawn from
the whole body in virtue of the heat which the movement produces,[e] is equivalent to saying that the
semen is a colliquescence. But colliquescence belongs to the class of things that are contrary to nature,
and from such things nothing that is in accordance
with nature is ever formed. Therefore the semen
must of necessity be a residue. Very well. Every
residue results either from useful or from useless
nourishment. By " useless nourishment " I mean
that which contributes nothing further to the natural
organism and which if too much of it is consumed
causes very great injury to the organism ; " useful
nourishment " is the opposite of this. It is obvious
that semen cannot be a residue resulting from useless
nourishment, for while residue of that sort is found in

ἔρχεται ἀπὸ παντὸς τοῦ ὑγροῦ τοῦ ἐν τῷ σώματι ἐόντος τὸ ἰσχυρότατον ἀποκριθέν· τούτου δὲ ἱστόριον τόδε, ὅτι ἀποκρίνεται τὸ
ἰσχυρότατον, ὅτι ἐπὴν λαγνεύσωμεν σμικρὸν οὕτω μεθέντες,
ἀσθενέες γινόμεθα with Aristotle's own statement at 725 b 6-8.

γὰρ κάκιστα διακειμένοις δι' ἡλικίαν ἢ νόσον[1]
πλεῖστον ἐνυπάρχει[2] τοιοῦτον, σπέρμα δὲ ἥκιστα·
ἢ γὰρ ὅλως οὐκ ἔχουσιν ἢ οὐ γόνιμον διὰ τὸ μί-
10 γνυσθαι ἄχρηστον περίττωμα καὶ νοσηματικόν.

Χρησίμου ἄρα περιττώματος μέρος τι ἐστὶ τὸ
σπέρμα. χρησιμώτατον δὲ τὸ ἔσχατον καὶ ἐξ οὗ
ἤδη γίνεται ἕκαστον τῶν μορίων. ἔστι γὰρ τὸ
μὲν πρότερον τὸ δ' ὕστερον. τῆς μὲν οὖν πρώτης
15 τροφῆς περίττωμα φλέγμα καὶ εἴ τι ἄλλο τοιοῦτον·
καὶ γὰρ τὸ φλέγμα τῆς χρησίμου τροφῆς περίττωμά ἐστιν· σημεῖον δ' ὅτι μιγνύμενον τροφῇ καθαρᾷ τρέφει καὶ πονοῦσι καταναλίσκεται. τὸ δὲ
τελευταῖον ἐκ πλείστης τροφῆς ὀλίγιστον.[3] ἐννοεῖν
δὲ δεῖ ὅτι μικρῷ αὐξάνεται τὰ ζῷα καὶ τὰ φυτὰ
20 τῷ[4] καθ' ἡμέραν· παμμικροῦ[5] γὰρ ἂν προστιθεμένου
τῷ αὐτῷ[6] ὑπερέβαλλε[7] τὸ μέγεθος.

Τοὐναντίον ἄρα ἢ οἱ ἀρχαῖοι ἔλεγον λεκτέον. οἱ
μὲν γὰρ τὸ ἀπὸ παντὸς ἀπιόν, ἡμεῖς δὲ τὸ πρὸς
ἅπαν ἰέναι πεφυκὸς σπέρμα ἐροῦμεν, καὶ οἱ μὲν
σύντηγμα, φαίνεται δὲ περίττωμα μᾶλλον. εὐ-
25 λογώτερον γὰρ ὅμοιον εἶναι τὸ προσιὸν ἔσχατον
καὶ τὸ περιττὸν γινόμενον τοῦ τοιούτου, οἷον τοῖς
γραφεῦσι τοῦ ἀνδρεικέλου πολλάκις περιγίνεται

---

[1] νόσον ἢ ἕξιν Z.
[2] ἐνυπάρχει PZ : ὑπάρχει vulg.
[3] γίνεται add. PZ.
[4] τῷ PY : τὸ vulg.
[5] παμμικροῦ A.-W. : πᾶν· μικροῦ vulg. : locus hic corruptus.
[6] τῷ αὐτῷ Platt : τοῦ αὐτοῦ vulg.
[7] ὑπερέβαλλε PY : ὑπερέβαλε vulg. : ὑπερβάλλοι Platt.

---

[a] See Introd. § 66.　　　　　　　　　[b] Cf. 728 a 31, n.
[c] Cf. 765 b 29 ff.
[d] Because it is the concocted residue of blood, the "ultimate nourishment" distributed to all the parts of the body.

considerable quantities in those who through age or disease are in a very bad state of health, the same is not true of semen ; such persons either have none at all, or if they have, it is infertile because of the useless and diseased residue that gets mixed with it.

Hence, semen is part of a useful residue ; and the most useful of the residues is that which is produced last, that from which each of the parts of the body is directly formed. I said "last," for of course some of the residues are produced earlier, some later. Nourishment in its first stage yields as its residue *phlegma* and any other such stuff.[a] Yes, *phlegma* too is a residue from the useful nourishment, as is shown by the fact that when it is mixed with pure nourishment it nourishes the body,[b] and that the body consumes it in cases of disease. The residue which comes last, however, is very small in bulk though the nourishment which yields it is very large[c] ; but we must bear in mind that it requires very little to supply the growth of animals and plants from day to day, since the continual addition of a very small amount to the same thing would make its size excessive.

Our own statement therefore must be the opposite of what the early people said. They said the semen is that which was drawn from the whole of the body ; we are going to say the semen is that whose nature it is to be distributed to the whole of the body.[d] And whereas they said it was a colliquescence, we see it is more correct to call it a residue. After all, it is more reasonable to suppose that the surplus residue of the final nourishment which is distributed all over the body resembles that nourishment, just as (to take a common instance) the paint left over on an artist's

Semen a "residue."

81

725 a

ὅμοιον τῷ ἀναλωθέντι. συντηκόμενον δὲ φθείρεται
πᾶν καὶ ἐξίσταται τῆς φύσεως. τεκμήριον δὲ τοῦ
μὴ σύντηγμα εἶναι ἀλλὰ περίττωμα μᾶλλον, τὸ
30 τὰ μεγάλα τῶν ζῴων ὀλιγοτόκα εἶναι, τὰ δὲ μικρὰ
πολύγονα. σύντηγμα μὲν γὰρ πλέον ἀναγκαῖον
εἶναι τοῖς μεγάλοις, περίττωμα δ' ἔλαττον· εἰς γὰρ
τὸ σῶμα μέγα ὂν ἀναλίσκεται τὸ πλεῖστον τῆς
τροφῆς, ὥστ' ὀλίγον γίνεται τὸ περίττωμα. ἔτι
τόπος συντήγματι μὲν οὐθεὶς ἀποδέδοται κατὰ
35 φύσιν, ἀλλὰ ῥεῖ ὅπου ἂν εὐοδήσῃ τοῦ σώματος,
725 b τοῖς δὲ κατὰ φύσιν περιττώμασι πᾶσιν, οἷον τῆς
τροφῆς τῆς ξηρᾶς ἡ κάτω κοιλία καὶ τῆς ὑγρᾶς
ἡ κύστις καὶ τῆς χρησίμης ἡ ἄνω κοιλία, καὶ τοῖς
σπερματικοῖς[1] ὑστέραι καὶ αἰδοῖα καὶ μαστοί· εἰς
τούτους γὰρ ἀθροίζεται καὶ συρρεῖ. καὶ μαρτύρια
5 τὰ συμβαίνοντα ὅτι τὸ εἰρημένον σπέρμα ἐστίν·
ταῦτα δὲ συμβαίνει διὰ τὸ τὴν φύσιν εἶναι τοῦ
περιττώματος τοιαύτην[2]· ἥ τε γὰρ ἔκλυσις ἐλα-
χίστου ἀπελθόντος τούτου γίνεται ἐπίδηλος, ὡς
στερισκόμενα τὰ σώματα τοῦ ἐκ τῆς τροφῆς γινομένου τέλους. (ὀλίγοις δέ τισιν ἐν μικρῷ χρόνῳ
10 κατὰ τὰς ἡλικίας κουφίζει τοῦτ' ἀπιόν, ὅταν πλεο-
νάσῃ, καθάπερ ἡ πρώτη τροφή, ἂν ὑπερβάλλῃ τῷ
πλήθει· καὶ γὰρ ταύτης ἀπιούσης τὰ σώματ' εὐ-

---

[1] τοῖς σπερματικοῖς PSZ : τῆς σπερματικῆς vulg.
[2] ὅτι . . . τοιαύτην fortasse secludenda (A.-W.), vel ὅτι τὸ
σπέρμα περίττωμά ἐστι χρήσιμον scribendum ; vertit Σ et
accidentia quae accidunt testificantur quod sperma est super-
fluum quo indigetur ad iuvamentum.

---

[a] For ἐξίστασθαι τῆς φύσεως, see 768 a 2, n.
[b] i.e., the large intestine.
[c] i.e., the small intestine.

palette resembles that which he has actually used; whereas everything that undergoes colliquescence gets destroyed and departs from its proper nature.[a] Here is a piece of evidence to show that semen is not a colliquescence but a residue : the large animals produce but few young, while the small ones are prolific. Now in the large animals there must of necessity be more colliquescence and less residue, because most of the nourishment is used up to maintain the large bulk of their body, so that but little residue is produced. Further, no place has been assigned by Nature for colliquescence, but it runs about in the body wherever it can find a clear way for itself ; whereas there is a proper place for all the *natural* residues—*e.g.*, the lower intestine [b] is set apart for the residue from the solid nourishment, the bladder for that from the fluid, the upper intestine [c] for that from the useful nourishment, the uterus, pudenda, and breasts for the seminal residues—they run into these places and collect there. As evidence of the truth of our statement about what semen is we can quote the actual facts, facts which directly result from this residue's being of the nature described by us. Thus (1) though only a very small quantity of semen be emitted, the exhaustion which follows is quite conspicuous,[d] which suggests that the body is being deprived of the final product formed out of the nourishment. (There are, I know, a few who for a short period during the heat of youth derive relief from the emission of the semen when it is superabundant. The same is true also of nourishment in its first stage, if there is an excessive quantity of it ;

[d] *Cf.* Hippocrates, π. γονῆς 1 (vii. 470 Littré), quoted in note on 725 a 1.

ἠμερεῖ μᾶλλον. ἔτι ὅταν συναπίῃ ἄλλα περιττώ-
ματα· οὐ γὰρ μόνον σπέρμα τὸ ἀπιόν, ἀλλὰ καὶ
ἔτεραι μεμιγμέναι δυνάμεις τούτῳ[1] συναπέρχονται,
15 αὗται δὲ νοσώδεις, διὸ ἐνίων γε καὶ ἄγονόν ποτε
γίνεται τὸ ἀποχωροῦν διὰ τὸ ὀλίγον ἔχειν τὸ σπερματικόν. ἀλλὰ τοῖς πλείστοις καὶ ὡς ἐπὶ τὸ πολὺ
εἰπεῖν συμβαίνει ἐκ τῶν ἀφροδισιασμῶν ἔκλυσις
καὶ ἀδυναμία μᾶλλον διὰ τὴν εἰρημένην αἰτίαν.)
ἔτι οὐκ ἐνυπάρχει σπέρμα οὔτ' ἐν τῇ πρώτῃ ἡλικίᾳ
20 οὔτ' ἐν τῷ γήρᾳ οὔτ' ἐν ταῖς ἀρρωστίαις, ἐν μὲν
τῷ κάμνειν διὰ τὴν ἀδυναμίαν, ἐν δὲ τῷ γήρᾳ διὰ
τὸ μὴ πέττειν τὸ ἱκανὸν τὴν φύσιν, νέοις δ' οὖσι
διὰ τὴν αὔξησιν· φθάνει γὰρ ἀναλισκόμενον πᾶν·
ἐν ἔτεσι γὰρ πέντε σχεδὸν ἐπί γε τῶν ἀνθρώπων
ἥμισυ λαμβάνειν δοκεῖ τὸ σῶμα τοῦ μεγέθους τοῦ[a]
25 ἐν τῷ ἄλλῳ χρόνῳ γιγνομένου ἅπαντος.

Πολλοῖς δὲ συμβαίνει καὶ ζῴοις καὶ φυτοῖς καὶ
γένεσι πρὸς γένη διαφορὰ περὶ ταῦτα κἂν τῷ γένει
τῷ αὐτῷ τοῖς ὁμοειδέσι πρὸς ἄλληλα, οἷον ἀνθρώπῳ
πρὸς ἄνθρωπον καὶ ἀμπέλῳ πρὸς ἄμπελον. τὰ
μὲν γὰρ πολύσπερμα τὰ δ' ὀλιγόσπερμά ἐστι, τὰ
30 δ' ἄσπερμα πάμπαν, οὐ δι' ἀσθένειαν, ἀλλ' ἐνίοις γε
διὰ[2] τοὐναντίον· καταναλίσκεται γὰρ εἰς τὸ σῶμα,
οἷον τῶν ἀνθρώπων ἐνίοις· εὐεκτικοὶ γὰρ ὄντες καὶ
γινόμενοι πολύσαρκοι[b] ἢ πιότεροι μᾶλλον ἧττον
προΐενται σπέρμα καὶ ἧττον ἐπιθυμοῦσι τοῦ ἀφ-

---

[1] τούτῳ Platt : τούτοις vulg.   [2] διὰ PZ : om. vulg.

---

[a] i.e., of nourishment.   [b] Or, muscle.

the body is more comfortable for having got rid of it. Relief is obtained too when other residues are got rid of in company with the semen : in such cases what is emitted is not merely semen, but there are other substances which come away at the same time mixed up with it, and these are morbid. This explains why at certain times with some persons the emission is infertile : it contains so small an amount of actual semen. However, speaking generally for the majority of men, the sequel to sexual intercourse is exhaustion and weakness rather than relief, and the cause is as I have described.) Besides (2), semen is absent during childhood, old age, and infirmity ; absent during infirmity on account of the weakness of the body, during old age because the organism does not concoct a sufficient amount [a] ; during childhood because the body is growing, and the concocted matter is all used up so soon that there is none left over : it is usually held that in about five years human beings, at any rate, grow to one-half of the complete size that they will attain in the rest of their lifetime.

In respect of semen we find that with many animals and plants one group differs from another group, and even within one and the same group individuals of the same kind differ from each other, e.g., one man from another, and one grape-vine from another. Some individuals have much semen, some little, some none at all ; and this is not due to any bodily weakness, but in some cases, at any rate, it is due to the opposite : the available supply gets used up to benefit the body ; as an example of this we have men in sound health putting on rather a lot of flesh [b] and getting a bit fat : these emit less semen and have less desire for sexual intercourse than is normal. A

725 b

ροδισιάζειν. ὅμοιον δὲ καὶ τὸ περὶ τὰς τραγώσας
35 ἀμπέλους πάθος, αἳ διὰ τὴν τροφὴν ἐξυβρίζουσιν
726 a (ἐπεὶ καὶ οἱ τράγοι πίονες ὄντες ἧττον ὀχεύουσιν,
διὸ καὶ προλεπτύνουσιν αὐτούς· καὶ τὰς ἀμπέλους
τραγᾶν ἀπὸ τοῦ πάθους τῶν τράγων καλοῦσιν).
καὶ οἱ πίονες δὲ ἀγονώτεροι φαίνονται εἶναι τῶν
μὴ πιόνων, καὶ γυναῖκες καὶ ἄνδρες, διὰ τὸ τοῖς
5 εὐτραφέσι πεττόμενον τὸ περίττωμα γίνεσθαι πι-
μελήν· ἔστι γὰρ καὶ ἡ πιμελὴ περίττωμα, δι'
εὐβοσίαν ὑγιεινόν.

Ἔνια δ' ὅλως οὐδὲ φέρει σπέρμα, οἷον ἰτέα καὶ
αἴγειρος. εἰσὶ μὲν οὖν ἑκάτεραι αἰτίαι[1] τούτου
τοῦ πάθους. καὶ γὰρ δι' ἀδυναμίαν οὐ πέττουσι
καὶ διὰ δύναμιν ἀναλίσκουσιν, ὥσπερ εἴρηται.
10 ὁμοίως δὲ καὶ πολύχοά[2] ἐστι καὶ πολύσπερμα[3] τὰ
μὲν διὰ δύναμιν τὰ δὲ δι' ἀδυναμίαν· πολὺ γὰρ καὶ
ἄχρηστον περίττωμα συμμίγνυται, ὥστ' ἐνίοις γίγνεσθαι καὶ ἀρρώστημα, ὅταν αὐτῶν μὴ εὐοδήσῃ
ἡ ἀποκάθαρσις. καὶ ἔνιοι μὲν ὑγιάζονται, οἱ δὲ
καὶ ἀναιροῦνται. συντήκονται γὰρ ταύτῃ ὥσπερ
15 καὶ εἰς τὸ οὖρον· ἤδη γὰρ καὶ τοῦτ' ἀσθένημα
συνέβη τισίν.

[⁴][Ἔτι ὁ πόρος ὁ αὐτὸς τῷ περιττώματι καὶ τῷ
σπέρματι· καὶ ὅσοις μὲν ἀμφοῖν γίγνεται περίτ-

---

[1] ἑκάτεραι αἰτίαι scripsi, post A.-W., qui αἰτίαι ἑκάτεραι :
αἰτίαι καὶ ἕτεραι P, καὶ ἕτεραι αἰτίαι vulg. (aliae Σ).
[2] πολύχοα Z¹ : πολυχρόνιά PSYZ².
[3] καὶ πολύσπερμα fort. secl.
[4] vv. 16-25 seclusit Platt ; 725 b 25—726 a 15 seclusit Sus.

---

[a] The former part of this interpolation seems to belong to
the interpolation connected with chh. 12 and 13 (cf. 719

86

similar phenomenon is that of grape-vines which
" go goaty," rampaging all over the place because
they are getting too much nourishment. (The reason
for the phrase " go goaty " is that they behave just
like he-goats, which when they get fat indulge less
in copulation, and incidentally this explains why goats
are made to slim before the breeding season comes
on.) And further it seems that fat people, men and
women alike, are less fertile than those who are not
fat, the reason being that when the body is too well
fed, the effect of concoction upon the residue is to
turn it into fat (since fat also is one of the residues,
a healthy one, because it results from good living).

Some living things actually produce no semen at
all : examples are the willow and the poplar. Both
reasons together are responsible for this state of
affairs ; in other words, on account of their weakness
the trees cannot concoct their nourishment, and on
account of their strength they use it all up, as
described above. Similarly, some animals are prolific and have abundance of semen because they are
strong, but others because they are weak ; the explanation being that in the latter case much useless
residue gets mixed up with the semen, and in some
instances, when there is no clear way open by which
the evacuated matter may leave, it actually produces
disease, from which some recover though others
succumb. Their semen is contaminated by the col-
liquescences which get into it, just as they do into
the urine—another malady by no means unknown.

^a [Further, the same passage serves both for the
residue and for the semen : (a) in those animals which

b 29 ff.) ; the latter part refers to the subjects discussed in
725 a—726 a.

τωμα, καὶ τῆς ὑγρᾶς καὶ τῆς ξηρᾶς τροφῆς, ᾗπερ
ἡ τοῦ ὑγροῦ, ταύτῃ καὶ ἡ τῆς γονῆς γίγνεται
ἀπόκρισις (ὑγροῦ γὰρ περίττωμά ἐστιν· ἡ γὰρ
20 τροφὴ πάντων ὑγρὰ μᾶλλον), οἷς δὲ μή ἐστιν αὕτη,
κατὰ τὴν τῆς ξηρᾶς ὑποστάσεως ἀποχώρησιν. ἔτι
ἡ μὲν σύντηξις ἀεὶ νοσώδης, ἡ δὲ τοῦ περιττώματος ἀφαίρεσις ὠφέλιμος· ἡ δὲ τοῦ σπέρματος
ἀποχώρησις ἀμφοτέρων[1] διὰ τὸ προσλαμβάνειν
τῆς μὴ χρησίμου τροφῆς. εἰ δέ γ᾽ ἦν σύντηξις,
25 ἀεὶ ἔβλαπτεν ἄν· νῦν δ᾽ οὐ ποιεῖ τοῦτο.]

Ὅτι μὲν οὖν περίττωμά ἐστι τὸ σπέρμα χρησίμου τροφῆς καὶ τῆς ἐσχάτης, εἴτε πάντα προΐεται
σπέρμα εἴτε μή, ἐν τοῖς προειρημένοις φανερόν.

XIX Μετὰ δὲ ταῦτα διοριστέον περίττωμά τε ποίας
30 τροφῆς, καὶ περὶ καταμηνίων· γίγνεται γάρ τισι
καταμήνια τῶν ζῳοτόκων. διὰ τούτων γὰρ φανερὸν ἔσται καὶ περὶ τοῦ θήλεος, πότερον προΐεται
σπέρμα ὥσπερ τὸ ἄρρεν καὶ ἔστιν ἐν[2] μίγμα τὸ
γινόμενον ἐκ δυοῖν σπερμάτοιν, ἢ οὐθὲν σπέρμα
ἀποκρίνεται ἀπὸ τοῦ θήλεος· καὶ εἰ μηθέν, πότερον
35 οὐδὲ ἄλλο οὐθὲν συμβάλλεται εἰς τὴν γένεσιν ἀλλὰ
726 b μόνον παρέχει τόπον, ἢ συμβάλλεταί τι, καὶ τοῦτο
πῶς καὶ τίνα τρόπον.

Ὅτι μὲν οὖν ἐστὶν ἐσχάτη τροφὴ τὸ αἷμα τοῖς

---

[1] haec non sana ; Aldus habet ἀεὶ νοσώδης, ἡ δὲ τοῦ σπέρματος ἀποχώρησις ὠφέλιμος διὰ τὸ προσλαμβ. κτλ.

[2] ἐν om. ΡΖΣ.

---

[a] See *P.A.* 650 a 34, 651 a 15, 678 a 8 ff. ; it has been
implied throughout the discussion in the preceding chapter
(ch. 18).

produce residue both from the fluid nourishment and from the solid, the semen is discharged by the same exit as the fluid residue, because it is itself a residue from a fluid, the nourishment of all animals tending to be fluid rather than solid ; (*b*) in those animals which produce no fluid residue, the semen leaves by the same way as the solid excrement. Further, colliquescence is always morbid, whereas the removal of residue is beneficial ; and the discharge of semen has both characteristics because it includes some of the useless nourishment. If it were just a colliquescence, it would always be injurious, whereas in fact it is not so.]

To conclude : the foregoing discussion makes it clear that, whether all animals discharge semen or not, semen is a residue derived from useful nourishment, and not only that, but from useful nourishment in its final form.

Our next task is to determine what is the character of the nourishment from which this residue is derived ; and we must discuss the menstrual discharge as well, because this occurs in some of the Vivipara. By this means we shall be able to give a clear answer to the following questions : Does the female discharge semen as the male does, which would mean that the object formed is a single mixture produced from two semens ; or is there no discharge of semen from the female ? And if there is none, then does the female contribute nothing whatever to generation, merely providing a place where generation may happen ; or does it contribute something else, and if so, how and in what manner does it do so ?

We have said before [a] that in blooded animals blood is the final form of the nourishment, and in

XIX

Menstrual discharge.

89

ἐναίμοις, τοῖς δ' ἀναίμοις τὸ ἀνάλογον, εἴρηται
πρότερον· ἐπεὶ δὲ καὶ ἡ γονὴ περίττωμά ἐστι
τροφῆς καὶ τῆς ἐσχάτης, ἤτοι αἷμα ἂν εἴη ἢ τὸ
5 ἀνάλογον ἢ ἐκ τούτων τι. ἐπεὶ δ' ἐκ τοῦ αἵματος
πεττομένου καὶ μεριζομένου πως γίνεται τῶν
μορίων ἕκαστον, τὸ δὲ σπέρμα πεφθὲν μὲν ἀλ-
λοιότερον ἀποκρίνεται τοῦ αἵματος, ἄπεπτον δ'
ὄν, καὶ ὅταν τις προσβιάζηται πλεονάκις χρώμενος
τῷ ἀφροδισιάζειν, ἐνίοις αἱματῶδες ἤδη προελή-
10 λυθεν, φανερὸν ὅτι τῆς αἱματικῆς ἂν εἴη περίττωμα
τροφῆς τὸ σπέρμα, τῆς εἰς τὰ μέρη διαδιδομένης
τελευταίας. καὶ διὰ τοῦτο μεγάλην ἔχει δύναμιν—
καὶ γὰρ ἡ τοῦ καθαροῦ καὶ ὑγιεινοῦ αἵματος ἀποχώρησις ἐκλυτικόν—καὶ τὸ ὅμοια γίγνεσθαι τὰ
ἔκγονα τοῖς γεννήσασιν εὔλογον· ὅμοιον γὰρ τὸ
15 προσελθὸν πρὸς τὰ μέρη τῷ ὑπολειφθέντι. ὥστε
τὸ σπέρμα ἐστὶ τὸ τῆς χειρὸς ἢ τὸ τοῦ προσώπου
ἢ ὅλου τοῦ ζῴου ἀδιορίστως χεὶρ ἢ πρόσωπον ἢ
ὅλον ζῷον· καὶ οἷον ἐκείνων ἕκαστον ἐνεργείᾳ,
τοιοῦτον τὸ σπέρμα δυνάμει, ἢ κατὰ τὸν ὄγκον τὸν
ἑαυτοῦ, ἢ ἔχει[1] τινὰ δύναμιν ἐν ἑαυτῷ (τοῦτο γὰρ
20 οὔπω δῆλον ἡμῖν ἐκ τῶν διωρισμένων, πότερον τὸ

---

[1] ἔχον A.-W.

---

[a] See Introd. § 18.　　[b] Cf. P.A. 678 a 8 ff.
[c] Dynamis : see below, b 19.
[d] And concocted into semen.　Cf. also 725 a 25 ff.
[e] Introd. § 36.
[f] See Introd. §§ 26 ff.　This is an important passage for
the meaning of dynamis in this particular connexion.　Cf.
727 b 14, and ch. 21.

bloodless animals the analogous substance. And since semen also is a residue from nourishment—from nourishment in its final form, surely it follows that semen will be either blood or the analogous substance, or something formed out of these. Now every one of the parts [a] is formed out of the blood as it becomes concocted and in some way divided up into portions ; and though semen which has been concocted is by the time of its secretion from it considerably different in character from blood, yet unconcocted semen, and semen emitted under strain due to excessively frequent intercourse, has been known in some cases to have a bloodlike appearance when discharged ; and this shows that semen is pretty certainly a residue from that nourishment which is in the form of blood and which, as being the final form of nourishment, is distributed to the various parts of the body.[b] This, of course, is the reason why semen has great potency [c] —the loss of it from the system is just as exhausting as the loss of pure healthy blood—and this, too, is why we should expect children to resemble their parents : because there is a resemblance between that which is distributed to the various parts of the body and that which is left over.[d] Thus, the semen of the hand or of the face or of the whole animal really *is* hand or face or a whole animal though in an undifferentiated way ; in other words, what each of those is *in actuality*, such the semen is *potentially*,[e] whether in respect of its own proper bulk, or because it has some *dynamis* [f] within itself (I mention both alternatives because from what we have said so far it is not clear which is the correct one,[g] *i.e.*, whether

[g] This will be settled during the remaining part of the Book ; see especially ch. 21.

σῶμα τοῦ σπέρματός ἐστι τὸ αἴτιον τῆς γενέσεως,
ἢ ἔχει τινὰ ἕξιν καὶ ἀρχὴν κινήσεως γεννητικήν)·
οὐδὲ γὰρ ἡ χεὶρ οὐδ' ἄλλο τῶν μορίων οὐδὲν ἄνευ
ψυχῆς[1] ἢ ἄλλης τινὸς δυνάμεώς ἐστι χεὶρ οὐδὲ
μόριον οὐθέν, ἀλλὰ μόνον ὁμώνυμον.

25 [2][Φανερὸν δὲ καὶ ὅτι ὅσοις σύντηξις γίνεται σπερ-
ματική, καὶ τοῦτο περίττωμά ἐστιν. συμβαίνει
δὲ τοῦτο ὅταν ἀναλύηται εἰς τὸ προελθόν,[3] ὥσπερ
ὅταν ἀποπέσῃ τὸ ἐναλειφθὲν[4] τοῦ κονιάματος εὐθύς·
ταὐτὸν γάρ ἐστι τὸ ἀπελθὸν τῷ πρώτῳ προσ-
τεθέντι. τὸν αὐτὸν τρόπον καὶ τὸ τελευταῖον περίττωμα τῷ πρώτῳ συντήγματι ταὐτόν ἐστιν.[5] καὶ
30 περὶ μὲν τούτων διωρίσθω τὸν τρόπον τοῦτον.]

Ἐπεὶ δ' ἀναγκαῖον καὶ τῷ ἀσθενεστέρῳ γίγνεσθαι περίττωμα πλεῖον καὶ ἧττον πεπεμμένον,
τοιοῦτον δ' ὂν ἀναγκαῖον εἶναι αἱματώδους ὑγρότητος πλῆθος, ἀσθενέστερον δὲ τὸ ἐλάττονος

---

[1] ψυχικῆς PSY (ἄνευ . . . οὐθέν om. Z).
[2] vv. 24-30 secluserunt A.-W., Sus., Platt.
[3] προελθόν Z, προσελθόν vulg. : in primo dissolvitur Σ, qui
et valde diversa hic habet.
[4] ἐναπολειφθὲν YZ : confer τὸ προσελθὸν . . . ὑπολειφθέντι
supra, vv. 14, 15.
[5] ἐστι τὸ ἀπελθὸν . . . ταὐτόν ἐστιν] ἐστι τὸ τελευταῖον
(τελευταῖον om. P) περίττωμα τῷ πρώτῳ περιττώματι PSY.

---

[a] ἕξις. See definition in Met. 1022 b 4 : οἷον ἐνέργειά τις
τοῦ ἔχοντος καὶ ἐχομένου, ὥσπερ πρᾶξίς τις ἢ κίνησις· ὅταν γὰρ
τὸ μὲν ποιῇ τὸ δὲ ποιῆται, ἔστι ποίησις μεταξύ.
[b] See Introd. §§ 41 ff.
[c] Aristotle often repeats this in the Generation of Animals
and the Parts of Animals; see also Met. 1035 b 24. For
ὁμώνυμον, cf. Cat. 1 a 1 ὁμώνυμα λέγεται ὧν ὄνομα μόνον

the physical substance of the semen is the cause of
generation, or whether it contains some disposition [a]
and some principle of movement which effects generation), since neither a hand nor any other part of
the body whatsoever is a hand or any other part of the
body if it lacks Soul [b] or some other *dynamis*; it has
the same name,[c] but that is all.[d]

[e] [It is clear also that in cases where seminal
colliquescence occurs, this too is a residue; and this
happens when ⟨a fresh secretion⟩ is decomposed into
that which preceded it; just as when a ⟨fresh⟩ layer
of plaster spread on a wall immediately drops away,
the reason being that the stuff which comes away is
identical with that which was applied in the first
instance. In just the same way, the final residue is
identical with the original colliquescence. Such then
are the lines on which we treat that subject.]

Now (1) the weaker creature too must of necessity produce a residue, greater in amount and less
thoroughly concocted; and (2) this, if such is its
character, must of necessity be a volume of bloodlike
fluid.[f] (3) That which by nature has a smaller share

κοινόν, ὁ δὲ κατὰ τοὔνομα λόγος τῆς οὐσίας ἕτερος. In this case,
the οὐσία required to be present is Soul (see following note,
and reference to *De anima* given in note on 738 b 26); but
it is absent. For συνώνυμον, see note on 721 a 3.

[d] Because Soul is the essence of any particular body (or of
any part of it). *Cf.* 738 b 26 and note there.

[e] This paragraph seems to be a continuation of the preceding interpolation, 726 a 25. There are variations in the
text. Thus, the mss. PSY replace "stuff which . . .
colliquescence" by "final residue is the same as the first
residue." Some of the words seem to echo lines 14 and 15
above.

[f] Semen of course has undergone a further stage of concoction, and has lost its bloodlike appearance.

θερμότητος κοινωνοῦν κατὰ φύσιν, τὸ δὲ θῆλυ ὅτι
35 τοιοῦτον εἴρηται πρότερον, ἀναγκαῖον καὶ τὴν ἐν
τῷ θήλει γινομένην αἱματώδη ἀπόκρισιν περίτ-

τωμα εἶναι. γίνεται δὲ τοιαύτη ἡ τῶν καλουμένων
καταμηνίων ἔκκρισις.

Ὅτι μὲν οὖν ἐστὶ τὰ καταμήνια περίττωμα, καὶ
ὅτι ἀνάλογον ὡς τοῖς ἄρρεσιν ἡ γονὴ οὕτω τοῖς
θήλεσι τὰ καταμήνια, φανερόν. ὅτι δ' ὀρθῶς
5 εἴρηται, σημεῖα τὰ συμβαίνοντα περὶ αὐτά. κατὰ
γὰρ τὴν αὐτὴν ἡλικίαν τοῖς μὲν ἄρρεσιν ἄρχεται
ἐγγίνεσθαι γονὴ καὶ ἀποκρίνεται, τοῖς δὲ θήλεσι
ῥήγνυται τὰ καταμήνια καὶ φωνήν τε μεταβάλλουσι
καὶ ἐπισημαίνει τὰ περὶ τοὺς μαστούς. καὶ παύ-
10 εται τῆς ἡλικίας ληγούσης τοῖς μὲν τὸ δύνασθαι
γεννᾶν, ταῖς δὲ τὰ καταμήνια. ἔτι δὲ καὶ τὰ
τοιάδε σημεῖα ὅτι περίττωμά ἐστιν αὕτη ἡ ἔκκρισις τοῖς θήλεσιν. ὡς γὰρ ἐπὶ τὸ πολὺ οὔθ'
αἱμορροΐδες γίνονται ταῖς γυναιξὶν οὔτ' ἐκ τῶν
ῥινῶν ῥύσις αἵματος οὔτε τι ἄλλο μὴ τῶν καταμηνίων ἱσταμένων· ἐάν τε συμβῇ τι τούτων, χεί-
15 ρους γίγνονται αἱ καθάρσεις ὡς μεθισταμένης[1] εἰς
ταῦτα τῆς ἀποκρίσεως. ἔτι δὲ οὔτε φλεβώδη[2]
ὁμοίως γλαφυρώτερά[3] τε καὶ λειότερα τὰ θήλεα
τῶν ἀρρένων ἐστὶ διὰ τὸ συνεκκρίνεσθαι τὴν εἰς
ταῦτα περίττωσιν ἐν τοῖς καταμηνίοις. τὸ δ' αὐτὸ
τοῦτο δεῖ νομίζειν αἴτιον εἶναι καὶ τοῦ τοὺς ὄγκους
20 ἐλάττους εἶναι τῶν σωμάτων τοῖς θήλεσιν ἢ τοῖς
ἄρρεσιν ἐν τοῖς ζῳοτοκοῦσιν· ἐν τούτοις γὰρ ἡ

---

[1] μεθισταμένης PZ : ἀναλισκομένης vulg.
[2] φλεβώδη Peck, φλεβώδεις vulg.   [3] ἀτριχώτερά Z.

of heat is weaker ; and (4) the female answers to this description, as we have said already. From which we conclude that the bloodlike secretion which occurs in the female must of necessity be a residue just as much ⟨as the secretion in the male⟩. Of such a character is the discharge of what is called the menstrual fluid.

Thus much then is evident : the menstrual fluid is a residue, and it is the analogous thing in females to the semen in males. Its behaviour shows that this statement is correct. At the same time of life that semen begins to appear in males and is emitted, the menstrual discharge begins to flow in females, their voice changes and their breasts begin to become conspicuous ; and similarly, in the decline of life the power to generate ceases in males and the menstrual discharge ceases in females. Here are still further indications that this secretion which females produce is a residue. Speaking generally, unless the menstrual discharge is suspended, women are not troubled by haemorrhoids or bleeding from the nose or any other such discharge, and if it happens that they are, then the evacuations fall off in quantity, which suggests that the substance secreted is being drawn off to the other discharges. Again, their blood-vessels are not so prominent as those of males ; and females are more neatly made [a] and smoother than males, because the residue which goes to produce those characteristics in males is in females discharged together with the menstrual fluid. We are bound to hold, in addition, that for the same cause the bulk of the body in female Vivipara is smaller than that of the males, as of course it is only in Vivipara that the

[a] Also implying "hairless," "delicate," "dainty."

τῶν καταμηνίων γίνεται ῥύσις θύραζε μόνοις, καὶ
τούτων ἐπιδηλότατα ἐν ταῖς γυναιξίν· πλείστην
γὰρ ἀφίησιν ἀπόκρισιν γυνὴ τῶν ζῴων. διόπερ
ἐπιδηλοτάτως ἀεὶ ὠχρόν τέ ἐστι καὶ ἀδηλόφλεβον,
25 καὶ τὴν ἔλλειψιν πρὸς τοὺς ἄρρενας ἔχει τοῦ σώματος φανεράν.

Ἐπεὶ δὲ τοῦτ' ἐστὶν ὃ γίγνεται τοῖς θήλεσιν ὡς
ἡ γονὴ τοῖς ἄρρεσιν, δύο δ' οὐκ ἐνδέχεται σπερ-
ματικὰς ἅμα γίνεσθαι ἀποκρίσεις, φανερὸν ὅτι τὸ
θῆλυ οὐ συμβάλλεται σπέρμα εἰς τὴν γένεσιν. εἰ
μὲν γὰρ σπέρμα ἦν, καταμήνια[1] οὐκ ἂν ἦν· νῦν
30 δὲ διὰ τὸ ταῦτα γίγνεσθαι ἐκεῖνο οὐκ ἔστιν.

Διότι μὲν οὖν, ὥσπερ τὸ σπέρμα, καὶ τὰ καταμήνια περίττωμά ἐστιν, εἴρηται· λάβοι δ' ἄν τις
εἰς τοῦτο μαρτύρια ἔνια τῶν συμβαινόντων τοῖς[2]
ζῴοις. τά τε γὰρ πίονα ἧττόν ἐστι σπερματικὰ
τῶν ἀπιμέλων, ὥσπερ εἴρηται πρότερον. αἴτιον
35 δ' ὅτι καὶ ἡ πιμελὴ περίττωμά ἐστι καθάπερ τὸ
σπέρμα, καὶ πεπεμμένον αἷμα, ἀλλ' οὐ τὸν αὐτὸν
τρόπον τῷ σπέρματι. ὥστ' εὐλόγως εἰς τὴν πι-
727 b μελὴν ἀνηλωμένης τῆς περιττώσεως ἐλλείπει τὰ
περὶ τὴν γονήν, οἷον τῶν τε ἀναίμων τὰ μαλάκια
καὶ τὰ μαλακόστρακα περὶ τὴν κύησίν ἐστιν ἄρι-
στα. διὰ τὸ ἄναιμα γὰρ εἶναι καὶ μὴ γίνεσθαι
5 πιμελὴν ἐν αὐτοῖς, τὸ ἀνάλογον αὐτοῖς τῇ πιμελῇ
ἀποκρίνεται εἰς τὸ περίττωμα τὸ σπερματικόν.[3]
σημεῖον δ' ὅτι οὐ τοιοῦτο σπέρμα προΐεται τὸ
θῆλυ οἷον τὸ ἄρρεν, οὐδὲ μιγνυμένων ἀμφοῖν γίνεται, ὥσπερ τινές φασιν, ὅτι πολλάκις τὸ θῆλυ
συλλαμβάνει οὐ γενομένης αὐτῇ τῆς ἐν τῇ ὁμιλίᾳ

---

[1] καταμήνια P : τὰ καταμήνια vulg.
[2] τοῖς ἄλλοις PZ.   [3] 727 a 31–b 6 secl. Sus.

menstrual discharge flows externally, and most conspicuously of all in women, who discharge a greater amount than any other female animals. On this account it is always very noticeable that the female is pale, and the blood-vessels are not prominent, and there is an obvious deficiency in physique as compared with males.

Now it is impossible that any creature should produce two seminal secretions at once, and as the secretion in females which answers to semen in males is the menstrual fluid, it obviously follows that the female does not contribute any semen to generation ; for if there were semen, there would be no menstrual fluid ; but as menstrual fluid is in fact formed, therefore there is no semen.

We have said why it is that the menstrual fluid as well as semen is a residue. In support of this, there are a number of facts concerning animals which may be adduced. (1) Fat animals produce less semen than lean ones, as we said before, and the reason is that fat is a residue just as semen is, *i.e.*, it is blood that has been concocted, only not in the same way as semen. Hence it is not surprising that when the residue has been consumed to make fat the semen is deficient. Take a parallel from the bloodless animals : Cephalopods and Crustacea are in their finest condition at the breeding season. Why ? Because, being bloodless, they produce no fat ; hence, what in them corresponds to fat is at this period secreted into the seminal residue. (2) Here is an indication that the female does not discharge semen of the same kind as the male, and that the offspring is not formed from a mixture of two semens, as some allege. Very often the female conceives although she has derived

ἡδονῆς· καὶ γιγνομένης πάλιν οὐδὲν ἧττον, καὶ
10 ἰσοδρομησάντων [παρὰ]¹ τοῦ ἄρρενος καὶ τοῦ θήλεος,
οὐ γεννᾷ,² ἐὰν μὴ ἡ τῶν καλουμένων καταμηνίων
ἰκμὰς ὑπάρχῃ σύμμετρος. διὸ οὔτε ὅλως μὴ
γιγνομένων αὐτῶν γεννᾷ τὸ θῆλυ, οὔτε γιγνομένων
ὅταν ἐξικμάζῃ ὡς ἐπὶ τὸ πολύ, ἀλλὰ μετὰ τὴν
κάθαρσιν. ὁτὲ μὲν γὰρ οὐκ ἔχει τροφὴν οὐδ’
15 ὕλην ἐξ ἧς δυνήσεται συστῆσαι τὸ ζῷον ἡ ἀπὸ
τοῦ ἄρρενος ἐνυπάρχουσα ἐν τῇ γονῇ δύναμις, ὁτὲ
δὲ συνεκκλύζεται διὰ τὸ πλῆθος. ὅταν δὲ γενομένων ἀπέλθῃ, τὸ ὑπολειφθὲν συνίσταται. ὅσαι
δὲ μὴ γιγνομένων τῶν καταμηνίων συλλαμβάνουσιν, ἢ μεταξὺ γιγνομένων ὕστερον δὲ μή, αἴτιον
20 ὅτι ταῖς μὲν τοσαύτη γίνεται ἰκμὰς ὅση μετὰ τὴν
κάθαρσιν ὑπολείπεται ταῖς γονίμοις, πλείων δ’ οὐ
γίγνεται περίττωσις ὥστε καὶ θύραζε ἀπελθεῖν,
ταῖς δὲ μετὰ τὴν κάθαρσιν συμμύει τὸ στόμα τῶν
ὑστερῶν. ὅταν οὖν πολὺ μὲν τὸ ἀπεληλυθὸς ᾖ,
ἔτι δὲ γίγνηται μὲν κάθαρσις, μὴ τοσαύτη δὲ ὥστε
25 συνεξικμάζειν³ τὸ σπέρμα, τότε πλησιάζουσαι συλλαμβάνουσι πάλιν.⁴ οὐδὲν δὲ ἄτοπον τὸ συνειλη-
φυίαις ἔτι γίγνεσθαι· καὶ γὰρ ὕστερον μέχρι τινὸς
φοιτᾷ τὰ καταμήνια, ὀλίγα δὲ καὶ οὐ διὰ παντός.

---

¹ seclusit Platt : τὸ παρά Z.
² γεννᾷ A.-W. : γίγνεται vulg. : γίγνεται ⟨σύλληψις⟩ Btf.
³ συνεξικμάζειν Z : ἐξικμάζειν vulg.
⁴ πάλιν om. PS.

ᵃ See above, 726 b 19.
ᵇ This really means ordinary individuals in which the
menstrual discharge takes place.

no pleasure from the act of coitus ; and, on the contrary side, when the female derives as much pleasure as the male, and they both keep the same pace, the female does not bear—unless there is a proper amount of menstrual liquid (as it is called) present. Thus, the female does not bear (a) if the menstrual fluid is completely absent, (b) if it is present and the discharge of moisture is in progress (in most instances) ; but only (c) after the evacuation is over. The reason is that in one case (a) the female has no nourishment, no material, for the *dynamis* [a] supplied by the male in the semen to draw upon and so to cause the living creature to take shape from it ; in the other case (b) it is washed right away owing to the volume of the menstrual fluid. When, however, (c) the discharge is over and most of it has passed off, then what remains begins to take shape as a fetus. There are instances of women who conceive without the occurrence of menstrual discharge ; others conceive during its occurrence but not after it. The reasons are these. The former produce only just so much liquid as remains in fertile individuals [b] after the evacuation is over, and there is no surplus residue to be discharged externally ; in the latter, the mouth of the uterus closes up after the evacuation is over. Therefore, when there has been a plentiful discharge and yet the evacuation still continues, though not so copiously that the discharge of moisture carries the semen away with it, that is the time when if they have intercourse women can conceive again. There is nothing odd about the menstrual fluid's continuing to flow after conception has taken place ; indeed it actually recurs afterwards up to a point, but it is scanty and does not last throughout gestation. How-

ARISTOTLE

ἀλλὰ τοῦτο μὲν νοσηματῶδες, διόπερ ὀλίγαις καὶ
ὀλιγάκις συμβαίνει· τὰ δ' ὡς ἐπὶ τὸ πολὺ γινόμενα
30 μάλιστα κατὰ φύσιν ἐστίν.

Ὅτι μὲν οὖν συμβάλλεται τὸ θῆλυ εἰς τὴν γένεσιν τὴν ὕλην, τοῦτο δ' ἐστὶν ἐν τῇ τῶν καταμηνίων συστάσει, τὰ δὲ καταμήνια περίττωμα,
δῆλον.

XX Ὁ δ' οἴονταί τινες σπέρμα συμβάλλεσθαι ἐν
35 τῇ συνουσίᾳ τὸ θῆλυ διὰ τὸ γίνεσθαι παραπλησίαν τε χαρὰν ἐνίοτε αὐταῖς τῇ τῶν ἀρρένων καὶ
ἅμα ὑγρὰν ἀπόκρισιν, οὐκ ἔστιν ἡ ὑγρασία αὕτη
σπερματικὴ ἀλλὰ τοῦ τόπου ἴδιος ἑκάσταις. ἔστι
γὰρ τῶν ὑστερῶν ἔκκρισις, καὶ ταῖς μὲν γίγνεται
ταῖς δ' οὔ. γίγνεται μὲν γὰρ¹ ταῖς λευκοχρόοις καὶ
θηλυκαῖς ὡς ἐπὶ τὸ πολὺ εἰπεῖν, οὐ γίνεται δὲ ταῖς
μελαίναις καὶ ἀρρενωποῖς. τὸ δὲ πλῆθος, αἷς γί-
5 γνεται, ἐνίοτε οὐ κατὰ σπέρματος πρόεσίν ἐστιν,
ἀλλὰ πολὺ ὑπερβάλλει. ἔτι δὲ καὶ ἐδέσματα ἕτερα

---

¹ γὰρ SZ: om. vulg.

ᵃ Aristotle's notion that the menstrual blood is the substance from which the embryo is formed reigned un-
questioned for many centuries. (It appears in the Wisdom
of Solomon, vii. 2, "In the womb of a mother was I moulded
into flesh in the time of ten months, being compacted in
blood of the seed of man and the pleasure that came with
sleep.") It can be seen pictured in 16th century obstetrical
books such as the *De conceptu et generatione hominis* of
Jacob Rueff (1554). Its falsity was decisively demonstrated
by William Harvey, who in his *Exercitationes de generatione
animalium* (1651) describes his dissections of the uteri of
does in King Charles the First's forests, at different stages
after coitus. The expected mass of blood and seed was
never found; a source of great perplexity to Harvey himself,
since the mammalian egg was not discovered until long after

ever, this is a morbid condition, and that is why it only occurs infrequently and in few subjects. It is what occurs generally that is most in accord with the course of Nature.

By now it is plain that the contribution which the female makes to generation is the *matter* used therein, that this is to be found in the substance constituting the menstrual fluid,[a] and finally that the menstrual fluid is a residue.

There are some who think that the female con- XX tributes semen[b] during coition because women sometimes derive pleasure from it comparable to that of the male and also produce a fluid secretion. This fluid, however, is not seminal; it is peculiar to the part from which it comes in each several individual; there is a discharge from the uterus, which though it happens in some women does not in others.[c] Speaking generally, this happens in fair-skinned women[d] who are typically feminine, and not in dark women of a masculine appearance. Where it occurs, this discharge is sometimes on quite a different scale from the semen discharged by the male, and greatly exceeds it in bulk. Furthermore, differences of food

his death. We know now that the menstrual bleeding is a phase in the sexual cycle, this phase being usually succeeded by the periodical liberation of the egg from the ovary, and by its attachment (if fertilized) to the wall of the uterus.
  [b] The view that the female also contributed semen was apparently adopted by the Epicureans; see Lucretius iv. 1229 *semper enim partus duplici de semine constat*; cf. 1247, 1257-1258.
  [c] This apparently refers to the so-called vaginal discharge, which is a natural secretion (*cf.* 739 a 37), but the latter part of the paragraph seems to describe leucorrhoea, which is pathological. The two have apparently been confused.
  [d] *Cf. H.A.* 583 a 11.

ἑτέρων ποιεῖ πολλὴν διαφορὰν τοῦ γίγνεσθαι τὴν
ἔκκρισιν ἢ ἐλάττω ἢ πλείω τὴν τοιαύτην, οἷον
ἔνια τῶν δριμέων ἐπίδηλον ποιεῖ εἰς πλῆθος τὴν
ἀπόκρισιν.

10 Τὸ δὲ συμβαίνειν ἡδονὴν ἐν τῇ συνουσίᾳ οὐ μόνον
τοῦ σπέρματος προϊεμένου ἐστίν, ἀλλὰ καὶ πνεύματος, ἐξ οὗ συνισταμένου ἀποσπερματίζει. δῆλον
δ᾽ ἐπὶ τῶν παίδων τῶν μήπω δυναμένων προΐεσθαι,
ἐγγὺς δὲ τῆς ἡλικίας ὄντων, καὶ τῶν ἀγόνων
ἀνδρῶν· γίνεται γὰρ πᾶσι τούτοις ἡδονὴ ξυομένοις.
15 καὶ τοῖς γε διεφθαρμένοις τὴν γένεσιν ἔστιν ὅτε
ἀναλύονται αἱ κοιλίαι διὰ τὸ ἀποκρίνεσθαι περίττωμα εἰς τὴν κοιλίαν οὐ δυνάμενον πεφθῆναι καὶ
γενέσθαι σπέρμα.

Ἔοικε δὲ καὶ τὴν μορφὴν γυναικὶ[1] παῖς, καὶ
ἔστιν ἡ γυνὴ ὥσπερ ἄρρεν ἄγονον· ἀδυναμίᾳ γάρ
τινι τὸ θῆλύ ἐστι, τῷ μὴ δύνασθαι πέττειν ἐκ τῆς
20 τροφῆς σπέρμα τῆς ὑστάτης (τοῦτο δ᾽ ἐστὶν ἢ
αἷμα ἢ τὸ ἀνάλογον ἐν τοῖς ἀναίμοις) διὰ ψυχρότητα
τῆς φύσεως. ὥσπερ οὖν ἐν ταῖς κοιλίαις διὰ τὴν
ἀπεψίαν γίνεται διάρροια, οὕτως ἐν ταῖς φλεψὶν αἵ
τ᾽ ἄλλαι αἱμορροΐδες καὶ ἡ τῶν καταμηνίων ῥύσις[2]·
καὶ γὰρ αὕτη αἱμορροΐς ἐστιν, ἀλλ᾽ ἐκεῖναι μὲν διὰ
25 νόσον, αὕτη δὲ φυσική.

Ὥστε φανερὸν ὅτι εὐλόγως γίνεται ἐκ τούτου ἡ
γένεσις. ἔστι γὰρ τὰ καταμήνια σπέρμα οὐ καθαρὸν ἀλλὰ δεόμενον ἐργασίας, ὥσπερ ἐν τῇ περὶ

---

[1] γυναικὶ Z : γυνὴ καὶ vulg.
[2] ἡ τῶν κ. ῥύσις Y : αἱ τῶν κ. vulg.

cause a great difference in the amount of this discharge which is produced : *e.g.*, some pungent foods cause a noticeable increase in the amount.

The pleasure which accompanies copulation is due to the fact that not only semen but also *pneuma* [a] is emitted : it is from this *pneuma* as it collects together that the emission of the semen results. This is plain in the case of boys who cannot yet emit semen, though they are not far from the age for it, and in infertile men, because all of them derive pleasure from attrition. Indeed, men whose generative organs have been destroyed sometimes suffer from looseness of the bowels caused by residue which cannot be concocted and converted into semen being secreted into the intestine.

Further, a boy actually resembles a woman in physique, and a woman is as it were an infertile male ; the female, in fact, is female on account of inability [b] of a sort, viz., it lacks the power to concoct semen out of the final state of the nourishment (this is either blood, or its counterpart in bloodless animals) because of the coldness of its nature. Thus, just as lack of concoction produces in the bowels diarrhoea, so in the blood-vessels it produces discharges of blood of various sorts, and especially the menstrual discharge (which has to be classed as a discharge of blood, though it is a natural discharge, and the rest are morbid ones).

Hence, plainly, it is reasonable to hold that generation takes place from this process ; for, as we see, the menstrual fluid is semen, not indeed semen in a pure condition, but needing still to be acted upon. It

[a] See 718 a 4, 738 a 1, etc.
[b] *Cf.* 765 b 9.

728 a

τοὺς καρποὺς γενέσει, ὅταν ᾖ μήπω διηθημένη,[1]
ἔνεστι μὲν ἡ τροφή, δεῖται δ' ἐργασίας πρὸς τὴν
30 κάθαρσιν. διὸ καὶ μιγνυμένη ἐκείνη μὲν τῇ γονῇ,
αὕτη δὲ καθαρᾷ τροφῇ, ἡ μὲν γεννᾷ, ἡ δὲ τρέφει.

Σημεῖον δὲ τοῦ τὸ θῆλυ μὴ προΐεσθαι σπέρμα
καὶ τὸ γίνεσθαι ἐν τῇ ὁμιλίᾳ τὴν ἡδονὴν τῇ ἁφῇ
κατὰ τὸν αὐτὸν τόπον τοῖς ἄρρεσιν· καίτοι οὐ
προΐενται τὴν ἰκμάδα ταύτην ἐντεῦθεν. ἔτι δ' οὐ
35 πᾶσι γίνεται τοῖς θήλεσιν αὕτη ἡ ἔκκρισις, ἀλλὰ
τοῖς αἱματικοῖς, καὶ οὐδὲ τούτοις πᾶσιν, ἀλλ' ὅσων
αἱ ὑστέραι μὴ πρὸς τῷ ὑποζώματί εἰσι μηδ' ᾠο-
728 b τοκοῦσιν, ἔτι δ' οὐδὲ τοῖς αἷμα μὴ ἔχουσιν ἀλλὰ
τὸ ἀνάλογον· ὅπερ γὰρ ἐν ἐκείνοις[2] τὸ αἷμα, ἐν τούτοις ἑτέρα ὑπάρχει σύγκρισις. τοῦ δὲ μήτε τούτοις
γίγνεσθαι κάθαρσιν μήτε τῶν αἷμα ἐχόντων τοῖς
εἰρημένοις, τοῖς κάτω ἔχουσι καὶ μὴ ᾠοτοκοῦσιν,[3]
5 αἰτία ἡ ξηρότης τῶν σωμάτων, ὀλίγον λείπουσα
τὸ περίττωμα, καὶ τοσοῦτον ὅσον εἰς τὴν γένεσιν
ἱκανὸν μόνον, ἔξω δὲ μὴ προΐεσθαι. ὅσα δὲ ζῳοτόκα ἄνευ ᾠοτοκίας (ταῦτα δ' ἐστὶν ἄνθρωπος καὶ
τῶν τετραπόδων ὅσα κάμπτει τὰ ὀπίσθια σκέλη ἐν-
10 τός[4]· ταῦτα μὲν γὰρ πάντα ζῳοτοκεῖ ἄνευ ᾠοτοκίας)
τούτοις δὲ γίγνεται μὲν πᾶσιν, πλὴν εἴ τι πεπήρω-

---

[1] διηθημένη Bonitz: διηττημένη Z, A.-W.: διητημένη vulg.:
cibus . . . incompletus Σ; cf. indicem Aristot.
[2] ἐν ἐκείνοις Platt: ἐνίοις vulg.
[3] τοῖς κάτω . . . ᾠοτοκοῦσιν om. Z.
[4] ἐκτός Z[1], Platt: τὰ ἐκτός Y.

---

[a] Cf. Pol. 1281 b 27. ἡ μὴ καθαρὰ τροφὴ μετὰ τῆς καθαρᾶς
τὴν πᾶσαν ποιεῖ χρησιμωτέραν τῆς ὀλίγης.—For the two sorts of
τροφή, see 744 b 32 ff. Cf. 725 a 17.
[b] Cf. 739 b 15.
[c] i.e., the extremity of the bent limb is moved towards the

is the same with fruit when it is forming. The nourishment is present right enough, even before it has been strained off, but it stands in need of being acted upon in order to purify it. That is why when the former is mixed with the semen, and when the latter is mixed with pure nourishment,[a] the one effects generation, and the other effects nutrition.

An indication that the female emits no semen is actually afforded by the fact that in intercourse the pleasure is produced in the same place as in the male by contact, yet this is not the place from which the liquid is emitted.[b] Further, this discharge does not occur in all females, but only in those which are blooded, and not in all of them, but only in those whose uterus is not close by the diaphragm and which are not oviparous ; nor again in those which have an analogous substance instead of blood (they have another composition which is for them what blood is for the others). Dryness of the body is the cause why neither these animals nor the blooded ones I mentioned (viz., those whose uterus is low down and which are not oviparous) produce this evacuation ; their dryness leaves over but little residue, only enough in fact for generation, not enough to be emitted externally. Take next the animals which are viviparous but not previously oviparous : this means man, and those quadrupeds which bend their hind legs inwards.[c] The menstrual discharge occurs in all of these ; though if they are deformed [d] in any respect

main bulk of the body and not away from it, so that the angle of the bent joint points away from the body. " Inwards " thus has no reference to " knock knees." See *I.A.* 704 a 19 ff., 711 a 8 ff. ; *H.A.* 498 a 3 ff. ; and my diagram in *Parts of Animals* (Loeb), p. 433.

[d] See Introd. § 12.

ται ἐν τῇ γενέσει, οἷον ὀρεύς, οὐ μὴν ἐπιπολάζουσί
γε αἱ καθάρσεις ὥσπερ ἀνθρώποις. δι' ἀκριβείας
δέ, πῶς συμβαίνει ταῦτα περὶ ἕκαστον τῶν ζώων,
γέγραπται ἐν ταῖς περὶ τὰ ζῷα ἱστορίαις. πλείστη
15 δὲ γίνεται κάθαρσις τῶν ζώων ταῖς γυναιξί, καὶ
τοῖς ἄρρεσι πλείστη τοῦ σπέρματος πρόεσις κατὰ
λόγον τοῦ μεγέθους. αἴτιον δ' ἡ τοῦ σώματος
σύστασις ὑγρὰ καὶ θερμὴ οὖσα· ἀναγκαῖον γὰρ ἐν
τῷ τοιούτῳ γίνεσθαι πλείστην περίττωσιν. ἔτι δὲ
οὐδὲ τὰ τοιαῦτ' ἔχει ἐν τῷ σώματι μέρη εἰς ἃ
20 τρέπεται ἡ περίττωσις, ὥσπερ ἐν τοῖς ἄλλοις· οὐ
γὰρ ἔχει οὔτε τριχῶν πλῆθος κατὰ τὸ σῶμα, οὔτε
ὀστῶν καὶ κεράτων καὶ ὀδόντων ἐκκρίσεις.

Σημεῖον δ' ὅτι ἐν τοῖς καταμηνίοις τὸ σπέρμα
ἐστίν· ἅμα γάρ, ὥσπερ εἴρηται πρότερον, τοῖς ἄρρεσι γίνεται τὸ περίττωμα τοῦτο καὶ τοῖς θήλεσι
25 τὰ καταμήνια ἐπισημαίνει ἐν τῇ αὐτῇ ἡλικίᾳ, ὡς
καὶ ἅμα διισταμένων τῶν τόπων τῶν δεκτικῶν
ἑκατέρου τοῦ περιττώματος· καὶ ἀραιουμένων
ἑκατέρων τῶν πλησίον τόπων ἐξανθεῖ ἡ τῆς ἥβης
τρίχωσις. μελλόντων δὲ διίστασθαι οἱ τόποι ἀν-
οιδοῦσιν ὑπὸ τοῦ πνεύματος, τοῖς μὲν ἄρρεσιν ἐπι-
δηλότερον περὶ τοὺς ὄρχεις, ἐπισημαίνει δὲ καὶ
30 περὶ τοὺς μαστούς, τοῖς δὲ θήλεσι περὶ τοὺς μαστοὺς μᾶλλον· ὅταν γὰρ δύο δακτύλους ἀρθῶσι,
τότε γίνεται τὰ καταμήνια ταῖς πλείσταις.[1]

Ἐν ὅσοις μὲν οὖν τῶν ζωὴν ἐχόντων μὴ κεχώρισται τὸ θῆλυ καὶ τὸ ἄρρεν, τούτοις μὲν τὸ

---

[1] vv. 22-32 secl. Sus.

---

[a] See Book II, ch. 8.   [b] H.A. 572 b 29 ff.

in their formation, as, *e.g.*, the mule,[a] the evacuation is not as obvious as it is in human beings. An exact account of this matter, as it concerns every sort of animal, is to be found in the *Researches upon Animals.*[b] A larger amount of evacuation is produced by women than by any other animal, and a larger amount of semen in proportion to their size is emitted by men ; the reason being that the composition of the human body is fluid and warm, and that is just the sort of organism which of necessity produces the greatest amount of residue ; further, the human body does not possess the sort of parts to which the residue gets diverted, as other animals do : it has no great coat of hair all over the body,[c] and no secretions in the form of bones, horns and tusks.

Here is an indication that the semen resides in the menstrual discharge. As I said before, this residue is formed in males at the same time of life as the menstrual discharge becomes noticeable in females ; which suggests that the places which are the receptacles of these residues also become differentiated at the same time in each sex ; and as the neighbouring places in each sex become less firm in their consistency, the pubic hair grows up too. Just before these places receive their differentiation, they are swelled up by the *pneuma* : in males, this is clearer in regard to the testes, but it is also to be noticed in the breasts ; whereas in females it is clearer in the breasts : it is when the breasts have risen a couple of fingers' breadth that the menstrual discharge begins in most women.

Now in those living creatures where male and female are not separate, the semen (seed) is as it

[c] Or, in proportion to the size of the body.

σπέρμα οἷον κύημά ἐστιν. λέγω δὲ κύημα τὸ
35 πρῶτον μίγμα[1] θήλεος καὶ ἄρρενος. διὸ καὶ ἐξ
ἑνὸς σπέρματος ἓν σῶμα γίνεται, οἷον ἐξ[2] ἑνὸς
πυροῦ εἷς πυθμήν, ὥσπερ ἐξ ἑνὸς ᾠοῦ ἓν ζῷον (τὰ
729 a γὰρ δίδυμα τῶν ᾠῶν δύο ᾠά ἐστιν). ἐν ὅσοις δὲ
τῶν γενῶν διώρισται τὸ θῆλυ καὶ τὸ ἄρρεν, ἐν[3] τούτοις ἀφ᾽ ἑνὸς σπέρματος ἐνδέχεται πολλὰ γίνεσθαι
ζῷα, ὡς διαφέροντος τῇ φύσει τοῦ σπέρματος ἐν
τοῖς φυτοῖς τε καὶ ζῴοις. σημεῖον δέ, ἀπὸ μιᾶς
5 γὰρ ὀχείας πλείω γίνεται ἐν τοῖς πλείω δυναμένοις
γεννᾶν ἑνός. ᾗ καὶ δῆλον ὅτι οὐκ ἀπὸ παντὸς
ἔρχεται ἡ γονή· οὔτε γὰρ ἂν κεχωρισμένα ἀπὸ τοῦ
αὐτοῦ μέρους εὐθὺς ἀπεκρίνετο, οὔτε ἅμα ἐλθόντα
εἰς τὰς ὑστέρας ἐκεῖ διεχωρίζετο· ἀλλὰ συμβαίνει
10 ὥσπερ εὔλογον, ἐπειδὴ τὸ μὲν ἄρρεν παρέχεται τό
τε εἶδος καὶ τὴν ἀρχὴν τῆς κινήσεως, τὸ δὲ θῆλυ
τὸ σῶμα καὶ τὴν ὕλην, οἷον ἐν τῇ τοῦ γάλακτος
πήξει τὸ μὲν σῶμα τὸ γάλα ἐστίν, ὁ δὲ ὀπὸς ἢ ἡ[4]
πυτία τὸ τὴν ἀρχὴν ἔχον τὴν συνιστᾶσαν, οὕτω
τὸ ἀπὸ τοῦ ἄρρενος ἐν τῷ θήλει μεριζόμενον. δι᾽
15 ἣν δ᾽ αἰτίαν μερίζεται ἔνθα μὲν εἰς πλείω ἔνθα δ᾽
εἰς ἐλάττω ἔνθα δὲ μοναχῶς, ἕτερος ἔσται λόγος.
ἀλλὰ διὰ τὸ μηθέν γε διαφέρειν τῷ εἴδει, ἀλλ᾽ ἐὰν

---

[1] ἐκ add. PSY, om. vulg., Z.　　[2] ἐξ Y*Z, om. vulg.
[3] ἐν Z : ἐν δὲ vulg. : ἐν δὴ Y.
[4] ἢ ἡ Z[2], Platt, Btf. : ἢ Z[1] : ἡ vulg.

---

[a] See Introd. §§ 56 ff.　　[b] Cf. 723 b 10, 728 a 27.
[c] Cf. above, 722 b 28, 723 b 14.
[d] The "Formal" Cause, and the "Motive" (or
"Efficient") Cause, i.e., sentient Soul.
[e] The "Material" Cause. See Introd. §§ 1 ff. With this
passage cf. Met. 1044 a 34 ἀνθρώπου τίς αἰτία ὡς ὕλη; ἆρα τὰ
καταμήνια; τί δ᾽ ὡς κινοῦν; ἆρα τὸ σπέρμα;

were a fetation.[a] (By fetation I mean the primary mixture of male and female.) This explains incidentally why one body only is formed from one seed—*e.g.*, one stalk from one grain of corn, just like one animal from one egg (double-yolked eggs of course count as two eggs). In those groups, however, where male and female are distinct, many animals may be formed from one semen, which suggests that the nature of semen in animals differs from that in plants.[b] We have as a proof of this those animals which are able to produce more offspring than one at a time, where more than one are formed as the result of one act of coitus. This shows also that the semen is not drawn from the whole body; because we cannot suppose (*a*) that at the moment of discharge it contains a number of *separate* portions from one and the same part of the body; nor (*b*) that these portions all enter the uterus *together* and separate themselves out when they have got there.[c] No; what happens is what one would expect to happen. The male provides the " form " and the " principle of the movement," [d] the female provides the body, in other words, the material.[e] Compare the coagulation of milk. Here, the milk is the body, and the fig-juice or the rennet contains the principle which causes it to set.[f] The semen of the male acts in the same way as it gets divided up into portions within the female. (Another part of the treatise [g] will explain the *Cause* why in some cases it gets divided into many portions, in others into few, while in others it is not divided up at all.) But as this semen which gets divided up exhibits no difference in kind, all that

[f] *Cf.* 739 b 23.
[g] 771 b 14 ff.

109

μόνον σύμμετρον ἦ τὸ διαιρούμενον πρὸς τὴν ὕλην,
καὶ μήτε ἔλαττον ὥστε μὴ πέττειν μηδὲ συνιστάναι, μήτε πλεῖον ὥστε ξηρᾶναι, πλείω οὕτω
20 γεννᾶται. ἐκ δὲ τοῦ συνιστάντος πρώτου, ἐξ ἑνὸς
ἤδη ἓν γίνεται μόνον.

Ὅτι μὲν οὖν τὸ θῆλυ εἰς τὴν γένεσιν γονὴν μὲν
οὐ συμβάλλεται, συμβάλλεται δέ τι, καὶ τοῦτ'
ἐστὶν ἡ τῶν καταμηνίων σύστασις καὶ τὸ ἀνάλογον
ἐν τοῖς ἀναίμοις, ἔκ τε τῶν εἰρημένων δῆλον καὶ
κατὰ τὸν λόγον καθόλου σκοπουμένοις. ἀνάγκη
25 γὰρ εἶναι τὸ γεννῶν καὶ ⟨τὸ⟩[1] ἐξ οὗ, καὶ ταῦτ'[2] ἂν
καὶ ἓν ἦ, τῷ γε εἴδει διαφέρειν καὶ τῷ τὸν λόγον
αὐτῶν εἶναι ἕτερον, ἐν δὲ τοῖς κεχωρισμένας ἔχουσι
τὰς δυνάμεις καὶ τὰ σώματα καὶ τὴν φύσιν ἑτέραν
εἶναι τοῦ τε ποιοῦντος καὶ τοῦ πάσχοντος. εἰ οὖν
τὸ ἄρρεν ἐστὶν ὡς κινοῦν καὶ ποιοῦν, τὸ δὲ θῆλυ,
30 ᾗ θῆλυ,[3] ὡς παθητικόν, εἰς τὴν τοῦ ἄρρενος γονὴν
τὸ θῆλυ ἂν συμβάλλοιτο οὐ γονὴν ἀλλ' ὕλην. ὅπερ
καὶ φαίνεται συμβαῖνον· κατὰ γὰρ τὴν πρώτην
ὕλην[4] ἐστὶν ἡ τῶν καταμηνίων φύσις.

---

[1] ⟨τὸ⟩ Rackham.
[2] ταῦτ' Peck : τοῦτ' vulg.
[3] ᾗ θῆλυ fort. secl. (ex 729 b 12 insertum ?).
[4] κατὰ . . . ὕλην] ἡ γὰρ πρώτη ὕλη Z[1].

---

[a] Cf. 772 a 12.   [b] In one individual.
[c] i.e., specifically, in "form."   [d] See Introd. § 10.
[e] At Met. 1015 a 8 (cf. 1014 b 27) Aristotle speaks
of "prime matter" in two senses: e.g., in the case of bronze
articles (a) the prime matter relatively to them is bronze, but
(b) generally it is water (because all things that can be melted,
according to Aristotle, consist of water). And "prime
matter" is one of the meanings of φύσις, both according to
Met. (loc. cit.) and Phys. 193 a 28 : "one meaning of φύσις
is ἡ πρώτη ἑκάστῳ ὑποκειμένη ὕλη τῶν ἐχόντων ἐν αὑτοῖς ἀρχὴν

is required in order to produce numerous offspring is that there should be the right amount of it to suit the material available—neither so little that it fails to concoct it or even to set it, nor so much that it dries it up.[a] If on the other hand this semen which causes the original setting remains single and undivided, then one single offspring only is formed from it.

The foregoing discussion will have made it clear that the female, though it does not contribute any semen to generation, yet contributes something, viz., the substance constituting the menstrual fluid (or the corresponding substance in bloodless animals). But the same is apparent if we consider the matter generally, from the theoretical standpoint. Thus : there must be that which generates, and that out of which it generates ; and even if these two be united in one,[b] at any rate they must differ in kind,[c] and in that the *logos*[d] of each of them is distinct. In those animals in which these two faculties are separate, the body—that is to say the physical nature—of the active partner and of the passive must be different. Thus, if the male is the active partner, the one which originates the movement, and the female *qua* female is the passive one, surely what the female contributes to the semen of the male will be not semen but material. And this is in fact what we find happening ; for the natural substance of the menstrual fluid is to be classed as " prime matter." [e]

κινήσεως καὶ μεταβολῆς." In its lowest phase, "prime matter" is that which, united with the prime contrarieties (hot, cold, solid, fluid), produces the "elements" Earth, Air, Fire, Water; but, as the term "prime" itself suggests, "matter" is altogether a relative conception, and in its highest phase matter is one and the same as "form" (*Met.* 1045 b 18).

729 a

XXI Καὶ περὶ μὲν τούτων διωρίσθω τὸν τρόπον τοῦ-
35 τον. ἅμα δ' ἐκ τούτων φανερόν, περὶ ὧν ἐχόμενόν
729 b ἐστιν ἐπισκέψασθαι, πῶς ποτε συμβάλλεται εἰς τὴν
γένεσιν τὸ ἄρρεν, καὶ πῶς αἴτιόν ἐστι τοῦ γινομένου τὸ σπέρμα τὸ ἀπὸ τοῦ ἄρρενος, πότερον ὡς
ἐνυπάρχον καὶ μόριον ὂν εὐθὺς τοῦ γινομένου σώματος, μιγνύμενον τῇ ὕλῃ τῇ παρὰ τοῦ θήλεος, ἢ
5 τὸ μὲν σῶμα οὐθὲν κοινωνεῖ τοῦ σπέρματος, ἡ δ'
ἐν αὐτῷ δύναμις καὶ κίνησις· αὕτη μὲν γάρ ἐστιν
ἡ ποιοῦσα, τὸ δὲ συνιστάμενον καὶ λαμβάνον τὴν
μορφὴν τὸ τοῦ ἐν τῷ θήλει περιττώματος λοιπόν.
κατά τε δὴ τὸν λόγον οὕτω φαίνεται καὶ ἐπὶ τῶν
ἔργων. καθόλου τε γὰρ ἐπισκοποῦσιν οὐ φαίνεται
10 γιγνόμενον ἓν ἐκ τοῦ παθητικοῦ καὶ τοῦ ποιοῦντος
ὡς ἐνυπάρχοντος ἐν τῷ γινομένῳ τοῦ ποιοῦντος,
οὐδ' ὅλως δὴ ἐκ τοῦ κινουμένου καὶ κινοῦντος.
ἀλλὰ μὴν τό γε θῆλυ, ᾗ θῆλυ, παθητικόν, τὸ δ'
ἄρρεν, ᾗ ἄρρεν, ποιητικὸν καὶ ὅθεν ἡ ἀρχὴ τῆς
κινήσεως. ὥστε ἂν ληφθῇ τὰ ἄκρα ἑκατέρων, ᾗ
15 τὸ μὲν ποιητικὸν καὶ κινοῦν, τὸ δὲ παθητικὸν καὶ
κινούμενον, οὐκ ἔστιν ἐκ τούτων τὸ γιγνόμενον ἕν,
ἀλλ' ἢ οὕτως ὡς ἐκ τοῦ τέκτονος καὶ ξύλου ἡ
κλίνη, ἢ ὡς ἐκ τοῦ κηροῦ καὶ τοῦ εἴδους
ἡ σφαῖρα. δῆλον ἄρα ὅτι οὔτ' ἀνάγκη ἀπιέναι τι

---

[a] Aristotle now comes to grips with deciding between the
alternatives stated at 726 b 18 ff.

[b] i.e., that portion of the menstrual fluid which is not discharged externally.

[c] Cf. 716 a 27 ff.

These then are the lines upon which that subject
should be treated. And what we have said indicates plainly at the same time how we are to answer the questions which we next have to consider, viz., how it is that the male makes its contribution to generation, and how the semen produced by the male is the cause of the offspring; that is to say, Is the semen inside the offspring to start with, from the outset a part of the body which is formed, and mingling with the material provided by the female; or does the physical part of the semen have no share nor lot in the business, only the *dynamis* and movement contained in it? [a] This, anyway, is the active and efficient ingredient; whereas the ingredient which gets set and given shape is the remnant [b] of the residue in the female animal. The second suggestion is clearly the right one, as is shown both by reasoning and by observed fact. (a) If we consider the matter on general grounds, we see that when some one thing is formed from the conjunction of an active partner with a passive one, the active partner is not situated within the thing which is being formed; and we may generalize this still further by substituting " moving " and " moved " for " active " and " passive." Now of course the female, *qua* female,[c] is passive, and the male, *qua* male, is active—it is that whence the principle of movement comes. Taking, then, the widest formulation of each of these two opposites, viz., regarding the male *qua* active and causing movement, and the female *qua* passive and being set in movement, we see that the one thing which is formed is formed *from them* only in the sense in which a bedstead is formed from the carpenter and the wood, or a ball from the wax and the form. It is plain, then,

729 b

ἀπὸ τοῦ ἄρρενος, οὔτ' εἴ τι ἀπέρχεται, διὰ τοῦτο
20 ἐκ τούτου ὡς ἐνυπάρχοντος τὸ γιγνόμενόν ἐστιν,
ἀλλ' ὡς ἐκ κινήσαντος καὶ τοῦ εἴδους, ὡς καὶ ἀπὸ
τῆς ἰατρικῆς ὁ ὑγιασθείς. συμβαίνει δ' ὁμολογούμενα τῷ λόγῳ καὶ ἐπὶ τῶν ἔργων. διὰ τοῦτο γὰρ
ἔνια τῶν ἀρρένων καὶ συνδυαζομένων τοῖς θήλεσιν
οὐδὲ μόριον οὐθὲν φαίνεται προϊέμενα εἰς τὸ θῆλυ,
25 ἀλλὰ τοὐναντίον τὸ θῆλυ εἰς τὸ ἄρρεν, ὃ συμβαίνει
ἐνίοις τῶν ἐντόμων. ὃ γὰρ¹ τοῖς προϊεμένοις ἀπεργάζεται τὸ σπέρμα ἐν τῷ θήλει, τούτοις² ἡ ἐν
τῷ ζῴῳ αὐτῷ θερμότης³ καὶ δύναμις ἀπεργάζεται,
εἰσφέροντος τοῦ θήλεος τὸ δεκτικὸν τοῦ περιττώματος μόριον. καὶ διὰ τοῦτο τὰ τοιαῦτα τῶν
30 ζῴων συμπλέκεται μὲν πολὺν χρόνον, διαλυθέντα δὲ
γεννᾷ ταχέως. συνδεδύασται μὲν οὖν⁴ μέχρις οὗ⁵
ἂν συστήσῃ, ὥσπερ ἡ γονή⁶· διαλυθέντα δὲ προΐεται
τὸ κύημα ταχέως· γεννᾷ γὰρ ἀτελές· σκωληκοτοκεῖ
γὰρ πάντα τὰ τοιαῦτα.

Μέγιστον δὲ σημεῖον τὸ συμβαῖνον περὶ τοὺς
ὄρνιθας καὶ τὸ τῶν ἰχθύων γένος τῶν ᾠοτόκων
35 τοῦ μήτε ἀπὸ πάντων ἰέναι⁷ τὸ σπέρμα τῶν μορίων,

730 a

μήτε προΐεσθαι τὸ ἄρρεν τοιοῦτόν τι μόριον ὃ
ἔσται ἐνυπάρχον τῷ γεννηθέντι, ἀλλὰ μόνον τῇ δυνάμει τῇ ἐν τῇ γονῇ ζῳοποιεῖν, ὥσπερ εἴπομεν ἐπὶ

---

¹ γὰρ Z : γὰρ ἐν vulg.   ² τούτοις Z¹ : τοῦτο vulg.
³ θερμότης Z¹ : ὑγρότης vulg.
⁴ μὲν οὖν] γὰρ Z.   ⁵ μέχρις οὗ PZ : μέχρι vulg.
⁶ hic locus haud sanus videtur ; fortasse συνδεδύασται . . .
ταχέως secludenda ; om. Σ.
⁷ ἰέναι A.-W., Z¹ : exit Σ : εἶναι vulg.

ᵃ See above, ch. 16.
ᵇ Probably the words " the copulation . . . discharge the
fetation " should be deleted.

that there is no necessity for any substance to pass from the male ; and if any does pass, this does not mean that the offspring is formed from it as from something situated within itself during the process, but as from that which has imparted movement to it, or that which is its " form." The relationship is the same as that of the patient who has been healed to the medical art. (b) This piece of reasoning is entirely borne out by the facts. It explains why certain of those males which copulate with the females are observed to introduce no part at all into the female, but on the contrary the female introduces a part into the male. This occurs in certain insects.ᵃ In those cases where the male introduces some part, it is the semen which produces the effect inside the female ; but in the case of these insects, the same effect is produced by the heat and *dynamis* inside the ⟨male⟩ animal itself when the female inserts the part which receives the residue. And that is why animals of this sort take a long time over copulation, and once they have separated the young are soon produced : the copulation lasts until ⟨the *dynamis* in the male⟩ has " set " ⟨the material in the female⟩, just as the semen does ; but once they have separated they soon discharge the fetation,ᵇ because the offspring they produce is imperfect ; all such creatures, in fact, produce larvae.

However, it is the behaviour of birds and the group of oviparous fishes which provides us with our strongest proof (a) that the semen is not drawn from all the parts of the body, and (b) that the male does not emit any part such as will remain situated within the fetus, but begets the young animal simply by means of the *dynamis* residing in the semen (just as

τῶν ἐντόμων, ἐν οἷς τὸ θῆλυ προΐεται εἰς τὸ ἄρρεν.
5 ἐάν τε γὰρ ὑπηνέμια τύχῃ κύουσα ἡ ὄρνις, ἐὰν
μετὰ ταῦτα ὀχεύηται, μήπω μεταβεβληκότος τοῦ
ᾠοῦ ἐκ τοῦ ὠχρὸν ὅλον εἶναι εἰς τὸ λευκαίνεσθαι,
γόνιμα γίνεται ἀντὶ ὑπηνεμίων· ἐάν τε ὑφ᾽ ἑτέρου
ὠχευμένη ⟨ᾖ⟩[1] καὶ ἔτι ὠχροῦ ὄντος, κατὰ τὸν ὕστερον
ὀχεύσαντα τὸ γένος ἀποβαίνει πᾶν τὸ τῶν νεοττῶν.
10 διὸ ἔνιοι τοῦτον τὸν τρόπον τῶν περὶ τὰς ὄρνιθας
τὰς γενναίας σπουδαζόντων ποιοῦσι, μεταβάλλοντες
τὰ πρῶτα ὀχεῖα καὶ τὰ ὕστερα, ὡς οὐ συμμιγνύ-
μενον καὶ ἐνυπάρχον, οὐδ᾽ ἀπὸ παντὸς ἐλθὸν τὸ
σπέρμα· ἀπ᾽ ἀμφοῖν γὰρ ἂν ἦλθεν, ὥστ᾽ εἶχεν ἂν
δὶς ταὐτὰ μέρη. ἀλλὰ τῇ δυνάμει τὸ τοῦ ἄρρενος
15 σπέρμα τὴν ἐν τῷ θήλει ὕλην καὶ τροφὴν ποιάν
τινα κατασκευάζει. τοῦτο γὰρ ἐνδέχεται ποιεῖν τὸ
ὕστερον ἐπεισελθὸν ἐκ τοῦ θερμᾶναι καὶ πέψαι·
λαμβάνει γὰρ τροφὴν τὸ ᾠὸν ἕως ἂν αὐξάνηται.

Τὸ δ᾽ αὐτὸ συμβαίνει καὶ περὶ τὴν τῶν ἰχθύων
γένεσιν τῶν ᾠοτοκουμένων. ὅταν γὰρ ἀποτέκῃ
20 τὰ ᾠὰ ἡ θήλεια, ὁ ἄρρην ἐπιρραίνει τὸν θορόν·
καὶ ὧν μὲν ἂν ἐφάψηται, γόνιμα ταῦτα γίνεται τὰ
ᾠά, ὧν δ᾽ ἂν μή, ἄγονα, ὡς οὐκ εἰς τὸ ποσὸν
συμβαλλομένου τοῖς ζῴοις τοῦ ἄρρενος, ἀλλ᾽ εἰς
τὸ ποιόν.

Ὅτι μὲν οὖν οὔτ᾽ ἀπὸ παντὸς ἀπέρχεται τὸ
25 σπέρμα τοῖς προϊεμένοις σπέρμα τῶν ζῴων, οὔτε

[1] ⟨ᾖ⟩ Peck.

[a] See below, 757 b 2 f.

we said happened with those insects where the female inserts a part into the male). Here is the evidence. Supposing a hen bird is in process of producing wind-eggs, and then that she is trodden by the cock while the egg is still completely yellow and has not yet started to whiten : the result is that the eggs are not wind-eggs but fertile ones. And supposing the hen has been trodden by another cock while the egg is still yellow,[a] then the whole brood of chickens when hatched out takes after the second cock. Some breeders who specialize in first-class strains act upon this, and change the cock for the second treading. The implication is (a) that the semen is not situated inside the egg and mixed up with it, and (b) that it is not drawn from the whole of the body of the male : if it were in this case, it would be drawn from both males, so the offspring would have every part twice over. No ; the semen of the male acts otherwise ; in virtue of the *dynamis* which it contains it causes the material and nourishment in the female to take on a particular character ; and this can be done by that semen which is introduced at a later stage, working through heating and concoction, since the egg takes in nourishment so long as it is growing.

The same thing occurs in the generation of oviparous fishes. When the female fish has laid her eggs, the male sprinkles his milt over them ; the eggs which it touches become fertile, but the others are infertile, which seems to imply that the contribution which the male makes to the young has to do not with bulk but with specific character.

What has been said makes it clear that, in the case of animals which emit semen, the semen is not drawn from the whole of the body, and also that in genera-

τὸ θῆλυ πρὸς τὴν γένεσιν οὕτω συμβάλλεται τοῖς
συνισταμένοις ὡς τὸ ἄρρεν, ἀλλὰ τὸ μὲν ἄρρεν
ἀρχὴν κινήσεως, τὸ δὲ θῆλυ τὴν ὕλην, δῆλον ἐκ
τῶν εἰρημένων. διὰ γὰρ τοῦτο οὔτ᾿ αὐτὸ καθ᾿
αὑτὸ γεννᾷ τὸ θῆλυ, δεῖται γὰρ ἀρχῆς καὶ τοῦ
30 κινήσοντος καὶ διοριοῦντος (ἀλλ᾿ ἐνίοις γε τῶν
ζῴων, οἷον τοῖς ὄρνισι, μέχρι τινὸς ἡ φύσις δύναται
γεννᾶν· αὗται γὰρ συνιστᾶσι μέν, ἀτελῆ δὲ συν-
XXII ιστᾶσι τὰ καλούμενα ὑπηνέμια ᾠά), ἥ τε γένεσις
ἐν τῷ θήλει συμβαίνει τῶν γινομένων, ἀλλ᾿ οὐκ
εἰς τὸ ἄρρεν οὔτ᾿ αὐτὸ τὸ[1] ἄρρεν προΐεται τὴν γονὴν
35 οὔτε τὸ θῆλυ, ἀλλ᾿ ἄμφω εἰς τὸ θῆλυ συμβάλλονται

τὸ παρ᾿ αὐτῶν γιγνόμενον, διὰ τὸ ἐν τῷ θήλει
εἶναι τὴν ὕλην ἐξ ἧς ἐστι τὸ δημιουργούμενον.
καὶ εὐθὺς τὴν μὲν ἀθρόον ὑπάρχειν ἀναγκαῖον ἐξ
ἧς συνίσταται τὸ κύημα τὸ πρῶτον, τὴν δ᾿ ἐπι-
γίνεσθαι ἀεὶ[2] τῆς ὕλης, ἵν᾿ αὐξάνηται τὸ κυούμενον.[3]
5 ὥστ᾿ ἀνάγκη ἐν τῷ θήλει ὑπάρχειν τὸν τόκον· καὶ
γὰρ πρὸς τῷ ξύλῳ ὁ τέκτων καὶ πρὸς τῷ πηλῷ ὁ
κεραμεύς, καὶ ὅλως πᾶσα ἡ ἐργασία καὶ ἡ κίνησις
ἡ ἐσχάτη πρὸς τῇ ὕλῃ, οἷον ἡ οἰκοδόμησις ἐν τοῖς
οἰκοδομουμένοις. λάβοι δ᾿ ἄν τις ἐκ τούτων καὶ
τὸ ἄρρεν πῶς συμβάλλεται πρὸς τὴν γένεσιν· οὐδὲ
10 γὰρ τὸ ἄρρεν ἅπαν προΐεται σπέρμα, ὅσα τε

---

[1] ἄρρεν οὔτ᾿ αὐτὸ τὸ Buss.-Platt (καὶ οὐκ αὐτὸ τὸ Z) : ἄρρεν
οὔτ᾿ αὖ τὸ vulg. (⟨καὶ⟩ οὔτ᾿ Sus.).
[2] ἀεὶ SYZ : δεῖ vulg.   [3] κυούμενον SZ : κυόμενον vulg.

---

[a] This is explained in the passage which follows (730 b 15 ff.).

tion the contribution which the female makes to the embryos when they are being " set " and constituted is on different lines from that of the male ; in other words, the male contributes the principle of movement [a] and the female contributes the material. This is why (*a*) on the one hand the female does not generate on its own : it needs some source or principle to supply the material with movement and to determine its character (though in some ⟨female⟩ animals, as in birds, Nature can generate up to a point : the females of these species do actually " set " a fetation, but what they " set " is imperfect, viz., what are known as wind-eggs) ; (*b*) on the other hand, the formation XXII of the young does in fact take place in the female, whereas neither the male himself nor the female emits semen into the male, but they both deposit together what they have to contribute in the female —it is because that is where the material is out of which the creature that is being fashioned is made. And as regards this material, a good quantity of it must of necessity be available immediately, out of which the fetation is " set " and constituted in the first place, and after that fresh supplies of it must be continually arriving to make its growth possible. Hence, of necessity, it is in the female that parturition takes place. After all, the carpenter is close by his timber, and the potter close by his clay ; and to put it in general terms, the working or treatment of any material, and the ultimate movement [a] which acts upon it, is in all cases close by the material, *e.g.*, the location of the activity of house-building is in the houses which are being built. These instances may help us to understand how the male makes its contribution to generation ; for not every male emits

προΐεται τῶν ἀρρένων, οὐθὲν μόριον τοῦτ' ἐστὶ
τοῦ γιγνομένου κυήματος, ὥσπερ οὐδ' ἀπὸ τοῦ
τέκτονος πρὸς τὴν τῶν ξύλων ὕλην οὔτ' ἀπέρχεται
οὐθὲν οὔτε μόριον οὐθέν ἐστιν ἐν τῷ γιγνομένῳ
τῆς τεκτονικῆς, ἀλλ' ἡ μορφὴ καὶ τὸ εἶδος ἀπ'
15 ἐκείνου ἐγγίνεται διὰ τῆς κινήσεως ἐν τῇ ὕλῃ, καὶ
ἡ μὲν ψυχὴ ἐν ᾗ τὸ εἶδος καὶ ἡ ἐπιστήμη κινοῦσι
τὰς χεῖρας ἤ τι μόριον ἕτερον ποιάν τινα κίνησιν,
ἑτέραν μὲν ἀφ' ὧν τὸ γιγνόμενον ἕτερον, τὴν αὐτὴν
δὲ ἀφ' ὧν τὸ αὐτό, αἱ δὲ χεῖρες τὰ ὄργανα τὰ δ'
ὄργανα τὴν ὕλην.[1] ὁμοίως δὲ καὶ ἡ φύσις[2] ἐν τῷ
20 ἄρρενι τῶν σπέρμα προϊεμένων χρῆται τῷ σπέρματι
ὡς ὀργάνῳ καὶ ἔχοντι κίνησιν ἐνεργείᾳ, ὥσπερ ἐν
τοῖς κατὰ τέχνην γινομένοις τὰ ὄργανα κινεῖται·
ἐν ἐκείνοις γάρ πως ἡ κίνησις τῆς τέχνης. ὅσα
μὲν οὖν προΐεται σπέρμα, συμβάλλεται τοῦτον τὸν
25 τρόπον εἰς τὴν γένεσιν· ὅσα δὲ μὴ προΐεται σπέρμα,
ἀλλ' ἐναφίησι τὸ θῆλυ εἰς τὸ ἄρρεν τῶν αὑτοῦ τι
μορίων, ὅμοιον ἔοικε ποιοῦντι ὥσπερ ἂν εἰ τὴν
ὕλην κομίσειέ τις πρὸς τὸν δημιουργόν. δι' ἀσθένειαν γὰρ τῶν τοιούτων ἀρρένων οὐθὲν δι' ἑτέρων
οἷά τε ποιεῖν ἡ φύσις, ἀλλὰ μόλις αὐτῆς προσ-
εδρευούσης ἰσχύουσιν αἱ κινήσεις, καὶ ἔοικε τοῖς
30 πλάττουσιν, οὐ τοῖς τεκταινομένοις· οὐ γὰρ δι'
ἑτέρου θιγγάνουσα δημιουργεῖ τὸ συνιστάμενον,
ἀλλ' αὐτὴ τοῖς αὑτῆς μορίοις.

XXIII   Ἐν μὲν οὖν τοῖς ζῴοις πᾶσι τοῖς πορευτικοῖς

---

[1] sic Z : αἱ δὲ χεῖρες καὶ τὰ ὄργανα τὴν ὕλην vulg.
[2] φύσις Z : φύσις ἡ vulg., Z².

---

[a] Cf. P.A. 639 b 16—641 a 14.

semen, and in the case of those which do, this semen is not a part of the fetation as it develops. In the same way, nothing passes from the carpenter into the pieces of timber, which are *his* material, and there is no part of the art of carpentry present in the object which is being fashioned : it is the shape and the form which pass from the carpenter, and they come into being by means of the movement in the material. It is his soul, wherein is the " form," and his knowledge, which cause his hands (or some other part of his body) to move in a particular way (different ways for different products, and always the same way for any one product) ; his hands move his tools and his tools move the material.[a] In a similar way to this, Nature acting in the male of semen-emitting animals uses the semen as a tool, as something that has movement in actuality ; just as when objects are being produced by any art the tools are in movement, because the movement which belongs to the art is, in a way, situated in them. Males, then, that emit semen contribute to generation in the manner described. Those which emit no semen, males into which the female inserts one of its parts, may be compared to a craftsman who has his material brought to him. Males of this sort are so weak that Nature is unable to accomplish anything at all through intermediaries : indeed, their movements are only just strong enough when Nature herself sits watching over the business ; the result is that here Nature resembles a modeller in clay rather than a carpenter ; she does not rely upon contact exerted at second hand when fashioning the object which is being given shape, but uses the parts of her own very self to handle it.

In all animals which can move about, male and XXIII

κεχώρισται τὸ θῆλυ τοῦ ἄρρενος, καὶ ἔστιν ἕτερον
35 ζῷον θῆλυ καὶ ἕτερον ἄρρεν, τῷ δὲ εἴδει ταὐτόν,

οἷον ἄνθρωπος ἢ ἵππος[1] ἀμφότερα· ἐν δὲ τοῖς φυτοῖς
μεμιγμέναι αὗται αἱ δυνάμεις εἰσί, καὶ οὐ κεχώρισται τὸ θῆλυ τοῦ ἄρρενος. διὸ καὶ γεννᾷ αὐτὰ
ἐξ αὑτῶν, καὶ προΐεται οὐ γονὴν ἀλλὰ κύημα τὰ
καλούμενα σπέρματα. καὶ τοῦτο καλῶς λέγει
5 Ἐμπεδοκλῆς ποιήσας

οὕτω δ᾽ ᾠοτοκεῖ μακρὰ δένδρεα· πρῶτον ἐλαίας ...

τό τε γὰρ ᾠὸν κύημά ἐστι, καὶ ἔκ τινος αὐτοῦ
γίγνεται τὸ ζῷον, τὸ δὲ λοιπὸν τροφή, καὶ τοῦ[2]
σπέρματος ἐκ[3] μέρους γίγνεται τὸ φυόμενον, τὸ δὲ
λοιπὸν[4] τροφὴ γίγνεται τῷ βλαστῷ καὶ τῇ ῥίζῃ
10 τῇ πρώτῃ. τρόπον δέ τινα ταὐτὰ[5] συμβαίνει καὶ
ἐν τοῖς κεχωρισμένον ἔχουσι ζῴοις τὸ θῆλυ καὶ τὸ
ἄρρεν. ὅταν γὰρ δεήσῃ γεννᾶν, γίνεται ἀχώριστον,
ὥσπερ ἐν τοῖς φυτοῖς, καὶ βούλεται ἡ φύσις αὐτῶν
ἓν γίνεσθαι· ὅπερ ἐμφαίνεται κατὰ τὴν ὄψιν μιγνυμένων καὶ συνδυαζομένων [ἔν τι ζῷον γίγνεσθαι
ἐξ ἀμφοῖν].[6]

15 Καὶ τὰ μὲν μὴ προϊέμενα σπέρμα πολὺν χρόνον
συμπεπλέχθαι πέφυκεν, ἕως ἂν συστήσῃ τὸ κύημα,
οἷον τὰ συνδυαζόμενα τῶν ἐντόμων· τὰ δ᾽, ἕως ἂν
ἀποπέμψῃ τι τῶν ἐπεισάκτων αὐτοῦ μορίων, ὃ
συστήσει τὸ κύημα ἐν πλείονι χρόνῳ, οἷον ἐπὶ τῶν
ἐναίμων. τὰ μὲν γὰρ ἡμέρας τι μόριον συνέχεται,

---

[1] ἢ ἵππος ΖΣ : om. vulg.   [2] A.-W. : τοῦ PSY : καὶ ἐκ τοῦ vulg.
[3] ἐκ A.-W., Diels : ἐκ (non καὶ) Ζ[1] : καὶ ἐκ vulg., Ζ[3].
[4] ἐν αὑτῷ addit Ζ.
[5] ταὐτὰ Y : ταὐτὸ A.-W. : ταῦτα vulg.
[6] secl. Rackham.

---

[a] Empedocles, fr. 79 (Diels).

female are separate ; one animal is male and another female, though they are identical in species, just as men and women are both human beings, and stallion and mare are both horses. In plants, however, these faculties are mingled together ; the female is not separate from the male ; and that is why they generate out of themselves, and produce not semen but a fetation—what we call their " seeds." Empedocles puts this well in his poem, when he says [a] :

So the great trees lay eggs ; the olives first . . .,

because just as the egg is a fetation from part of which [b] the creature is formed while the remainder is nourishment, so from part of the seed is formed the growing plant, while the remainder is nourishment for the shoot and the first root. And in a sort of way the same happens even in those animals where male and female are separate ; for when they have need to generate they cease to be separate and are united as they are in plants : their nature desires that they should become one. And this is plain to see when they are uniting and copulating [that one animal is produced out of the two of them].

The natural practice of those animals which emit no semen is to remain united for a long time, until ⟨the male⟩ has " set " the fetation : those Insects which copulate are an example of this. Other animals, however, remain united until the male has introduced from those " parts " [c] of himself which he inserts one which will " set " the fetation but will take a longer time to do so : the blooded animals illustrate this. The former sort remain in copulation

[b] See 732 a 29.
[c] The use of " part " here to refer to semen is a good illustration of the meaning of this term in Aristotle.

20 ἡ δὲ γονὴ ἐν ἡμέραις συνίστησι πλείοσιν· προέ-
μενα¹ δὲ τὸ τοιοῦτον ἀπολύεται. καὶ ἀτεχνῶς ἔοικε
τὰ ζῷα ὥσπερ φυτὰ εἶναι διαιρετά, οἷον εἴ τις
κἀκεῖνα, ὅτε σπέρμα ἐξενέγκειεν, διαλύσειε καὶ
χωρίσειεν εἰς τὸ ἐνυπάρχον θῆλυ καὶ ἄρρεν.

Καὶ ταῦτα πάντα εὐλόγως ἡ φύσις δημιουργεῖ.
25 τῆς μὲν γὰρ τῶν φυτῶν οὐσίας οὐθέν ἐστιν ἄλλο
ἔργον οὐδὲ πρᾶξις οὐδεμία πλὴν ἡ τοῦ σπέρματος
γένεσις, ὥστ᾽ ἐπεὶ τοῦτο διὰ τοῦ θήλεος γίνεται
καὶ τοῦ ἄρρενος συνδεδυασμένων, μίξασα ταῦτα
διέθηκε μετ᾽ ἀλλήλων· διὸ ἐν τοῖς φυτοῖς ἀχώριστον τὸ θῆλυ καὶ τὸ ἄρρεν. ἀλλὰ περὶ μὲν φυτῶν
30 ἐν ἑτέροις ἐπέσκεπται, τοῦ δὲ ζῴου οὐ μόνον τὸ
γεννῆσαι ἔργον (τοῦτο μὲν γὰρ κοινὸν τῶν ζώντων
πάντων), ἀλλὰ καὶ γνώσεώς τινος πάντα μετέχουσι,
τὰ μὲν πλείονος, τὰ δ᾽ ἐλάττονος, τὰ δὲ πάμπαν
μικρᾶς. αἴσθησιν γὰρ ἔχουσιν, ἡ δ᾽ αἴσθησις
γνῶσίς τις. ταύτης δὲ τὸ τίμιον καὶ ἄτιμον πολὺ
35 διαφέρει σκοποῦσι πρὸς φρόνησιν καὶ πρὸς τὸ τῶν

ἀψύχων γένος. πρὸς μὲν γὰρ τὸ φρονεῖν ὥσπερ
οὐδὲν εἶναι δοκεῖ τὸ κοινωνεῖν ἁφῆς καὶ γεύσεως
μόνον, πρὸς δὲ ἀναισθησίαν² βέλτιστον· ἀγαπητὸν
γὰρ ἂν δόξειε καὶ ταύτης τυχεῖν τῆς γνώσεως
ἀλλὰ μὴ κεῖσθαι τεθνεὸς καὶ μὴ ὄν. διαφέρει δ᾽

---

¹ προέμενα coniecit Platt : προϊέμενα vulg.
² ἀναισθησίαν] φυτὸν ἢ λίθον Z, unde φυτοῦ ἢ λίθου addunt
A.-W. : pro πρὸς . . . βέλτιστον *inter ergo istud animal et necib
est differentia mirabilis* Σ (θαύμασιον pro βέλτιστον Z).

ᵃ Cf. above, 717 a 20.

for a fair part of a day ; whereas semen takes several days to " set " fetations, and when the creatures have emitted this they free themselves. Indeed, animals seem to be just like divided plants : as though you were to pull a plant to pieces when it was bearing its seed and separate it into the male and female present in it.

In all her workmanship herein Nature acts in every particular as reason would expect. A plant, in its essence, has no function or activity to perform other than the production of its seed [a] ; and since this is produced as the result of the union of male with female, Nature has mixed the two and placed them together, so that in plants male and female are not separate. Plants, however, have been dealt with in another treatise ; here we are concerned with animals, and generation is not the only function which an animal has—that is a function common to all things living. All animals have, in addition, some measure of knowledge of a sort (some have more, some less, some very little indeed), because they have sense-perception,[b] and sense-perception is, of course, a sort of knowledge. The value we attach to this knowledge varies greatly according as we judge it by the standard of human intelligence or the class of lifeless objects. Compared with the intelligence possessed by man, it seems as nothing to possess the two senses of touch and taste only ; but compared with entire absence of sensibility it seems a very fine thing indeed. We should much prefer to have even this sort of knowledge to a state of death and non-existence. Now it is by sense-perception that animals

[b] See 732 a 13, n. With this passage (731 a 29-b 3) *cf.* the whole *Protrepticus* passage there referred to.

5 αἰσθήσει τὰ ζῷα τῶν ζώντων μόνον. ἐπεὶ δ'
ἀνάγκη καὶ ζῆν, ἐὰν¹ ᾖ ζῷον, ὅταν δεήσῃ ἀποτελεῖν
τὸ τοῦ ζῶντος ἔργον, τότε συνδυάζεται καὶ μίγνυται καὶ γίγνεται ὡσπερανεὶ φυτόν, καθάπερ
εἴπομεν.

Τὰ δ' ὀστρακόδερμα τῶν ζῴων μεταξὺ ὄντα τῶν
ζῴων καὶ τῶν φυτῶν, ὡς ἐν ἀμφοτέροις ὄντα τοῖς
10 γένεσιν, οὐδετέρων ποιεῖ τὸ ἔργον· ὡς μὲν γὰρ
φυτὸν² οὐκ ἔχει τὸ θῆλυ καὶ τὸ ἄρρεν καὶ οὐ
γεννᾷ εἰς ἕτερον, ὡς δὲ ζῷον οὐ φέρει ἐξ αὑτοῦ
καρπὸν ὥσπερ τὰ φυτά, ἀλλὰ συνίσταται καὶ
γεννᾶται ἔκ τινος συστάσεως γεοειδοῦς καὶ ὑγρᾶς.
ἀλλὰ περὶ μὲν τῆς τούτων γενέσεως ὕστερον
λεκτέον.

---

¹ ἐὰν] ὃ ἂν A.-W.   ² φυτὸν Z : φυτὸν ὂν vulg.

ᵃ *i.e.*, to reproduce itself, because τὸ θρεπτικόν, which all differ from the creatures which are merely alive ; since, however, if it be an animal, its attributes must of necessity include that of being alive, when the time comes for it to accomplish the function proper to that which is alive,[a] then it copulates and unites and becomes as it were a plant, just as we have said.

The Testacea stand midway between animals and plants and so, as being in both groups, perform the function of neither : as plants, they do not have male and female and so they do not generate by pairing ; as animals they bear no fruit externally like that borne by plants ; but they take shape and are generated out of a certain earthy and fluid coagulation. The manner of generation of these creatures, however, must be described later.[b]

living things must possess, is also τὸ γεννητικὸν ἑτέρου οἷον αὐτό (735 a 17, 18).

[b] In Bk. III, ch. 11.

B

731 b 18 I Τὸ δὲ θῆλυ καὶ τὸ ἄρρεν ὅτι μέν εἰσιν ἀρχαὶ
γενέσεως εἴρηται πρότερον, καὶ τίς ἡ δύναμις καὶ
20 ὁ λόγος τῆς οὐσίας αὐτῶν· διὰ τί δὲ γίνεται καὶ
ἔστι τὸ μὲν θῆλυ τὸ δ' ἄρρεν, ὡς μὲν ἐξ ἀνάγκης
καὶ τοῦ[1] πρώτου κινοῦντος καὶ ὁποίας ὕλης[2]
προϊόντα πειρᾶσθαι δεῖ φράζειν τὸν λόγον, ὡς δὲ
διὰ τὸ βέλτιον καὶ τὴν αἰτίαν τὴν ἕνεκά τινος,
ἄνωθεν[3] ἔχει τὴν ἀρχήν· ἐπεὶ γάρ ἐστι τὰ μὲν
25 ἀΐδια καὶ θεῖα τῶν ὄντων, τὰ δ' ἐνδεχόμενα καὶ
εἶναι καὶ μὴ εἶναι, τὸ δὲ καλὸν καὶ τὸ θεῖον αἴτιον
ἀεὶ κατὰ τὴν αὐτοῦ φύσιν τοῦ βελτίονος ἐν τοῖς
ἐνδεχομένοις, τὸ δὲ μὴ ἀΐδιον ἐνδεχόμενόν ἐστι καὶ

---

[1] το τοῦ Z[1].   [2] καὶ . . . ὕλης fortasse secludenda.
[3] ἀπὸ τοῦ παντὸς addit P, Aldus (=ἄνωθεν).

---

[a] See Introd. §§ 25, 30, etc.   [b] See Introd. § 10.
[c] The sense, though perhaps not the syntax, of the
following sentence is clear. The contrast is between (a)
causes ἐξ ἀνάγκης (i.e., mechanical causes, viz., the " motive "
and " material " causes, the operation of which in the production of male and female individuals Aristotle describes in
detail in Bk. IV. 765 b 5—766 b 26 ; cf. 767 a 36—768 b 36) ;
and (b) the " final " cause, the better purpose or " end " for
the sake of which male and female individuals are produced.
[d] See Introd. § 7 (ii).

BOOK II

I HAVE already said that the male and the female are
" principles " of generation, and I have also said
what is their *dynamis* ^a and the *logos* ^b of their essence.
^c As for the reason why one comes to be formed, and
is, male, and another female, (*a*) in so far as this
results from *necessity*,^d *i.e.*, from the proximate
motive cause and from what sort of matter, our
argument as it proceeds must endeavour to explain ;
(*b*) in so far as this occurs on account of what is
better, *i.e.*, on account of the final cause (the Cause
" for the sake of which "), the principle is derived
from the upper cosmos.^e   What I mean is this.   Of
the things which are, some are eternal and divine,
others admit alike of being and not-being, and the
beautiful and the divine acts always, in virtue of its
own nature, as a cause which produces that which
is *better* in the things which admit of it ^f ; while

^e And this principle Aristotle proceeds to explain at once,
since it is really beyond the normal scope of the present
treatise which is concerned chiefly with the " motive " and
" material " causes of generation.   ἄνωθεν (*cf*. τὸ ἄνω σῶμα,
App. B § 26)=*via* the " heavens " from the Unmoved Mover,
" God."   The best commentary on the passage which follows
is afforded by Aristotle's own statements in other treatises,
of which the pertinent passages will be found in App. A
(esp. §§ 12-18), and I have therefore thought it unnecessary
to provide full annotations here.

^f *Cf*. *Met*. 1013 a 22 πολλῶν γὰρ καὶ τοῦ γνῶναι καὶ τῆς
κινήσεως ἀρχὴ τἀγαθὸν καὶ τὸ καλόν.

731 b

εἶναι ⟨καὶ μὴ εἶναι⟩¹ καὶ μεταλαμβάνειν καὶ τοῦ
χείρονος καὶ τοῦ βελτίονος, βέλτιον δὲ ψυχὴ μὲν
σώματος, τὸ δ' ἔμψυχον τοῦ ἀψύχου διὰ τὴν
30 ψυχήν, καὶ τὸ εἶναι τοῦ μὴ εἶναι καὶ τὸ ζῆν τοῦ
μὴ ζῆν, διὰ ταύτας τὰς αἰτίας γένεσις ζῴων ἐστίν·
ἐπεὶ γὰρ ἀδύνατος ἡ φύσις τοῦ τοιούτου γένους
ἀΐδιος εἶναι, καθ' ὃν ἐνδέχεται τρόπον, κατὰ τοῦτόν
ἐστιν ἀΐδιον τὸ γινόμενον. ἀριθμῷ μὲν οὖν ἀδύνατον, ἡ γὰρ οὐσία τῶν ὄντων ἐν τῷ καθ' ἕκαστον·
35 τοιοῦτον δ' εἴπερ ἦν, ἀΐδιον ἂν ἦν· εἴδει δ' ἐν-

732 a

δέχεται. διὸ γένος ἀεὶ ἀνθρώπων καὶ ζῴων ἐστὶ
καὶ φυτῶν. ἐπεὶ δὲ τούτων ἀρχὴ τὸ θῆλυ καὶ τὸ
ἄρρεν, ἕνεκα τῆς γενέσεως ἂν εἴη τὸ θῆλυ καὶ τὸ
ἄρρεν ἐν τοῖς οὖσιν ἑκάτερον τούτων.² βελτίονος

---

¹ supplevit Platt.
² ἑκάτερον τούτων Z : om. vulg.

ᵃ *i.e.,* this is the Final Cause, which can be equated with
" the better," as opposed to the mere mechanical sort of
causation. See above 731 b 23.

ᵇ The reader may at first be confused in this passage
owing to the fact that Aristotle uses ἀΐδιος in two senses:
(a) in the true and full sense, as applicable to the ἄφθαρτα and
θεῖα, as in line 731 b 25, in which sense it can be applied
only to the things which οὐκ ἐνδέχεται εἶναι καὶ μὴ εἶναι,
i.e., which always *are*; but then he goes on to use it in a
modified sense (b), and applies it to that which ἐνδέχεται εἶναι
καὶ μὴ εἶναι, *i.e.,* to τὸ γιγνόμενον, and says that τὸ γίγνομενον
is ἀΐδιον *in the way which is open to it.* (Aristotle seems to
regard this extension of the use of ἀΐδιος as justifiable, since,
as he states in the passage of *De anima* quoted in App. A
(§ 17), τὰ γιγνόμενα, although they are not eternal, do *partake
in* eternity.) These two modes of being ἀΐδιον he then
describes more exactly as ἀΐδιον ἀριθμῷ (the eternity of
individual identity) and ἀΐδιον εἴδει (the eternity of specific

that which is not eternal admits of being ⟨and not-being⟩, and of acquiring a share both in the better and in the worse ; also, Soul is better than body, and a thing which has Soul in it is better than one which has not, in virtue of that Soul ; and being is better than not-being, and living than not living. These are the causes on account of which generation of animals takes place,ᵃ because since the nature of a class of this sort is unable to be eternal, that which comes into being is eternal in the manner that is open to it. Now it is impossible for it to be so *numerically*, since the " being " of things is to be found in the particular, and if it really were so, then it would be eternal ᵇ ; it is, however, open to it to be so *specifically*. That is why there is always a *class* of men, of animals, of plants ; and since the principle of these is " the male " and " the female," it will surely be for the sake of generation that " the male " and " the female " are present in the individuals which are male and female. And as the

form).   Hence, in the present sentence τοιοῦτον means ἀριθμῷ ἀίδιον ; and the sense of the statement is that if an animal really were ἀριθμῷ ἀίδιον, its οὐσία would be ἀίδιος, *i.e.*, ἄφθαρτος ; in other words, it would no longer be a φθαρτόν or a γιγνόμενον.   The translation might be expanded as follows to bring out the meaning : " Now it is impossible for it to be so *numerically*, since the " being " of things is in the particular ⟨*i.e.*, in the individual concrete object consisting of matter and form ; and obviously no such particular φθαρτόν —animal or plant—can be *numerically* eternal⟩ ; and if it really were so, then it would be eternal ⟨in the full and proper sense of the term, viz., it would be ἄφθαρτον, and no longer a γιγνόμενον at all⟩ ; it is, however, open to it to be eternal *specifically*."   It is useful to note that at *Met.* 999 b 33 Aristotle states that there is no difference between the terms ἀριθμῷ ἕν and καθ' ἕκαστον (τὸ ἀριθμῷ ἕν ἢ τὸ καθ' ἕκαστον λέγειν διαφέρει οὐδέν).—See further, App. A §§ 15-18.

732 a

δὲ καὶ θειοτέρας τὴν φύσιν οὔσης τῆς αἰτίας τῆς
5 κινούσης πρώτης, ᾗ ὁ λόγος ὑπάρχει καὶ τὸ εἶδος,
τῆς ὕλης, βέλτιον καὶ τὸ κεχωρίσθαι τὸ κρεῖττον
τοῦ χείρονος. διὰ τοῦτ᾽ ἐν ὅσοις ἐνδέχεται καὶ
καθ᾽ ὅσον ἐνδέχεται, κεχώρισται τοῦ θήλεος τὸ
ἄρρεν· βέλτιον γὰρ καὶ θειότερον ᾗ[1] ἀρχὴ τῆς
κινήσεως [ᾗ ἄρρεν ὑπάρχει][2] τοῖς γινομένοις· ὕλη
10 δὲ τὸ[3] θῆλυ. συνέρχεται δὲ καὶ μίγνυται πρὸς
τὴν ἐργασίαν τῆς γενέσεως τῷ θήλει τὸ ἄρρεν·
αὕτη γὰρ κοινὴ ἀμφοτέροις.

[4][Κατὰ μὲν οὖν τὸ μετέχειν τοῦ θήλεος καὶ τοῦ
ἄρρενος ζῇ, διὸ καὶ τὰ φυτὰ μετέχει ζωῆς· κατὰ
δὲ τὴν αἴσθησιν τὸ τῶν ζῴων ἐστὶ γένος. τούτων
δὲ σχεδὸν ἐν πᾶσι τοῖς πορευτικοῖς κεχώρισται τὸ
15 θῆλυ καὶ τὸ ἄρρεν διὰ τὰς εἰρημένας αἰτίας· καὶ
τούτων τὰ μέν, ὥσπερ ἐλέχθη, προΐεται σπέρμα,
τὰ δ᾽ οὐ προΐεται ἐν τῷ συνδυασμῷ. τούτου δ᾽
αἴτιον ὅτι τὰ τιμιώτερα καὶ αὐταρκέστερα τὴν
φύσιν ἐστίν, ὥστε μεγέθους μετειληφέναι. τοῦτο
δ᾽ οὐκ ἄνευ θερμότητος ψυχικῆς· ἀνάγκη γὰρ τὸ
20 μεῖζον ὑπὸ πλείονος κινεῖσθαι δυνάμεως, τὸ δὲ
θερμὸν κινητικόν. διόπερ, ὡς ἐπὶ τὸ πᾶν βλέ-

---

[1] ᾗ Peck : ἡ vulg. (ἡ Z² in ras. ; fuerat καὶ Z¹).
[2] om. S.
[3] τὸ Υ : τὸ ᾗ vulg. : ὕλης ᾗ τὸ θῆλυ coni. A.-W., τὸ ⟨θῆλυ⟩
ᾗ θῆλυ Btf. ; sed fortasse haec verba secludenda.   scr. Platt
ᾗ τὸ ἄρρεν ὑπάρχει τοῖς γινομένοις ᾗ ἡ ὕλη ᾗ τὸ θῆλυ.
[4] vv. 11-23 secludenda.

---

[a] Cf. 716 a 5.
[b] i.e., the Material Cause. Cf. 716 a 5.
[c] See Introd. §§ 1 ff., 10, 50.
[d] This paragraph seems to be out of place, consisting of
various remarks which are irrelevant here. Cf. 715 a 18 ff.,
and parts of Bk. I, ch. 23.

proximate motive cause,[a] to which belong the *logos*
and the Form, is *better* and more divine in its nature
than the Matter,[b] it is *better* also that the superior one
should be separate from the inferior one. That
is why wherever possible and so far as possible the
male is separate from the female, since it is something *better* and more divine in that it is the principle
of movement[c] for generated things, while the female
serves as their matter. The male, however, comes
together with the female and mingles with it for the
business of generation, because this is something that
concerns both of them.

[d] [Thus things are alive in virtue of having in them
a share of the male and of the female, and that is why
even plants have life. The class of animals, however,
is ⟨what it is⟩ in virtue of its power of sense-perception.[e] In practically all animals which can move
about the male and the female are found separate;
and the causes are the ones which have been stated;
and, as was said,[f] some of them emit semen during
copulation, some do not. The reason for this is that
the higher animals are more self-sufficient in their
nature, and so are large in size: this cannot be so
without heat of Soul, since of necessity the larger a
thing is, the greater the power required to move it,
and heat acts as a motive power. Hence, if we take

---

[e] *Cf. P.A.* 666 a 34 τὸ μὲν γὰρ ζῷον αἰσθήσει ὥρισται, and
651 b 4, 653 b 22. Aristotle seems to have perceived early
the importance of this point, as it occurred in his early work
Protrepticus. See Iamblichus, *Protrepticus* 7 (44. 9 Pistelli;
37. 9 Walzer, *Aristot. Dial. Frag.*), a passage which according to Jaeger (*Aristotle*, 69) comes from Aristotle's *Protrepticus*: ἀλλὰ μὴν τό γε ζῆν τῷ αἰσθάνεσθαι διακρίνεται τοῦ μὴ ζῆν,
and with that whole passage *cf.* 731 a 29–b 3 above.

[f] Bk. I, ch. 17.

ARISTOTLE

ψαντας εἰπεῖν, τὰ ἔναιμα μείζω τῶν ἀναίμων καὶ
τὰ πορευτικὰ τῶν μονίμων ζῴων· ἅπερ προΐεται
σπέρμα διὰ τὴν θερμότητα καὶ τὸ μέγεθος.]

Καὶ περὶ μὲν ἄρρενος καὶ θήλεος, δι' ἣν αἰτίαν
25 ἐστὶν ἑκάτερον, εἴρηται.

Τῶν δὲ ζῴων τὰ μὲν τελεσιουργεῖ καὶ ἐκπέμπει θύραζε ὅμοιον ἑαυτῷ, οἷον ὅσα ζῳοτοκεῖ εἰς
τοὐμφανές, τὰ δὲ ἀδιάρθρωτον ἐκτίκτει καὶ οὐκ ἀπ-
ειληφὸς τὴν αὑτοῦ μορφήν. τῶν δὲ τοιούτων τὰ
μὲν ἔναιμα ᾠοτοκεῖ, τὰ δ' ἄναιμα ⟨ἢ ᾠοτοκεῖ ἢ⟩[1]
σκωληκοτοκεῖ. διαφέρει δ' ᾠὸν καὶ σκώληξ· ᾠὸν
30 μὲν γάρ ἐστιν ἐξ οὗ γίνεται τὸ γινόμενον ἐκ
μέρους, τὸ δὲ λοιπόν ἐστι τροφὴ τῷ γινομένῳ,
σκώληξ δ' ἐξ οὗ τὸ γινόμενον ὅλου ὅλον γίνεται.
τῶν δὲ εἰς τὸ φανερὸν ὅμοιον ἀποτελούντων ζῷον
καὶ ζῳοτοκούντων τὰ μὲν εὐθὺς ἐν αὑτοῖς ζῳοτοκεῖ, οἷον ἄνθρωπος καὶ ἵππος καὶ βοῦς καὶ τῶν
35 θαλαττίων δὲ[2] δελφὶς καὶ τἆλλα τὰ τοιαῦτα, τὰ
732 bδ' ἐν αὑτοῖς ᾠοτοκήσαντα πρῶτον οὕτω ζῳοτοκεῖ
θύραζε, οἷον τὰ σελάχη καλούμενα. τῶν δ' ᾠοτοκούντων τὰ μὲν τέλειον προΐεται τὸ ᾠόν, οἷον
ὄρνιθες καὶ ὅσα τετράποδα ᾠοτοκεῖ καὶ ὅσα ἄποδα,
οἷον σαῦραι καὶ χελῶναι καὶ τῶν ὄφεων τὸ πλεῖστον
5 γένος (τὰ γὰρ τούτων ᾠὰ ὅταν ἐξέλθῃ, οὐκέτι
λαμβάνει αὔξησιν), τὰ δ' ἀτελῆ, οἷον οἵ τ' ἰχθύες

---

[1] Platt.　　　　[2] δὲ om. PSY.

---

[a] See Introd §§ 74 ff.
[b] Cf. 752 a 27, 758 b 10 ff., and H.A. 489 b 6 ff. The
distinction wh‘ch Aristo!le makes here is that between the
utilization of yolk as the raw material of embryonic develop-

134

a general view, we may say that blooded animals are larger than bloodless, and mobile ones larger than stationary ; and they are the ones which emit semen on account of their heat and their size.]

We have now stated the Cause why each of the two, male and female, is.

Some animals bring their young to perfection, and bring forth externally a creature similar to themselves—*e.g.*, those which are externally viviparous ; others produce something which is unarticulated and has not yet assumed its proper shape. In the latter class those which are blooded lay eggs, those which are bloodless produce ⟨either eggs or⟩ larvae.[a] The difference between an egg and a larva is this : an egg is something from *part* of which the new creature is formed, while the remainder is nourishment for it ; whereas in the case of the larva, the *whole* of it is used to form the whole of the offspring.[b] Of the animals which produce externally a perfected creature similar to themselves, *i.e.*, the Vivipara, some are internally viviparous from the outset (as man, horse, ox ; and of sea-creatures, the dolphin and the other animals of that sort), others are internally oviparous at the first stage, and thereafter are externally viviparous (as what are called Selachia). Of oviparous animals, some lay their eggs in a perfected state (as birds, oviparous quadrupeds and footless animals, *e.g.*, lizards and tortoises, and the great majority of the serpents [c])—eggs which once they are laid do not grow any more ; others lay their eggs in an imper-

The various methods of generation.

ment, and the utilization of tissue-disintegration products in metamorphosis. The embryo feeds upon its yolk, but the pupa feeds upon itself.

[c] The viper is the exception ; see below, line 21.

καὶ τὰ μαλακόστρακα καὶ τὰ μαλάκια καλούμενα·
τούτων γὰρ τὰ ᾠὰ αὐξάνεται ἐξελθόντα.

Πάντα δὲ τὰ ζωοτοκοῦντα [ἢ ᾠοτοκοῦντα][1] ἔναιμά
ἐστιν, καὶ τὰ ἔναιμα ἢ ζωοτοκεῖ ἢ ᾠοτοκεῖ, ὅσα
10 μὴ ὅλως ἄγονά ἐστιν. τῶν δ' ἀναίμων τὰ ἔντομα
σκωληκοτοκεῖ, ὅσα ἢ ἐκ συνδυασμοῦ γίνεται ἢ
αὐτὰ συνδυάζεται. ἔστι γὰρ ἔνια τοιαῦτα τῶν
ἐντόμων ἃ γίνεται μὲν αὐτόματα, ἔστι δὲ θήλεα
καὶ ἄρρενα, καὶ ἐκ συνδυαζομένων γίνεταί τι
αὐτῶν, ἀτελὲς μέντοι τὸ γιγνόμενον· ἡ δ' αἰτία
εἴρηται πρότερον ἐν ἑτέροις.

15 Συμβαίνει δὲ πολλὴ ἐπάλλαξις τοῖς γένεσιν.
οὔτε γὰρ τὰ δίποδα πάντα ζωοτοκεῖ (οἱ γὰρ ὄρνιθες
ᾠοτοκοῦσιν) οὔτ' ᾠοτοκεῖ πάντα (ὁ γὰρ ἄνθρωπος
ζωοτοκεῖ), οὔτε τὰ τετράποδα πάντα ᾠοτοκεῖ
(ἵππος γὰρ καὶ βοῦς καὶ ἄλλα μυρία ζωοτοκεῖ)
οὔτε ζωοτοκεῖ πάντα (σαῦροι[2] γὰρ καὶ κροκόδειλοι
20 καὶ ἄλλα πολλὰ ᾠοτοκοῦσιν). οὐδ' ἐν τῷ πόδας
ἔχειν ἢ μὴ ἔχειν διαφέρει· καὶ γὰρ ἄποδα ζωοτοκεῖ,
οἷον οἱ ἔχεις καὶ τὰ σελάχη, τὰ δ' ᾠοτοκεῖ, οἷον
τὸ τῶν ἰχθύων γένος καὶ τὸ τῶν ἄλλων ὄφεων·
καὶ τῶν πόδας ἐχόντων καὶ ᾠοτοκεῖ πολλὰ καὶ
ζωοτοκεῖ, οἷον τὰ εἰρημένα τετράποδα. καὶ ἐν
25 αὑτοῖς δὲ ζωοτοκεῖ καὶ πόδας ἔχοντα, οἷον ἄνθρωπος, καὶ ἄποδα, οἷον φάλαινα καὶ δελφίς. ταύτῃ
μὲν οὖν οὐκ ἔστι διελεῖν, οὐδ' αἴτιον τῆς διαφορᾶς

---

[1] seclusit Platt (idem Sus.).
[2] σαῦροι PSY*Z : σαύραι O[b], vulg.

---

[a] Cf. 718 b 8 and note there.　　　　　[b] See 721 a 3 ff.
[c] Aristotle may have in mind the method of dichotomy,
against which he inveighs elsewhere (see P.A. 642 b 5 ff.,

fect state, as the Fishes, and the Crustacea and the Cephalopods as they are called, whose eggs do grow in size after they are laid.[a]

All animals that are viviparous [or oviparous] are blooded, and animals that are blooded are either viviparous or oviparous, apart from those which are completely infertile. Of bloodless animals, Insects produce a larva ; this holds good both for those which are formed as a result of copulation and those which themselves copulate.[b] (A note of explanation : there are certain Insects which, although formed by spontaneous generation, nevertheless are male and female, and as a result of their copulation something is formed, though it is imperfect : the cause of this has already been stated elsewhere.)

Actually there is a good deal of overlapping between the various classes. Bipeds are not all viviparous (birds are oviparous) nor all oviparous (man is viviparous) ; quadrupeds are not all oviparous (the horse and ox and heaps of others are viviparous), nor all viviparous (lizards and crocodiles and many others are oviparous). Nor does the difference lie even in having or not having feet : some footless animals are viviparous (as vipers, and the Selachia), some are oviparous (as the class of fishes, and the rest of the serpents) ; and of the footed animals many are oviparous, many viviparous (e.g., the quadrupeds already mentioned). There are footed animals which are internally viviparous (as man), and footless ones also (as the whale and dolphin). So we find no means here for making a division [c] : the cause of this difference

Classification of animals.

and my note there), as used, though for a different purpose, by Plato in *Sophist* and *Politicus* (*e.g.*, the division into τὸ πεζόν and τὸ νευστικόν at *Sophist* 220 A).

732 b

ταύτης οὐθὲν τῶν πορευτικῶν ὀργάνων, ἀλλὰ ζωοτοκεῖ μὲν¹ τὰ τελεώτερα τὴν φύσιν τῶν ζῴων καὶ
30 μετέχοντα καθαρωτέρας ἀρχῆς· οὐθὲν γὰρ ζωοτοκεῖ
ἐν αὑτῷ, μὴ δεχόμενον τὸ πνεῦμα καὶ ἀναπνέον.
τελεώτερα δὲ τὰ θερμότερα τὴν φύσιν καὶ ὑγρότερα
καὶ μὴ γεώδη. τῆς δὲ θερμότητος τῆς φυσικῆς
ὅρος ὁ πλεύμων, ὅσων ἔναιμός ἐστιν· ὅλως μὲν γὰρ
τὰ ἔχοντα πλεύμονα τῶν μὴ ἐχόντων θερμότερα,
35 τούτων δ' αὐτῶν τὰ μὴ σομφὸν ἔχοντα μηδὲ στι-
733 a φρὸν μηδ' ὀλίγαιμον ἀλλ' ἔναιμον καὶ μαλακόν.
ὥσπερ δὲ τὸ μὲν ζῷον τέλειον,² ὁ δὲ σκώληξ καὶ
τὸ ᾠὸν ἀτελές, οὕτως τὸ τέλειον ἐκ τοῦ τελειοτέρου
γίνεσθαι πέφυκεν. τὰ δὲ θερμότερα μὲν διὰ τὸ
ἔχειν πλεύμονα, ξηρότερα δὲ τὴν φύσιν, ἢ τὰ ψυ-
5 χρότερα μὲν ὑγρότερα δέ, τὰ μὲν ᾠοτοκεῖ τέλειον
ᾠόν, τὰ δ' ᾠοτοκήσαντα ζῳοτοκεῖ ἐν αὑτοῖς. οἱ
μὲν γὰρ ὄρνιθες καὶ τὰ φολιδωτὰ διὰ μὲν θερμότητα τελεσιουργοῦσι, διὰ δὲ ξηρότητα ᾠοτοκοῦσι, τὰ δὲ σελάχη θερμὰ μὲν ἧττον τούτων,
ὑγρὰ δὲ μᾶλλον, ὥστε μετέχει ἀμφοτέρων· καὶ γὰρ
10 ᾠοτοκεῖ καὶ ζῳοτοκεῖ ἐν αὑτοῖς, ᾠοτοκεῖ μὲν ὅτι
ψυχρά, ζῳοτοκεῖ δ' ὅτι ὑγρά· ζωτικὸν γὰρ τὸ
ὑγρόν, πορρωτάτω δὲ τοῦ ἐμψύχου τὸ ξηρόν. ἐπεὶ
δ' οὔτε πτερωτὰ οὔτε φολιδωτὰ οὔτε λεπιδωτά
ἐστιν, ἃ σημεῖα ξηρᾶς μᾶλλον καὶ γεώδους φύσεως,

---

¹ ἐν αὑτοῖς add. Z.   ² τέλειον PSYZ² : τέλεον vulg.

ᵃ See Introd. § 38.   ᵇ Not a living creature.

does not lie in any of the organs of locomotion. No ; those animals are viviparous which are more perfect in their nature, which partake of a purer " principle " ; in other words, no animal is internally viviparous unless it draws in breath—respires. The more perfect animals are those which are by their nature hotter and more fluid and are not earthy. (The test of natural heat is the presence of the lung, provided it has blood in it. Speaking generally, animals which have a lung are hotter than those that have none, and of the former those are hotter whose lung is not spongy nor compact nor poorly supplied with blood, but well supplied with blood and soft.) And since an actual animal is something perfect whereas larvae and eggs are something imperfect, Nature's rule is that the perfect offspring shall be produced by the more perfect sort of parent. Those animals which are hotter (as their having a lung indicates), though of a more solid ^{a} consistency, or are colder but more fluid, either (a) are oviparous and lay a perfect egg, or (b) first lay an egg and then are viviparous internally. Thus, birds and the animals with horny scales, on account of their heat, produce something perfect, but on account of their solidity it is an egg only ^{b} ; the Selachia are less hot than these are, but more fluid ; hence they share in the characteristics of both—they are oviparous because they are cold creatures, and internally viviparous because they are fluid (the reason being that fluid matter is conducive to life, whereas solid matter and the living organism are at opposite poles) ; and as they have neither feathers nor horny plates nor scales, which are signs of a constitution that tends to be solid and earthy, the egg which they produce is

μαλακὸν τὸ ᾠὸν γεννῶσιν· ὥσπερ γὰρ οὐδ' ἐν
15 αὐτῷ, οὐδ' ἐν τῷ ᾠῷ ἐπιπολάζει τὸ γεηρόν. καὶ
διὰ τοῦτο εἰς αὐτά[1] ᾠοτοκεῖ· θύραζε γὰρ ἂν ἰὸν
διεφθείρετο τὸ ᾠόν, οὐκ ἔχον προβολήν.

Τὰ δὲ ψυχρὰ καὶ ξηρὰ μᾶλλον ᾠοτοκεῖ μέν,
ἀτελὲς δὲ τὸ ᾠόν, καὶ σκληρόδερμον δὲ διὰ τὸ
γεηρὰ εἶναι καὶ ἀτελὲς προΐεσθαι, ἵνα σῴζηται
20 φυλακὴν ἔχον τὸ ὀστρακῶδες. οἱ μὲν οὖν ἰχθύες
λεπιδωτοὶ ὄντες καὶ τὰ μαλακόστρακα γεηρὰ ὄντα
σκληρόδερμα τὰ ᾠὰ γεννᾷ. τὰ δὲ μαλάκια, ὥσπερ
αὐτὰ γλίσχρα τὴν τοῦ σώματός ἐστι φύσιν, οὕτως
σῴζει ἀτελῆ προϊέμενα τὰ ᾠά· προΐεται γὰρ γλισχρότητα περὶ τὸ κύημα πολλήν. τὰ δ' ἔντομα
25 πάντα σκωληκοτοκεῖ. ἔστι δ' ἅπαντα ἄναιμα τὰ
ἔντομα, διὸ καὶ[2] σκωληκοτοκοῦντα θύραζε. τὰ δ'
ἄναιμα οὐ πάντα σκωληκοτοκεῖ ἁπλῶς· ἐπαλλάτ-
τουσι γὰρ ἀλλήλοις [τά τ' ἔντομα][3] τὰ σκωληκοτοκοῦντα[4] καὶ τὰ ἀτελὲς τίκτοντα τὸ ᾠόν, οἷον
οἵ τ' ἰχθύες οἱ λεπιδωτοὶ καὶ τὰ μαλακόστρακα
30 καὶ τὰ μαλάκια. τούτων μὲν γὰρ τὰ ᾠὰ σκωληκώδη ἐστίν (αὔξησιν γὰρ λαμβάνει θύραζε),
ἐκείνων δ' οἱ σκώληκες γίνονται προϊόντες ᾠο-
ειδεῖς· ὃν δὲ τρόπον, ἐν τοῖς ὕστερον διοριοῦμεν.

Δεῖ δὲ νοῆσαι ὡς εὖ καὶ ἐφεξῆς τὴν γένεσιν
ἀποδίδωσιν ἡ φύσις. τὰ μὲν γὰρ τελεώτερα καὶ
θερμότερα[5] τῶν ζῴων τέλειον ἀποδίδωσι τὸ τέκνον
κατὰ τὸ ποιόν (κατὰ δὲ τὸ ποσὸν ὅλως οὐθὲν τῶν

---

[1] αὐτὰ P : αὐτὸ vulg.
[2] καὶ τὰ PSYΣ.   [3] seclusi.
[4] τὰ ἔντομα καὶ τὰ σκωληκοτοκοῦντα ZΣ : τά τ' ἔντομα vulg.
[5] τελειότατα καὶ θερμότατα P.

---

[a] Bk. III, ch. 9.

a soft one : the earthy substance does not come to the surface in the egg any more than it does in the creature which lays it. And that is why they lay their eggs internally : if the eggs emerged they would be destroyed through lack of protection.

Animals that tend to be cold and solid lay eggs, it is true, but their egg is imperfect, and it has a hard covering (a) because the animals themselves are earthy and (b) because it is in an imperfect state when laid, and the shelly exterior serves as a protection to keep it safe. Thus fishes, being scaly, and Crustacea, being earthy, produce eggs with a hard covering ; while the Cephalopods, which also lay imperfect eggs, keep them safe by a method in accordance with the sticky nature of their own bodies ; they exude a large amount of sticky substance over the fetation. Insects all produce larvae. Now all Insects are bloodless, and that actually is why they are externally larva-producing. But it is not true that all bloodless animals are larva-producing without qualification, because there is overlapping as between the larva-producing animals and those that produce imperfect eggs (e.g., the scaly fishes, the Crustacea and the Cephalopods), since the eggs of the latter are larva-like, in that they grow bigger after they have been laid externally, while the larvae of the former, as they develop, become egg-like : we shall explain later how this happens.[a]

We should notice how well Nature brings generation about in its several forms : they are arranged in a regular series, thus : (1) The more perfect and hotter of the animals produce their young in a perfect state so far as their quality is concerned (no animal brings forth young that are perfect in *size*, because

141

733 b

ζῴων· πάντα γὰρ γενόμενα λαμβάνει αὔξησιν), καὶ
γεννᾷ δὴ ταῦτα ζῷα ἐν αὑτοῖς εὐθύς. τὰ δὲ δεύ-
5 τερα ἐν αὑτοῖς μὲν οὐ γεννᾷ τέλεια εὐθύς (ζῳοτοκεῖ
γὰρ ᾠοτοκήσαντα πρῶτον), θύραζε δὲ ζῳοτοκεῖ.
τὰ δὲ ζῷον μὲν οὐ τέλειον γεννᾷ, ᾠὸν δὲ γεννᾷ,
καὶ τοῦτο τέλειον τὸ ᾠόν. τὰ δ' ἔτι τούτων ψυχροτέραν ἔχοντα τὴν φύσιν ᾠὸν μὲν γεννᾷ οὐ
τέλειον δὲ ᾠόν, ἀλλ' ἔξω τελειοῦται, καθάπερ τὸ
10 τῶν λεπιδωτῶν ἰχθύων γένος καὶ τὰ μαλακόστρακα
καὶ τὰ μαλάκια. τὸ δὲ πέμπτον γένος καὶ ψυχρότατον οὐδ' ᾠοτοκεῖ ἐξ αὑτοῦ, ἀλλὰ καὶ τὸ[1] τοιοῦτον
ἔξω συμβαίνει πάθος αὐτῷ, ὥσπερ εἴρηται· τὰ
γὰρ ἔντομα σκωληκοτοκεῖ τὸ πρῶτον· προελθὼν δ'
ὠώδης γίνεται ὁ σκώληξ (ἡ γὰρ χρυσαλλὶς κα-
15 λουμένη δύναμιν ᾠοῦ ἔχει)· εἶτ' ἐκ τούτου γίνεται
ζῷον, ἐν τῇ τρίτῃ μεταβολῇ λαβὸν τὸ τῆς γενέσεως
τέλος.

Τὰ μὲν οὖν οὐ γίνεται τῶν ζῴων ἀπὸ σπέρματος,
ὥσπερ ἐλέχθη καὶ πρότερον· τὰ δ' ἔναιμα πάντα
γίνεται ἀπὸ σπέρματος, ὅσα ἐκ συνδυασμοῦ γί-
20 νεται, προϊεμένου τοῦ ἄρρενος εἰς τὸ θῆλυ γονήν,
ἧς εἰσελθούσης τὰ ζῷα συνίσταται καὶ λαμβάνει
τὴν οἰκείαν μορφήν, τὰ μὲν ἐν αὑτοῖς τοῖς ζῴοις
ὅσα ζῳοτοκεῖ, τὰ δ' ἐν ᾠοῖς [καὶ σπέρμασι καὶ
τοιαύταις ἄλλαις ἀποκρίσεσιν].[2]

Περὶ ὧν ἐστὶν ἀπορία πλείων, πῶς ποτε γίνεται ἐκ

---

[1] τοῦ Bekker per typoth. err.
[2] seclusit Platt (om. Σ), sed monet quaedam de plantis
fortasse excidisse.

---

[a] Above, 733 a 31.

they all grow in size after they have been produced), and these young which they generate are living creatures inside them from the outset. (2) The second class do not generate perfect animals within themselves from the outset : although they are viviparous, they lay eggs first of all ; externally however they are viviparous. (3) Others produce not a perfect animal, but an egg, which is perfect. (4) Those whose constitution is still colder than this produce an egg, but it is not a perfect one : it reaches its perfection outside the parent. Examples are the scaly fishes, the Crustacea and the Cephalopods. (5) The fifth class of creatures, which are the coldest of all, do not even lay an egg directly themselves, but the formation of their egg takes place outside the parent, as has been said.[a] What happens is that Insects first produce a larva, then the larva develops till it becomes egg-like (what is called the chrysalis is really equivalent to an egg [b]) ; then out of this an animal is formed, and it is not until this third stage in its series of changes that it reaches the end and perfection of its generation.

There are, then, some animals which are not formed from semen, as I have in fact said already. All blooded ones, however, are formed from semen, so many as are formed as the result of copulation, that is to say, the male emits semen into the female, and upon the entry of the semen the young animals are " set " and constituted and assume their proper shape ; with the viviparous animals this stage takes place within the parent, with others in the eggs [and seeds and other such secretions].

And on this subject we are confronted by no small How is the

[b] Lit., " has the *dynamis* of an egg " : see Introd. § 26.

733 b

τοῦ σπέρματος τῶν φυτῶν¹ ἢ τῶν ζῴων ὁτιοῦν.
25 ἀνάγκη γὰρ τὸ γιγνόμενον καὶ ἔκ τινος γίνεσθαι
καὶ ὑπό τινος καί τι. ἐξ οὗ μὲν οὖν ἐστιν ὕλη, ἣν
ἔνια μὲν ζῷα ἔχει πρώτην ἐν αὑτοῖς, λαβόντα² ἐκ τοῦ
θήλεος, οἷον ὅσα μὴ ζῳοτοκεῖται ἀλλὰ σκωληκοτοκεῖται ἢ ᾠοτοκεῖται, τὰ δὲ μέχρι πόρρω ἐκ τοῦ
30 θήλεος λαμβάνει διὰ τὸ θηλάζειν, ὥσπερ ὅσα ζῳοτοκεῖται μὴ μόνον ἐκτὸς ἀλλὰ καὶ ἐντός. ἐξ οὗ μὲν
οὖν γίνεται, ἡ τοιαύτη ὕλη ἐστίν· ζητεῖται δὲ νῦν
οὐκ ἐξ οὗ ἀλλ' ὑφ' οὗ γίνεται τὰ μόρια. ἤτοι γὰρ
τῶν ἔξωθέν τι ποιεῖ, ἢ ἐνυπάρχον³ τι ἐν τῇ γονῇ καὶ
734 a σπέρματι· καὶ τοῦτ' ἐστὶν ἢ μέρος τι ψυχῆς ἢ
ψυχὴ ἢ ἔχον ἂν εἴη ψυχήν. τῶν μὲν οὖν ἔξωθέν
τι ποιεῖν ἕκαστον ἢ τῶν σπλάγχνων ἢ τῶν ἄλλων
μερῶν ἄλογον ἂν δόξειεν· κινεῖν τε γὰρ μὴ ἁπτόμενον ἀδύνατον καὶ μὴ κινοῦντος πάσχειν τι ὑπὸ
5 τούτου. ἐν αὐτῷ ἄρα τῷ κυήματι ἐνυπάρχει τι
ἤδη ἢ⁴ αὐτοῦ μόριον ἢ κεχωρισμένον. τὸ μὲν οὖν

---

¹ τῶν φυτῶν Z : τὸ φυτὸν vulg.
² λαβόντα S : λαμβάνοντα vulg.
³ ἐνυπάρχον Peck : ἐνυπάρχει vulg.
⁴ ἤδη ἢ Υ : ἤδη Z : ἢ δὴ vulg.

ᵃ The discussion which follows shows that Aristotle fully
appreciated the greatest problem of embryological theory, a
problem which gave rise to centuries of controversy. Does
the embryo contain all its parts in little from the beginning,
unfolding like a Japanese paper flower in water (" preformation "), or is there a true formation of new structures
as it develops (" epigenesis ")? Aristotle was an epigenesist,
but he was not vindicated till the time of C. F. Wolff and
K. E. von Baer, at the end of the 18th and the beginning of
the 19th century. The history of the controversy will be
found in J. Needham's *History of Embryology* and A. W.

puzzle.[a] How, we ask, is any plant formed out of the seed, or any animal out of the semen ? That which is formed by means of a process must of necessity be formed (a) out of something (b) by something (c) into something. " Out of something." This of course is the material or matter. Some animals have their primary matter [b] within themselves, having derived it from the female parent, e.g., those animals which are produced not viviparously but out of larvae or eggs. Others derive it from the mother for a considerable time by being suckled. These are the animals which are produced viviparously not externally only but also internally.[c] So then, that " out of which " the parts are formed is material of this sort. The problem now before us however is not Out of what, but, By what, are they formed ? Either something external fashions them, or else something present in the semen or seminal fluid ; and this is either some part of Soul, or Soul, or something which possesses Soul. Now it would appear unreasonable to suppose that anything external fashions all the individual parts, whether they be the viscera or any others, because unless it is in contact [d] it cannot set up any movement, and unless it sets up a movement no effect can be produced upon anything by it. Hence it follows that there must be something already present inside the fetation itself, which is either a part of it or separate from it.

embryo formed ?

Meyer's *The Rise of Embryology*. Like many erroneous theories, preformationism contained some truth, for we know to-day that the course of the embryo's development is predetermined by its genetic constitution.

[b] *Cf.* 729 a 33 note.

[c] This excludes the Selachia.

[d] *Cf.* Bk. I. 730 b 5 ff., and see App. B § 22, n.

ARISTOTLE

ἄλλο τι εἶναι κεχωρισμένον ἄλογον· γεννηθέντος
γὰρ τοῦ ζῴου πότερον φθείρεται ἢ ἐμμένει; ἀλλ'
οὐδὲν τοιοῦτον φαίνεται ἐνὸν ὃ οὐ μόριον τοῦ ὅλου
ἢ φυτοῦ ἢ ζῴου ἐστίν. ἀλλὰ μὴν καὶ τὸ φθείρεσθαί
10 γε ποιῆσαν εἴτε πάντα τὰ μέρη εἴτε τινὰ ἄτοπον·
τὰ λοιπὰ γὰρ τί ποιήσει; εἰ γὰρ ἐκεῖνο μὲν τὴν
καρδίαν, εἶτ' ἐφθάρη, αὕτη δ' ἕτερον, τοῦ αὐτοῦ
λόγου ἢ πάντα φθείρεσθαι ἢ πάντα μένειν. σῴζεται
ἄρα. αὐτοῦ ἄρα μόριόν ἐστιν, ὃ εὐθὺς ἐνυπάρχει
ἐν τῷ σπέρματι. εἰ δὲ δὴ μή ἐστι τῆς ψυχῆς
15 μηθὲν ὃ μὴ τοῦ σώματός ἐστιν ἔν τινι μορίῳ, καὶ
ἔμψυχον ἄν τι εἴη μόριον εὐθύς.

Τὰ οὖν ἄλλα πῶς; ἢ γάρ τοι ἅμα πάντα γίγνεται τὰ μόρια, οἷον καρδία πλεύμων ἧπαρ ὀφθαλμὸς καὶ τῶν ἄλλων ἕκαστον, ἢ ἐφεξῆς, ὥσπερ
ἐν τοῖς καλουμένοις Ὀρφέως ἔπεσιν· ἐκεῖ γὰρ
20 ὁμοίως φησὶ γίγνεσθαι τὸ ζῷον τῇ τοῦ δικτύου
πλοκῇ. ὅτι μὲν οὖν οὐχ ἅμα, καὶ τῇ αἰσθήσει ἐστὶ
φανερόν· τὰ μὲν γὰρ φαίνεται ἐνόντα ἤδη τῶν
μορίων, τὰ δ' οὔ. ὅτι δ' οὐ διὰ μικρότητα οὐ
φαίνεται, δῆλον· μείζων γὰρ τὸ μέγεθος ὢν ὁ
πνεύμων τῆς καρδίας ὕστερον φαίνεται τῆς καρδίας
25 ἐν τῇ ἐξ ἀρχῆς γενέσει. ἐπεὶ δὲ τὸ μὲν πρότερον
τὸ δ' ὕστερον, πότερον θάτερον ποιεῖ θάτερον, καὶ

---

[a] It would be inconsistent to say that the disappearance
was arrested at some arbitrary stage in the process.
[b] Apart from rational Soul, the connexion is reciprocal;
and Aristotle often remarks that there is no part of the body
which has no Soul in it ; see 726 b 22 and 735 a 6 ff.
146

To suppose it is some other thing, and separate from it, is not reasonable. If it were, the question arises: When the animal's generation is completed, does this something disappear, or does it remain within the animal? We cannot detect any such thing, something which is in the plant or the animal and yet is no part of the organism as a whole. And again, to say that it fashions all the parts or some parts of the organism and then disappears is ridiculous. If it fashions only some of the parts, what will fashion the rest? Supposing it fashions the heart, and then disappears, and the heart fashions some other part: to be consistent we must say that either all the parts disappear or all the parts remain.[a] It must, then, persist. And therefore it must be a part of the whole, existing in the semen from the outset. And if it is true that there is no part of the Soul which is not in some part of the body,[b] then it must also be a part which contains Soul from the outset.

How, then, are the other parts formed? Either they are all formed simultaneously—heart, lung, liver, eye, and the rest of them—or successively, as we read in the poems ascribed to Orpheus, where he says that the process by which an animal is formed resembles the plaiting of a net. As for simultaneous formation of the parts, our senses tell us plainly that this does not happen: some of the parts are clearly to be seen present in the embryo while others are not. And our failure to see them is not because they are too small; this is certain, because although the lung is larger in size than the heart it makes its appearance later in the original process of formation. Since one part, then, comes earlier and another later, is it the case that A fashions B and that it is there on

734 a

ἔστι διὰ τὸ ἐχόμενον, ἢ μᾶλλον μετὰ τόδε γίνεται
τόδε; λέγω δ' οἷον οὐχ ἡ καρδία γενομένη ποιεῖ
τὸ ἧπαρ, τοῦτο δ' ἕτερόν τι, ἀλλὰ τόδε μετὰ τόδε,
[ὥσπερ μετὰ τὸ παῖς ἀνὴρ γίνεται],[1] ἀλλ' οὐχ ὑπ'
30 ἐκείνου. λόγος δὲ τούτου, ὅτι ὑπὸ τοῦ ἐντελεχείᾳ
ὄντος τὸ δυνάμει ὂν γίνεται ἐν τοῖς φύσει ἢ τέχνῃ
γινομένοις, ὥστε δέοι ἂν τὸ εἶδος καὶ τὴν μορφὴν
ἐν ἐκείνῳ εἶναι, οἷον ἐν τῇ καρδίᾳ τὸ τοῦ ἥπατος.
καὶ ἄλλως δ' ἄτοπος καὶ πλασματίας ὁ λόγος.
ἀλλὰ μὴν καὶ τὸ ἐν τῷ σπέρματι εὐθὺς ἐνυπάρχειν
35 τι μόριον τοῦ ζῴου ἢ φυτοῦ γεγενημένον, εἴτε
δυνάμενον ποιεῖν τἆλλα εἴτε μή, ἀδύνατον, εἰ πᾶν
ἐκ σπέρματος καὶ γονῆς γίγνεται. δῆλον γὰρ ὅτι
ὑπὸ τοῦ τὸ σπέρμα ποιήσαντος ἐγένετο, εἴπερ

734 b

εὐθὺς ἐνυπάρχει. ἀλλὰ σπέρμα δεῖ γενέσθαι πρότερον, καὶ τοῦτ' ἔργον τοῦ γεννῶντος. οὐθὲν ἄρα
οἷόν τε μόριον ὑπάρχειν. οὐκ ἄρα ἔχει τὸ ποιοῦν
τὰ μόρια ἐν αὑτῷ. ἀλλὰ μὴν οὐδ' ἔξω· ἀνάγκη δὲ
τούτων εἶναι θάτερον.
5 Πειρατέον δὴ ταῦτα λύειν· ἴσως γάρ τι τῶν
εἰρημένων ἐστὶν οὐχ ἁπλοῦν, οἷον πῶς ποτε ὑπὸ
τοῦ ἔξω οὐκ ἐνδέχεται γίνεσθαι. ἔστι μὲν γὰρ ὡς
ἐνδέχεται, ἔστι δ' ὡς οὔ. τὸ μὲν οὖν τὸ σπέρμα

---

[1] seclusi ; velit secludere Platt.

---

[a] As argued already, 734 a 2 ff.

148

account of B which is next to it, or is it rather the
case that B is formed after A ? I mean, for instance,
not that the heart, once it is formed, fashions the
liver, and then the liver fashions something else ;
but that the one is formed after the other [just as a
man is formed after a child], not by it. The reason
of this is that, so far as the things formed by nature
or by human art are concerned, the formation of
that which is *potentially* is brought about by that
which is *in actuality* ; so that the Form, or conformation, of B would have to be contained in A,
e.g., the Form of the liver would have to be in the
heart—which is absurd. And there are other ways
too in which the theory is absurd and fondly in-
vented. But besides, for any part of the animal
or plant to be present from the outset ready formed
within the semen or seed, whether it has the power
to fashion the other parts or not—even this is impossible if everything is formed out of semen or seed ;
because it is plain that it was formed by that which
fashioned the semen if it is present within the semen
from the outset ; but semen must be formed before
⟨any part⟩, and that is the business of the parent.
Therefore no part can be present within the semen.
Therefore it does not contain in itself that which
fashions the parts. And yet this cannot be external
to the semen either [a]: and it must be either external to it or inside it.

Well, we must endeavour to solve this difficulty.
Maybe there is some statement of ours, made without
qualification, which ought to be qualified : *e.g.*, if we
ask, *in what sense* exactly is it impossible for the parts
to be formed by something external ? we see that
in one sense it is possible, though in another it is not.

λέγειν ἢ ἀφ' οὗ τὸ σπέρμα, οὐθὲν διαφέρει ἢ ἔχει
τὴν κίνησιν ἐν ἑαυτῷ ἣν ἐκεῖνο ἐκίνει. ἐνδέχεται
10 δὲ τόδε μὲν τόδε κινῆσαι, τόδε δὲ τόδε, καὶ εἶναι
οἷον τὰ αὐτόματα τῶν θαυμάτων. ἔχοντα γάρ πως
ὑπάρχει δύναμιν τὰ μόρια ἠρεμοῦντα· ὧν τὸ πρῶτον
ὅταν τι κινήσῃ τῶν ἔξωθεν, εὐθὺς τὸ ἐχόμενον
γίγνεται[1] ἐνεργείᾳ. ὥσπερ[2] οὖν[3] ἐν τοῖς αὐτομάτοις,
τρόπον μέν τινα ἐκεῖνο κινεῖ οὐχ ἁπτόμενον νῦν
15 οὐθενός, ἁψάμενον μέντοι, ὁμοίως [δὲ][4] καὶ ⟨τὸ⟩[5]
ἀφ' οὗ τὸ σπέρμα ἢ τὸ ποιῆσαν τὸ σπέρμα, ἁψάμενον μέν τινος, οὐχ ἁπτόμενον δ' ἔτι· τρόπον δέ
τινα ἡ ἐνοῦσα κίνησις, ὥσπερ ἡ οἰκοδόμησις τὴν
οἰκίαν.

Ὅτι μὲν οὖν ἔστι τι ὃ ποιεῖ, οὐχ οὕτως δὲ
ὡς τόδε τι, οὐδ' ἐνυπάρχον[6] ὡς τετελεσμένον τὸ
πρῶτον, δῆλον.

20 Πῶς δέ ποτε ἕκαστον γίγνεται, ἐντεῦθεν δεῖ
λαβεῖν, ἀρχὴν ποιησαμένους πρῶτον μὲν ὅτι ὅσα

---

[1] κινεῖται coni. A.-W. [2] ὥστε P.
[3] καθάπερ PS, om. Z. [4] secl. A.-W. [5] ⟨τὸ⟩ Peck.
[6] sic A.-W. : οὐδὲν ὑπάρχον P : οὐδ' ἐνυπάρχει vulg.

---

[a] It will be noticed that the passage which follows sounds
surprisingly modern ; this is largely due to the great emphasis
which Aristotle here gives to the rôle played by the Efficient
(or Motive) Cause.—See however App. B § 5.

[b] Cf. 741 b 9 ; and G. & C. II, chh. 10 and 11. At Mech.
848 a, there is a description of the mechanism by which these
may have been worked.

[c] κινεῖται (" is set in movement ") has been suggested for
γίγνεται (" comes to be "). But perhaps γίγνεσθαι ἐνεργείᾳ
is the inceptive form of εἶναι ἐνεργείᾳ, as in the phrase
ὄντος ἐνεργείᾳ, line 21 below.

Now it makes no difference whether we say " the semen " or " that from which the semen comes," in so far as the semen has within itself the movement which the generator set going. [a] And it is possible that A should move B, and B move C, and that the process should be like that of the " miraculous " automatic puppets [b] : the parts of these automatons, even while at rest, have in them somehow or other a *potentiality*, and when some external agency sets the first part in movement, then immediately the adjacent part comes to be [c] *in actuality*. The cases then are parallel : just as with the automaton (1) in one way it is the external agency which is causing the thing's movement—viz., not by being in contact with it anywhere now, but by having at one time been in contact with it, so too that from which the semen originally came, or that which fashioned the semen, ⟨causes the embryo's movement⟩ [d]—viz., not by being in contact with it still, but by having once been in contact with it at some point ; (2) in another way, it is the movement resident within ⟨which causes it to move⟩, just as the activity of building causes the house to get built.[e]

It is clear by now that there is something which fashions the parts of the embryo, but that this agent is not by way of being a definite individual thing,[f] nor is it present in the semen as something already perfected to begin with.

To answer the question, How exactly is each of the parts formed ? we must take first of all as our

---

[d] *i.e.*, development ; see Introd. §§ 47 ff.

[e] *Cf.* above, 730 b 8.

[f] τόδε τι: cf. Met. 1030 a 7 τὸ τόδε τι ταῖς οὐσίαις ὑπάρχει μόνον. A τόδε τι is often.equated with an οὐσία. Also cf. P.A. 641 b 31 γένεσις μὲν γὰρ τὸ σπέρμα, οὐσία δὲ τὸ τέλος.

734 b

φύσει γίγνεται ἢ τέχνη, ὑπ' ἐνεργείᾳ ὄντος γίνεται
ἐκ τοῦ δυνάμει τοιούτου. τὸ μὲν οὖν σπέρμα
τοιοῦτον, καὶ ἔχει κίνησιν καὶ ἀρχὴν τοιαύτην,
ὥστε παυομένης[1] τῆς κινήσεως γίνεσθαι ἕκαστον
25 τῶν μορίων καὶ ἔμψυχον. οὐ γάρ ἐστι πρόσωπον
μὴ ἔχον ψυχήν, οὐδὲ σάρξ, ἀλλὰ φθαρέντα ὁμωνύμως λεχθήσεται τὸ μὲν εἶναι πρόσωπον τὸ δὲ
σάρξ, ὥσπερ κἂν εἰ ἐγίγνετο λίθινα ἢ ξύλινα. ἅμα
δὲ τὰ ὁμοιομερῆ γίνεται καὶ τὰ ὀργανικά· καὶ
ὥσπερ οὐδ' ἂν πέλεκυν οὐδ' ἄλλο ὄργανον φή-
σαιμεν ἂν ποιῆσαι τὸ πῦρ μόνον, οὕτως οὐδὲ πόδα
30 οὐδὲ χεῖρα. τυν αὐτὸν δὲ τρόπον οὐδὲ σάρκα· καὶ
γὰρ ταύτης ἔργον τί ἐστιν. σκληρὰ μὲν οὖν καὶ
μαλακὰ καὶ γλίσχρα καὶ κραῦρα, καὶ ὅσα ἄλλα
τοιαῦτα[2] πάθη ὑπάρχει τοῖς ἐμψύχοις μορίοις, θερμότης καὶ ψυχρότης ποιήσειεν ἄν, τὸν δὲ λόγον ᾧ
ἤδη τὸ μὲν σάρξ τὸ δ' ὀστοῦν, οὐκέτι, ἀλλ' ἡ κίνησις
35 ἡ ἀπὸ τοῦ γεννήσαντος τοῦ ἐντελεχείᾳ ὄντος ὅ ἐστι
δυνάμει τὸ[3] ἐξ οὗ γίνεται, ὥσπερ καὶ ἐπὶ τῶν γινο-

[1] quieverit Σ : λυομένης coni. Platt.
[2] τοιαῦτα P, om. vulg.
[3] τὸ Y : ἢ Z : ἡ vulg. : om. P, A.-W., Platt.

---

[a] Cf. below, 734 b 36 and 735 a 4. Also see Introd.
§§ 34 ff.

[b] i.e., the principle of movement.

[c] If the text is sound, this can only refer to the original
"movement" imparted by the generating parent which
produced the semen ; and this would be comparable with
the initial movement imparted to the automaton mentioned
above.

starting-point this principle. Whatever is formed either by Nature or by human Art, say X, is formed by something which is X *in actuality* out of something which is X *potentially*.[a] Now semen, and the movement and principle [b] which it contains, are such that, as the movement ceases [c] each one of the parts gets formed and acquires Soul. (I add " acquires Soul," because there is no such thing as face, or flesh either, without Soul in it ; and though they are still said to be " face " and " flesh " after they are dead, these terms will be names merely ("homonyms "),[d] just as if the things were to turn into stone or wooden ones.) And the formation of the "uniform" parts [e] and of the instrumental parts goes on simultaneously. And as in speaking of an axe or any other instrument, we should not say that it was made solely by fire, so we should not say this about a foot or a hand ⟨in the embryo⟩, nor, similarly, of flesh either, because this too is an instrument with a function to perform. As for hardness, softness, toughness, brittleness and the rest of such qualities which belong to the parts that have Soul in them —heat and cold may very well produce these, but they certainly do not produce the *logos* [f] in direct consequence of which one thing is flesh and another bone ; this is done by the movement which derives from the generating parent, who is *in actuality* what the material out of which the offspring is formed is *potentially*. Exactly the same happens with things

[d] See note on 726 b 24 (and 721 a 3). They have merely the name in common with the living face and flesh, but not the essential nature. *Cf.* line 34 below.

[e] See Introd. § 19. Note that the non-uniform parts are here called the instrumental parts.

[f] See Introd. § 10.

ARISTOTLE

μένων κατὰ τέχνην· σκληρὸν μὲν γὰρ καὶ μαλακὸν
τὸν σίδηρον ποιεῖ τὸ θερμὸν καὶ τὸ ψυχρόν, ἀλλὰ
ξίφος ἡ κίνησις ἡ τῶν ὀργάνων, ἔχουσα λόγον τὸν
τῆς τέχνης. ἡ γὰρ τέχνη ἀρχὴ καὶ εἶδος τοῦ
γινομένου, ἀλλ᾽ ἐν ἑτέρῳ· ἡ δὲ τῆς φύσεως κίνησις
ἐν αὐτῷ ἀφ᾽ ἑτέρας οὖσα φύσεως τῆς ἐχούσης τὸ
5 εἶδος ἐνεργείᾳ. πότερον δ᾽ ἔχει ψυχὴν τὸ σπέρμα
ἢ οὔ; ὁ αὐτὸς λόγος καὶ περὶ τῶν μορίων· οὔτε
γὰρ ψυχὴ ἐν ἄλλῳ οὐδεμία ἔσται πλὴν ἐν ἐκείνῳ
οὗ γ᾽ ἐστίν, οὔτε μόριον ἔσται μὴ μετέχον ἀλλ᾽ ἢ
ὁμωνύμως, ὥσπερ τεθνεῶτος ὀφθαλμός. δῆλον
οὖν ὅτι καὶ ἔχει καὶ ἔστι δυνάμει. ἐγγυτέρω δὲ
10 καὶ πορρωτέρω αὐτὸ αὑτοῦ ἐνδέχεται εἶναι δυνάμει,
ὥσπερ ὁ καθεύδων γεωμέτρης τοῦ ἐγρηγορότος
πορρωτέρω, καὶ οὗτος τοῦ θεωροῦντος. ταύτης μὲν
οὖν οὐθὲν μόριον αἴτιον τῆς γενέσεως, ἀλλὰ τὸ
πρῶτον κινῆσαν ἔξωθεν. οὐθὲν γὰρ αὐτὸ ἑαυτὸ
γεννᾷ· ὅταν δὲ γένηται, αὔξει ἤδη αὐτὸ ἑαυτό.
15 διόπερ πρῶτόν τι γίγνεται, καὶ οὐχ ἅμα πάντα.
τοῦτο δὲ γίγνεσθαι ἀνάγκη πρῶτον, ὃ αὐξήσεως
ἀρχὴν ἔχει· εἴτε γὰρ φυτὸν εἴτε ζῷον, ὁμοίως τοῦτο
πᾶσιν ὑπάρχει τὸ θρεπτικόν. τοῦτο δ᾽ ἔστι τὸ

a See Introd. § 11.   *b* See above, 734 b 25.
c See note, 726 b 24.
d The argument now resumes from line 4 above.
e Cf. De anima 416 b 16, and context.

formed by the processes of the arts. Heat and cold
soften and harden the iron, but they do not produce
the sword ; this is done by the movement of the
instruments employed, which contains the *logos* of the
Art ; since the Art is both the principle [a] and Form
of the thing which is produced ; but it is located
elsewhere than in that thing, whereas Nature's
movement is located in the thing itself which is
produced, and it is derived from another natural
organism which possesses the Form *in actuality*. As
for the question whether the semen possesses Soul or
not, the same argument [b] holds as for the parts of
the body, viz., (*a*) no Soul will be present elsewhere
than in that of which it is the Soul ; (*b*) no part of
the body will be such in more than name [c] unless it
has some Soul in it (*e.g.*, the eye of a dead person).
Hence it is clear both that semen possesses Soul, and
that it is Soul, *potentially*. And there are varying
degrees in which it may be *potentially* that which it
is capable of being—it may be nearer to it or further
removed from it (just as a sleeping geometer is at a
further remove than one who is awake, and a waking
one than one who is busy at his studies). So [d] then,
the cause of this process of formation is not any part
of the body, but the external agent which first set
the movement going—for of course nothing generates itself,[e] though as soon as it has been formed a
thing makes itself grow.[f] That is why one part is
formed first, not all the parts simultaneously. And
the part which must of necessity be formed first is
the one which possesses the principle of growth : be
they plants or animals, this, the nutritive, faculty is
present in all of them alike (this also is the faculty

[f] *Cf.* below, 735 a 22, 740 a 19 ff.

γεννητικὸν ἑτέρου οἷον αὐτό· τοῦτο γὰρ παντὸς
φύσει τελείου ἔργον καὶ ζῴου καὶ φυτοῦ. ἀνάγκη
20 δὲ διὰ τόδε, ὅτι ὅταν τι γένηται, αὐξάνεσθαι ἀνάγκη. ἐγέννησε μὲν τοίνυν τὸ συνώνυμον, οἷον
ἄνθρωπος ἄνθρωπον, αὔξεται δὲ δι' ἑαυτοῦ. ἑαυτὸ[1]
ἄρα τι ὂν αὔξει.[2] εἰ δὴ ἕν τι καὶ τοῦτο πρῶτον,[3]
τοῦτο ἀνάγκη γίγνεσθαι πρῶτον. ὥστ' εἰ ἡ καρδία
πρῶτον ἔν τισι ζῴοις γίγνεται, ἐν δὲ τοῖς μὴ ἔχουσι
25 καρδίαν τὸ ταύτῃ ἀνάλογον, ἐκ ταύτης ἂν εἴη ἡ
ἀρχὴ τοῖς ἔχουσι, τοῖς δ' ἄλλοις ἐκ τοῦ ἀνάλογον.

Τί μὲν οὖν ἐστιν αἴτιον ὡς ἀρχὴ τῆς περὶ ἕκαστον γενέσεως, κινοῦν πρῶτον καὶ δημιουργοῦν,
εἴρηται πρὸς τὰ διαπορηθέντα πρότερον·

II 30   Περὶ δὲ τῆς τοῦ σπέρματος φύσεως ἀπορήσειεν
ἄν τις. τὸ γὰρ σπέρμα ἐξέρχεται μὲν ἐκ τοῦ ζῴου
παχὺ καὶ λευκόν, ψυχόμενον δὲ γίνεται ὑγρὸν
ὥσπερ ὕδωρ, καὶ τὸ χρῶμα ὕδατος. ἄτοπον δὴ ἂν

---

[1] ἑαυτὸ Peck : αὐτὸ vulg.
[2] ἑαυτὸ . . . αὔξει] ἔστιν ἄρά τί ὃ αὔξει Z[1].
[3] πρῶτον om. PS : A.-W. coni. ἔν τι τοῦτο, καὶ τοῦτο ἀνάγκη.

---

[a] Cf. De anima 415 a 26 ff., and for identity of nutritive
and generative faculty, 416 a 18 ff., and note on 744 b 36
below.
[b] See note on 721 a 3.
[c] This seems to be the meaning of this phrase ; cf. the
twice-repeated remark above, that once a thing has been
brought into being, it makes itself grow : Aristotle now says,
" now that it is making itself grow, it is something—but
what ? Some one thing—it is so far just that one thing which
is able to cause growth, which contains the principle of
156

of generating another creature like itself, since this
is a function which belongs to every animal and plant
that is perfect in its nature).[a] The reason why this
must of necessity be so is that once a thing has been
formed, it must of necessity grow. And though it
was generated by another thing bearing the same
name [b] (*e.g.*, a man is generated by a man), it grows
by means of itself. So then, since it makes itself
grow, it *is* something [c] : and if indeed it is some *one*
thing, and if it is this first of all, then this must of
necessity be formed first. Thus, if the heart is
formed first in certain animals (or the part analogous
to the heart, in those animals which have no heart),
we may suppose that it is the heart (or its analogue)
which supplies the principle.[d]

The queries raised earlier have now been dealt
with. We have answered the question, What is the
cause (in the sense of principle) of the generation of
each individual—what is that which first sets it in
movement and fashions it ?

A puzzle which may now be propounded is, What II
is the nature of Semen ? Semen when it leaves the Semen.
animal is thick and white, but when it cools it becomes
fluid like water and is of the colour of water. This

nutritive Soul, viz., the heart. And that is why the heart is
the first thing to be formed." *Cf.* 740 a 21 (where there is no
need to alter the text).

[d] The meaning of this passage seems to be that.the semen,
though it must have (and be) Soul, can have (and be) Soul
potentially only ; and the realizing of this potentiality,
which is the process of formation or generation (of which the
parent is the agent), goes on gradually—thus, the first part
of the Soul to be formed, generated, or realized, is the part
which produces growth (τὸ θρεπτικόν), and with it the part of
the body in which that part of the Soul resides, viz., the
heart. (See 763 b 25, n.)

735 a

δόξειεν· οὐ γὰρ παχύνεται ὕδωρ θερμῷ, τὸ δ'
ἔσωθεν ἐκ θερμοῦ ἐξέρχεται παχύ, ψυχόμενον δὲ
γίνεται ὑγρόν. καίτοι πήγνυταί γε τὰ ὑδατώδη·
35 τὸ δὲ σπέρμα οὐ πήγνυται τιθέμενον ἐν τοῖς πάγοις
ὑπαίθριον, ἀλλ' ὑγραίνεται, ὡς ὑπὸ τοῦ ἐναντίου
παχυνθέν. ἀλλὰ μὴν οὐδ' ὑπὸ θερμοῦ παχύνεσθαι

735 b

εὔλογον. ὅσα γὰρ γῆς πλεῖον ἔχει, ταῦτα συνίσταται καὶ παχύνεται ἑψόμενα, οἷον καὶ τὸ γάλα.
ἔδει οὖν ψυχόμενον στερεοῦσθαι. νῦν δ' οὐθὲν
γίνεται στερεόν, ἀλλὰ πᾶν ὥσπερ ὕδωρ. ἡ μὲν
οὖν ἀπορία αὕτη ἐστίν. εἰ μὲν γὰρ ὕδωρ, τὸ ὕδωρ
5 οὐ φαίνεται παχυνόμενον ὑπὸ τοῦ θερμοῦ, τὸ δ'
ἐξέρχεται παχὺ καὶ θερμὸν καὶ ἐκ θερμοῦ τοῦ
σώματος· εἰ δ' ἐκ γῆς[1] ἢ μικτὸν γῆς καὶ ὕδατος,
οὐκ ἔδει ὑγρὸν πᾶν γίνεσθαι καὶ ὕδωρ. ἢ οὐ
πάντα τὰ συμβαίνοντα διηρήκαμεν; οὐ γὰρ μόνον
παχύνεται τὸ ἐξ ὕδατος καὶ γεώδους συνιστάμενον
10 ὑγρόν, ἀλλὰ καὶ τὸ ἐξ ὕδατος καὶ πνεύματος, οἷον
καὶ ὁ ἀφρὸς γίνεται παχύτερος καὶ λευκός, καὶ
ὅσῳ ἂν ἐλάττους καὶ ἀδηλότεραι αἱ πομφόλυγες
ὦσι, τοσούτῳ καὶ λευκότερος καὶ στιφρότερος ὁ
ὄγκος φαίνεται. τὸ δ' αὐτὸ καὶ τὸ ἔλαιον πάσχει·
παχύνεται γὰρ τῷ πνεύματι μιγνύμενον· διὸ καὶ
15 τὸ λευκαινόμενον παχύτερον γίνεται, τοῦ ἐνόντος
ὑδατώδους ὑπὸ τοῦ θερμοῦ διακρινομένου καὶ γι-

___

[1] εἰ δὲ γῆς P, A.-W.

may seem strange, because water is not thickened by heat, yet semen is thick when it leaves the inside of the animal, which is hot, and becomes fluid when it cools. Moreover, watery substances freeze, but semen does not freeze when exposed to frost in the open air; it becomes fluid, which suggests that it was heat that thickened it. And yet it is not very probable that it is thickened by heat, because it is substances that contain a large proportion of earth which " set " and thicken when boiled—milk, for example; hence it ought to solidify when it cools, but in fact it does not solidify at all; the whole of it becomes fluid like water. This then is the puzzle. Suppose that semen is water. Water is never observed to be thickened by heat; whereas semen is both thick and hot, and the body it comes from is hot. Or suppose it consists of earth, or is a mixture of earth and water. In that case the whole of it ought not to become fluid and turn to water. Perhaps then after all we have not distinguished all the cases that occur. Other fluids thicken beside those which are composed of water and earthy matter, viz., those composed of water and *pneuma*,[a] for instance, foam, which becomes thicker, and white; and the smaller and more microscopic the bubbles are, the whiter and more compact is the appearance of the bulk. Oil behaves in the same way; it thickens when it gets mixed with *pneuma*; and that is why ⟨oil⟩ when it becomes whiter is thickening, since the watery substance in it is separated out from

---

[a] *Pneuma* is defined below (736 a 1) as " hot air "; see, however, 736 b 35 ff. below. Rather than attempt a misleading or inadequate translation of the word (*e.g.*, spirit, breath), I have decided to keep the original term, as elsewhere. See further, Appendix B.

νομένου πνεύματος. καὶ ἡ μολύβδαινα μιγνυμένη
ὕδατι καὶ[1] ἐλαίῳ ἐξ ὀλίγου τε πολὺν ὄγκον ποιεῖ
καὶ ἐξ ὑγροῦ στιφρὸν καὶ ἐκ μέλανος λευκόν.
αἴτιον δ' ὅτι ἐγκαταμίγνυται πνεῦμα, ὃ τόν τε
20 ὄγκον ποιεῖ καὶ τὴν λευκότητα διαφαίνει, ὥσπερ
ἐν τῷ ἀφρῷ καὶ τῇ χιόνι· καὶ γὰρ ἡ χιών ἐστιν
ἀφρός. καὶ αὐτὸ τὸ ὕδωρ[2] ἐλαίῳ μιγνύμενον
γίνεται παχὺ καὶ λευκόν· καὶ γὰρ ὑπὸ τῆς τρίψεως
ἐγκατακλείεται πνεῦμα, καὶ αὐτὸ τὸ ἔλαιον ἔχει
25 πνεῦμα πολύ· ἔστι γὰρ οὔτε γῆς οὔτε ὕδατος ἀλλὰ
πνεύματος τὸ λιπαρόν. διὸ καὶ ἐπὶ τῷ ὕδατι
ἐπιπολάζει· ὁ γὰρ ἐν αὐτῷ ὢν ἀήρ, ὥσπερ ἐν ἀγ-
γείῳ, φέρει ἄνω καὶ ἐπιπολάζει καὶ αἴτιος τῆς
κουφότητός ἐστιν. καὶ ἐν τοῖς ψύχεσι δὲ καὶ
πάγοις παχύνεται τὸ ἔλαιον, πήγνυται δ' οὔ· διὰ
30 μὲν γὰρ θερμότητα οὐ πήγνυται (ὁ γὰρ ἀὴρ θερμὸν
καὶ ἄπηκτον), διὰ δὲ τὸ συνίστασθαι αὐτὸν καὶ
πυκνοῦσθαι [ὥσπερ][3] ὑπὸ τοῦ ψύχους παχύτερον
γίνεται τὸ ἔλαιον. διὰ ταύτας τὰς αἰτίας καὶ τὸ
σπέρμα ἔσωθεν μὲν ἐξέρχεται στιφρὸν καὶ λευκόν,
ὑπὸ τῆς ἐντὸς θερμότητος πνεῦμα πολὺ ἔχον θερ-

[1] καὶ ΡΣ : ἢ καὶ vulg.   [2] ὕδωρ Ρ : ὕδωρ τῷ vulg.
[3] secl. A.-W. ; fortasse αὐτὸν πυκνοῦται [ὥσπερ . . . ἔλαιον]
scribendum (haec om. Σ).   lac. post ὥσπερ stat. Sus.

---

[a] This is no doubt galena (lead sulphide), the chief ore
found in the Attic mines at Laurium, although these were
more famous for their silver output. The reference to the
mixing of the ore with water *and oil*, which heretofore seems
to have passed unnoticed, must imply an early process of
" flotation," a stage which follows the mechanical crushing
of the ore and precedes the metallurgical extracting of the
metal, its object being to separate the metalliferous from
the non-metalliferous constituents of the ore by means of the
production of a froth. The first practically successful

it by the heat and becomes *pneuma*. Lead ore,[a] too, when it gets mixed with water and oil, increases its bulk, and whereas it was fluid and black it becomes thick and coherent and white. The reason is that *pneuma* gets mixed in with it, and this produces the increase of bulk and lets the whiteness show through, precisely as it does with foam, and also with snow (because snow too is a foam). Even water itself when it gets mixed with oil becomes thick and white, the reason being that some *pneuma* is left behind in it owing to the friction of mixing, and also that oil itself contains a good deal of *pneuma*—for of course shininess is a quality of *pneuma*, not of earth or water. And that too is why oil floats on the surface of water ; air is contained in it, as though in a vessel, and this air buoys it up and causes it to float ; thus the air is the cause of its lightness. Further, in time of cold and frost, oil thickens, but does not freeze. Its failure to freeze is due to its heat—because the air is hot and is impervious to frost. But it thickens because the air is coagulated and compressed [as] by the cold. These reasons explain the behaviour of semen as well. It is coherent and white when it comes forth from within, because it contains a good deal of hot *pneuma* owing to the internal heat of the animal.

attempt at flotation in modern times was made by the brothers Elmore at the Glasdir gold-mine in Wales (patent 1898), though suggestions for the use of oil had been made by William Haynes of Holywell some years earlier (patent 1860). For details see S. J. Truscott, *Text-book of Ore-dressing* ; T. A. Rickard, *Man and Metals*, id., *Concentration by Flotation* (which includes two essays on the flotation of galena at Broken Hill, N.S.W.). The term στιφρός corresponds exactly to the " thick coherent froth " mentioned by Truscott (*op. cit.* 392, etc.).—For a full account of the mines at Laurium see E. Ardaillon, *Les Mines du Laurion* (1897).

735 b

35 μόν, ἐξελθὸν¹ δὲ ὅταν ἀποπνεύσῃ τὸ θερμὸν καὶ
ὁ ἀὴρ ψυχθῇ, ὑγρὸν γίνεται καὶ μέλαν· λείπεται
γὰρ τὸ ὕδωρ καὶ εἴ τι μικρὸν γεῶδες, ὥσπερ ἐν
φλέγματι, καὶ ἐν τῷ σπέρματι ξηραινομένῳ.

736 a

Ἔστι μὲν οὖν τὸ σπέρμα κοινὸν πνεύματος καὶ
ὕδατος, τὸ δὲ πνεῦμά ἐστι θερμὸς ἀήρ· διὸ ὑγρὸν
τὴν φύσιν, ὅτι ἐξ ὕδατος. Κτησίας γὰρ ὁ Κνί-
διος ἃ περὶ τοῦ σπέρματος τῶν ἐλεφάντων εἴρηκε, φανερός ἐστιν ἐψευσμένος. φησὶ γὰρ οὕτω
5 σκληρύνεσθαι ξηραινόμενον ὥστε γίνεσθαι ἠλέκτρῳ
ὅμοιον. τοῦτο δ' οὐ γίνεται· μᾶλλον μὲν γὰρ
ἕτερον ἑτέρου σπέρμα γεωδέστερον ἀναγκαῖον
εἶναι, καὶ μάλιστα τοιοῦτον ὅσοις πολὺ γεῶδες
ὑπάρχει κατὰ τὸν ὄγκον τὸν τοῦ σώματος. παχὺ
δὲ καὶ λευκὸν διὰ τὸ μεμῖχθαι πνεῦμα. καὶ γὰρ
10 λευκόν ἐστι τὸ σπέρμα πάντων· Ἡρόδοτος γὰρ οὐκ
ἀληθῆ λέγει, φάσκων μέλαιναν εἶναι τὴν τῶν
Αἰθιόπων γονήν, ὥσπερ ἀναγκαῖον ὂν τῶν τὴν
χρόαν μελάνων εἶναι πάντα μέλανα, καὶ ταῦθ' ὁρῶν
καὶ τοὺς ὀδόντας αὐτῶν ὄντας λευκούς. αἴτιον δὲ
τῆς λευκότητος τοῦ σπέρματος ὅτι ἐστὶν ἡ γονὴ
15 ἀφρός, ὁ δ' ἀφρὸς λευκόν, καὶ μάλιστα τὸ ἐξ

---

¹ ἐξελθὸν Peck : ἐξελθόντος vulg.

ᵃ See 725 a 15 ff.

ᵇ Ktesias of Knidos in Caria, a contemporary of Xenophon,
belonged to an old medical family, and was physician to the
Persian king Artaxerxes Mnemon (405–362 B.C.). His chief
work was his Περσικά, in 23 books, containing the history
of the East down to 398–397 B.C. Most of his zoological
matter, however, seems to have been contained in his Ἰνδικά,
judging from this reference and three others in the *History
of Animals*. Abridgements of both these works by Photius
are extant.   ᶜ Herodotus III. 101.

ᵈ The view that semen was foam was held by Diogenes of

162

Later, when it has lost its heat by evaporation and the air has cooled, it becomes fluid and dark, because the water and whatever tiny quantity of earthy matter it may contain stay behind in the semen as it solidifies, just as happens with *phlegma.*[a]

Semen, then, is a compound of *pneuma* and water (*pneuma* being hot air), and that is why it is fluid in its nature ; it is made of water. Ktesias of Knidos [b] is obviously mistaken in his statement about the semen of elephants : he says that it gets so hard when it solidifies that it becomes like amber. It does not. It is, of course, true that one semen must of necessity be earthier than another, and the earthiest will be in those animals which, for their bodily bulk, contain a large amount of earthy matter ; but semen is thick and white because there is *pneuma* mixed with it. What is more, it is white in all cases. Herodotus [c] is incorrect when he says that the semen of Ethiopians is black, as though everything about a person with a black skin were bound to be black—and this too in spite of their teeth being white, as he could see for himself. The cause of the whiteness of semen is that it is foam,[d] and foam is white, the whitest being that

Apollonia ; see Vindicianus, § 1 (Diels, *Vorsokr.*[5] 64 B 6) *Alexander Amator veri* (= Φιλαλήθης) . . . *libro primo De semine spumam sanguinis eius essentiam dixit Diogenis placitis consentiens* ; and *cf.* § 3. See Jaeger's discussion of the subject in *Diokles von Karystos*, 198-211. *Cf.* also Hippocrates, π. γονῆς κτλ. 1 (vii. 470 Littré) ἀποκρίνεται ἀπὸ τοῦ ὑγροῦ ἀφρεόντος τὸ ἰσχυρότατον. In modern times a similar idea has been put forward, *e.g.*, by Bütschli (*Untersuchungen über mikroskopische Schäume und das Protoplasma*, Leipzig, 1892), who " thought of protoplasm as a foam, or rather as an emulsion composed of two liquids, one in the form of droplets, the other as lamellae [*i.e.*, films] between the droplets " (Heilbrunn, *An Outline of General Physiology*, 1938, p. 25).

ὀλιγίστων συγκείμενον μορίων καὶ οὕτω μικρῶν
ὥσπερ ἑκάστης ἀοράτου τῆς πομφόλυγος οὔσης,
ὅπερ συμβαίνει καὶ ἐπὶ τοῦ ὕδατος καὶ τοῦ ἐλαίου
μιγνυμένων καὶ τριβομένων, καθάπερ ἐλέχθη
πρότερον.

Ἔοικε δὲ οὐδὲ τοὺς ἀρχαίους λανθάνειν ἀφρώδης
20 ἡ τοῦ σπέρματος οὖσα φύσις· τὴν γοῦν κυρίαν
θεὸν τῆς μίξεως ἀπὸ τῆς δυνάμεως ταύτης
προσηγόρευσαν.

Ἡ μὲν οὖν αἰτία τῆς λεχθείσης ἀπορίας εἴρηται,
φανερὸν δὲ ὅτι διὰ τοῦτ᾽ οὐδὲ πήγνυται· ὁ γὰρ
ἀὴρ ἄπηκτος.

III Τούτου δ᾽ ἐχόμενόν ἐστιν[1] ἀπορῆσαι καὶ εἰπεῖν,
25 εἰ τῶν προϊεμένων εἰς τὸ θῆλυ γονὴν μηθέν μόριόν
ἐστι τὸ εἰσελθὸν τοῦ γιγνομένου κυήματος, ποῦ[2]
τρέπεται τὸ σωματῶδες αὐτοῦ, εἴπερ ἐργάζεται τῇ
δυνάμει τῇ ἐνούσῃ ἐν αὐτῷ. διορίσαι δὲ[3] δεῖ πότερον μεταλαμβάνει τὸ συνιστάμενον ἐν τῷ θήλει
ἀπὸ τοῦ εἰσελθόντος τι ἢ οὐθέν, καὶ περὶ ψυχῆς
30 καθ᾽ ἣν λέγεται ζῷον (ζῷον δ᾽ ἐστὶ κατὰ τὸ μόριον
τῆς ψυχῆς τὸ αἰσθητικόν) πότερον ἐνυπάρχει τῷ
σπέρματι καὶ τῷ κυήματι ἢ οὔ, καὶ πόθεν. οὔτε
γὰρ ὡς ἄψυχον ἂν θείη τις τὸ κύημα κατὰ πάντα
τρόπον ἐστερημένον ζωῆς· οὐδὲν γὰρ ἧττον τά τε

---

[1] ἐστιν καί PSY, Galen.   [2] ποῖ Btf.   [3] δὲ P : τε vulg.

---

[a] Lit., "called after this substance (*dynamis*)." Aphrodite,
after *aphros*. *Cf.* Galen, π. σπέρματος I. 5 (iv. 531 Kühn);
and Clem. Paedag. I. 6. 48 (Diels, *Vorsokr.*[5] 64 A 24) τινὲς
δὲ καὶ τὸ σπέρμα τοῦ ζῴου ἀφρὸν εἶναι τοῦ αἵματος κατ᾽ οὐσίαν ὑποτίθενται . . . ἐντεῦθεν γὰρ ὁ Ἀπολλωνιάτης Διογένης τὰ
ἀφροδίσια κεκλῆσθαι βούλεται. *Cf.* preceding note.
[b] See note on meaning of κύημα, Introd. § 56.

which consists of the tiniest particles, so small that each individual bubble cannot be detected by the eye. An instance of such a foam, mentioned earlier, is that produced by the mechanical mixing of water and oil.

That the natural substance of semen is foam-like was, so it seems, not unknown even in early days ; at any rate, the goddess who is supreme in matters of sexual intercourse was called after foam.[a]

We have now given the reason which solves the puzzle that was stated. And this also shows, incidentally, why semen does not freeze : it is because air is impervious to frost.

The next puzzle to be stated and solved is this. III Take the case of those groups of animals in which Semen and semen is emitted into the female by the male. Suppose Soul. posing it is true that the semen which is so introduced is not an ingredient in the fetation [b] which is formed, but performs its function simply by means of the *dynamis* [c] which it contains. Very well ; if so, what becomes of the physical part of it ? First of all we shall have to decide (a) whether that which takes shape within the female does or does not incorporate into itself any portion of that which was introduced ⟨from the male⟩ ; and (b) whether Soul—and it is in virtue of Soul that an animal has the name of " animal " : it is in fact in virtue of the sentient part [d] of Soul that it is an animal [e]—whether Soul is or is not in the semen and in the fetation to begin with, and if so where it comes from. No one, of course, would maintain that the fetation is quite without Soul, completely devoid of life in every sense,

[c] See also 726 b 18 ff., 727 b 15, 16, 738 b 12, and Bk. I, ch. 21.   [d] See Introd. § 43.   [e] See 732 a 13, n.

736 a

σπέρματα καὶ τὰ κυήματα τῶν ζῴων ζῇ τῶν
35 φυτῶν, καὶ γόνιμα μέχρι τινός ἐστιν. ὅτι μὲν οὖν
τὴν θρεπτικὴν ἔχουσι ψυχήν, φανερόν (δι' ὅτι δὲ
ταύτην πρῶτον ἀναγκαῖόν ἐστι λαβεῖν, ἐκ τῶν περὶ

736 b

ψυχῆς διωρισμένων ἐν ἄλλοις φανερόν)· προϊόντα
δὲ καὶ τὴν αἰσθητικήν, καθ' ἣν ζῷον. οὐ γὰρ ἅμα
γίνεται ζῷον καὶ ἄνθρωπος οὐδὲ ζῷον καὶ ἵππος,
ὁμοίως δὲ καὶ ἐπὶ τῶν ἄλλων ζῴων· ὕστατον[1] γὰρ
γίνεται τὸ τέλος, τὸ δ' ἴδιόν ἐστι τὸ ἑκάστου τῆς
5 γενέσεως τέλος. διὸ καὶ περὶ νοῦ, πότε καὶ πῶς
μεταλαμβάνει καὶ πόθεν τὰ μετέχοντα ταύτης τῆς
ἀρχῆς, ἔχει τ' ἀπορίαν πλείστην, καὶ δεῖ προ-
θυμεῖσθαι κατὰ δύναμιν λαβεῖν καὶ καθ' ὅσον
ἐνδέχεται.

Τὴν μὲν οὖν θρεπτικὴν ψυχὴν τὰ σπέρματα καὶ
τὰ κυήματα τὰ ⟨ἀ⟩χώριστα[2] δῆλον ὅτι δυνάμει μὲν
10 ἔχοντα θετέον, ἐνεργείᾳ δ' οὐκ ἔχοντα, πρὶν ἢ[3]
καθάπερ τὰ χωριζόμενα τῶν κυημάτων ἕλκει τὴν
τροφὴν καὶ ποιεῖ τὸ τῆς τοιαύτης ψυχῆς ἔργον·
πρῶτον μὲν γὰρ ἅπαντ' ἔοικε ζῆν τὰ τοιαῦτα

---

[1] ὕστατον P: ὕστερον vulg.
[2] Buss.: ὄντα χωριστὰ Platt.   [3] πλὴν εἰ Platt.

---

[a] e.g., wind-eggs, Bk. III.
[b] De anima, Bk. II, ch. 4 ; and see 735 a 13 ff. above.
[c] These are two instances of the rule that there are definite
stages in the development or formation of living things.
Nutritive Soul (the mark of a living thing) is acquired before
sentient Soul (the mark of an animal), just as the formation
of an animal precedes the formation of any particular species
of animal. *Cf.* von Baer's " biogenetic law," that the character of the class is acquired before that of the genus, and that
of the genus before that of the species. (K. E. von Baer,

for the semens and the fetations of animals are just as much alive as plants are, and up to a point they are fertile.[a] Thus it is clear that they possess nutritive Soul (*vide* my remarks on Soul in another treatise[b] for an explanation of why nutritive Soul must of necessity be acquired first). It is while they develop that they acquire sentient Soul as well, in virtue of which an animal is an animal—I say, " while they develop," for it is not the fact that when an animal is formed at that same moment a human being, or a horse, or any other particular sort of animal is formed, because the end or completion is formed last of all, and that which is peculiar to each thing is the end of its process of formation.[c] That is why it is a very great puzzle to answer another question, concerning Reason. At what moment, and in what manner, do those creatures which have this principle of Reason acquire their share in it, and where does it come from ? This is a very difficult problem which we must endeavour to solve, so far as it may be solved, to the best of our power.

As regards nutritive Soul, then,[d] it is clear that we must posit that semens and fetations which are not separated ⟨from the parent⟩ possess it *potentially*, though not *in actuality*—*i.e.*, not until they begin to draw the nourishment to themselves and perform the function of nutritive Soul, as fetations which get separated[e] ⟨from the parent⟩ do ; for to begin with it seems that all things of this sort live the life of a

Über Entwicklungsgeschichte der Thiere, Beobachtung und Reflexion (1828), i. 224, Scholion V (1) Dass das Gemeinsame einer grössern Thiergruppe sich früher im Embryo bildet, als das Besondere, *et seqq.*)

[d] The solution begins by resuming the argument from 736 a 32-34.   [e] *e.g.*, seeds of plants.

φυτοῦ βίον. ἑπομένως δὲ δῆλον ὅτι καὶ περὶ τῆς
αἰσθητικῆς λεκτέον ψυχῆς καὶ περὶ τῆς νοητικῆς·
15 πάσας γὰρ ἀναγκαῖον δυνάμει πρότερον ἔχειν ἢ
ἐνεργείᾳ. ἀναγκαῖον δὲ ἤτοι μὴ οὔσας πρότερον
ἐγγίνεσθαι πάσας, ἢ πάσας προϋπαρχούσας, ἢ τὰς
μὲν τὰς δὲ μή, καὶ ἐγγίνεσθαι ἢ ἐν τῇ ὕλῃ μὴ
εἰσελθούσας ἐν τῷ τοῦ ἄρρενος σπέρματι, ἢ ἐνταῦθα μὲν ἐκεῖθεν ἐλθούσας, ἐν δὲ τῷ ἄρρενι ἢ
20 θύραθεν ἐγγινομένας ἁπάσας ἢ μηδεμίαν ἢ τὰς μὲν
τὰς δὲ μή. ὅτι μὲν τοίνυν οὐχ οἷόν τε πάσας
προϋπάρχειν, φανερόν ἐστιν ἐκ τῶν τοιούτων. ὅσων
γὰρ ἐστιν ἀρχῶν[1] ἡ ἐνέργεια σωματική, δῆλον ὅτι
ταύτας ἄνευ σώματος ἀδύνατον ὑπάρχειν, οἷον βαδίζειν ἄνευ ποδῶν· ὥστε καὶ θύραθεν εἰσιέναι
25 ἀδύνατον· οὔτε γὰρ αὐτὰς καθ᾽ αὑτὰς εἰσιέναι
οἷόν τε ἀχωρίστους οὔσας, οὔτ᾽ ἐν σώματι εἰσιέναι·

---

[1] πράξεων coniecerunt A.-W.

---

[a] This elaborate scheme of possibilities is not really so over-
whelming as it looks, though the argument would have been
more lucid if Aristotle had explicitly named the several sorts
of Soul involved. It will be seen, however, that of the first
three possibilities, the last, (c), is the operative one ; in fact,
it is nutritive Soul which the material of the female (more
specifically, the fetation) possesses (see 736 a 32 ff., 737 a 23 ff);
thus it remains for the other two, sentient and rational Souls,
to be supplied by the male (Aristotle explains in ch. 5 below
that the reason why a fetation can grow yet is unable to
develop fully into an animal is that it lacks sentient Soul,
which only the male can supply). Hence in the second series
of possibilities it is again the last one, (c), which is the operative
one: sentient Soul is present inside the male (i.e., the semen),
and it remains that rational Soul comes into being inside the
male (i.e., the semen) from some outside source, for it alone
is not affected by the two considerations which preclude the
entry from outside of the other parts of Soul, whose activity

plant. And it is clear we should follow a similar line also in our statements about sentient Soul and rational Soul, since a thing must of necessity possess every one of the sorts of Soul *potentially* before it possesses them *in actuality*. And necessity requires either (*a*) that none of them exists previously, and that they all come to be formed in ⟨the fetation⟩ ; or (*b*) that they are all there beforehand ; or (*c*) that some of them are there and some are not ; and further, that they come to be formed in the material supplied by the female either (*a*) without having entered in the semen of the male or (*b*) after having so entered—that is, having come from the male, and if so, then that either (*a*) all of them or (*b*) none of them or (*c*) some of them come to be formed within the male from some outside source.[a] Now the following considerations plainly show that they cannot all be present beforehand. Clearly, those principles whose activity is physical cannot be present without a physical body—there can, for example, be no walking without feet[b] ; and this also rules out the possibility of their entering from outside, since it is impossible either that they enter by themselves, because they are inseparable ⟨from a physical body⟩, or that they enter by transmission in some body, because the

is essentially physical (see also below, 737 a 9 f.). Thus, sentient Soul, and *a fortiori* rational Soul, are supplied by the male, through the semen, to the material provided by the female. Aristotle does not, however, give any fuller solution than this to his own "very difficult puzzle" how and when rational Soul, which is thus supplied in a *potential* state by the male, is *actualized* in the offspring.

[b] Aristotle takes the "locomotive Soul," the highest of the "parts" or "faculties" of Soul apart from "rational Soul," and shows that this cannot enter by itself ; *a fortiori* therefore none of the lower "parts" can do so.

736 b

τὸ γὰρ σπέρμα περίττωμα μεταβαλλούσης τῆς
τροφῆς ἐστίν. λείπεται δὴ¹ τὸν νοῦν μόνον θύραθεν
ἐπεισιέναι καὶ θεῖον εἶναι μόνον· οὐθὲν γὰρ αὐτοῦ
τῇ ἐνεργείᾳ κοινωνεῖ σωματικὴ ἐνέργεια.

30 Πάσης μὲν οὖν ψυχῆς δύναμις ἑτέρου σώματος
ἔοικε κεκοινωνηκέναι καὶ θειοτέρου τῶν καλουμένων στοιχείων· ὡς δὲ διαφέρουσι τιμιότητι αἱ
ψυχαὶ καὶ ἀτιμίᾳ ἀλλήλων, οὕτω καὶ ἡ τοιαύτη
διαφέρει φύσις. πάντων μὲν γὰρ ἐν τῷ σπέρματι
ἐνυπάρχει, ὅπερ ποιεῖ γόνιμα εἶναι τὰ σπέρματα,
35 τὸ καλούμενον θερμόν. τοῦτο δ' οὐ πῦρ οὐδὲ
τοιαύτη δύναμίς ἐστιν, ἀλλὰ τὸ ἐμπεριλαμβανό-
μενον ἐν τῷ σπέρματι καὶ ἐν τῷ ἀφρώδει πνεῦμα
καὶ ἡ ἐν τῷ πνεύματι φύσις, ἀνάλογον οὖσα τῷ

737 a

τῶν ἄστρων στοιχείῳ. διὸ πῦρ μὲν οὐθὲν γεννᾷ
ζῷον, οὐδὲ φαίνεται συνιστάμενον ἐν² πυρουμένοις
οὔτ' ἐν ὑγροῖς οὔτ' ἐν ξηροῖς οὐθέν· ἡ δὲ τοῦ ἡλίου
θερμότης καὶ ἡ τῶν ζῴων οὐ μόνον ἡ διὰ τοῦ

---

¹ δὴ Platt, Zeller, Btf. : δὲ vulg.   ² ἐν P : om. vulg.

ᵃ *i.e.*, it is not a body possessing the parts necessary in
order to give effect to the activities involved, such as legs for
walking. *Cf. P.A.* 641 b 31 γένεσις μὲν γὰρ τὸ σπέρμα, οὐσία
δὲ τὸ τέλος.
ᵇ *Cf. De anima* 413 a 4 ff.
ᶜ *Cf.* 762 a 20.   ᵈ See 736 a 13 ff.
ᵉ This is the so-called " fifth element," (*i.e.*, over and
above the four " elements " found in the sublunary regions,
viz., earth, air, fire, and water), though Aristotle's own name
for it is "the first of the elements " (τὸ πρῶτον τῶν στοιχείων,
De caelo 298 b 6, τὸ πρῶτον σῶμα, 270 b 21), owing to its
pre-eminent qualities. The arguments for its existence will
be found in *De caelo*, Bk. I ; it is ungenerated, indestruct-
170

semen is a residue of the nourishment that is undergoing change.[a] It remains, then, that Reason alone
enters in, as an additional factor, from outside, and
that it alone is divine, because physical activity has
nothing whatever to do with the activity of Reason.[b]

Now so far as we can see, the faculty of Soul of
every kind has to do with some physical substance
which is different from the so-called " elements " and
more divine than they are ; and as the varieties of
Soul differ from one another in the scale of value, so
do the various substances concerned with them differ
in their nature. In all cases the semen contains
within itself that which causes it to be fertile—what
is known as " hot " substance,[c] which is not fire nor
any similar substance, but the *pneuma* which is enclosed within the semen or foam-like stuff,[d] and the
natural substance which is in the *pneuma* ; and this
substance is analogous to the element which belongs
to the stars.[e] That is why fire does not generate
any animal,[f] and we find no animal taking shape either
in fluid or solid substances while they are under the
influence of fire ; whereas the heat of the sun [g] does
effect generation, and so does the heat of animals,

ible, and divine (269 a 31 ff., 270 a 12 ff., 270 b 10 ff.).   Aristotle claims that it was vaguely recognized by the ancients, as
is suggested by the name (*aither*) they gave to " the uppermost place " (270 b 16 ff.) : ἀπὸ τοῦ θεῖν ἀεὶ τὸν ἀΐδιον χρόνον
θέμενοι τὴν ἐπωνυμίαν αὐτῷ. (*Cf.* Hippocrates, π. σαρκῶν 2
(viii. 584 Littré) δοκέει δέ μοι ὃ καλέομεν θερμόν, ἀθάνατόν τε
εἶναι . . . τοῦτο οὖν . . . ἐξεχώρησεν εἰς τὴν ἀνωτάτω περι-
φορὴν καὶ αὐτό μοι δοκέει αἰθέρα τοῖς παλαιοῖς εἰρῆσθαι.)   Its
motion is circular ; so is that of the stars, which are composed of it (289 a 15). It is not found in the sublunary
regions, but *pneuma* is its " counterpart " (see Introd. §§ 70 ff.,
App. A §§ 7 ff., and B).   [f] But see 761 b 15 ff., and note.
    [g] See App. A §§ 7 ff., B §§ 7-17.

σπέρματος, ἀλλὰ κἄν τι περίττωμα τύχῃ τῆς φύ-
5 σεως ὂν ἕτερον, ὅμως ἔχει καὶ τοῦτο ζωτικὴν
ἀρχήν. ὅτι μὲν οὖν ἡ ἐν τοῖς ζώοις θερμότης οὔτε
πῦρ οὔτε ἀπὸ πυρὸς ἔχει τὴν ἀρχήν, ἐκ τῶν τοιούτων ἐστὶ φανερόν.

Τὸ δὲ τῆς γονῆς σῶμα, ἐν ᾧ συναπέρχεται [τὸ
σπέρμα]¹ τὸ τῆς ψυχικῆς ἀρχῆς, τὸ μὲν χωριστὸν ὂν
10 σώματος, ὅσοις ἐμπεριλαμβάνεταί τι² θεῖον (τοιοῦτος δ' ἐστὶν ὁ καλούμενος νοῦς), τὸ δ' ἀχώριστον,
τοῦτο τὸ σῶμα³ τῆς γονῆς διαλύεται καὶ πνευμα-
τοῦται, φύσιν ἔχον ὑγρὰν καὶ ὑδατώδη. διόπερ
οὐ δεῖ ζητεῖν ἀεὶ θύραζε αὐτὸ ἐξιέναι, οὐδὲ μόριον
οὐθὲν εἶναι τῆς συστάσης μορφῆς, ὥσπερ οὐδὲ τὸν
15 ὀπὸν τὸν τὸ γάλα συνιστάντα· καὶ γὰρ οὗτος
μεταβάλλει καὶ μόριον οὐθέν ἐστι τῶν συνισταμένων ὄγκων.

Περὶ μὲν οὖν ψυχῆς, πῶς ἔχει τὰ κυήματα καὶ ἡ
γονὴ καὶ πῶς οὐκ ἔχει, διώρισται· δυνάμει μὲν γὰρ
ἔχει, ἐνεργείᾳ δ' οὐκ ἔχει.⁴

Τοῦ δὲ σπέρματος ὄντος περιττώματος καὶ κινουμένου κίνησιν τὴν αὐτὴν καθ' ἥνπερ τὸ σῶμα
20 αὐξάνεται μεριζομένης τῆς ἐσχάτης τροφῆς,ᵃ ὅταν
ἔλθῃ εἰς τὴν ὑστέραν, συνίστησι καὶ κινεῖ τὸ περίττωμα τὸ τοῦ θήλεος τὴν αὐτὴν κίνησιν ἥνπερ αὐτὸ
τυγχάνει κινούμενον κἀκεῖνο. καὶ γὰρ ἐκεῖνο

---

¹ τὸ σπέρμα om. P, secl. A.-W. : τὸ πνεῦμα Platt, Σ.
² τι P : τὸ vulg.
³ σῶμα A.-W. : σπέρμα vulg.   ⁴ haec seclusit Platt.

ᵃ The " ultimate nourishment." *Cf.* 726 b 1 ff., and *P.A.*
650 a 34, 651 a 15, 678 a 8 ff. This is nourishment in its
final form, viz., blood.

and not only the heat of animals which operates through the semen, but also any other natural residue which there may be has within it a principle of life. Considerations of this sort show us that the heat which is in animals is not fire and does not get its origin or principle from fire.

Consider now the physical part of the semen. (This it is which, when it is emitted by the male, is accompanied by the portion of soul-principle and acts as its vehicle. Partly this soul-principle is separable from physical matter—this applies to those animals where some divine element is included, and what we call Reason is of this character—partly it is inseparable.) This physical part of the semen, being fluid and watery, dissolves and evaporates ; and on that account we should not always be trying to detect it leaving the female externally, or to find it as an ingredient of the fetation when that has set and taken shape, any more than we should expect to trace the fig-juice which sets and curdles milk. The fig-juice undergoes a change ; it does not remain as a part of the bulk which is set and curdled ; and the same applies to the semen.

We have now determined in what sense fetations and semen have Soul and in what sense they have not. They have Soul *potentially*, but not *in actuality*.

As semen is a residue, and as it is endowed with the same movement as that in virtue of which the body grows through the distribution of the ultimate nourishment,[a] when the semen has entered the uterus it " sets " the residue produced by the female and imparts to it the same movement with which it is itself endowed. The female's contribution, of course, is a residue too, just as the male's is, and

737 a

περίττωμα, καὶ πάντα τὰ μόρια ἔχει δυνάμει,
ἐνεργείᾳ δ' οὐθέν. καὶ γὰρ τὰ τοιαῦτ' ἔχει μόρια
25 δυνάμει, ᾗ διαφέρει τὸ θῆλυ τοῦ ἄρρενος. ὥσπερ
γὰρ καὶ ἐκ πεπηρωμένων ὁτὲ μὲν γίνεται πεπηρω-
μένα ὁτὲ δ' οὔ, οὕτω καὶ ἐκ θήλεος ὁτὲ μὲν θῆλυ
ὁτὲ δ' οὔ, ἀλλ' ἄρρεν. τὸ γὰρ θῆλυ ὥσπερ ἄρρεν
ἐστὶ πεπηρωμένον, καὶ τὰ καταμήνια σπέρμα, οὐ
καθαρὸν δέ. ἐν γὰρ οὐκ ἔχει μόνον, τὴν τῆς ψυχῆς
30 ἀρχήν. καὶ διὰ τοῦτο ὅσοις ὑπηνέμια γίνεται τῶν
ζῴων, ἀμφοτέρων ἔχει τὰ μέρη τὸ συνιστάμενον
ᾠόν, ἀλλὰ τὴν ἀρχὴν οὐκ ἔχει, διὸ οὐ γίνεται
ἔμψυχον· ταύτην γὰρ τὸ τοῦ ἄρρενος ἐπιφέρει
σπέρμα. ὅταν δὲ μετάσχῃ τοιαύτης ἀρχῆς τὸ
περίττωμα τὸ τοῦ θήλεος, κύημα γίνεται.
35 [1][Τοῖς δ' ὑγροῖς μὲν σωματώδεσι δὲ θερμαινο-
μένοις περιίσταται, καθάπερ ἐν τοῖς ἐψήμασι ψυ-
737 b χομένοις τὸ περίξηρον. πάντα δὲ τὰ σώματα
συνέχει τὸ γλίσχρον· ὅπερ καὶ προϊοῦσι καὶ μείζοσι
γιγνομένοις ἡ τοῦ νεύρου λαμβάνει φύσις, ἥπερ
συνέχει τὰ μόρια τῶν ζῴων, ἐν μὲν τοῖς οὖσα
νεῦρον, ἐν δὲ τοῖς τὸ ἀνάλογον. τῆς δ' αὐτῆς
5 μορφῆς ἐστι καὶ δέρμα καὶ φλὲψ καὶ ὑμὴν καὶ πᾶν

---

[1] vv. 34–b 7 secluserunt A.-W.

[a] Other attempts to bring out the meaning of this
word would include " imperfectly developed," " under-
developed," " malformed," " mutilated," " congenitally
disabled."

[b] i.e., as appears later, sentient Soul (ch. 5).

[c] i.e., as above (ll. 23-25), potentially.

contains all the parts of the body *potentially*, though
none *in actuality* ; and " all " includes those parts
which distinguish the two sexes. Just as it sometimes happens that deformed [a] offspring are produced
by deformed parents, and sometimes not, so the offspring produced by a female are sometimes female,
sometimes not, but male. The reason is that the
female is as it were a deformed male ; and the
menstrual discharge is semen, though in an impure
condition ; *i.e.*, it lacks one constituent, and one only,
the principle of Soul.[b] This explains why, in the
case of the wind-eggs produced by some animals, the
egg which takes shape contains the parts of both
sexes,[c] but it has not this principle, and therefore it
does not become a living thing with Soul in it ; this
principle has to be supplied by the semen of the
male, and it is when the female's residue secures this
principle that a fetation is formed.[d]

[e] [When substances which are fluid but also corporeal are heated, an outer layer forms round them,
just as we find a solid layer forming round things that
have been boiled, as they cool. All bodies depend on
something glutinous to hold them together ; and as
their development proceeds and they become larger,
this glutinous character is acquired by the substance
known as sinew, which holds the parts of animals
together (in some it is actual sinew which does
this, in others its counterpart).[f] Skin, blood-vessels,
membrane and all that class of substances are of the

[d] Or, " it becomes a fetation," *i.e.*, a perfect fetation ; see
737 a 10.
[e] The following paragraph, which consists partly of remarks taken from elsewhere, is irrelevant here.
[f] Sometimes, as here, " counterpart " could be represented by the modern term " analogue " ; *cf. P.A.* 653 b 36.

τὸ τοιοῦτον γένος· διαφέρει γὰρ ταῦτα τῷ μᾶλλον
καὶ ἧττον καὶ ὅλως[1] ὑπεροχῇ καὶ ἐλλείψει.]

IV Τῶν δὲ ζῴων τὰ μὲν ἀτελεστέραν ἔχοντα τὴν
φύσιν, ὅταν γένηται κύημα τέλειον ζῷον δὲ μήπω
10 τέλειον, θύραζε προΐεται· δι᾽ ἃς δ᾽ αἰτίας εἴρηται
πρότερον. τέλειον δ᾽ ἤδη τότ᾽ ἐστίν, ὅταν τὸ μὲν
ἄρρεν ᾖ τὸ δὲ θῆλυ τῶν κυημάτων, ἐν ὅσοις ἐστὶν
αὕτη ἡ διαφορὰ τῶν γινομένων· ἔνια γὰρ οὔτε θῆλυ
γεννᾷ οὔτ᾽ ἄρρεν, ὅσα μηδ᾽ αὐτὰ γίνεται ἐκ θήλεος
καὶ ἄρρενος μηδ᾽ ἐκ ζῴων μιγνυμένων. καὶ περὶ
15 μὲν τῆς τούτων γενέσεως ὕστερον ἐροῦμεν·

Τὰ δὲ ζῳοτοκοῦντα ἐν αὑτοῖς τὰ τέλεια τῶν
ζῴων, μέχρι περ ἂν οὗ γεννήσῃ ζῷον καὶ θύραζε
ἐκπέμψῃ, ἔχει συμφυὲς ἐν αὑτοῖς[2] τὸ γιγνόμενον
ζῷον.

Ὅσα δὲ θύραζε μὲν ζῳοτοκεῖ, ἐν αὑτοῖς δ᾽ ᾠοτοκεῖ τὸ πρῶτον, ὅταν γεννήσῃ τὸ ᾠὸν τέλειον,
20 τούτων ἐνίων μὲν ἀπολύεται τὸ ᾠὸν ὥσπερ τῶν
θύραζε ᾠοτοκούντων, καὶ τὸ ζῷον ἐκ τοῦ ᾠοῦ
γίνεται ἐν τῷ θήλει, ἐνίων δ᾽ ὅταν καταναλωθῇ ἡ
ἐκ τοῦ ᾠοῦ τροφή, τελειοῦται ἀπὸ τῆς ὑστέρας, καὶ
διὰ τοῦτο οὐκ ἀπολύεται τὸ ᾠὸν ἀπὸ τῆς ὑστέρας.
ταύτην δ᾽ ἔχουσι τὴν διαφορὰν οἱ σελαχώδεις ἰχ-
25 θύες, περὶ ὧν ὕστερον καθ᾽ αὑτὰ λεκτέον.

Νῦν δ᾽ ἀπὸ τῶν πρώτων ἀρκτέον πρῶτον. ἔστι

---

[1] ὅλως PS : ὅλως ἐν vulg.
[2] αὑτοῖς Rackham : αὑτῷ vulg.

---

[a] Cf. P.A. 644 a 17, and note there; also Introd. § 70.
[b] For the meaning of " perfect " animals, see below,
737 b 15, 16, and the fuller definition given at 732 b 28 ff.
[c] i.e., a " perfect " egg; for another sense, see 776 b 1.
[d] For Selachia, see Bk. III, ch. 3.

same stamp ; they differ only by the " more and less," or putting it generally, by excess and deficiency.[a]]

So far as those animals whose nature is more im- IV perfect are concerned,[b] as soon as a perfect fetation[c] has been formed, though it is not so far a perfect animal, they expel it. The reasons for this I have already stated. A fetation is perfect by the time it is either male or female. (This applies to those animals whose offspring have this distinction of sex, for there are some which generate offspring that are neither male nor female ; these are the animals which are not themselves produced by male and female parents—not produced in fact as the result of the copulation of a pair of animals. We will speak later of the way in which these are generated.)

The perfect animals, the ones which are internally viviparous, retain within themselves the animal which is forming, and it remains joined to them until it is brought to birth and expelled.

With regard to those which are internally oviparous in the first stage although they are externally viviparous, the egg, when it has been perfectly formed, in some cases (a) is released, just as it is in the externally oviparous animals, and the animal is produced out of the egg inside the female ; in other cases (b), when the nourishment in the egg has been used up, the supply for the creature's perfecting is derived from the uterus ; and that is why the egg is not released from the uterus. This distinguishing feature belongs to the Selachian fishes, which will have to receive special mention later.[d]

For the present, however, we must begin first of all with the animals that come first. These are the Generation in Vivipara.

737 b

δὲ τὰ τέλεια ζῷα πρῶτα, τοιαῦτα δὲ τὰ ζῳοτοκοῦντα, καὶ τούτων ἄνθρωπος πρῶτον.

Ἡ μὲν οὖν ἀπόκρισις γίνεται πᾶσι τοῦ σπέρματος ὥσπερ ἄλλου τινὸς περιττώματος. φέρεται
γὰρ ἕκαστον εἰς τὸν οἰκεῖον τόπον οὐθὲν ἀποβια-
30 ζομένου τοῦ πνεύματος, οὐδ' ἄλλης αἰτίας τοιαύτης
ἀναγκαζούσης, ὥσπερ τινές φασιν, ἕλκειν τὰ αἰδοῖα
φάσκοντες ὥσπερ τὰς σικύας, τῷ τε πνεύματι βια-
ζομένων, ὥσπερ ἐνδεχόμενον ἄλλοθί που πορευ-
θῆναι μὴ βιασαμένων ἢ¹ ταύτην τὴν περίττωσιν
ἢ² τὴν τῆς ὑγρᾶς ἢ ξηρᾶς τροφῆς, ὅτι τὰς ἐξόδους
35 αὐτῶν ἠθροισμένῳ τῷ πνεύματι συνεκκρίνουσιν.

738 a

τοῦτο δὲ κοινὸν κατὰ πάντων ὅσα δεῖ κινῆσαι, διὰ
γὰρ τοῦ τὸ πνεῦμα κατασχεῖν ἡ ἰσχὺς ἐγγίνεται·
ἐπεὶ καὶ ἄνευ ταύτης τῆς βίας ἐκκρίνεται τὰ περιττώματα καὶ καθεύδουσι, ἂν ἄνετοί τε καὶ πλήρεις
περιττώματος οἱ τόποι τύχωσιν ὄντες. ὅμοιον δὲ
5 κἂν εἴ τις φαίη τοῖς φυτοῖς ὑπὸ τοῦ πνεύματος
ἑκάστοτε τὰ σπέρματα ἀποκρίνεσθαι πρὸς τοὺς
τόπους πρὸς οὓς εἴωθε φέρειν τὸν καρπόν. ἀλλὰ
τούτου μὲν αἴτιον, ὥσπερ εἴρηται, τὸ πᾶσιν εἶναι
μόρια δεκτικὰ τοῖς περιττώμασι τοῖς τ' ἀχρήστοις
⟨καὶ τοῖς χρησίμοις⟩³ [οἷον τῇ τε ξηρᾷ καὶ τῇ
ὑγρᾷ, καὶ τῷ αἵματι τὰς καλουμένας φλέβας].

10 Τοῖς μὲν οὖν θήλεσι περὶ τὸν τῶν ὑστερῶν τόπον,
σχιζομένων ἄνωθεν τῶν δύο φλεβῶν, τῆς τε με-

¹ ἢ P : om. vulg.   ² ἢ P : om. vulg.
³ supplevi, cetera seclusi ; vid. p. 562, infra.

ᵃ Cf. Hippocrates, π. ἀρχ. ἰητρικῆς 22 (i. 626-628 Littré),
where the action of the bladder, the head and the uterus in
drawing fluid to themselves is compared to the action of
σικύαι.

178

perfect animals, which means the viviparous ones ; and the first of these is Man.

In all of them the semen is secreted in precisely the same way as any other residue. Each of the residues is carried to its proper place without the exertion of any force from the *pneuma* and without compulsion by any other cause of that sort, although some people assert this, alleging that the sexual parts draw the residue like cupping-glasses [a] and that we exert force by means of the *pneuma,* as though it were possible for the seminal residue or for the residue of the liquid or of the solid nourishment to take any other course unless such force were exerted. The reason given for this view is that our discharge of these residues is accompanied by the collecting of the *pneuma* (the holding of the breath). But this is a phenomenon which is common to all cases where something has to be moved, because holding the breath is the way in which the required strength is obtained. Besides, even without the exertion of this force residues are actually discharged during sleep, if the places concerned are relaxed and full of residue. Such statements are on a par with saying that the seeds of plants are on each occasion secreted to the places where they commonly bear their fruit by means of *pneuma.* No, the real reason for this, as has been said, is that in all animals there are parts for the reception of the residues, both for the useless ⟨and for the useful ones⟩ [*e.g.*, both for the solid and the fluid ; and for the blood there are the blood-vessels as they are called].[b]

The region of the uterus in females.—Higher up in the body the two blood-vessels, the Great Blood-

(a) The generative residues.

---

[b] This phrase is an interpolation. See p. 562.

γάλης καὶ τῆς ἀορτῆς, πολλαὶ καὶ λεπταὶ φλέβες
τελευτῶσιν εἰς τὰς ὑστέρας, ὧν ὑπερπληρουμένων
ἐκ τῆς τροφῆς, καὶ τῆς φύσεως διὰ ψυχρότητα
πέττειν οὐ δυναμένης, ἐκκρίνεται διὰ λεπτοτάτων
15 φλεβῶν εἰς τὰς ὑστέρας, οὐ δυναμένων διὰ τὴν
στενοχωρίαν δέχεσθαι τὴν ὑπερβολὴν τοῦ πλήθους,
καὶ γίνεται τὸ πάθος οἷον αἱμορροΐς. ἀκριβῶς μὲν
οὖν ἡ περίοδος οὐ τέτακται ταῖς γυναιξί, βούλεται
δὲ φθινόντων γίνεσθαι τῶν μηνῶν εὐλόγως· ψυχρότερα γὰρ τὰ σώματα τῶν ζῴων ὅταν καὶ τὸ
20 περιέχον συμβαίνῃ γίγνεσθαι τοιοῦτον, αἱ δὲ τῶν
μηνῶν σύνοδοι ψυχραὶ διὰ τὴν τῆς σελήνης ἀπό-
λειψιν, διόπερ καὶ χειμερίους συμβαίνει τὰς συν-
όδους εἶναι τῶν μηνῶν μᾶλλον ἢ τὰς μεσότητας.
μεταβεβληκότος μὲν οὖν εἰς αἷμα τοῦ περιττώματος
βούλεται γίγνεσθαι τὰ καταμήνια κατὰ τὴν εἰρη-
25 μένην περίοδον, μὴ πεπεμμένου δὲ κατὰ μικρὸν ἀεί
τι ἀποκρίνεται· διὸ τὰ λευκὰ μικροῖς ἔτι[1] καὶ παιδίοις οὖσι γίνεται τοῖς θήλεσιν. μετριάζουσαι μὲν
οὖν ἀμφότεραι αὗται αἱ ἀποκρίσεις τῶν περιττωμάτων τὰ σώματα σώζουσιν, ἅτε γιγνομένης καθάρσεως τῶν περιττωμάτων ἃ τοῦ νοσεῖν αἴτια
30 τοῖς σώμασιν· μὴ γινομένων δὲ ἢ πλειόνων γιγνομένων βλάπτει· ποιεῖ γὰρ ἢ νόσους ἢ τῶν σωμάτων
καθαίρεσιν, διὸ καὶ τὰ λευκὰ συνεχῶς γινόμενα καὶ
πλεονάζοντα τὴν αὔξησιν ἀφαιρεῖται τῶν παιδίων.

Ἐξ ἀνάγκης μὲν οὖν ἡ περίττωσις αὕτη γίνεται

---

[1] μικροῖς ἔτι] μικρὰ σημεῖα Z¹ : μικροῖς οὖσι Z².

---

[a] i.e., the *vena cava* and the whole venous system, and the
aorta and the whole arterial system.

[b] The moon has no real connexion with menstruation.
Various notions on this subject will be found in H. M. Fox,

vessel and the Aorta,[a] branch out into many fine blood-vessels, which terminate in the uterus. When these are overfull of nourishment (which owing to its own coldness the female system is unable to concoct), it passes through these extremely fine blood-vessels into the uterus ; but owing to their being so narrow they cannot hold the excessive quantity of it, and so a sort of haemorrhage takes place. In women the period is not accurately fixed, but it tends to happen when the moon is waning,[b] which is what we should expect, since the bodies of animals are colder when their environment is colder, and the time of new moon is a cold time on account of the disappearance [c] of the moon : the same thing explains why the end of the month is stormier than the middle.[d] When the residue has changed into blood, the menstrual discharge tends to occur in accordance with the period just mentioned ; but when the residue has not been concocted, small quantities are secreted from time to time, and this is why " whites " occur in females, even while they are still quite small children. These two secretions of residue, if moderate in amount, keep the body in a sound condition, because they constitute an evacuation of the residues which cause disease. If they fail to occur, or occur too plenteously, they are injurious, producing either diseases or a lowering of the body ; and that is why continuous and abundant discharge of " whites " prevents young girls from growing.

Thus the production of this residue by females is,

Selene. For other references see F. H. A. Marshall, " Sexual Periodicity," in *Phil. Trans. Royal Soc.* (B), CCXXVI (No. 539), p. 442, n.   [c] *i.e.*, complete waning.
    [d] See 777 b 35, n.

738 a

τοῖς θήλεσι διὰ τὰς εἰρημένας αἰτίας· μὴ δυναμένης
35 τε γὰρ πέττειν τῆς φύσεως ἀνάγκη περίττωμα γίγνεσθαι μὴ μόνον τῆς ἀχρήστου τροφῆς, ἀλλὰ καὶ
ἐν ταῖς φλεψίν, ὑπερβάλλειν τε πληθύοντα[1] κατὰ

738 b

τὰς λεπτοτάτας φλέβας. ἕνεκα δὲ τοῦ βελτίονος
καὶ τοῦ τέλους ἡ φύσις καταχρῆται πρὸς τὸν τόπον
τοῦτον τῆς γενέσεως χάριν, ὅπως οἷον ἔμελλε
τοιοῦτον γένηται ἕτερον· ἤδη γὰρ ὑπάρχει δυνάμει
γε ὂν τοιοῦτον οἷον πέρ ἐστι σώματος ἀπόκρισις.

5 Τοῖς μὲν οὖν θήλεσιν ἅπασιν ἀναγκαῖον γίγνεσθαι
περίττωμα, τοῖς μὲν αἱματικοῖς πλεῖον, καὶ τούτων
ἀνθρώπῳ πλεῖστον· ἀνάγκη δὲ καὶ τοῖς ἄλλοις
ἀθροίζεσθαί τινα σύστασιν εἰς τὸν ὑστερικὸν τόπον.
τὸ δ' αἴτιον, ὅτι τοῖς θ' αἱματικοῖς πλεῖον καὶ
τούτων ὅτι πλεῖστον τοῖς ἀνθρώποις, εἴρηται
πρότερον.

10 Τοῦ δ' ἐν μὲν τοῖς θήλεσι πᾶσιν ὑπάρχειν περίττωμα τοιοῦτον, ἐν δὲ τοῖς ἄρρεσι μὴ πᾶσιν, ἔνια
γὰρ οὐ προΐεται γονήν, ἀλλ' ὥσπερ τὰ προϊέμενα[2]
τῇ ἐν τῇ γονῇ κινήσει δημιουργεῖ τὸ συνιστάμενον
ἐκ τῆς ἐν τοῖς θήλεσιν ὕλης, οὕτω τὰ τοιαῦτα [ἐν][3]
τῇ ἐν αὑτοῖς κινήσει ἐν τῷ μορίῳ τούτῳ, ὅθεν
15 ἀποκρίνεται τὸ σπέρμα, ταὐτὸ ποιεῖ καὶ συνίστησιν.
τοῦτο δ' ἐστὶν ὁ τόπος ὁ περὶ τὸ ὑπόζωμα πᾶσι
τοῖς ἔχουσιν· ἀρχὴ γὰρ τῆς φύσεως ἡ καρδία καὶ

---

[1] πληθύοντα Z : πληθύνοντα vulg.
[2] προϊέμενα PS : προειρημένα vulg.
[3] secluserunt A.-W.

---

[a] Sc., from the useful nourishment, viz., blood.
[b] At 727 a 21 ff., and 728 a 30 ff.
[c] This sentence has been remodelled in the translation, since in the Greek the construction is not carried through.

on the one hand, the result of *necessity*, and the reasons have been given : The female system cannot effect concoction, and therefore of necessity residue must be formed not only from the useless nourishment, but also [a] in the blood-vessels, and when there is a full complement of it in those very fine blood-vessels, it must overflow. On the other hand, in order to serve the *better* purpose, the End, Nature diverts it to this place and employs it there for the sake of generation, in order that it may become another creature of the same kind as it would have become, since even as it is, it is *potentially* the same in character as the body whose secretion it is.

In all female animals, then, some residue must of necessity be formed : a greater amount of it in the blooded ones, and the greatest of all in human beings, though some substance must of necessity collect in the region of the uterus in the other animals too. The reason why a larger amount is produced in the blooded animals, and the largest amount of all in human beings, has already been stated.[b]

But although a residue of this sort occurs in all females, it does not occur in all males. Why is this ? [c] Some males do not emit semen, but, just as the ones which emit semen fashion the creature that is taking shape out of the material supplied by the female by the agency of the movement resident in the semen, so these fashion it into shape by the agency of the movement which resides in that part of themselves whence the semen is secreted ; they produce this same effect of causing the material to set.[d] (The part to which I refer is the region around the diaphragm in all those animals which have one, because

[a] *Cf.* above, 736 a 27 and references there given.

τὸ ἀνάλογον, τὸ δὲ κάτω προσθήκη καὶ τούτου
χάριν. αἴτιον δὴ τοῦ τοῖς μὲν ἄρρεσι μὴ πᾶσιν
εἶναι περίττωμα γεννητικόν, τοῖς δὲ θήλεσι πᾶσιν,
20 ὅτι τὸ ζῷον σῶμα ἔμψυχόν ἐστιν. ἀεὶ δὲ παρέχει
τὸ μὲν θῆλυ τὴν ὕλην, τὸ δ' ἄρρεν τὸ δημιουργοῦν.
ταύτην γὰρ αὐτῶν φαμὲν ἔχειν τὴν δύναμιν ἑκάτερον, καὶ τὸ εἶναι τὸ μὲν θῆλυ τὸ δ' ἄρρεν τοῦτο.
ὥστε τὸ μὲν θῆλυ ἀναγκαῖον παρέχειν σῶμα καὶ
ὄγκον, τὸ δ' ἄρρεν οὐκ ἀναγκαῖον· οὔτε γὰρ τὰ
25 ὄργανα ἀνάγκη ἐνυπάρχειν ἐν τοῖς γιγνομένοις οὔτε
τὸ ποιοῦν. ἔστι δὲ τὸ μὲν σῶμα ἐκ τοῦ θήλεος, ἡ
δὲ ψυχὴ ἐκ τοῦ ἄρρενος· ἡ γὰρ ψυχὴ οὐσία σώματός τινός ἐστιν. καὶ διὰ τοῦτο ὅσα τῶν μὴ
ὁμογενῶν μίγνυται θῆλυ καὶ ἄρρεν (μίγνυται δὲ ὧν
ἴσοι οἱ χρόνοι καὶ ἐγγὺς αἱ κυήσεις, καὶ τὰ μεγέθη
30 τῶν σωμάτων μὴ πολὺ διέστηκεν), τὸ μὲν πρῶτον
κατὰ τὴν ὁμοιότητα γίγνεται κοινὸν ἀμφοτέρων,
οἷον τὰ γιγνόμενα ἐξ ἀλώπεκος καὶ κυνὸς καὶ
πέρδικος καὶ ἀλεκτρυόνος, προϊόντος δὲ τοῦ χρόνου
καὶ ἐξ ἑτέρων ἕτερα γιγνόμενα τέλος ἀποβαίνει
κατὰ τὸ θῆλυ τὴν μορφήν, ὥσπερ τὰ σπέρματα τὰ
35 ξενικὰ κατὰ τὴν χώραν. αὕτη γὰρ ἡ τὴν ὕλην

ᵃ Or " reality." *Cf. De anima* 415 b 7 ff., where the Soul
is said to be the Cause and principle of the body (a) as the
source of its movement, (b) as its Final Cause, that " for the
sake of which " the body exists, (c) as being the essence of
living bodies. The last is explained thus : the cause (or
ground) of the being of anything is its essence ; the being
of living things is to live ; and the Cause and principle of
their being and living is Soul. *Cf.* also Aristotle's repeated

the first principle of any natural creature's system is the heart or its counterpart, while the lower parts are an appendage added for the sake of that.) Why does this generative residue, then, not occur in all males, although it occurs in all females ? The answer is that an animal is a living body, a body with Soul in it. The female always provides the material, the male provides that which fashions the material into shape ; this, in our view, is the specific characteristic of each of the sexes : that is what it means to be male or to be female. Hence, necessity requires that the female should provide the physical part, *i.e.*, a quantity of material, but not that the male should do so, since necessity does not require that the tools should reside in the product that is being made, nor that the agent which uses them should do so. Thus the physical part, the body, comes from the female, and the Soul from the male, since the Soul is the essence [a] of a particular body. On this account, when a male and a female of different species copulate (which happens in the case of animals whose periods are equal and whose times of gestation run close, and which do not differ widely in physical size), the first generation, so far as resemblance goes, takes equally after both parents (examples are the offspring of fox and dog,[b] and of partridge and common fowl), but as time goes on and successive generations are produced, the offspring finish up by taking after the female as regards their bodily form, just as happens when seeds are introduced into a strange locality— the plants take after the soil, the reason being that

statements that no part of the body can be such in anything but name unless it has Soul in it ; see also *P.A.* 641 a 25 ff.

[b] Viz., the so-called Laconian hound ; see *H.A.* 607 a 3.

παρέχουσα καὶ τὸ σῶμα τοῖς σπέρμασίν ἐστιν. καὶ
διὰ τοῦτο τοῖς μὲν θήλεσι τὸ μόριον τὸ δεκτικὸν οὐ
πόρος ἐστίν, ἀλλ᾽ ἔχουσι διάστασιν αἱ ὑστέραι·
τοῖς δ᾽ ἄρρεσι πόροι τοῖς σπέρμα προϊεμένοις,
ἄναιμοι δ᾽ οὗτοι.

Τῶν δὲ περιττωμάτων ἕκαστον ἅμα ἔν τε τοῖς
οἰκείοις τόποις ἐστὶ καὶ γίγνεται περίττωμα· πρότερον δ᾽ οὐθέν, ἂν μή τι βίᾳ πολλῇ καὶ παρὰ φύσιν.
5 Δι᾽ ἣν μὲν οὖν αἰτίαν ἀποκρίνεται τὰ περιττώματα τὰ γεννητικὰ τοῖς ζῴοις, εἴρηται.

Ὅταν δ᾽ ἔλθῃ τὸ σπέρμα ἀπὸ τοῦ ἄρρενος τῶν
σπέρμα προϊεμένων, συνίστησι τὸ καθαρώτατον τοῦ
περιττώματος—τὸ γὰρ πλεῖστον ἄχρηστον καὶ ἐν
τοῖς καταμηνίοις ἐστὶν ὑγρὸν ⟨ὄν⟩[1], ὥσπερ καὶ τῆς
10 τοῦ ἄρρενος γονῆς τὸ ὑγρότατον· καὶ τῆς εἰσάπαξ
προέσεως [καὶ][2] ἡ προτέρα τῆς ὑστέρας ἄγονος
μᾶλλον τοῖς πλείστοις· ἐλάττω γὰρ ἔχει θερμότητα
ψυχικὴν διὰ τὴν ἀπεψίαν, τὸ δὲ πεπεμμένον πάχος
ἔχει καὶ σεσωμάτωται μᾶλλον.

Ὅσαις δὲ μὴ γίνεται θύραζέ τις πρόεσις, ἢ τῶν
γυναικῶν ἢ τῶν ἄλλων ζῴων, διὰ τὸ μὴ ἐνυπάρχειν
15 ἄχρηστον περίττωμα πολὺ ἐν τῇ ἀποκρίσει τῇ
τοιαύτῃ, τοσοῦτόν ἐστι τὸ ἐγγινόμενον ὅσον τὸ
ὑπολειπόμενον τοῖς θύραζε προϊεμένοις ζῴοις, ὃ
συνίστησιν ἡ τοῦ ἄρρενος δύναμις ἡ ἐν τῷ σπέρματι

---

[1] ⟨ὄν⟩ supplevi. [2] seclusi.

[a] See Bk. I. 718 a 10 ff.
[b] Cf. Hippocrates, π. σαρκῶν 13 (viii. 600 Littré) ἡ δὲ
τροφὴ ἐπειδὰν ἀφίκηται ἐς ἕκαστον, τοιαύτην ἀπέδωκε τὴν εἰδέην
ἑκάστου ὁκοία περ ἦν.
[c] The "concoction" of the semen in viviparous land-
animals takes place actually during copulation (see 717 b 24

the soil provides the material—*i.e.*, the physical body —for the seeds. And on this account the part in females which receives the semen is not a passage, but it—*i.e.*, the uterus—is fairly wide, whereas the males that emit semen have passages only, and these have no blood in them.[a]

It is only when it occupies its own proper place that each of the residues becomes that particular residue[b]; before that time none of them can do so without great violence exerted contrary to nature.

We have now given the reason for the secretion of the generative residues in animals.

In those species which emit semen, when the semen from the male has entered, it causes the purest portion of the residue to " set "—I say " purest portion," because the most part of the menstrual discharge is useless, being fluid, just as the most fluid portion of the male semen is, and in most cases the earlier discharge during any one emission is less fertile than the later, because it has less soul-heat owing to its being unconcocted, whereas that which has been concocted is thicker and has more body in it.[c]

In those cases (whether women or other female animals) where there is no external discharge (due to there being no large amount of useless residue in the generative secretion), the amount of stuff which is produced within them corresponds in quantity to that which remains behind in those animals which discharge externally. This stuff gets " set " by the *dynamis* of the male (*a*) present in the semen which

and 718 a 5 above), which explains the phenomenon here mentioned. In fishes and serpents the semen is already concocted before the time of copulation (*ibid.*).

739 a

τῷ ἀποκρινομένῳ, ἢ εἰς τὸ ἄρρεν ἐλθόντος τοῦ
ἀνάλογον μορίου ταῖς ὑστέραις, ὥσπερ ἔν τισι τῶν
20 ἐντόμων φαίνεται συμβαῖνον.

Ὅτι δ᾽ ἡ γινομένη ὑγρότης μετὰ τῆς ἡδονῆς τοῖς
θήλεσιν οὐδὲν συμβάλλεται εἰς τὸ κύημα, εἴρηται
πρότερον. μάλιστα δ᾽ ἂν δόξειεν, ὅτι καθάπερ τοῖς
ἄρρεσι, γίγνεται καὶ ταῖς γυναιξὶ νύκτωρ ὃ καλοῦσιν ἐξονειρώττειν. ἀλλὰ τοῦτο σημεῖον οὐθέν· γί-
25 νεται γὰρ καὶ τοῖς νέοις τῶν ἀρρένων τοῖς μέλλουσι
μὲν μηθὲν δὲ προϊεμένοις, ἢ τοῖς ἔτι[1] προϊεμένοις
ἄγονον.

Ἄνευ μὲν οὖν τῆς τοῦ ἄρρενος προέσεως ἐν τῇ
συνουσίᾳ ἀδύνατον συλλαβεῖν, καὶ ἄνευ τῆς τῶν
γυναικείων περιττώσεως ἢ θύραζε προελθούσης
ἢ ἐντὸς ἱκανῆς οὔσης. οὐ συμβαινούσης μέντοι τῆς
30 εἰωθυίας γίγνεσθαι τοῖς θήλεσιν ἡδονῆς περὶ τὴν
ὁμιλίαν τὴν τοιαύτην συλλαμβάνουσιν, ἂν τύχῃ ὁ
τόπος ⟨γ᾽⟩ ὀργῶν[2] καὶ καταβεβηκυῖαι αἱ ὑστέραι
ἐντός.[3] ἀλλ᾽ ὡς ἐπὶ τὸ πολὺ συμβαίνει ἐκείνως
διὰ τὸ μὴ συμμεμυκέναι τὸ στόμα γινομένης τῆς
ἐκκρίσεως, μεθ᾽ ἧς εἴωθε γίγνεσθαι καὶ τοῖς ἄρρεσιν
35 ἡ ἡδονὴ καὶ ταῖς γυναιξίν· οὕτω δ᾽ ἐχόντος εὐοδεῖται
μᾶλλον καὶ τῷ τοῦ ἄρρενος σπέρματι.

Ἡ δ᾽ ἄφεσις οὐκ ἐντὸς γίγνεται, καθάπερ οἴονταί
τινες (στενὸν γὰρ τὸ στόμα τῶν ὑστερῶν), ἀλλ᾽ εἰς
739 b τὸ πρόσθεν, οὗπερ τὸ θῆλυ προΐεται τὴν ἐν ἐνίαις
αὐτῶν ἰκμάδα γινομένην, ἐνταῦθα καὶ τὸ ἄρρεν
προΐεται [ἐάν τις ἐξικμάσῃ].[4] ὁτὲ μὲν οὖν μένει

---

[1] ἔτι προϊεμένοις corr. P : ἐπιπροϊεμένοις vulg.
[2] τόπος γ᾽ ὀργῶν A.-W.: τόπος ὁ γεωργῶν P : γ᾽ om. vulg.
[3] ἐντός P : ἐγγύς vulg.
[4] secl. A.-W., Platt.

is secreted, or (b) when the part of the female analogous to the uterus is inserted into the male (as is observed to take place in certain insects).[a]

I have said already [b] that the fluid which is produced in females and accompanies sexual excitement contributes nothing at all to the fetation. The strongest reason for believing that it does is that the phenomenon of night effusions occurs in women just as in men; but this is no proof at all, because it occurs with young men who come almost to the point but in fact emit nothing, and also with those who as yet emit infertile semen.

Conception cannot occur without (a) an emission from the male during copulation and without (b) the presence of the menstrual residue either externally discharged or available in sufficient quantity internally. Conception takes place, however, even if the pleasure which women usually experience during sexual intercourse fails to occur, if the part concerned happens to be in heat and the uterus has descended within. Generally, however, pleasure does occur, because when the secretion, which is usually accompanied by pleasure in man and woman alike, takes place, the *os uteri* has not closed, and in these conditions a better passage is afforded for the semen of the male.

The discharge does not (as some suppose) take place within the uterus, because the *os uteri* is narrow. The discharge of the male takes place in front of it, at precisely the same spot where the female discharges the moisture which is produced in some instances.[c] Sometimes it remains in this place,

[a] Cf. 738 b 12.   [b] Bk. I, ch. 20.
[c] Cf. 727 b 33 ff.

τοῦτον ἔχον[1] τὸν τόπον,[2] ὁτὲ δέ, ἂν τύχῃ συμμέτρως ἔχουσα καὶ θερμὴ διὰ τὴν κάθαρσιν ἡ ὑ-
5 στέρα, εἴσω σπᾷ. σημεῖον δέ· καὶ γὰρ τὰ πρόσθετα[3]
ὑγρὰ προστεθέντα ἀφαιρεῖται ξηρά. ἔτι δὲ ὅσα
τῶν ζῴων πρὸς τῷ ὑποζώματι ἔχει τὰς ὑστέρας,
καθάπερ ὄρνις καὶ τῶν ἰχθύων οἱ ζωοτοκοῦντες,
ἀδύνατον ἐκεῖ μὴ σπᾶσθαι τὸ σπέρμα, ἀλλ' ἀφεθὲν
ἐλθεῖν. ἕλκει δὲ τὴν γονὴν ὁ τόπος διὰ τὴν θερ-
10 μότητα τὴν ὑπάρχουσαν. καὶ ἡ τῶν καταμηνίων
δὲ ἔκκρισις καὶ συνάθροισις ἐμπυρεύει θερμότητα
ἐν τῷ μορίῳ τούτῳ, [ὥστε][4] καθάπερ τὰ κωνικὰ[5]
τῶν ἀγγείων, ὅταν θερμῷ διακλυσθῇ, σπᾷ τὸ ὕδωρ
εἰς αὑτὰ καταστρεφομένου τοῦ στόματος. καὶ
τοῦτον μὲν τὸν τρόπον γίγνεται σπάσις, ὡς δέ τινες
15 λέγουσι, τοῖς ὀργανικοῖς πρὸς τὴν συνουσίαν μορίοις οὐ γίνεται κατ' οὐθένα τρόπον. ἀνάπαλιν δὲ
συμβαίνει καὶ τοῖς λέγουσι προΐεσθαι καὶ τὴν γυ-
ναῖκα σπέρμα. προϊεμέναις γὰρ ἔξω συμβαίνει
ταῖς ὑστέραις πάλιν εἴσω σπᾶν, εἴπερ μιχθήσεται
τῇ γονῇ τῇ τοῦ ἄρρενος. τὸ δ' οὕτω γίγνεσθαι
20 περίεργον, ἡ δὲ φύσις οὐδὲν ποιεῖ περίεργον.

Ὅταν δὲ συστῇ ἡ ἐν ταῖς ὑστέραις ἀπόκρισις
τοῦ θήλεος ὑπὸ τῆς τοῦ ἄρρενος γονῆς, παραπλήσιον ποιούσης ὥσπερ ἐπὶ τοῦ γάλακτος τῆς
πυετίας· καὶ γὰρ ἡ πυετία γάλα ἐστὶ θερμότητα
ζωτικὴν ἔχον, ἣ τὸ ὅμοιον εἰς ἓν ἄγει καὶ συνίστησι,

---

[1] ἔχον Y : ἔχοντα vulg.   [2] τόπον Platt : τρόπον vulg.
[3] πρόσθετα P : πρόσθεν vulg.   [4] ὥστε seclusi.
[5] κωνικὰ Platt : ἀκόνιτα vulg. : vas quod non est plenum
Σ (= κενὰ ?).

---

[a] Cf. 728 a 31 ff.

sometimes, if the uterus happens to be in a suitable condition and hot owing to the evacuation of the menses, the uterus draws it in. Evidence for this is the fact that pessaries though wet when applied are dry when removed. Also, in those animals (such as birds and viviparous fishes) whose uterus is close by the diaphragm there is no alternative : the semen must be drawn in ; it cannot enter at the moment of discharge. This region, in virtue of the heat present in it (the discharge and aggregation of the menstrual fluid also produce fiery heat in this part) draws up the semen in the same way that conical vessels which have been washed out with something warm draw water up into themselves when they are turned mouth downwards. And that is the way in which the semen is drawn in ; it is certainly not done, as some allege, by the parts that are instrumental in copulation.[a] We find the situation reversed in the theory that the woman as well as the man emits semen, since if the uterus emits any semen outside itself, it will have to draw it back inside again if it is to mingle with the semen of the male. Such a performance is superfluous, and Nature does nothing which is superfluous.

The action of the semen of the male in " setting " the female's secretion in the uterus is similar to that of rennet upon milk.[b] Rennet is milk which contains vital heat, as semen does, and this integrates the homogeneous substance and makes it " set." As the

[b] *Cf.* 755 a 18. This is a remarkable intuition of the essential rôle played by fermentation in embryonic development. *Cf.* also Job x. 10 " Hast thou not poured me out as milk, and curdled me like cheese ? Thou hast clothed me with skin and flesh, and knit me together with bones and sinews " (R.V.).

191

739 b

25 καὶ ἡ γονὴ πρὸς τὴν τῶν καταμηνίων φύσιν ταὐτὸ[1]
πέπονθεν· ἡ γὰρ αὐτὴ φύσις ἐστὶ γάλακτος καὶ
καταμηνίων. συνιόντος δὴ[2] τοῦ σωματώδους ἐκκρίνεται τὸ ὑγρόν, καὶ περίστανται κύκλῳ ξηραινο-
μένων τῶν γεηρῶν ὑμένες, καὶ ἐξ ἀνάγκης καὶ
ἔνεκά τινος· καὶ γὰρ θερμαινομένων ξηραίνεσθαι
30 ἀναγκαῖον τὰ ἔσχατα καὶ ψυχομένων, καὶ δεῖ μὴ
ἐν ὑγρῷ τὸ ζῷον εἶναι ἀλλὰ κεχωρισμένον. κα-
λοῦνται δὲ τούτων οἱ μὲν ὑμένες τὰ δὲ χόρια,
διαφέροντα τῷ μᾶλλον καὶ ἧττον· ὁμοίως δ' ἐνυπάρ-
χουσιν ἔν τε τοῖς ᾠοτόκοις ταῦτα καὶ τοῖς ζῳοτόκοις.

Ὅταν δὲ συστῇ τὸ κύημα ἤδη, παραπλήσιον
35 ποιεῖ τοῖς σπειρομένοις. ἡ μὲν γὰρ ἀρχὴ καὶ ἐν
τοῖς σπέρμασιν ἐν αὐτοῖς ἐστὶν ἡ πρώτη· ὅταν δ'
αὕτη ἀποκριθῇ ἐνοῦσα δυνάμει πρότερον, ἀπὸ ταύτης ἀφίεται ὅ τε βλαστὸς καὶ ἡ ῥίζα. αὕτη δ'

740 a
ἐστὶν ᾗ τὴν τροφὴν λαμβάνει· δεῖται γὰρ αὐξήσεως
τὸ φυτόν. οὕτω καὶ ἐν τῷ κυήματι τρόπον τινὰ
πάντων ἐνόντων τῶν μορίων δυνάμει ἡ ἀρχὴ πρὸ
ὁδοῦ μάλιστα ἐνυπάρχει. διὸ ἀποκρίνεται πρῶτον
ἡ καρδία ἐνεργείᾳ. καὶ τοῦτο οὐ μόνον ἐπὶ τῆς
5 αἰσθήσεως δῆλον (συμβαίνει γὰρ οὕτως), ἀλλὰ καὶ
ἐπὶ τοῦ λόγου. ὅταν γὰρ ἀπ' ἀμφοῖν ἀποκριθῇ,
δεῖ αὐτὸ αὐτὸ διοικεῖν τὸ γενόμενον, καθάπερ ἀπ-

---

[1] ταὐτὸ P : τοῦτο vulg. [2] δὴ A.-W., O[b]* : δὲ vulg.

---

[a] φύσις, as often, refers specially to the substance of the
thing. The substance of milk and the menstrual fluid is
identical, because they are both residues of the useful
nourishment.

nature [a] of milk and the menstrual fluid is one and the same, the action of the semen upon the substance of the menstrual fluid is the same as that of rennet upon milk. Thus when the " setting " is effected, *i.e.*, when the bulky portion " sets," the fluid portion comes off ; and as the earthy portion solidifies membranes form all round its outer surface. (This is the result of *necessity* ; but also it is to serve a *purpose* : (*a*) Necessity ordains that the extreme surface of a thing should solidify when heated as well as when cooled ; (*b*) it is requisite that the young animal should not be situated in fluid but well away from it.) Some of these are called membranes ; some *choria* [b] : and they differ by the " more and less." [c] They are found in Ovipara and Vivipara alike.

Once the fetation has " set," it behaves like seeds sown in the ground. The first principle ⟨of growth⟩ is present in the seeds themselves too, and as soon as this, which at first was present *potentially*, has become distinct, a shoot and a root are thrown out from it, the root being the channel by which nourishment is obtained, for of course the plant needs material for growth. So too in the fetation, in a way all the parts are present *potentially*, but the first principle has made the most headway, and on that account the first to become distinct in *actuality* is the heart. This is plain not only to the senses (for after all it is a matter of fact), but also to the reason. Once the fetation which has been formed is separate and distinct from both the parents, it must manage for itself, just like a son who has set up a house of his own independently of his

(b) The development of the embryo

Heart.

---

[b] See also *H.A.* Bk. VI, ch. 3.
[c] See 737 b 7, n., and Introd. § 70.

οικισθὲν τέκνον ἀπὸ πατρός. ὥστε δεῖ ἀρχὴν ἔχειν,
ἀφ᾽ ἧς καὶ ὕστερον ἡ διακόσμησις τοῦ σώματος
γίνεται τοῖς ζῴοις. εἰ γὰρ ἔξωθέν ποτ᾽ ἔσται καὶ
10 ὕστερον ἐνεσομένη, οὐ μόνον διαπορήσειεν ἄν τις
τὸ πότε, ἀλλ᾽ ὅτι ἀνάγκη, ὅταν ἕκαστον χωρίζηται
τῶν μορίων, ταύτην ὑπάρχειν πρῶτον, ἐξ ἧς καὶ
ἡ αὔξησις ὑπάρχει καὶ ἡ κίνησις τοῖς ἄλλοις μορίοις. διόπερ ὅσοι λέγουσιν, ὥσπερ Δημόκριτος,
15 τὰ ἔξω πρῶτον διακρίνεσθαι τῶν ζῴων, ὕστερον
δὲ τὰ ἐντός, οὐκ ὀρθῶς λέγουσιν, ὥσπερ ξυλίνων
ἢ λιθίνων ζῴων. τὰ μὲν γὰρ τοιαῦτ᾽ οὐκ ἔχει
ἀρχὴν ὅλως, τὰ δὲ ζῷα πάντ᾽ ἔχει καὶ ἐντὸς ἔχει.
διὸ πρῶτον ἡ καρδία φαίνεται διωρισμένη πᾶσι
τοῖς ἐναίμοις· ἀρχὴ γὰρ αὕτη καὶ τῶν ὁμοιομερῶν
καὶ τῶν ἀνομοιομερῶν. ἤδη γὰρ ἀρχὴν ταύτην
20 ἄξιον ἀκοῦσαι τοῦ ζῴου καὶ τοῦ συστήματος, ὅταν
δέηται τροφῆς· τὸ γὰρ δὴ ὂν¹ αὐξάνεται. τροφὴ
δὲ ζῴου ἡ ἐσχάτη αἷμα καὶ τὸ ἀνάλογον. τούτων
δ᾽ ἀγγεῖον αἱ φλέβες· διὸ ἡ καρδία καὶ τούτων

¹ ὄν] ζῷον Y.

ᵃ See Diels, *Vorsokr.*⁵ 68 A 145.
ᵇ See Introd. § 19.
ᶜ The point is that by this time the fetation is definitely
constituted—it is an individual—it *exists*, and that which
exists can correctly be said to have an ἀρχή. Also, that
which exists needs nourishment, and in animals nourishment
means blood, of which the heart is the ἀρχή. (As Aristotle
says elsewhere, 735 a, the heart supplies the principle of
growth, and the nutritive faculty of Soul operates through
the heart.) This, then, is why, as soon as the fetation is
definitely constituted, the heart is formed—otherwise no
growth could take place.
ᵈ It is unnecessary to read ζῷον for ὄν : ὄν gives better
point to the argument, with which compare the passage
194

father. That is why it must have a first principle, from which also the subsequent ordering of the animal's body is derived. Otherwise, supposing this principle is to come in at some moment from outside and take up its position inside later on, then we may well be puzzled at what moment this is to happen, and also we may point out that of necessity the first principle must be present at the outset, at the time when each of the parts is being separated from the rest, since the growth and movement of the other parts are derived from it. That is why those people are wrong who, like Democritus,[a] hold that the external parts of animals become distinct first, and then the internal ones. They might be speaking of animals carved out of wood or stone, the sort of things which have no first principle at all, whereas living animals all have such a principle, and it is inside them. On this account in all blooded animals it is the heart which can first be seen as something distinct, as this is the first principle both of the " uniform " and of the " non-uniform " parts [b]—since this is justifiably designated as first principle of the animal or organism from the moment when it begins to need nourishment,[c] for of course that which exists grows,[d] and, for an animal, the ultimate form of nourishment is blood or its counterpart. Of these fluids the blood-vessels are the receptacle,[e] and therefore

735 a 13-26 (where again the reading with ὄν should be kept in 735 a 22). Here the point is clearly made that, once a thing has come into being (γένηται), it must of necessity grow. See also note on 744 b 36.

[e] The blood-vessels distribute the " ultimate nourishment " to the parts of the body, which, as Aristotle says (743 a 1), are moulded round them like a wax figure round a core or foundation, and are formed out of them.

740 a

ἀρχή. δῆλον δὲ τοῦτο ἐκ τῶν ἱστοριῶν καὶ τῶν ἀνατομῶν.[a]

Ἐπεὶ δὲ δυνάμει μὲν ἤδη ζῷον ἀτελὲς δέ, ἄλ-
25 λοθεν ἀναγκαῖον λαμβάνειν τὴν τροφήν· διὸ χρῆται
τῇ ὑστέρᾳ καὶ τῇ ἐχούσῃ, ὥσπερ γῇ φυτόν, τοῦ
λαμβάνειν τροφήν, ἕως ἂν τελεωθῇ πρὸς τὸ εἶναι
ἤδη ζῷον δυνάμει πορευτικόν. διὸ ἐκ τῆς καρδίας
τὰς δύο φλέβας πρῶτον[1] ἡ φύσις ὑπέγραψεν·[b] ἀπὸ
δὲ τούτων φλέβια ἀπήρτηται πρὸς τὴν ὑστέραν ὁ
30 καλούμενος ὀμφαλός. ἔστι γὰρ ὁ ὀμφαλὸς φλέψ,
τοῖς μὲν μία, τοῖς δὲ πλείους τῶν ζῴων. περὶ δὲ
ταύτας κέλυφος δερματικὸν [ὁ καλούμενος ὀμφα-
λός][2] διὰ τὸ δεῖσθαι σωτηρίας καὶ σκέπης τὴν τῶν
φλεβῶν ἀσθένειαν. αἱ δὲ φλέβες οἷον ῥίζαι πρὸς
35 τὴν ὑστέραν συνάπτουσι, δι᾿ ὧν λαμβάνει τὸ κύημα
τὴν τροφήν. τούτου γὰρ χάριν ἐν ταῖς ὑστέραις
μένει τὸ ζῷον, ἀλλ᾿ οὐχ ὡς Δημόκριτός φησιν,[d] ἵνα
διαπλάττηται τὰ μόρια κατὰ τὰ μόρια τῆς ἐχούσης.

740 b
τοῦτο γὰρ ἐπὶ τῶν ᾠοτοκούντων φανερόν· ἐκεῖνα
γὰρ ἐν τοῖς ᾠοῖς λαμβάνει τὴν διάκρισιν, κεχωρισμένα τῆς μήτρας.

Ἀπορήσειε δ᾿ ἄν τις, εἰ τὸ αἷμα μὲν τροφή ἐστιν,
ἡ δὲ καρδία πρώτη γίνεται ἔναιμος οὖσα, [τὸ δ᾿
αἷμα τροφή,][3] ἡ δὲ τροφὴ θύραθεν, πόθεν εἰσῆλθεν
5 ἡ πρώτη τροφή; ἢ τοῦτ᾿ οὐκ ἀληθές, ὡς πᾶσα

---

[1] πρώτας P.   [2] seclusit Bekker.
[3] secluserunt A.-W.; pro τὸ δ᾿ αἷμα . . . θύραθεν et sanguis
est ex extrinseco Σ.

---

[a] H.A. Bk. III, ch. 3.
[b] Or, "sketches in," "traces out."   Cf. 743 b 20, and
a different metaphor at 743 a 2.
[c] Cf. 745 b 25 ff.   [d] See Diels, Vorsokr.⁵ 68 A 144.

the heart is the first principle of them as well. This is clearly brought out in the *Researches* [a] and in the *Dissections*.

Now since the fetation is already an animal *potentially*, though an imperfect one, it must get its nourishment from elsewhere ; and that is why it makes use of the uterus, *i.e.*, of the mother, just as a plant makes use of the earth, in order to get its nourishment, until such time as it is sufficiently perfected to be a *potentially* locomotive animal. That is why Nature prescribes [b] first of all the two blood-vessels that run from the heart ; and attached to these are some small blood-vessels which run to the uterus, forming what is known as the umbilicus, the umbilicus [c] being of course a blood-vessel—a single blood-vessel in some animals, and consisting of more numerous ones in others. Round these blood-vessels there is a skin-like integument, because the blood-vessels being weak need a protective covering to keep them safe and sound. The blood-vessels join on to the uterus as though they were roots, and through them the fetation gets its nourishment. And that of course is the reason why the young animal stays in the uterus (not as Democritus [d] alleges, in order that its parts may be moulded after the fashion of the parts of its mother). This is manifest in the case of the Ovipara, whose parts become distinct in the egg, *i.e.*, after they have been separated from the matrix.

Here is a puzzle which may be raised. If (1) the blood is nourishment, (2) the heart is the first thing to be formed, and when formed contains blood, and (3) the nourishment comes from outside, from whence did the first nourishment [e] enter ? Well, perhaps

---

[e] *i.e.*, the blood which is in the heart to begin with.

θύραθεν, ἀλλ' εὐθύς, ὥσπερ ἐν τοῖς τῶν φυτῶν
σπέρμασιν ἔνεστί τι τοιοῦτον τὸ φαινόμενον πρῶτον γαλακτῶδες, οὕτω καὶ ἐν τῇ ὕλῃ τῶν ζῴων
τὸ περίττωμα τῆς συστάσεως τροφή ἐστιν.

Ἡ μὲν οὖν αὔξησις τῷ κυήματι γίνεται διὰ τοῦ
10 ὀμφαλοῦ τὸν αὐτὸν τρόπον ὅνπερ διὰ τῶν ῥιζῶν
τοῖς φυτοῖς [καὶ τοῖς ζῴοις αὐτοῖς, ὅταν ἀπο-
λυθῶσιν, ἐκ τῆς ἐν αὐτοῖς τροφῆς]¹· περὶ ὧν ὕστερον
λεκτέον κατὰ τοὺς οἰκείους τῶν λόγων καιρούς. ἡ
δὲ διάκρισις γίγνεται τῶν μορίων οὐχ ὡς τινες
ὑπολαμβάνουσι, διὰ τὸ πεφυκέναι φέρεσθαι τὸ
15 ὅμοιον πρὸς τὸ ὅμοιον (πρὸς γὰρ πολλαῖς ἄλλαις
αἷς ὁ λόγος οὗτος ἔχει δυσχερείαις, συμβαίνει
χωρὶς ἕκαστον γίνεσθαι τῶν μορίων τῶν ὁμοιομερῶν, οἷον ὀστᾶ καθ' αὑτὰ καὶ νεῦρα, καὶ τὰς
σάρκας καθ' αὑτάς, εἴ τις ἀποδέξαιτο ταύτην τὴν
αἰτίαν), ἀλλ' ὅτι τὸ περίττωμα τὸ τοῦ θήλεος
20 δυνάμει τοιοῦτόν ἐστιν οἷον φύσει τὸ ζῷον, καὶ
ἔνεστι δυνάμει τὰ μόρια, ἐνεργείᾳ δ' οὐθέν, διὰ
ταύτην τὴν αἰτίαν γίνεται ἕκαστον αὐτῶν, καὶ ὅτι
τὸ ποιητικὸν καὶ τὸ παθητικόν, ὅταν θίγωσιν, ὃν
τρόπον ἐστὶ τὸ μὲν ποιητικὸν τὸ δὲ παθητικόν (τὸν
δὲ τρόπον λέγω τὸ ὡς καὶ οὗ καὶ ὅτε), εὐθὺς τὸ
25 μὲν ποιεῖ τὸ δὲ πάσχει. ὕλην μὲν οὖν παρέχει τὸ

---

¹ seclusi: suspicatus est Platt: τὸν αὐτὸν ... τροφῆς om. Σ.

ᵃ This phrase seems to be an interpolation, connected
perhaps with ll. 29-31 below.
ᵇ This commonplace of thought in Greek philosophy and
medicine is a pseudo-scientific form of a proverbial maxim
(cf. " birds of a feather "), specially alluring to the Greeks.
Cf. especially Hippocrates, π. φύσιος παιδίου, ch. 17 init. and
fin. (vii. 496-498 Littré).   See quotation in note on 742 a 1.
198

after all it is not true to say that all the nourishment comes from outside. In the seeds of plants there is some nutritive matter, which at first has a milky appearance ; and it may be that in the same way, in the material of the animal, the residue left over from its construction is present as nourishment for it from the outset.

So then, the fetation's growth is supplied through the umbilicus in the same way that a plant's growth is supplied through its roots [and also as that of animals is, when they have been separated, from the nourishment which is in themselves].[a] Of these matters we shall have to speak later at the appropriate occasions in our discussions. As for the differentiation of the various parts : this is not due, as some suppose, to any natural law that " like makes its way to like." [b] This theory involves quite a number of difficulties, one being that if you accept it as stating a valid reason, it follows that each of the " uniform " parts, such as bones, and sinews, and flesh, is formed separately, each one all on its own. The true reason why each of these parts is formed is that the residue provided by the female is *potentially* the same in character as the future animal will be, according to its nature ; and although none of the parts is present *in actuality* in that residue, they are all there *potentially*. A further reason is this. When a pair of factors, the one active and the other passive, come into contact in the way in which one is active and the other passive (by " way " I mean the manner, the place, and the time of the contact), then immediately both are brought into play, the one acting, the other being acted upon. In this case, it is the female which provides the matter, and the male which provides the

θῆλυ, τὴν δ' ἀρχὴν τῆς κινήσεως τὸ ἄρρεν. ὥσπερ
δὲ τὰ ὑπὸ τῆς τέχνης γινόμενα γίνεται διὰ τῶν
ὀργάνων, ἔστι δ' ἀληθέστερον εἰπεῖν διὰ τῆς κινήσεως αὐτῶν, αὕτη δ' ἐστὶν ἡ ἐνέργεια τῆς τέχνης,
ἡ δὲ τέχνη μορφὴ τῶν γινομένων ἐν ἄλλῳ, οὕτως
30 ἡ τῆς θρεπτικῆς ψυχῆς δύναμις, ὥσπερ καὶ ἐν
αὐτοῖς τοῖς ζῴοις καὶ τοῖς φυτοῖς ὕστερον ἐκ τῆς
τροφῆς ποιεῖ τὴν αὔξησιν, χρωμένη οἷον ὀργάνοις
θερμότητι καὶ ψυχρότητι (ἐν γὰρ τούτοις ἡ κίνησις
ἐκείνης, καὶ λόγῳ τινὶ ἕκαστον γίνεται), οὕτω καὶ
ἐξ ἀρχῆς συνίστησι τὸ φύσει γιγνόμενον· ἡ γὰρ
35 αὐτή ἐστιν ὕλη ᾗ αὐξάνεται καὶ ἐξ ἧς συνίσταται
τὸ πρῶτον, ὥστε καὶ ἡ ποιοῦσα δύναμις ταὐτὸ [τῷ
ἐξ ἀρχῆς· μείζων δὲ αὕτη ἐστίν][1]· εἰ οὖν αὕτη
ἐστὶν ἡ θρεπτικὴ ψυχή, αὕτη ἐστὶ καὶ ἡ γεννῶσα·
καὶ τοῦτ' ἐστιν ἡ φύσις ἡ ἑκάστου, ἐνυπάρχουσα
καὶ ἐν φυτοῖς καὶ ἐν ζῴοις πᾶσιν. τὰ δ' ἄλλα
μόρια τῆς ψυχῆς τοῖς μὲν ὑπάρχει τοῖς δ' οὐχ
ὑπάρχει τῶν ζώντων.[2]

Ἐν μὲν οὖν τοῖς φυτοῖς οὐ κεχώρισται τὸ θῆλυ
5 τοῦ ἄρρενος· ἐν δὲ τοῖς ζῴοις ἐν οἷς κεχώρισται,
V προσδεῖται τὸ θῆλυ τοῦ ἄρρενος.[3] καίτοι τις ἀπο-

---

[1] seclusi. μείζων . . . ἐστίν secl. A.-W., qui et ταὐτὸ τῷ
ἐξ ἀρχῆς γεννήσαντί ἐστιν: Btf. τῇ ἐξ ἀρχῆς [μ. . . . . ἐ.].
[2] ζώντων Peck : ζῴων vulg.: cf. 731 a 31, ubi PY ζῴων
pro ζώντων : in quibusdam corporibus quae vivunt Σ.
[3] τὸ θῆλυ τοῦ ἄρρενος Peck, docente Platt : τοῦ θήλεος τὸ
ἄρρεν vulg.

---

[a] Cf. 734 b 36 ff. and *P.A.* 640 a 32.
[b] See App. B §§ 6, 9, 15.
[c] Cf. *Phys.* 192 b 21 ff. ὡς οὔσης τῆς φύσεως ἀρχῆς τινος καὶ
αἰτίας τοῦ κινεῖσθαι καὶ ἠρεμεῖν . . . 32 φύσιν δὲ ἔχει ὅσα τοιαύτην

principle of movement. Now the products which are formed by human art are formed by means of instruments, or rather it would be truer to say they are formed by means of the movement of the instruments, and this movement is the activity, the actualization, of the art, for by " art " we mean the shape of the products which are formed, though it is resident elsewhere than in the products themselves.[a] The *dynamis* of the nutritive Soul behaves in the same way. Just as, in the independently existing animal or plant, this Soul, which uses heat and cold as its instruments (for it is in these that its movement subsists,[b] each several thing being formed according to some definite *logos*), at a later stage produces growth out of the nourishment supplied, so in precisely the same way at the very outset, this Soul, while the natural object is being formed, causes it to be set and constituted ; since, as the matter from which the object derives its growth is identical with that out of which it was originally set and constituted, so too the *dynamis* which fashions the object is identical. If, then, this is the nutritive Soul, this it is which also generates the object. And this part of Soul it is which is the " nature " of each several object,[c] being present alike in plants and in animals one and all, whereas the other parts of Soul,[d] while present in some living things, are absent from others.

Now in plants the female is not separate from the male ; in certain of the animals, however, it is separate, and here, in addition, it has need of the male. And yet anyone might well raise the puzzle,

(c) Why the female cannot generate alone.

V

ἔχει ἀρχήν, and *De caelo* 301 b 17 φύσις μέν ἐστιν ἡ ἐν αὐτῷ ὑπάρχουσα κινήσεως ἀρχή. See also Introd. § 42.
    [d] See Introd. § 43.

ρήσειεν ἂν διὰ τίν' αἰτίαν. εἴπερ ἔχει τὸ θῆλυ
τὴν αὐτὴν ψυχὴν καὶ ἡ ὕλη τὸ περίττωμα τὸ τοῦ
θήλεός ἐστι, τί προσδεῖται τοῦ ἄρρενος, ἀλλ' οὐκ
αὐτὸ ἐξ αὑτοῦ γεννᾷ τὸ θῆλυ; αἴτιον δ' ὅτι
10 διαφέρει τὸ ζῷον τοῦ φυτοῦ αἰσθήσει· ἀδύνατον
δὲ πρόσωπον ἢ χεῖρα ἢ σάρκα εἶναι ἢ ἄλλο τι
μόριον μὴ ἐνούσης αἰσθητικῆς ψυχῆς, ἢ ἐνεργείᾳ
ἢ δυνάμει, καὶ ἤ πη ἢ ἁπλῶς· ἔσται γὰρ οἷον
νεκρὸς ἢ νεκροῦ μόριον. εἰ οὖν τὸ ἄρρεν ἐστὶ τὸ
τῆς τοιαύτης ποιητικὸν ψυχῆς, ὅπου κεχώρισται
15 τὸ θῆλυ καὶ τὸ ἄρρεν, ἀδύνατον τὸ θῆλυ αὐτὸ ἐξ
αὑτοῦ γεννᾶν ζῷον· τὸ γὰρ εἰρημένον ἦν τὸ ἄρρενι[1]
εἶναι· ἐπεὶ ὅτι γ' ἔχει λόγον ἡ λεχθεῖσα ἀπορία,
φανερὸν ἐπὶ τῶν ὀρνίθων τῶν τὰ ὑπηνέμια τικτόν-
των, ὅτι δύναται μέχρι γέ τινος τὸ θῆλυ γεννᾶν.
ἔτι δ' ἔχει καὶ τοῦτο ἀπορίαν, πῶς τις αὐτῶν τὰ
20 ᾠὰ φύσει ζῆν. οὔτε γὰρ οὕτως ὡς τὰ γόνιμα ᾠὰ
ἐνδέχεται (ἐγίγνετο γὰρ ἂν ἐξ αὐτῶν ἐνεργείᾳ
ἔμψυχον) οὔθ' οὕτως ὥσπερ ξύλον ἢ λίθος. ἔστι
γὰρ καὶ τούτων τῶν ᾠῶν φθορά τις ὡς μετεχόντων
τρόπον τινὰ ζωῆς πρότερον. δῆλον οὖν ὅτι ἔχει
τινὰ δυνάμει ψυχήν. ποίαν οὖν ταύτην; ἀνάγκη
25 δὴ τὴν ἐσχάτην. αὕτη δ' ἐστὶν ἡ θρεπτική· αὕτη

---

[1] ἄρρενι S : ἄρρεν vulg.

to what cause this is due. Granted that the female possesses the same Soul ⟨as the male⟩ and that the residue provided by the female is the material ⟨for the fetation⟩, why has the female any need of the male in addition? Why does not the female accomplish generation all by itself and from itself? The reason is that there is a difference between animal and plant : the animal possesses sense-perception.[a] It is impossible for any part of the body whatever (face, hand, flesh, etc.) to exist unless sentient Soul is present in it, whether *in actuality* or *potentially*, whether in some qualified sense or without qualification. Otherwise what we have will be on a par with a dead body or a dead limb. Thus, if the male is the factor which produces the sentient Soul in cases where male and female are separate, it is impossible for the female all by itself and from itself to generate an animal ; because the faculty just mentioned [b] is the essence of what is meant by "male." Still, it is not at all unreasonable to raise the puzzle we have stated, as is shown by the instance of those birds which lay wind-eggs : this proves that up to a point the female is able to generate. But there is a puzzle here too : In what sense are we to say that these eggs are alive? We cannot say that they are alive in the same sense as fertile eggs, for in that case an *actual* living creature would hatch out from them ; nor are they on a par with wood and stone, because these eggs go bad just as fertile ones do, and this seems to indicate that to start with they were in some way alive. Hence it is clear that *potentially* they possess Soul of a sort. What sort, then? The lowest, it must be, obviously ; and this is nutritive Soul, because this it is which is present

741 a

γὰρ ὑπάρχει πᾶσιν ὁμοίως ζῴοις τε καὶ φυτοῖς. διὰ τί οὖν οὐκ ἀποτελεῖ τὰ μόρια καὶ τὸ ζῷον; ὅτι δεῖ αἰσθητικὴν αὐτὰ ἔχειν ψυχήν· οὐ γάρ ἐστιν ὥσπερ φυτοῦ τὰ μόρια τῶν ζῴων. διὸ δεῖται τῆς τοῦ ἄρρενος κοινωνίας· κεχώρισται γὰρ ἐν τούτοις
30 τὸ ἄρρεν. ὅπερ καὶ συμβαίνει· τὰ γὰρ ὑπηνέμια γίνεται γόνιμα, ἐὰν ἔν τινι καιρῷ τὸ ἄρρεν ἐπο-χεύσῃ. ἀλλὰ περὶ μὲν τῆς τούτων αἰτίας ὕστερον διορισθήσεται.

Εἰ δ' ἔστι τι γένος ὃ θῆλυ μέν ἐστιν, ἄρρεν δὲ μὴ ἔχει κεχωρισμένον, ἐνδέχεται τοῦτο[1] ζῷον ἐξ αὑτοῦ γεννᾶν. ὅπερ ἀξιοπίστως μὲν οὐ συνῶπται
35 μέχρι γε τοῦ νῦν, ποιεῖ δὲ διστάζειν ⟨ἔνια⟩[2] ἐν τῷ γένει τῷ τῶν ἰχθύων· τῶν γὰρ καλουμένων ἐρυ-θρίνων ἄρρην μὲν οὐθεὶς ὦπταί πω, θήλειαι δὲ καὶ κυημάτων πλήρεις. ἀλλὰ τούτων μὲν οὔπω πεῖραν ἔχομεν ἀξιόπιστον, οὔτε δὲ θήλεα οὔτε ἄρρενα καὶ

741 b

ἐν τῷ τῶν ἰχθύων γένει ἐστίν, οἷον αἵ τ' ἐγχέλεις καὶ γένος τι κεστρέων περὶ τοὺς τελματιαίους πο-ταμούς. ἐν ὅσοις δὲ κεχώρισται τὸ θῆλυ καὶ τὸ ἄρρεν, ἀδύνατον αὐτὸ καθ' αὑτὸ τὸ θῆλυ γεννᾶν εἰς

---

[1] ἄνευ ὀχείας addit P.   [2] ⟨ἔνια⟩ Hackforth.

---

[a] See 750 b 3 ff., 757 b 1 ff., also 730 a 5 ff. and *H.A.* 539 b 1.

[b] Probably some species of *Serranus*, perhaps *S. anthias* (a sea-perch). *Cf. H.A.* 538 a 21, 567 a 27. Actually the majority of species of *Serranus* are hermaphrodite (see E. S. Goodrich, *Cyclostomes and Fishes*, 430), as was discovered by Cavolini in the latter part of the 18th cent. See A.-W., *Introduction*, pp. 32 ff.

[c] *i.e.,* roe.

[d] Eels do not develop generative organs except in deep water, whither they go in order to breed. This is taken to

204

alike in all animals and plants. Why then does this
Soul fail to bring the parts to their completion and so
produce an animal ? Because the parts of an animal
are bound to possess sentient Soul, since they are not
on a par with those of a plant ; and that is why the
male is required to take its share in the business (the
male being separate from the female in such animals).
The facts bear this out : wind-eggs become fertile
if the male treads the female within a certain period.
However, the cause of these things will be fully determined later on.[a]

If there is any class of animal which is female and
has no separate male, it is possible that this generates
offspring from itself. This has not so far been
reliably observed, it is true, but some instances in the
class of fishes give cause to suspect that it may be
the case. Thus, of the fish known as *erythrinus* [b]
not a single male specimen has so far been observed, whereas female ones have been, full of
fetations.[c] But although with regard to these we
have no reliable proof so far, there are also in the
class of fishes some which are neither male nor female : *e.g.*, eels,[d] and one sort of *cestreus* [e] which
frequents marshland rivers. In all animals, however, where the male and female are separate,
the female is unable by itself to generate offspring

---

indicate that they are descended from an original deep-
water fish. See additional note, p. 565.

Cestreus seems to be a generic name for the grey mullet,
Mugil (Thompson, *Glossary*, p. 108). In *P.A.* 696 a 5, Aristotle speaks of a *cestreus* found in the lake at Siphae in
Boeotia, on the south coast, near Thespiae (now Tipha). *Cf.*
also the reference at 763 b 1 to Pyrrha, where there was a
lagoon which was apparently one of Aristotle's favourite
spots for studying animals.

τέλος· τὸ γὰρ ἄρρεν μάτην ἂν ἦν, ἡ δὲ φύσις οὐδὲν
5 ποιεῖ μάτην. διόπερ ἐν τοῖς τοιούτοις ἀεὶ τὸ ἄρρεν
ἐπιτελεῖ τὴν γένεσιν. ἐμποιεῖ γὰρ τοῦτο τὴν αἰσθητικὴν ψυχήν, ἢ δι᾽ αὑτοῦ ἢ διὰ τῆς γονῆς.
ἐνυπαρχόντων δ᾽ ἐν τῇ ὕλῃ δυνάμει τῶν μορίων,
ὅταν ἀρχὴ γένηται κινήσεως, ὥσπερ ἐν τοῖς αὐτομάτοις θαύμασι, συνείρεται τὸ ἐφεξῆς· καὶ ὃ βού-
10 λονται λέγειν τινὲς τῶν φυσικῶν, τὸ " φέρεσθαι εἰς
τὸ ὅμοιον," λεκτέον οὐχ ὡς τόπον μεταβάλλοντα
τὰ μόρια κινεῖσθαι, ἀλλὰ μένοντα καὶ ἀλλοιούμενα
μαλακότητι καὶ σκληρότητι καὶ χρώμασι καὶ ταῖς
ἄλλαις ταῖς τῶν ὁμοιομερῶν διαφοραῖς, γινόμενα
15 ἐνεργείᾳ ἃ ὑπῆρχεν ὄντα δυνάμει πρότερον. γίγνεται δὲ πρῶτον ἡ ἀρχή. αὕτη δ᾽ ἐστὶν ἡ καρδία
τοῖς ἐναίμοις, τοῖς δ᾽ ἄλλοις τὸ ἀνάλογον, ὥσπερ
εἴρηται πολλάκις. καὶ τοῦτο φανερὸν οὐ μόνον
κατὰ τὴν αἴσθησιν, ὅτι γίνεται πρῶτον, ἀλλὰ καὶ
περὶ τὴν τελευτήν· ἀπολείπει γὰρ τὸ ζῆν ἐντεῦθεν
20 τελευταῖον, συμβαίνει δ᾽ ἐπὶ πάντων τὸ τελευταῖον
γινόμενον¹ πρῶτον ἀπολείπειν, τὸ δὲ πρῶτον τελευταῖον, ὥσπερ τῆς φύσεως διαυλοδρομούσης καὶ

---

¹ γενόμενον P.

ᵃ i.e., the matter provided by the female.
ᵇ See note on 734 b 10.
ᶜ φυσικοί, sometimes φυσιολόγοι, a term used by Aristotle
to describe the early writers on φύσις, i.e., nature, or the
nature (stuff) of the universe and its contents. They include
the so-called " early philosophers," and apparently also
Hippocrates, as here (see note on 740 b 14). Several of the
pre-Socratic philosophers had made use of this principle in
various connexions.—See also pp. xvi f.
ᵈ Cf. above, 740 b 14.   ᵉ See Introd. § 48.
ᶠ See App. B §§ 4-6, 9-10.

and bring it to completion : if it could, the existence of the male would have no purpose, and Nature
does nothing which lacks purpose. Hence in such
animals the male always completes the business of
generation—it implants sentient Soul, either acting
by itself directly or by means of semen. As the
parts of the animal to be formed are present *potentially* in the matter,[a] once the principle of movement has been supplied, one thing follows on after
another without interruption, just as it does in the
" miraculous " automatic puppets.[b] And the meaning of the statement, made by some of the physiologers,[c] about like " making its way to like,"[d] must be
taken to be not that the parts of the body " move "[e]
in the sense of changing their position, but that while
remaining in the same position they undergo " alteration "[f] as regards softness, hardness, colour, and the
other differences which belong to the uniform parts ;
that is, they become *in actuality* what previously all
along they had been *potentially*. The first to be formed
is the " principle," which in blooded animals is the
heart and in the others the counterpart of the heart,
as I have said many times over. There can be no
doubt about this, because our senses tell us that it
is the first thing formed ; but the truth of it is confirmed by what happens when the creature dies : the
heart is the place where life fails last of all ; and we
find universally that what is the last to be formed is
the first to fail, and the first to be formed is the last
to fail.[g] It is as though Nature were a runner, covering a double course there and back, and retracing her

---

[g] *Cor primum vivens ultimum moriens*: cf. Ebstein *et al.*,
Mitt. zur Gesch. der Medizin und Naturw. 19 (1920), 102,
219, 305.

741 b

ἀνελιττομένης ἐπὶ τὴν ἀρχὴν ὅθεν ἦλθεν. ἔστι γὰρ
ἡ μὲν γένεσις ἐκ τοῦ μὴ ὄντος εἰς τὸ ὄν, ἡ δὲ
φθορὰ ἐκ τοῦ ὄντος πάλιν εἰς τὸ μὴ ὄν.

VI 25 Γίνεται δὲ μετὰ τὴν ἀρχήν, ὥσπερ ἐλέχθη, τὰ
ἐντὸς πρότερον τῶν ἐκτός. φαίνεται δὲ πρότερα
τὰ μέγεθος ἔχοντα τῶν ἐλαττόνων, οὐδ' ἔνια γιγνόμενα πρότερον. πρῶτον δὲ τὰ ἄνω διαρθροῦται
τοῦ διαζώματος, καὶ διαφέρει μεγέθει· τὸ δὲ κάτω
καὶ ἔλαττον καὶ ἀδιοριστότερον. καὶ τοῦτο γίγνε-
30 ται ἐν πᾶσιν, ὅσοις τὸ ἄνω καὶ τὸ κάτω διώρισται,
πλὴν ἐν τοῖς ἐντόμοις· τούτων δ' ἐν τοῖς σκωληκοτοκουμένοις ἐπὶ τὸ ἄνω ἡ αὔξησις γίνεται· τὸ
γὰρ ἄνω ἐξ ὑπαρχῆς ἔλαττον. ἀδιόριστον δὲ καὶ
τὸ ἄνω καὶ τὸ κάτω τοῖς μαλακίοις τῶν πορευτικῶν μόνοις. τὸ δὲ λεχθὲν συμβαίνει καὶ ἐπὶ τῶν
35 φυτῶν, τὸ προτερεῖν τῇ γενέσει τὸ ἄνω κύτος τοῦ
κάτωθεν· τὰς γὰρ ῥίζας πρότερον ἀφιᾶσι τὰ σπέρματα τῶν πτόρθων.

Διορίζεται δὲ τὰ μέρη τῶν ζῴων πνεύματι, οὐ
μέντοι οὔτε τῷ τῆς γεννώσης οὔτε τῷ αὐτοῦ,

---

[a] See 740 a 12 ff.

[b] Aristotle's observations are quite correct. *Cf.* the
theories of C. M. Child on axial gradients, physiological
dominance (*cf.* Aristotle's own use of κύριος, 742 a 34 below),
etc., conveniently discussed by J. Huxley and G. R. de Beer
in *Elements of Experimental Embryology.* See also 742 b 14.

[c] According to Aristotle (*I.A.* 705 a 29 ff.), the distinction
between the upper and lower portions of animals and plants
is determined by function, and not by position relative to the
earth and the sky. The " upper " portion is that from which
is received the distribution of nourishment and material for
growth ; and the extremity towards which the nourishment
and growth penetrate is the " lower " extremity. Thus, as

208

steps towards the starting-point whence she set out. The process of formation, genesis, starts from not-being and advances till it reaches being ; that of decay starts from being and goes back again till it reaches not-being.

After the " principle " is formed, the other parts are formed, the internal ones earlier than the external, as I have said.[a] The larger parts become visible, however, earlier than the smaller ones, although some of them are not in fact formed earlier. First the parts above the diaphragm become articulated, and these are larger in size, whereas that which is below is smaller and less clearly defined.[b] This happens in all cases where the upper and the lower portions [c] are definite and distinct, except Insects : in those Insects which are produced as larvae, the increase occurs towards the upper part, as this is smaller to begin with. The only locomotive animals in which there is no definite distinction between the upper and lower portions are the Cephalopods.[d] What has been said here applies to plants as well : the formation of the upper portion precedes that of the lower : seeds send out their roots before their shoots.[e]

Now the parts of animals are differentiated by means of *pneuma* [f] ; but this is not the *pneuma* of the mother, nor that of the creature itself, as some of

VI
(d) Development of the embryo *(continued)*.

he says (705 b 6), in plants, the roots are the " upper " portion, since it is through their roots that plants get their nourishment, just as animals do through the mouth. *Cf.* the end of the present paragraph, 741 b 34 ff. ; also the passage in *P.A.* 686 b 21 ff.

[d] Because (720 b 18, *P.A.* 684 b 15, 685 a 1) their back-part is drawn up on to the front-part, their tail-end is bent right over to meet the front, and in consequence the residual vent is brought close to the mouth.

[e] See note on 741 b 30.   [f] See App. B §§ 7 ff.

742 a καθάπερ τινὲς τῶν φυσικῶν φασίν. φανερὸν δὲ[1]
τοῦτο ἐπὶ τῶν ὀρνίθων καὶ τῶν ἰχθύων καὶ τῶν
ἐντόμων. τὰ μὲν γὰρ χωρισθέντα τῆς γεννώσης
γίνεται ἐξ ᾠοῦ, ἐν ᾧ λαμβάνει τὴν διάρθρωσιν· τὰ
δ' ὅλως οὐκ ἀναπνεῖ τῶν ζῴων, σκωληκοτοκεῖται δὲ
5 καὶ ᾠοτοκεῖται· τὰ δ' ἀναπνέοντα καὶ ἐν τῇ μήτρα
λαμβάνοντα τὴν διάρθρωσιν οὐκ ἀναπνεῖ πρὶν ἢ ὁ
πλεύμων λάβῃ τέλος· διαρθροῦται δὲ καὶ οὗτος καὶ
τὰ ἔμπροσθεν μόρια πρὶν ἀναπνεῖν. ἔτι δ' ὅσα
πολυσχιδῆ τῶν τετραπόδων, οἷον κύων λέων λύκος
10 ἀλώπηξ θώς, πάντα τυφλὰ γεννᾷ, καὶ διίσταται τὸ
βλέφαρον γενομένων ὕστερον. ὥστε δῆλον ὅτι τὸν
αὐτὸν τρόπον καὶ ἐν τοῖς ἄλλοις πᾶσι, καθάπερ καὶ
τὸ ποιόν, καὶ τὸ ποσὸν γίνεται δυνάμει προϋπάρχον,
ἐνεργείᾳ δ' ὕστερον, ὑπὸ τῶν αὐτῶν αἰτίων ὑφ'
ὧνπερ καὶ τὸ ποιὸν διορίζεται, καὶ γίγνεται δύο ἐξ
15 ἑνός. πνεῦμα δ' ὑπάρχειν ἀναγκαῖον, ὅτι ὑγρὸν καὶ
θερμόν, τοῦ μὲν ποιοῦντος, τοῦ δὲ πάσχοντος.

Τῶν δ' ἀρχαίων τινὲς φυσιολόγων τί μετὰ τί
γίγνεται τῶν μορίων ἐπειράθησαν λέγειν, οὐ λίαν
ἐμπειρικῶς ἔχοντες τῶν συμβαινόντων. τῶν γὰρ

---

[1] δὲ P : γὰρ vulg.

[a] See note on 741 b 10. *e.g.*, Hippocrates, π. φύσιος
παιδίου 17 (vii. 496-498 Littré) ἡ δὲ σὰρξ αὐξομένη ὑπὸ τοῦ
πνεύματος ἀρθροῦται, καὶ ἔρχεται ἐν αὐτέῃ ἕκαστον τὸ ὅμοιον ὡς
τὸ ὅμοιον . . . διαρθροῦται ὑπὸ τῆς πνοῆς ἕκαστα, φυσώμενα γὰρ
διίσταται ξύμπαντα κατὰ συγγένειαν. *Cf.* also ch. 19. According to this treatise the embryo both received nourishment
and breathed through the umbilicus (*cf.* chh. 13, 15).
[b] Viz., birds.   [c] Viz., fishes and insects.
[d] Viz., Vivipara.

the physiologers [a] allege. This point is clear in the case of birds, fishes, and insects : thus, some [b] of these are formed out of an egg, after separation from the mother, and it is in the egg that they get their articulation ; and some animals [c] do not breathe at all, but are produced as larvae or as eggs ; others, [d] which both breathe and get their articulation within the uterus, do not however breathe until their lungs have reached completion : with them, both the lungs and the preceding parts become articulated before they breathe. Further, the polydactylous quadrupeds (such as the dog, the lion, the wolf, the fox and the jackal) all bring forth their young blind, and the eyelid does not separate until some time after birth. Hence it is clear that, with regard to all the other parts as well, the same holds : just as the characteristics of quality are there *potentially* to begin with and later on are formed *in actuality*, so too those of quantity are formed—by the same causes as those by which the characteristics of quality are differentiated, and two things are formed out of a single one. [e] As for *pneuma*, its presence is the result of necessity, because liquid substance and hot substance are present, one being active and the other being acted upon. [f]

Some of the early physiologers endeavoured to describe the order in which the various parts are formed, but they were none too well acquainted with what actually happens. As with everything else, so

---

[e] *e.g.*, two eyelids ; an example of a potential duality being actualized.—See also App. B § 7, n.

[f] *i.e.*, the *pneuma* is not ἐπείσακτον, but σύμφυτον, derived from within, and hence can serve as an " instrument " (see 789 b 3 ff.) charged with a specific " movement " (see Introd. § 68, and App. B, esp. § 32).

μορίων, ὥσπερ καὶ ἐπὶ τῶν ἄλλων, πέφυκεν ἕτερον
20 ἑτέρου πρότερον. τὸ δὲ πρότερον ἤδη πολλαχῶς
ἐστιν. τό τε γὰρ οὗ ἕνεκα καὶ τὸ τούτου ἕνεκα
διαφέρει, καὶ τὸ μὲν τῇ γενέσει πρότερον αὐτῶν
ἐστι, τὸ δὲ τῇ οὐσίᾳ. δύο δὲ διαφορὰς ἔχει καὶ
τὸ τούτου¹ ἕνεκα· τὸ μὲν γάρ ἐστιν ὅθεν ἡ κίνησις,
τὸ δὲ ᾧ χρῆται τὸ οὗ ἕνεκα. λέγω δ' οἷον τό τε
25 γεννητικὸν καὶ τὸ ὀργανικὸν τῷ γεννωμένῳ²· τούτων
γὰρ τὸ μὲν ὑπάρχειν δεῖ πρότερον, τὸ ποιητικόν,
οἷον τὸ διδάξαν³ τοῦ μανθάνοντος, τοὺς δ' αὐλοὺς
ὕστερον τοῦ μανθάνοντος αὐλεῖν· περίεργον γὰρ μὴ
ἐπισταμένοις αὐλεῖν ὑπάρχειν αὐλούς. τριῶν δ'
ὄντων, ἑνὸς μὲν τοῦ τέλους, ὃ λέγομεν εἶναι οὗ
ἕνεκα, δευτέρου δὲ τῶν τούτου ἕνεκα τῆς ἀρχῆς
30 τῆς κινητικῆς καὶ γεννητικῆς (τὸ γὰρ ποιητικὸν
καὶ γεννητικόν, ᾗ τοιαῦτα, πρὸς τὸ ποιούμενόν
ἐστι καὶ γεννώμενον), τρίτου δὲ τοῦ χρησίμου καὶ
ᾧ χρῆται τὸ τέλος, πρῶτον μὲν τοῦ ὑπάρχειν ἀναγκαῖόν
τι μόριον ἐν ᾧ ἡ ἀρχὴ τῆς κινήσεως (καὶ γὰρ
εὐθὺς τοῦτο⁴ μόριόν ἐστι τοῦ τέλους ἓν καὶ κυ-
35 ριώτατον), ἔπειτα μετὰ τοῦτο τὸ ὅλον καὶ τὸ τέλος,
τρίτον δὲ καὶ τελευταῖον τὰ ὀργανικὰ τούτοις μέρη
πρὸς ἐνίας χρήσεις. ὥστ' εἴ τι τοιοῦτόν ἐστιν,

---

¹ τούτου PS : οὗ vulg.
² γεννωμένῳ Z¹ : γενομένῳ vulg.
³ διδάξον Richards.
⁴ fort. τοῦτο τὸ (Z¹) scribendum, et mox ἕν⟨εκα⟩.

ᵃ Cf. *Met.* 1035 b 18 ff.
ᵇ This will be modified in a moment, when Aristotle subdivides this heading. Some of the things which are for the
sake of the End are posterior to it in point of formation.
ᶜ By this, as appears from 742 b 13, 14 below, is meant the
" upper portion," the head and trunk.

with the parts of the body : one is, by nature, prior to another.[a] But the term "prior" at once comprises a variety of meanings. *E.g.*, take the difference between (*a*) that *for the sake of which* a thing is, and (*b*) that thing which is *for its sake* : of these, one (*b*) is prior in point of formation,[b] while the other (*a*) is prior in point of being or reality. Further, "that which is *for the sake of* the End" comprises two divisions : (i) that whence the movement is derived and (ii) that which is employed by the End ; or, in other words, (i) something which generates, and (ii) something which serves as an instrument for what is generated. Of the two, the productive factor must exist prior to the other : *e.g.*, a teacher must exist prior to a learner, while pipes are posterior to the person who is learning to play them : it is superfluous for people who cannot play pipes to possess them. So we have these three things : (1) the End, which we describe as being that *for the sake of which* ⟨other things are⟩ ; (2) the things which are *for the sake of the End*, viz., the activating and generative principle (second, because the existence of that which is productive and generative, *qua* such, is relative to what it produces and generates) ; (3) the things which are serviceable, which can be and are employed by the End. Thus, first of all there must of necessity exist some part in which the principle of movement resides (for of course this is a part of the End, and the supreme controlling part of it) ; after that comes the animal as a whole, *i.e.*, the End [c] ; third and last of all come the parts which serve these [d] as instruments for various employments. If it is true, then, that there is a part

---

[d] Or perhaps "this," referring only to the "End."

742 b ὅπερ ἀναγκαῖον ὑπάρχειν ἐν τοῖς ζῴοις, τὸ πάσης
ἔχον τῆς φύσεως ἀρχὴν καὶ τέλος, τοῦτο γίνεσθαι
πρῶτον ἀναγκαῖον, ᾗ μὲν κινητικόν, πρῶτον, ᾗ δὲ
μόριον τοῦ τέλους, μετὰ τοῦ ὅλου. ὥστε τῶν
μορίων τῶν ὀργανικῶν ὅσα μέν ἐστι γεννητικὰ τὴν
5 φύσιν, ἀεὶ πρότερον δεῖ ὑπάρχειν αὐτά (ἄλλου γὰρ
ἕνεκά ἐστιν ὡς¹ ἀρχή), ὅσα δὲ μὴ τοιαῦτα τῶν
ἄλλου ἕνεκα, ὕστερον. διὸ οὐ ῥᾴδιον διελεῖν πότερα
πρότερα τῶν μορίων, ὅσα ἄλλου ἕνεκα, ἢ οὗ² ἕνεκα
ταῦτα. παρεμπίπτει γὰρ τὰ κινητικὰ τῶν μορίων
πρότερον ὄντα τῇ γενέσει τοῦ τέλους, τὰ δὲ κινη-
10 τικὰ πρὸς τὰ ὀργανικὰ διελεῖν οὐ ῥᾴδιον. καίτοι
κατὰ ταύτην τὴν μέθοδον δεῖ ζητεῖν τί γίνεται μετὰ
τι· τὸ γὰρ τέλος ἐνίων μὲν ὕστερον, ἐνίων δὲ πρότερον. καὶ διὰ τοῦτο πρῶτον μὲν τὸ ἔχον τὴν
ἀρχὴν γίνεται μόριον, εἶτ' ἐχόμενον τὸ ἄνω κύτος.
15 διὸ τὰ περὶ τὴν κεφαλὴν καὶ τὰ ὄμματα μέγιστα
κατ' ἀρχὰς φαίνεται τοῖς ἐμβρύοις, τὰ δὲ κάτω
τοῦ ὀμφαλοῦ, οἷον τὰ κῶλα, μικρά· τοῦ γὰρ ἄνω
τὰ κάτω ἕνεκεν, καὶ οὔτε μόρια τοῦ τέλους οὔτε
γεννητικὰ αὐτοῦ.

Οὐ καλῶς δὲ λέγουσιν οὐδὲ τοῦ διὰ τί τὴν

¹ ὡς P : ὡς ἡ vulg.   ² οὗ] ὧν P.

---

[a] *i.e.*, generative of other parts, as the heart is.
[b] Or, reading ἡ ἀρχή, "just as the first principle is for the sake of the End."

214

of this kind—a part which contains the first principle and the End of the animal's whole nature—which must of necessity be present in an animal, then this part must of necessity be formed first of all—formed first, *qua* activating, though formed along with the whole creature, *qua* being a part of the End. Thus, those instrumental parts which are in their nature generative [a] must always be there themselves prior to the rest, because they are *for the sake of* something else, as being a first principle [b] ; those parts which, although they are *for the sake of* something else, are not generative, come later. That is why it is not easy to determine whether those parts are " prior " which are *for the sake of* something else, or that part *for whose sake* these others are present. The activating parts intrude themselves into the picture, because in formation they are prior to the End ; and it is not easy to determine as between the activating and the instrumental parts. Still, this is the line we must follow in trying to find out the order in which they are formed ; for the End, though it comes after some of them, is prior to others. And on this account the part which contains the first principle is the first to be formed ; then follows the upper portion of the body ; and that is why in embryos we see that the parts round the head and eyes are the largest at the outset, while the parts below the umbilicus, for instance the legs, are small. The reason is that the lower portions are for the sake of the upper portion, and they are not parts of the End [c] nor are they concerned in generating it.

People who say, like Democritus of Abdera, that

[c] See above, 742 a 35, 743 b 13, 14. They are merely useful adjuncts, enabling it to move about, etc.

ἀνάγκην, ὅσοι λέγουσιν ὅτι οὕτως ἀεὶ γίνεται, καὶ
20 ταύτην εἶναι νομίζουσιν ἀρχὴν ἐν αὐτοῖς, ὥσπερ
Δημόκριτος ὁ Ἀβδηρίτης, ὅτι τοῦ μὲν [ἀεὶ καὶ]¹
ἀπείρου οὐκ ἔστιν ἀρχή, τὸ δὲ διὰ τί ἀρχή, τὸ δ᾽
ἀεὶ ἄπειρον, ὥστε τὸ ἐρωτᾶν τὸ διὰ τί περὶ τῶν
τοιούτων τινὸς τὸ ζητεῖν εἶναί φησι τοῦ ἀπείρου
ἀρχήν. καίτοι κατὰ τοῦτον τὸν λόγον, καθ᾽ ὃν
25 ἀξιοῦσι τὸ διὰ τί μὴ ζητεῖν, οὐθενὸς ἀπόδειξις
ἔσται τῶν ἀιδίων· φαίνεται δ᾽ οὖσα πολλῶν, τῶν
μὲν γινομένων ἀεὶ τῶν δ᾽ ὄντων, ἐπεὶ καὶ τὸ τρί-
γωνον ἔχειν δυσὶν ὀρθαῖς ἴσας ἀεὶ καὶ τὸ τὴν
διάμετρον ἀσύμμετρον εἶναι πρὸς τὴν πλευρὰν
ἀίδιον, ἀλλ᾽ ὅμως ἐστὶν αὐτῶν αἴτιόν τι καὶ ἀπό-
30 δειξις. τὸ μὲν οὖν μὴ πάντων ἀξιοῦν ζητεῖν ἀρχὴν
λέγεται καλῶς, τὸ δὲ τῶν ὄντων ἀεὶ καὶ γινομένων
πάντων οὐ καλῶς, ἀλλ᾽ ὅσαι τῶν ἀιδίων ἀρχαὶ
τυγχάνουσιν οὖσαι· τῆς γὰρ ἀρχῆς ἄλλη γνῶσις καὶ
οὐκ ἀπόδειξις. ἀρχὴ δ᾽ ἐν μὲν τοῖς ἀκινήτοις τὸ

¹ secl. Platt.

ᵃ Cf. Met. 1011 a 13 ἀποδείξεως γὰρ ἀρχὴ οὐκ ἀπόδειξίς
ἐστιν. Also Anal. Post. 90 b 24 ff. αἱ ἀρχαὶ τῶν ἀποδείξεων
ὁρισμοί, ὧν ὅτι οὐκ ἔσονται ἀποδείξεις δέδεικται πρότερον· ἢ
ἔσονται αἱ ἀρχαὶ ἀποδεικταὶ καὶ τῶν ἀρχῶν ἀρχαί . . . ὁρισμὸς
μὲν γὰρ τοῦ τί ἐστι καὶ οὐσίας. See also 72 b 20 ff.; also
Met. 1013 a 15 (one of the definitions of ἀρχή) ἔτι ὅθεν
γνωστὸν τὸ πρᾶγμα πρῶτον, καὶ αὕτη ἀρχὴ λέγεται τοῦ πράγ-
ματος, οἷον τῶν ἀποδείξεων αἱ ὑποθέσεις. In Eth. N. 1142 a 26
it is said to be " intelligence " (νοῦς) which apprehends definitions that cannot be proved by reasoning. Aristotle also
speaks there of " the sort of intuition " (αἴσθησις) where-
216

" this is how they are always formed," and regard
this as a starting-point (first principle) in these cases,
make a mistake, nor do they even succeed in stating
the necessity involved in the cause. Their argument
is this : What is limitless has no starting-point ; but the
cause is a starting-point, and what is *always* is limitless ; therefore (says Democritus) to ask for a cause
in connexion with anything of this kind (*sc.*, anything
that *always* is) is the same as trying to discover a
starting-point in something that is limitless. Yet on
this line of argument, on the strength of which they
undertake to dispense with trying to discover the
cause, there will be no demonstration of any single
one of the " eternal " things. It is obvious, however,
that demonstrations of many of these (some of them
things which *always* come to be, some things which
always are) do in fact exist. For instance, the angles
of a triangle are *always* equal to two right angles,
and the diagonal of a square is *always* incommensur-
able with the side ; in both of these cases we have
something " eternal," yet there is a cause for them
and they are demonstrable. Thus it is right to say
that we cannot undertake to try to discover a starting-
point (a first principle) in all things and everything ;
but it is not right to deny the possibility in the case
of all the things that *always* are and that *always* come
to be ; it is impossible only with the first principles
of the eternal things, for of course the first principle
does not admit of demonstration, but is apprehended
by another mode of cognition.[a] Now with those
things that are " immutable," the first principle is

by we perceive that the ultimate figure in mathematics is a
triangle. Again (1143 b 1) in demonstrations, νοῦς apprehends the immutable (ἀκίνητα) and primary definitions.

ARISTOTLE

τί ἐστιν, ἐν δὲ τοῖς γινομένοις ἤδη πλείους, τρόπον
35 δ᾽ ἄλλον καὶ οὐ πᾶσαι τὸν αὐτόν· ὧν μία τὸν
ἀριθμόν, ὅθεν ἡ κίνησίς ἐστιν. διὸ πάντα τὰ
ἔναιμα καρδίαν ἔχει πρῶτον, ὥσπερ ἐλέχθη κατ᾽
ἀρχάς· ἐν δὲ τοῖς ἄλλοις τὸ ἀνάλογον γίνεται τῇ
καρδίᾳ πρῶτον.

Ἐκ δὲ τῆς καρδίας αἱ φλέβες διατέτανται[1] καθάπερ οἱ τοὺς κανάβους γράφοντες ἐν τοῖς τοίχοις·
τὰ γὰρ μέρη περὶ ταύτας ἐστίν, ἅτε γινόμενα
ἐκ τούτων. ἡ δὲ γένεσίς ἐστιν [ἐκ][2] τῶν ὁμοιο-
5 μερῶν ὑπὸ ψύξεως καὶ θερμότητος· συνίσταται
γὰρ καὶ πήγνυται τὰ μὲν ψυχρῷ τὰ δὲ θερμῷ.
περὶ δὲ τῆς τούτων διαφορᾶς εἴρηται πρότερον ἐν
ἑτέροις, ποῖα λυτὰ ὑγρῷ καὶ πυρί, καὶ ποῖα ἄλυτα
ὑγρῷ καὶ ἄτηκτα πυρί. διὰ μὲν οὖν τῶν φλεβῶν
καὶ τῶν ἐν ἑκάστοις πόρων διαπιδύουσα ἡ τροφή,
10 καθάπερ ἐν τοῖς ὠμοῖς κεραμίοις τὸ ὕδωρ, γίνονται

---

[1] Peck : διατεταμέναι vulg.
[2] om. SΣ, Platt : ἡ coni. A.-W.

---

[a] The term " immutable " is often used by Aristotle in
connexion with mathematics, as here.—" Essence," lit., " the
'what is it?'," the essential definition or nature of the thing.
Cf. quotation from *Anal. Post.* in preceding note, and *Phys.*
198 a 16 f. " in the case of the immutable things, *e.g.*, in
mathematics, where ultimately all is referred back to definitions, τὸ διὰ τί ('why ') is referred back to τὸ τί ἐστι ('what,'
the essence of the thing)." The essence is directly perceived,
not demonstrated. (See previous note.)
[b] This is one of the definitions given in *Met.* 1013 a 4—
that from which, being present within it, a thing first comes
into being (ὅθεν πρῶτον γίγνεται ἐνυπάρχοντος).
[c] He has repeated it almost continuously.

218

the essence [a] ; but as soon as we begin to deal with those things that come into being through a process of formation, we find there are several first principles —principles, however, of a different kind and not all of the same kind. Among them the source whence the movement comes [b] must be reckoned as one, and that is why the heart is the first part which all blooded animals have, as I said at the beginning [c] ; in the other animals it is the counterpart of the heart that is formed first.

Beginning at the heart, the blood-vessels extend all over the body. They may be compared to the skeleton models which are traced out on the walls of buildings,[d] since the parts are situated around the blood-vessels, because they are formed out of them. The formation of the uniform parts is effected by the agency of cooling and heat ; some things are " set " and solidified by the cold and some by the hot. I have spoken previously elsewhere [e] of the difference between these, and I have stated what sort of things are dissoluble by fluid and by fire, and what sorts are not dissoluble by fluid and cannot be melted by fire. Resuming then : As the nourishment oozes through the blood-vessels and the passages in the several parts (just as water does when it stands in unbaked

The uniform parts.

[d] *Cf. H.A.* 515 a 35. Hesychius's and Photius's definitions of κάναβοι describe them as the woodwork around which modellers, when they begin their modelling, mould the wax or plaster. There is a similar passage in *Parts of Animals,* though without mention of this term (654 b 29) ; there Aristotle speaks of a " hard and solid core or foundation " round which the figure is modelled ; though in that case he is speaking of the bones. There seems to be no justification for interpreting κάναβοι as a mere outline or sketch ; nor would such a meaning fit the passage. *Cf.* 764 b 31.

[e] *Meteorologica,* Bk. IV, chh. 7-10. *Cf.* also 762 a 31.

219

σάρκες ἢ τὸ ταύταις ἀνάλογον, ὑπὸ τοῦ ψυχροῦ
συνιστάμεναι, διὸ καὶ λύονται ὑπὸ πυρός. ὅσα δὲ
γεηρὰ λίαν τῶν ἀνατελλόντων, ὀλίγην ἔχοντα ὑγρότητα καὶ θερμότητα, ταῦτα δὲ ψυχόμενα ἐξατμίζοντος τοῦ ὑγροῦ μετὰ τοῦ θερμοῦ γίνεται σκληρὰ
15 καὶ γεώδη τὴν μορφήν, οἷον ὄνυχες καὶ κέρατα καὶ
ὁπλαὶ καὶ ῥύγχη· διὸ μαλάττεται μὲν πυρί, τήκεται
δ' οὐθέν, ἀλλ' ἔνια τοῖς ὑγροῖς, οἷον τὰ κελύφη
τῶν ᾠῶν.

Ὑπὸ δὲ τῆς ἐντὸς θερμότητος τά τε νεῦρα καὶ
τὰ ὀστᾶ γίνεται, ξηραινομένης τῆς ὑγρότητος. διὸ
καὶ ἄλυτά ἐστι τὰ ὀστᾶ ὑπὸ τοῦ πυρός, καθάπερ
20 κέραμος· οἷον γὰρ ἐν καμίνῳ, ὠπτημένα ἐστὶν[1] ὑπὸ
τῆς ἐν τῇ γενέσει θερμότητος. αὕτη δὲ οὔτε ὅ τι
ἔτυχε ποιεῖ σάρκα ἢ ὀστοῦν, οὔθ' ὅπου[2] ἔτυχεν, οὔθ'
ὁπότε ἔτυχεν,[3] ἀλλὰ τὸ πεφυκὸς καὶ οὗ[4] πέφυκε καὶ
ὅτε πέφυκεν. οὔτε γὰρ τὸ δυνάμει ὂν ὑπὸ τοῦ μὴ
τὴν ἐνέργειαν ἔχοντος κινητικοῦ ἔσται, οὔτε τὸ τὴν
25 ἐνέργειαν ἔχον ποιήσει ἐκ τοῦ τυχόντος, ὥσπερ
οὔτε κιβωτὸν μὴ ἐκ ξύλου ὁ τέκτων ποιήσειεν ἄν,
οὔτ' ἄνευ τούτου κιβωτὸς ἔσται ἐκ τῶν ξύλων.

Ἡ δὲ θερμότης ἐνυπάρχει ἐν τῷ σπερματικῷ
περιττώματι τοσαύτην καὶ τοιαύτην ἔχουσα τὴν
κίνησιν καὶ τὴν ἐνέργειαν, ὅση σύμμετρος εἰς
ἕκαστον τῶν μορίων. καθ' ὅσον δ' ἂν ἐλλείπῃ
30 ἢ ὑπερβάλλῃ, ἢ χεῖρον ἀποτελεῖ ἢ ἀνάπηρον τὸ
γινόμενον, παραπλησίως τοῖς ἔξω συνισταμένοις

---

[1] ἐστὶν P : om. vulg. [2] ὅπου P : ὅπῃ vulg.
[3] οὔθ' ὁπότε ἔτυχεν P : om. vulg. [4] οὗ P : ᾗ vulg.

earthenware), flesh, or its counterpart, is formed : it is the cold which " sets " the flesh, and that is why fire dissolves it. As the nourishment wells up, the excessively earthy stuff in it, which contains but little fluidity and heat, becomes cooled while the fluid is evaporating together with the hot substance, and is formed into parts that are hard and earthy in appearance, *e.g.*, nails, horns, hoofs and bills ; hence, these Nails, etc. can be softened, but not one of them can be melted, by fire ; though some, *e.g.*, eggshell, can be melted by fluids.

The sinews and bones are formed, as the fluidity Sinews and solidifies, by the agency of the internal heat ; hence bones. bones (like earthenware) cannot be dissolved by fire ; they have been baked as it were in an oven by the heat present at their formation. This heat, however, to produce flesh or bone, does not work on some casual material in some casual place at some casual time ; material, place and time must be those ordained by Nature : that which is *potentially* will not be brought into being by a motive agent which lacks the appropriate *actuality* ; so, equally, that which possesses the *actuality* will not produce the article out of any casual material. No more could a carpenter produce a chest out of anything but wood ; and, equally, without the carpenter no chest will be produced out of the wood.

This heat resides in the seminal residue, and the movement and the activity which it possesses are in amount and character correctly proportioned to suit each several part. If they are at all deficient or excessive, to that extent they cause the forming product to be inferior or deformed. The same is true of things that are " set " by heat elsewhere than in

221

διὰ τῆς ἑψήσεως πρὸς τροφῆς ἀπόλαυσιν ἤ τινα
ἄλλην ἐργασίαν. ἀλλ' ἐνταῦθα μὲν ἡμεῖς τὴν τῆς
θερμότητος συμμετρίαν εἰς τὴν κίνησιν παρασκευά-
ζομεν, ἐκεῖ δὲ δίδωσιν ἡ φύσις ἡ τοῦ γεννῶντος.
35 τοῖς δὲ αὐτομάτως γινομένοις ἡ τῆς ὥρας αἰτία
κίνησις καὶ θερμότης.

Ἡ δὲ ψύξις στέρησις θερμότητός ἐστιν. χρῆται
δ' ἀμφοτέροις ἡ φύσις ἔχουσι μὲν δύναμιν ἐξ

743 b ἀνάγκης ὥστε τὸ μὲν τοδὶ τὸ δὲ τοδὶ ποιεῖν, ἐν
μέντοι τοῖς γινομένοις ἕνεκά τινος συμβαίνει τὸ
μὲν ψύχειν αὐτῶν τὸ δὲ θερμαίνειν, καὶ γίνεσθαι
τῶν μορίων ἕκαστον, τὴν μὲν σάρκα μαλακὴν τῇ
μὲν ἐξ ἀνάγκης ποιούντων τοιαύτην τῇ δ' ἕνεκά
5 τινος, τὸ δὲ νεῦρον ξηρὸν καὶ ἑλκτόν, τὸ δ' ὀστοῦν
ξηρὸν καὶ θραυστόν. τὸ δὲ δέρμα ξηραινομένης
τῆς σαρκὸς γίνεται, καθάπερ ἐπὶ τοῖς ἑψήμασιν ἡ
καλουμένη γραῦς. οὐ μόνον δὲ διὰ τὸ ἔσχατον
συμβαίνει αὐτοῦ ἡ γένεσις, ἀλλὰ καὶ διότι ἐπιπολάζει τὸ γλίσχρον διὰ τὸ μὴ δύνασθαι ἐξατμίζειν.
10 ἐν μὲν οὖν τοῖς ἄλλοις αὐχμηρὸν τὸ γλίσχρον (διὸ
ὀστρακόδερμα καὶ μαλακόστρακα τὰ ἔσχατά ἐστι
τῶν ἀναίμων ζώων), ἐν δὲ τοῖς ἐναίμοις τὸ γλίσχρον λιπαρώτερόν ἐστιν. καὶ τούτων ὅσα μὴ γεώδη
τὴν φύσιν ἔχει λίαν, ἀθροίζεται τὸ πιμελῶδες ὑπὸ
τὴν τοῦ δέρματος σκέπην, ὡς τοῦ δέρματος γι-
15 νομένου ἐκ τῆς τοιαύτης γλισχρότητος· ἔχει γάρ
τινα γλισχρότητα τὸ λιπαρόν. πάντα δὲ ταῦτα,
καθάπερ εἴπομεν, λεκτέον γίνεσθαι τῇ μὲν ἐξ
ἀνάγκης τῇ δ' οὐκ ἐξ ἀνάγκης ἀλλ' ἕνεκά τινος.

---

^a Cf. 767 a 17 ff.
^b i.e., the change required to be effected; see Introd. § 48,
κίνησις.   ^c See Introd. § 8.

the uterus ; *e.g.*, things which we boil to make them pleasant for food, or for any other practical purpose. The only difference is that in this case the correct proportion of heat[a] to suit the movement[b] is supplied by us, whereas in the other, it is supplied by the nature of the generating parent. With those animals that are formed spontaneously the cause responsible is the movement and heat of the climatic conditions.

Heat and cooling (which is deprivation of heat) are both employed by Nature. Each has the faculty, grounded in *necessity*, of making one thing into this and another thing into that ; but in the case of the forming of the embryo it is for a *purpose* that their power of heating and cooling is exerted and that each of the parts is formed, flesh being made soft—as Flesh. heating and cooling make it such, partly owing to *necessity*, partly *for a purpose*,—sinew solid and elastic, bone solid and brittle. Skin is formed as the flesh Skin. solidifies, just as scum or " mother " forms on boiled liquids. Its formation is due not merely to its being on the outside, but also to the fact that glutinous substance remains on the surface because it cannot evaporate. In blooded animals the glutinous substance is more fatty than in bloodless ones, in which it is dry, and on this account the outer parts of the latter are testaceous or crustaceous. In those blooded animals whose nature is not excessively earthy, the fat collects under the protective covering, the skin, which seems to indicate that the skin is formed out of this sort of glutinous substance, since of course grease is to some extent glutinous. We are to say, then, as already stated, that all these things are formed partly as a result of *necessity*, partly also not of *necessity* but *for a purpose*.[c]

Πρῶτον μὲν οὖν τὸ ἄνω κύτος ἀφορίζεται κατὰ
τὴν γένεσιν, τὸ δὲ κάτω προϊόντος τοῦ χρόνου
20 λαμβάνει τὴν αὔξησιν ἐν τοῖς ἐναίμοις. ἅπαντα δὲ
ταῖς περιγραφαῖς διορίζεται πρότερον, ὕστερον δὲ
λαμβάνει τὰ χρώματα καὶ τὰς μαλακότητας καὶ
τὰς σκληρότητας, ἀτεχνῶς ὥσπερ ἂν ὑπὸ ζωγράφου
τῆς φύσεως δημιουργούμενα· καὶ γὰρ οἱ γραφεῖς
ὑπογράψαντες ταῖς γραμμαῖς οὕτως ἐναλείφουσι
25 τοῖς χρώμασι τὸ ζῷον.

Διὰ μὲν οὖν τὸ τὴν ἀρχὴν ἐν τῇ καρδίᾳ τῶν
αἰσθήσεων εἶναι καὶ τοῦ ζῴου παντὸς αὕτη γίνεται
πρῶτον· διὰ δὲ τὴν θερμότητα τὴν ταύτης, ᾗ
τελευτῶσιν αἱ φλέβες ἄνω, τὸ ψυχρὸν συνίστησιν
ἀντίστροφον τῇ θερμότητι τῇ περὶ τὴν καρδίαν τὸν
30 ἐγκέφαλον. διόπερ τὰ περὶ τὴν κεφαλὴν λαμβάνει
συνεχῆ τὴν γένεσιν μετὰ τὴν καρδίαν, καὶ μεγέθει
τῶν ἄλλων διαφέρει· πολὺς γὰρ καὶ ὑγρὸς ἐξ
ἀρχῆς ὁ ἐγκέφαλος.

Ἔχει δ' ἀπορίαν τὸ περὶ τοὺς ὀφθαλμοὺς συμβαῖνον τῶν ζῴων. μέγιστοι μὲν γὰρ ἐξ ἀρχῆς
φαίνονται καὶ πεζοῖς καὶ πλωτοῖς καὶ πτηνοῖς,
35 τελευταῖοι δὲ γίνονται τῶν μορίων· ἐν τῷ μεταξὺ
γὰρ χρόνῳ συμπίπτουσιν. αἴτιον δ' ὅτι τὸ τῶν
ὀφθαλμῶν αἰσθητήριον ἐστὶ μέν, ὥσπερ καὶ τὰ
ἄλλα αἰσθητήρια,[1] ἐπὶ πόρων· ἀλλὰ τὸ μὲν τῆς ἁφῆς

καὶ γεύσεως εὐθύς ἐστιν ἢ σῶμα ἢ τοῦ σώματός
τι τῶν ζῴων, ἡ δ' ὄσφρησις καὶ ἡ ἀκοὴ πόροι
συνάπτοντες πρὸς τὸν ἀέρα τὸν θύραθεν, πλήρεις
συμφύτου πνεύματος, περαίνοντες δὲ πρὸς τὰ

---

[1] ὥσπερ . . . αἰσθητήρια fort. secludenda ; suspic. est Platt.
μέν ἐστιν Z[1] pro ἐστὶ μέν ; πολλὰ P pro τὰ ἄλλα.

Now the upper portion of the body is the first to be marked off in the course of the embryo's formation; the lower portion receives its growth as time goes on. (This applies to the blooded animals.) In the early stages the parts are all traced out in outline ; later on they get their various colours and softnesses and hardnesses, for all the world as if a painter were at work on them, the painter being Nature. Painters, as we know, first of all sketch in [a] the figure of the animal in outline, and after that go on to apply the colours.

As the source of the sensations is in the heart, the heart is the first part of the whole animal to be formed ; and, on account of the heat of the heart, and to provide a corrective to it, the cold causes the brain to " set," where the blood-vessels terminate above. That is why the regions around the head Brain. begin to form immediately after the heart and are bigger than the other parts, the brain being large and fluid from the outset.

The development of the eyes is something of a Eyes. puzzle to the student. In birds, beasts, and fishes alike, the eyes are from the outset very large in appearance, yet they are the last of all the parts to be completely formed, since they shrink up in the meantime.[b] The reason is that the sense-organ of the eyes is indeed, like the other sense-organs, set upon passages ; but whereas the sense-organ of touch and of taste is just the animal's body or some portion of the body, and smell and hearing are passages full of connate *pneuma*,[c] connecting with the outer air and terminating at the small blood-vessels around

---

[a] *Cf.* note, 740 a 28.   [b] *Cf. H.A.* 561 a 19 ff.
    [c] See App. B §§ 26 ff.

744 a

φλέβια τὰ περὶ τὸν ἐγκέφαλον τείνοντα ἀπὸ τῆς
5 καρδίας· ὁ δ' ὀφθαλμὸς σῶμα μόνον ἴδιον ἔχει τῶν
αἰσθητηρίων. ἔστι δ' ὑγρὸν καὶ ψυχρόν, καὶ οὐ
προϋπάρχον ἐν τῷ τόπῳ καθάπερ καὶ τὰ ἄλλα
μόρια δυνάμει, ἔπειτα ἐνεργείᾳ γινόμενα ὕστερον·
ἀλλ' ἀπὸ τῆς περὶ τὸν ἐγκέφαλον ὑγρότητος ἀπο-
10 κρίνεται τὸ καθαρώτατον διὰ τῶν πόρων οἳ φαίνονται φέροντες ἀπ' αὐτῶν πρὸς τὴν μήνιγγα τὴν
περὶ τὸν ἐγκέφαλον. τούτου δὲ τεκμήριον· οὔτε
γὰρ ἄλλο μόριον ὑγρὸν καὶ ψυχρόν ἐστιν ἐν τῇ
κεφαλῇ παρὰ τὸν ἐγκέφαλον, τό τ' ὄμμα ψυχρὸν
καὶ ὑγρόν. ἐξ ἀνάγκης οὖν ὁ τόπος λαμβάνει
15 μέγεθος τὸ πρῶτον, συμπίπτει δ' ὕστερον. καὶ
γὰρ περὶ τὸν ἐγκέφαλον συμβαίνει τὸν αὐτὸν τρό-
πον· τὸ πρῶτον ὑγρὸς καὶ πολύς, ἀποπνέοντος δὲ
καὶ πεττομένου σωματοῦταί τε μᾶλλον καὶ συμπίπτει καὶ ὁ ἐγκέφαλος [καὶ τὰ σώματα]¹ καὶ τὸ
μέγεθος τὸ τῶν ὀμμάτων. ἐξ ἀρχῆς δὲ διὰ μὲν τὸν
20 ἐγκέφαλον ἡ κεφαλὴ μεγίστη, διὰ δὲ τὸ ὑγρὸν τὸ
ἐν τοῖς ὄμμασιν οἱ ὀφθαλμοὶ μεγάλοι φαίνονται.
τελευταῖοι δὲ λαμβάνουσι τέλος διὰ τὸ καὶ τὸν
ἐγκέφαλον συνίστασθαι μόλις· ὀψὲ γὰρ παύεται τῆς
ψυχρότητος καὶ τῆς ὑγρότητος ἐπὶ πάντων μὲν τῶν
ἐχόντων,² μάλιστα δ' ἐπὶ τῶν ἀνθρώπων. διὰ γὰρ
25 τοῦτο καὶ τὸ βρέγμα τῶν ὀστῶν γίνεται τελευ-
ταῖον· ἤδη γὰρ γεγενημένων θύραζε τῶν ἐμβρύων

---

¹ om. S, seclusit Bekker : καὶ τὰ ὄμματα Platt, om. καὶ τὸ
μέγεθος τὸ τῶν ὀμμάτων.
² τῶν ἐχόντων P : habentibus magnum cerebrum Σ : om.
vulg.

the brain which extend thither from the heart, the eye, by way of contrast, is the only one of the sense-organs which has a special " body " of its own. It is fluid and cold ; and, unlike the other parts, which are present in their places *potentially* to begin with and later on come to be formed *in actuality*, this one is not there at the start,[a] but it is produced by the purest part of the liquid around the brain being secreted off through those passages [b] which are to be observed leading from the eyes to the membrane around the brain. A sure sign of this is that beside the brain there is no part in the head except the eye which is cold and fluid. Hence it is due to *necessity* that this region gets large at first but shrinks later on ; because the same happens to the brain : at first this is fluid and large, but as evaporation and concoction proceed it becomes more solid and shrinks ; so does the size of the eyes. From the outset the head is very large, on account of the brain, and the eyes, as we see, are large on account of the fluid in them. But the eyes are the last of all to reach their completion, because the brain (on which they depend) does not " set " at all easily ; it is quite late before it ceases to be so cold and fluid ; and this is true of all animals that have a brain, especially of man. That is why the *bregma* [c] is the last of the bones to be formed : even after the embryos are brought to birth, this

[a] Aristotle's knowledge that the eye is an offshoot from the brain, and does not originate in the position which it finally occupies, is indeed remarkable.

[b] These are no doubt the optic nerves.

[c] *Cf. P.A.* 653 a 34 and *H.A.* 491 a 31. This is the bone which finally grows over the space at the top of the skull known as the " anterior fontanelle."

744 a

μαλακόν ἐστι τοῦτο τὸ ὀστοῦν τοῖς παιδίοις.[1]
αἴτιον δὲ τοῦ μάλιστ’ ἐπὶ τῶν ἀνθρώπων τοῦτο
συμβαίνειν, ὅτι τὸν ἐγκέφαλον ὑγρότατον ἔχουσι
καὶ πλεῖστον τῶν ζῴων, τούτου δ’ αἴτιον ὅτι καὶ
30 τὴν ἐν τῇ καρδίᾳ θερμότητα καθαρωτάτην. δηλοῖ
δὲ τὴν εὐκρασίαν ἡ διάνοια· φρονιμώτατον γάρ
ἐστι τῶν ζῴων ἄνθρωπος. ἀκρατῆ δὲ καὶ τὰ
παιδία μέχρι πόρρω τῆς κεφαλῆς ἐστι διὰ τὸ
βάρος τὸ περὶ τὸν ἐγκέφαλον. ὁμοίως δὲ καὶ
τῶν μορίων ὅσα δεῖ κινεῖν· ἡ γὰρ ἀρχὴ τῆς
κινήσεως ὀψὲ κρατεῖ τῶν ἄνωθεν καὶ τελευ-
35 ταῖον, ὅσων ἡ κίνησις μὴ συνήρτηται πρὸς αὐτήν,
ὥσπερ τῶν κώλων. τοιοῦτον δ’ ἐστὶ μόριον τὸ
βλέφαρον. ἐπεὶ δ’ οὐθὲν ποιεῖ περίεργον οὐδὲ
μάτην ἡ φύσις, δῆλον ὡς οὐδ’ ὕστερον οὐδὲ πρό-
τερον· ἔσται γὰρ τὸ γεγονὸς ἢ[2] μάτην ἢ περίεργον.

744 b

ὥσθ’ ἅμ’ ἀνάγκη τὰ βλέφαρα διαχωρίζεσθαί τε[3]
καὶ δύνασθαι κινεῖν. ὀψὲ μὲν οὖν διὰ τὸ πλῆθος
τῆς περὶ τὸν ἐγκέφαλον πέψεως τελειοῦται τὰ
ὄμματα τοῖς ζῴοις, τελευταῖα δὲ διὰ τὸ σφόδρα
κρατούσης τῆς κινήσεως εἶναι τὸ κινεῖν καὶ τὰ
5 οὕτως πόρρω τῆς ἀρχῆς καὶ ἀπεψυγμένα τῶν
μορίων. δηλοῖ δὲ τὰ βλέφαρα τοιαύτην ἔχοντα τὴν
φύσιν· ἂν γὰρ καὶ ὁποσονοῦν βάρος γένηται περὶ
τὴν κεφαλὴν δι’ ὕπνον ἢ μέθην ἢ ἄλλο τι τῶν
τοιούτων, οὐ δυνάμεθα τὰ βλέφαρα αἴρειν, οὕτω
βάρος αὐτῶν ἐχόντων μικρόν.

[1] τοῖς παιδίοις P : τῶν παιδίων vulg.
[2] ἢ P : om. vulg.   [3] τε PS : om. vulg.

[a] εὐκρασία. For κρᾶσις see Introd. § 40; and cf. P.A.
673 b 26 and Hippocrates, π. διαίτης I. 35.
[b] See Introd. §§ 11, 51.

bone is still soft in the case of children. The reason why this occurs especially in man is that in man the brain is more fluid and greater in volume than in any other animal, and the reason of this, in its turn, is that the heat in the heart is purest in man. The fineness of the blend [a] in man is shown by his possession of intellect : there is no other animal which is so intelligent. Even children however for a considerable period lack full control over their heads. This is due to the weight of the brain, and the same may be said of those parts of the body which have to be moved. It is quite late before the principle of movement gets control over the upper parts ; and its control over those parts (such as the legs) whose movement is not closely connected with it is achieved last of all. Another such part is the eyelid. Now, as Nature does nothing that is superfluous or pointless, it is plain that she will not do anything too late or too soon, for in that case what was done would be either pointless or superfluous. Therefore the separation of the eyelids and the ability to move them must coincide in time. Thus the completion of the formation of the eyes comes late, because of the large amount of concoction required by the brain, and it comes last, after all the other parts, because the movement [b] must be very strong and powerful in order to move parts which are so far away from the first principle,[c] and so much subjected to cold. That such is the nature of the eyelids is shown by the fact that even if a very little heaviness affects the head through sleep or intoxication or anything of that sort, we are unable to raise the eyelids although their weight is very slight.

[c] Viz., of movement, *i.e.*, the heart.

744 b

10 Περὶ μὲν οὖν ὀφθαλμῶν εἴρηται πῶς γίνονται καὶ
δι' ὅ τι, καὶ διὰ τίν' αἰτίαν τελευταίαν λαμβάνουσι
τὴν διάρθρωσιν.

Τῶν δ' ἄλλων γίνεται μορίων ἕκαστον ἐκ τῆς
τροφῆς, τὰ μὲν τιμιώτατα καὶ μετειληφότα τῆς κυ-
ριωτάτης ἀρχῆς ἐκ τῆς πεπεμμένης καὶ καθαρωτά-
της καὶ πρώτης τροφῆς, τὰ δ' ἀναγκαῖα μόρια καὶ
15 τούτων ἕνεκεν ἐκ τῆς χείρονος καὶ τῶν ὑπολειμμάτων καὶ περιττωμάτων. ὥσπερ γὰρ οἰκονόμος
ἀγαθός, καὶ ἡ φύσις οὐθὲν ἀποβάλλειν εἴωθεν ἐξ
ὧν ἔστι ποιῆσαί τι χρηστόν. ἐν δὲ ταῖς οἰκο-
νομίαις τῆς γινομένης τροφῆς ἡ μὲν βελτίστη τέτακται τοῖς ἐλευθέροις, ἡ δὲ χείρων καὶ τὸ πε-
20 ρίττωμα ταύτης ⟨τοῖς⟩[1] οἰκέταις, τὰ δὲ χείριστα καὶ
τοῖς συντρεφομένοις διδόασι ζῴοις. καθάπερ οὖν
εἰς τὴν αὔξησιν ὁ θύραθεν ταῦτα ποιεῖ νοῦς, οὕτως
ἐν τοῖς γινομένοις αὐτοῖς ἡ φύσις ἐκ μὲν τῆς καθ-
αρωτάτης ὕλης σάρκας καὶ τῶν ἄλλων αἰσθητηρίων τὰ σώματα συνίστησιν, ἐκ δὲ τῶν περιτ-
25 τωμάτων ὀστᾶ καὶ νεῦρα καὶ τρίχας, ἔτι δ' ὄνυχας
καὶ ὁπλὰς καὶ πάντα τὰ τοιαῦτα· διὸ τελευταῖα
ταῦτα λαμβάνει τὴν σύστασιν, ὅταν ἤδη γίγνηται
περίττωμα τῆς φύσεως.

Ἡ μὲν οὖν τῶν ὀστῶν φύσις ἐν τῇ πρώτῃ συστάσει γίνεται τῶν μορίων ἐκ τῆς σπερματικῆς
30 περιττώσεως, καὶ τῶν ζῴων αὐξανομένων ἐκ τῆς
φυσικῆς τροφῆς λαμβάνει τὴν αὔξησιν, ἐξ ἧσπερ
τὰ μόρια τὰ κύρια, ταύτης μέντοι αὐτῆς τὰ ὑπο-

---

[1] supplevit Richards.

---

[a] *i.e.*, blood.
[b] *Cf.* the regular distinction between " the better " and
" necessity." [c] The sense-organ of touch.

This concludes our discussion about the eyes. We have said how they are formed, and why, and what is the reason that they are the last of all the parts to be articulated.

Each of the remaining parts is formed out of the nourishment. The most honourable ones, those which have a share in the supreme controlling principle, are formed out of the first of the nourishment,[a] which has been concocted and is purest; the " necessary " parts,[b] which exist for the sake of those just mentioned, are formed out of inferior nourishment, out of the leavings and the residues. Like a good housekeeper, Nature is not accustomed to throw anything away if something useful can be made out of it. In housekeeping the best of the food available is reserved for the freemen; the residue left over from this as well as the inferior food goes to the servants, and the worst of all goes to the domestic animals. Here then is an instance of a mind, external to them, acting so as to provide for their growth. In the same way Nature is at work within the creatures themselves that are being formed, and constructs flesh[c] and the bodily parts of the other sense-organs out of the purest of the material, whereas out of the residues she constructs bones and sinews and hair, and also nails and hoofs and all such things, which means that they have to wait till Nature has some residue to hand, and that is why they are the last to be constructed.

The bones, then, are formed during the first stage **Bones, etc.** of construction out of the seminal residue, and as the animal grows they grow too. Their growth is derived from the natural nourishment, which is the same as that which supplies the supreme parts; only they

744 b

λείμματα καὶ τὰ περιττωματικά. γίνεται γὰρ ἐν
παντὶ τὸ πρῶτον καὶ τὸ δεύτερον τῆς¹ τροφῆς
τὸ μὲν θρεπτικὸν τὸ δ' αὐξητικόν, θρεπτικὸν μὲν
35 ὃ τὸ εἶναι παρέχεται τῷ τε ὅλῳ καὶ τοῖς μορίοις,
αὐξητικὸν δὲ τὸ εἰς μέγεθος ποιοῦν τὴν ἐπίδοσιν·
περὶ ὧν ὕστερον διοριστέον μᾶλλον. τὸν αὐτὸν δὲ
τρόπον τοῖς ὀστοῖς καὶ τὰ νεῦρα συνίσταται καὶ
ἐκ τῶν αὐτῶν, ἐκ τῆς σπερματικῆς περιττώσεως
745 a καὶ τῆς θρεπτικῆς. ὄνυχες δὲ καὶ τρίχες καὶ ὁπλαὶ
καὶ κέρατα καὶ ῥύγχη καὶ τὰ πλῆκτρα τῶν ὀρνίθων, καὶ εἴ τι τοιοῦτον ἕτερόν ἐστι μόριον, ἐκ τῆς
ἐπικτήτου τροφῆς καὶ τῆς αὐξητικῆς, ἥν τε παρὰ
τοῦ θήλεος ἐπικτᾶται καὶ [τῆς]² θύραθεν. διὰ τοῦτο
5 τὰ μὲν ὀστᾶ μέχρι τινὸς λαμβάνει τὴν αὔξησιν·
ἔστι γάρ τι πᾶσι τοῖς ζῴοις πέρας τοῦ μεγέθους,
διὸ καὶ τῆς τῶν ὀστῶν αὐξήσεως. εἰ γὰρ ταῦτ'
εἶχεν αὔξησιν ἀεί, καὶ τῶν ζῴων ὅσα ἔχει ὀστοῦν
ἢ τὸ ἀνάλογον, ηὐξάνετ' ἂν ἕως ἔζη· τοῦ γὰρ
μεγέθους ὅρος ἐστὶ ταῦτα τοῖς ζῴοις. δι' ἣν μὲν
10 οὖν αἰτίαν οὐκ ἀεὶ λαμβάνουσιν αὔξησιν λεκτέον
ὕστερον· τρίχες δὲ καὶ τὰ συγγενῆ τούτοις, ἕως ἂν

¹ τῆς Z : καὶ τῆς vulg.   ² seclusi.

---

[a] The functions of " nutritive Soul " (see above, 735 a 17,
and *De anima* 415 a 25) are to generate, and to make use of
nourishment; it is the same δύναμις of the Soul which generates
and which nourishes (*De anima* 416 a 19). In the passage
which there follows, a distinction is made between being
nourished (τρέφεσθαι) and growing (αὐξάνεσθαι). At 416 b 11,
Aristotle says that " nourishment " is not identical with
" that which is growth-promoting " ; thus, in so far as the
living thing (the creature " with Soul in it ") is of a certain
quantity, the food is " growth-promoting " (*i.e.*, increases its
quantity) : but in so far as the creature is a particular thing,
an individual " being," the food is " nourishment," *because*

232

get merely the leavings and the residues of it. In every instance, of course, there is nourishment of two grades present : (1) " nutritive," that is to say, which provides both the whole and the parts with being ; (2) " growth-promoting," that is to say, which causes increase of bulk. These will have to be more particularly distinguished later on.[a] The sinews are constructed in the same way as the bones, and out of the same materials, viz., the seminal or " nutritive " residue. As for nails, hair, hoofs, horns, bills, cocks' spurs and any other such part, these are formed out of the supplementary or " growth-promoting " nourishment, this additional nourishment being obtained from the female, and from outside. On this account, the bones continue growing only up to a certain point, for as all animals have a limit to their size, this involves a limit to the growth of the bones. If the bones continued growing for ever, then every animal which contains any bone or the counterpart of bone [b] would go on growing as long as it lived, because the bones set the limit for an animal's size. We shall have to explain later on why the bones do not continue growing for ever. Hair and similar things, on the other hand, continue growing so long

it maintains the creature's being. And it is also " productive of generation "—not, of course, of the generation of the creature which is getting the nourishment, for *its* " being " is already there, but of another creature similar to it (416 b 15-17). It thus appears that the business of " nutrition " is concerned with the maintenance of a living creature's *being*, and with the generation of new ones' *being*; " growth-promotion " is concerned with *increasing the bulk* of that which already has being—and this is precisely the distinction which Aristotle employs in the present passage.

[b] *e.g.*, the *os sepiae*, the " pen " of calamaries, the cartilaginous spines of Selachia (sharks, etc.) (*P.A.* 654 a 20, 655 a 23).

ὑπάρχωσιν, αὐξάνονται, καὶ μᾶλλον ἐν νόσοις καὶ
τῶν σωμάτων γηρασκόντων καὶ φθινόντων διὰ τὸ
λείπεσθαι περίττωμα πλεῖον ἐλάττονος εἰς τὰ
κύρια δαπανωμένου διὰ τὸ γῆρας καὶ τὰς νόσους,
15 ἐπεί γ᾽ ὅταν ὑπολείπῃ καὶ τοῦτο διὰ τὴν ἡλικίαν,
καὶ αἱ τρίχες ὑπολείπουσιν. τὰ δ᾽ ὀστᾶ τοὐναντίον·
συμφθίνει γὰρ τῷ σώματι καὶ τοῖς μέρεσιν. αὐξάνονται δ᾽ αἱ τρίχες καὶ τεθνεώτων, οὐ μέντοι
γίνονταί γ᾽ ἐξ ὑπαρχῆς.

Περὶ δ᾽ ὀδόντων ἀπορήσειεν ἄν τις. εἰσὶ γὰρ τὴν
20 μὲν φύσιν τὴν αὐτὴν ἔχοντες τοῖς ὀστοῖς, καὶ γίνονται ἐκ τῶν ὀστῶν, ὄνυχες δὲ καὶ τρίχες καὶ
κέρατα καὶ τὰ τοιαῦτα ἐκ τοῦ δέρματος, διὸ καὶ
συμμεταβάλλουσι τῷ δέρματι τὰς χρόας· λευκά τε
γὰρ καὶ μέλανα γίνονται καὶ παντοδαπὰ κατὰ τὴν
τοῦ δέρματος χρόαν, οἱ δ᾽ ὀδόντες οὐθέν· ἐκ γὰρ
τῶν ὀστῶν εἰσιν, ὅσα τῶν ζῴων ἔχει ὀδόντας καὶ
25 ὀστᾶ. αὐξάνονται δὲ διὰ βίου μόνοι τῶν ἄλλων
ὀστῶν· τοῦτο δὲ δῆλον ἐπὶ τῶν παρακλινόντων
ὀδόντων τὴν ἁφὴν τὴν ἀλλήλων. αἴτιον δὲ τῆς
αὐξήσεως, ὡς μὲν ἕνεκά του, διὰ τὸ ἔργον· ταχὺ
γὰρ ἂν κατετρίβοντο μὴ γινομένης τινὸς ἐπιρρύ-
σεως, ἐπεὶ καὶ νῦν ἐνίοις γηράσκουσι, τοῖς βρω-
30 τικοῖς μὲν μὴ μεγάλους δ᾽ ἔχουσι, κατατρίβονται
πάμπαν· πλείονι γὰρ λόγῳ καθαιροῦνται τῆς αὐξήσεως. διὸ καὶ τοῦτο εὖ μεμηχάνηται πρὸς τὸ

---

[a] In the case of rabbits, etc., it may happen that a tooth
in the upper jaw and one in the lower grow outwards and
thus continue growing indefinitely, so that finally the animal
is unable to eat at all.

as they are there at all, and they grow more during diseases, and when old age advances, and when the body is wasting. This is because old age and diseases mean that less ⟨nourishment⟩ is expended on the supreme parts of the body and therefore more residue is left over ; though when even this begins to fail through age, the hair follows suit. With the bones, the reverse occurs : they waste away along with the body and its parts. Hair actually continues to grow after life is extinct, though it will not begin growing where it does not already exist.

Teeth may present a puzzle. They possess the Teeth. same nature as the bones and are formed out of the bones ; nails, hair, horns and the like, however, are formed out of the skin, and that is why they change their colour along with the skin : they turn white and black and all shades according to the colour of the skin. The teeth do none of this, because they are formed out of the bones (this applies of course only to such animals as have both teeth and bones). They are unique among bones in that they continue grow-ing all through life, as is clear in the case of teeth which take an oblique direction and fail to come into contact with each other.[a] The reason for their growth, the purpose *for the sake of which* they grow, is to discharge their special function : they would soon be worn down unless the loss were made good in some way,[b] since even as it is, in some aged animals which eat a great deal but have small teeth, they are quite worn away, because their growth is not proportionate to their loss. And so here too Nature has produced

[b] L. & S. translate " unless there were some means of saving them " ; but Scot translates *si non crescerent con-sumerentur cito nisi esset materia ex qua crescunt.*

745 a

συμβαῖνον ἡ φύσις· συνάγει γὰρ εἰς τὸ γῆρας καὶ
τὴν τελευτὴν τὴν ὑπόλειψιν τῶν ὀδόντων. εἰ δ᾽
ἦν μυριετὴς ὁ βίος ἢ χιλιετής, παμμεγέθεις τ᾽ ἂν
35 ἔδει γίνεσθαι τοὺς ἐξ ἀρχῆς καὶ φύεσθαι πολλάκις·

745 b

καὶ γὰρ εἰ συνεχῆ τὴν αὔξησιν εἶχον, ὅμως ἂν
ἄχρηστοι λεαινόμενοι πρὸς τὴν ἐργασίαν ἦσαν. οὗ
μὲν οὖν ἕνεκα λαμβάνουσι τὴν αὔξησιν, εἴρηται·
συμβαίνει δὲ μηδὲ τὴν αὐτὴν ἔχειν φύσιν τοῖς
ἄλλοις ὀστοῖς τοὺς ὀδόντας· τὰ μὲν γὰρ ἐν τῇ
5 πρώτῃ συστάσει γίνεται πάντα καὶ οὐθὲν ὕστερον,
οἱ δ᾽ ὀδόντες ὕστερον. διὸ καὶ πάλιν δύνανται
φύεσθαι ἐκπεσόντες· ἅπτονται γάρ, ἀλλ᾽ οὐ συμ-
πεφύκασι τοῖς ὀστοῖς. ἐκ μέντοι τῆς τροφῆς τῆς
εἰς τὰ ὀστᾶ διαδιδομένης γίνονται, διὸ τὴν αὐτὴν[1]
ἔχουσι φύσιν, καὶ τότε ὅταν ἐκεῖνα ἔχῃ ἤδη τὸν
10 ἀριθμὸν τὸν αὐτῶν. τὰ μὲν οὖν ἄλλα ζῷα ἔχοντα
γίνεται ὀδόντας καὶ τὸ ἀνάλογον τοῖς ὀδοῦσιν, ἐὰν
μή τι γίγνηται παρὰ φύσιν, διὰ τὸ ἀπολύεσθαι τῆς
γενέσεως τετελεσμένα τοῦ ἀνθρώπου μᾶλλον· ὁ δ᾽
ἄνθρωπος, ἂν μή τι συμβῇ παρὰ φύσιν, οὐκ ἔχων.
δι᾽ ἣν δ᾽ αἰτίαν οἱ μὲν γίνονται τῶν ὀδόντων καὶ
15 ἐκπίπτουσιν, οἱ δ᾽ οὐκ ἐκπίπτουσιν, ὕστερον λεχθήσεται.

Διότι δ᾽ ἐκ περιττώματός ἐστι τὰ τοιαῦτα τῶν
μορίων, διὰ τοῦτ᾽ ἄνθρωπος ψιλότατόν τε κατὰ τὸ
σῶμα τῶν ζῴων πάντων ἐστὶ καὶ ὄνυχας ἐλαχί-
στους ἔχει ὡς κατὰ μέγεθος· ἐλάχιστον γὰρ ἔχει

---

[1] αὐτὴν Bekker, per typothetae errorem.

---

[a] Bk. V, ch. 8. [b] i.e., hair, nails, etc.

an excellent device to suit the case, in making the failure of the teeth coincide with the time of old age and the close of life. If life went on for 10,000 or even 1000 years, the teeth would have had to be quite enormous to begin with, and they would have had to grow afresh many times over ; not even continuous growth would have sufficed to prevent them being ground down and becoming useless for their work. We have now described the purpose *for the sake of which* the teeth grow. And yet as a matter of fact the teeth do not possess the same nature as the rest of the bones, because the bones, without exception, are all formed during the first stage of the embryo's construction, whereas the teeth are formed later ; and that, too, is why a fresh set of teeth is able to grow after the old ones have fallen out : although they are in touch with the bones, they are not all of a piece with them. Still, they are formed out of the nourishment which is distributed to the bones (which is why they possess the same nature), and at a time when the bones have already attained their full complement. All the animals except man already have their teeth (or the counterpart of teeth) when they are born—unless it be that something unnatural occurs—because when they are released from their process of formation they are more fully perfected than man ; man however when born has no teeth— unless something unnatural occurs. We shall explain later on [a] why some of the teeth are formed and fall out and why some do not fall out.

The reason why man's body is more naked than that of any single one of the other animals, and why he has the smallest nails in proportion to his size, is this. Parts of this sort [b] are made of residue ; now

745 b

περίττωμα γεῶδες, ἔστι δὲ περίττωμα μὲν τὸ
20 ἄπεπτον, τὸ δὲ γεηρὸν ἐν τοῖς σώμασι πάντων
ἀπεπτότατον.

Πῶς μὲν οὖν ἕκαστον συνίσταται τῶν μορίων,
εἴρηται, καὶ τί τῆς γενέσεως αἴτιον.

VII Ἔχει δὲ τὴν αὔξησιν τὰ ζῳοτοκούμενα τῶν
ἐμβρύων, ὥσπερ ἐλέχθη πρότερον, διὰ τῆς τοῦ
ὀμφαλοῦ προσφύσεως. ἐπεὶ γὰρ ἔνεστιν ἐν τοῖς
25 ζῴοις καὶ ἡ θρεπτικὴ δύναμις τῆς ψυχῆς, ἀφίησιν
εὐθὺς οἷον ῥίζαν τὸν ὀμφαλὸν εἰς τὴν ὑστέραν.
ἔστι δὲ ὁ ὀμφαλὸς ἐν κελύφει φλέβες, τοῖς μὲν
μείζοσι πλείους, οἷον βοΐ καὶ τοῖς τοιούτοις, τοῖς
δὲ μέσοις δύο, μία δὲ τοῖς ἐσχάτοις. διὰ δὲ τούτου
λαμβάνει τὴν τροφὴν αἱματικήν. αἱ γὰρ ὑστέραι
30 πέρατα φλεβῶν πολλῶν εἰσιν. τὰ μὲν οὖν μὴ
ἀμφώδοντα πάντα, καὶ τῶν ἀμφωδόντων ὅσων ἡ
ὑστέρα μὴ μίαν φλέβα μεγάλην ἔχει διατείνουσαν
ἀλλ᾽ ἀντὶ μιᾶς πυκνὰς πολλάς, ταῦτα ἐν ταῖς
ὑστέραις ἔχει τὰς καλουμένας κοτυληδόνας, πρὸς
ἃς¹ ὁ ὀμφαλὸς συνάπτει καὶ προσπέφυκεν· ἀποτέ-
τανται γὰρ αἱ φλέβες αἱ διὰ τοῦ ὀμφαλοῦ ἔνθεν
καὶ ἔνθεν καὶ σχίζονται πάντη κατὰ τὴν ὑστέραν· ᾗ
δὲ περαίνουσι, ταύτῃ γίγνονται αἱ κοτυληδόνες,²
τὸ μὲν περιφερὲς ἔχουσαι³ πρὸς τὴν ὑστέραν, τὸ
35 δὲ κοῖλον πρὸς τὸ ἔμβρυον. μεταξὺ δὲ τῆς ὑστέρας

746 a   καὶ τοῦ ἐμβρύου τὸ χόριον καὶ οἱ ὑμένες εἰσίν. αἱ

---

¹ ἃς Platt, Oᵇ*: ἇ P.
² πρὸς ἃς ὁ ὀμφαλὸς . . . γίγνονται αἱ κοτυληδόνες POᵇ*Σ:
om. vulg.   ³ ἔχουσαι Z¹ et corr. P : ἐχούσας vulg., Z².

ᵃ See 740 a 24 ff.
ᵇ Not quite the same as the modern use of the term.
Aristotle uses it to mean the pits in the modified wall of the
238

it is unconcocted substance which constitutes residue, and the most unconcocted substance in animals' bodies is the earthy substance, and man has a smaller amount of earthy residue than the other animals.

We have now described how each of the parts takes shape, and what is the cause of their formation.

In viviparous animals, as stated earlier,[a] the embryo obtains its growth through the umbilical attachment. Since the nutritive faculty of the Soul, as well as the others, is present in animals, it immediately sends off the umbilicus, like a root, to the uterus. The umbilicus consists of blood-vessels in a sheath. In the larger animals, such as the ox and the like, it contains numerous blood-vessels, in medium-sized animals, two, and in the smallest, one. Through this the embryo gets its nourishment, *i.e.*, blood ; the uterus being the terminus of many blood-vessels. The cotyledons [b] (as they are called) are present in the uterus (*a*) of all those animals which have no front teeth in the upper jaw, and (*b*) of those which have teeth in both jaws and also have a cluster of blood-vessels running right through the uterus instead of a single large one. The umbilicus is connected up to these cotyledons and firmly attached to them ; for the blood-vessels which pass through the umbilicus extend in both directions and branch out all over the uterus, and it is at their terminal points that the cotyledons are formed. Their convex side is towards the uterus, their hollow side towards the embryo. Between the uterus and the embryo are the *chorion* and the membranes. As the embryo grows and

VII
(e) Nutri-
tion of the
embryo.

uterus into which the villi of the outer membrane of the embryo fit. For the meaning attached to the term by Diocles, see Wellmann, reference in note on 746 a 19 below.

δὲ κοτυληδόνες αὐξανομένου καὶ τελεουμένου τοῦ
ἐμβρύου γίνονται ἐλάττους, καὶ τέλος ἀφανίζονται
τελεωθέντος. εἰς τοῦτο γὰρ προεκτίθεται τοῖς
ἐμβρύοις ἡ φύσις τὴν αἱματικὴν τροφὴν τῆς ὑστέρας ὥσπερ εἰς μαστούς, καὶ διὰ τὸ ἀθροίζε-
5 σθαι κατὰ[1] μικρὸν ἐκ πολλῶν οἷον ἐξάνθημα καὶ
φλεγμασία γίνεται τὸ σῶμα τὸ τῆς κοτυληδόνος.
ἕως μὲν ἂν οὖν ἔλαττον ᾖ τὸ ἔμβρυον, οὐ δυνάμενον
πολλὴν λαμβάνειν τροφήν, δῆλαί εἰσι καὶ μείζονες,
αὐξηθέντος δὲ συμπίπτουσιν.

Τὰ δὲ πολλὰ τῶν κολοβῶν ζῴων καὶ ἀμφωδόντων
10 οὐκ ἔχει κοτυληδόνας[2] ἐν ταῖς ὑστέραις, ἀλλ' ὁ
ὀμφαλὸς εἰς φλέβα τείνει μίαν, αὕτη δὲ τέταται διὰ
τῆς ὑστέρας ἔχουσα μέγεθος. ἐπεὶ δὲ τὰ μὲν
μονοτόκα τὰ δὲ πολυτόκα τῶν τοιούτων ἐστὶ
ζῴων, καὶ τὰ πλείω τῶν ἐμβρύων τὸν αὐτὸν ἔχει
τρόπον τῷ ἑνί. δεῖ δὲ ταῦτα θεωρεῖν ἔκ τε τῶν
15 παραδειγμάτων τῶν ἐν ταῖς ἀνατομαῖς καὶ τῶν ἐν
ταῖς ἱστορίαις γεγραμμένων. πεφύκασι γὰρ τὰ
ζῷα ἐκ τοῦ ὀμφαλοῦ, ὁ δ' ὀμφαλὸς ἐκ τῆς φλεβός,
ἐφεξῆς ἀλλήλοις, ὡσπερανεὶ παρ' ὀχετὸν τὴν
φλέβα ῥέουσαν· περὶ δὲ ἕκαστον τῶν ἐμβρύων οἵ
θ' ὑμένες καὶ τὸ χόριόν ἐστιν.

Οἱ δὲ λέγοντες τρέφεσθαι τὰ παιδία ἐν ταῖς
20 ὑστέραις διὰ τοῦ σαρκίδιόν τι βδάλλειν οὐκ ὀρθῶς

---

[1] κατὰ P : καὶ κατὰ vulg.
[2] κοτυληδόνας P : κοτυληδόνα vulg.

---

[a] Here seems to mean " hornless."
[b] Aëtius ascribes a similar theory to Democritus and
Epicurus (Aët. 5. 16 ; see Diels, *Vorsokr.*[5] 68 A 144);
Censorinus (*De die natali* 6. 3 ; Diels 38 A 17) to Diogenes
and Hippocrates. *Cf.* Hippocrates, π. σαρκῶν 6 (viii. 592

approaches its completion the cotyledons become smaller, and finally when it is completed they disappear. Nature lays in a store of the blood-like nourishment for the embryos in this part of the uterus, as it were into breasts, and the body of the cotyledon becomes as it were an eruption or an inflammation owing to the fact that the numerous cotyledons gradually get compacted together. While the embryo is fairly small, and unable to take much nourishment, they are large and plainly visible, but when it has grown they shrink up.

The great majority of the " stunted " [a] animals, and of those that have front teeth in both jaws, have no cotyledons in their uterus, but the umbilicus extends to meet a single blood-vessel, which is a large one and extends throughout the uterus. Some of these animals produce one at a birth, others several ; but what occurs when there is only one embryo occurs also when there are more. All this should be studied with the help of the illustrative diagrams given in the *Dissections* and *Researches*. The embryos are attached each to its umbilicus, and the umbilicus is attached to the blood-vessel ; they are arranged one after the other along the stream of the blood-vessel as it might be along a runnel in the garden ; and there are membranes and a *chorion* around each embryo.

Those people [b] who say that children are nourished in the uterus by means of sucking a bit of flesh are

Littré).   The view that the embryo sucked the " cotyledons " was held by Diocles of Carystus (Wellmann, *Fragmentsammlung der sikelischen Ärzte*, Diocles fr. 27, 10 ff.) ; and according to Jaeger (*Diokles von Karystos*, 166), Aristotle's detailed treatment of the subject of cotyledons here is due to the fact that Diocles was associated with him in the Lyceum.

λέγουσιν· ἐπί τε γὰρ τῶν ἄλλων ζῴων ταὐτὸν
συνέβαινεν ἄν, νῦν δ᾽ οὐ φαίνεται (θεωρῆσαι γὰρ
τοῦτο ῥάδιον διὰ τῶν ἀνατομῶν)· καὶ περὶ ἅπαντα
τὰ ἔμβρυα καὶ τὰ πτηνὰ καὶ τὰ πλωτὰ καὶ τὰ
τῶν πεζῶν ὁμοίως λεπτοὶ περιέχουσιν ὑμένες χω-
25 ρίζοντες ἀπό τε[1] τῆς ὑστέρας καὶ τῶν ἐγγινομένων
ὑγρῶν, ἐν οἷς οὔτ᾽ αὐτοῖς ἔνεστι τοιοῦτον οὐθέν,
οὔτε διὰ τούτων οὐθενὸς ἐνδέχεται ποιεῖσθαι τὴν
ἀπόλαυσιν· τὰ δ᾽ ᾠοτοκούμενα πάντα ὅτι λαμβάνει τὴν αὔξησιν χωρισθέντα τῆς μήτρας ἔξω,
φανερόν.

Γίνεται δὲ ὁ συνδυασμὸς τοῖς ζῴοις κατὰ φύσιν
30 μὲν τοῖς ὁμογενέσιν, οὐ μὴν ἀλλὰ καὶ τοῖς μὲν
σύνεγγυς[2] τὴν φύσιν ἔχουσιν, οὐκ ἀδιαφόροις δὲ
τῷ εἴδει, ἐὰν τά τε μεγέθη παραπλήσια ᾖ καὶ οἱ
χρόνοι ἴσοι ὦσι τῆς κυήσεως. σπάνια μὲν οὖν
γίνεται τὰ τοιαῦτα ἐπὶ τῶν ἄλλων, γίνεται δὲ καὶ
ἐπὶ κυνῶν καὶ ἀλωπέκων καὶ λύκων ⟨καὶ θώων⟩[3]·
35 καὶ οἱ Ἰνδικοὶ δὲ κύνες ἐκ θηρίου τινὸς κυνώδους

γεννῶνται καὶ κυνός. καὶ ἐπὶ τῶν ὀρνίθων δὲ τῶν
ὀχευτικῶν ὦπται τοῦτο συμβαῖνον, οἷον ἐπὶ περ-
δίκων καὶ ἀλεκτορίδων· καὶ τῶν γαμψωνύχων οἱ
ἱέρακες δοκοῦσιν οἱ διαφέροντες τῷ εἴδει μίγνυσθαι πρὸς ἀλλήλους· καὶ ἐπ᾽ ἄλλων δέ τινων
5 ὀρνέων ἔχει τὸν αὐτὸν τρόπον. ἐπὶ δὲ τῶν θαλαττίων οὐθὲν ἀξιόλογον ἑώραται, δοκοῦσι δὲ μάλιστα

---

[1] τε P : om. vulg.   [2] σύνεγγυς SZ : ἐγγὺς vulg.
[3] Btf. ; vid. p. 563.

---

[a] Cf. H.A. 607 a 4 ff. " they say too that the ' Indian dog '
is the offspring of a tiger and a bitch ; not the first cross,
but the offspring at the third generation." There seems to

mistaken. If this were true, the same would occur in the other animals, but it is not found to do so, as can be easily observed by means of dissections. Also, all embryos alike, whether they be of animals that fly or swim or walk, have round them fine membranes which separate them from the uterus and from the fluids which are formed there ; and there is nothing of the sort in these membranes nor can the embryos get the benefit of anything whatever through them. As for embryos that are produced by means of eggs, it is of course obvious that in all cases *their* growth takes place outside the uterus, after they have been separated from it.

The partners in copulation are naturally and ordi- Hybrids, narily animals of the same kind ; but beside that, etc. animals that are closely allied in their nature, and are not very different in species, copulate, if they are comparable in size and if their periods of gestation are equal in length. Although such crossing is infrequent among the majority of animals, it occurs among dogs, foxes, wolves ⟨and jackals⟩ ; the Indian dog [a] also is produced from the union of a dog with some wild doglike beast. It has also been observed to occur among those birds that are salacious, *e.g.*, partridges and common fowls. A case among the crook-taloned birds is that of the hawks, different species of which copulate, as it appears ; and the same occurs among certain other birds. We have no trustworthy observation of its occurrence among sea-animals ; but there is a strong suspicion that the *rhinobates* as it is called is produced by the copu-

be no general agreement as to what this animal was ; see Platt's note, *C.Q.* III (1909), 241 ff. *Cf.* too the " Laconian hound," 738 b 31.

οἱ ῥινοβάται καλούμενοι γίνεσθαι ἐκ ῥίνης καὶ
βάτου συνδυαζομένων. λέγεται δὲ καὶ τὸ περὶ
τῆς Λιβύης παροιμιαζόμενον, ὡς ἀεί τι τῆς
Λιβύης τρεφούσης καινόν, διὰ τὸ μίγνυσθαι καὶ
10 τὰ μὴ ὁμόφυλα ἀλλήλοις λεχθῆναι τοῦτο· διὰ γὰρ
τὴν σπάνιν τοῦ ὕδατος ἀπαντῶντα πάντα πρὸς
ὀλίγους τόπους τοὺς ἔχοντας νάματα μίγνυσθαι
καὶ τὰ μὴ ὁμογενῆ.

Τὰ μὲν οὖν ἄλλα τῶν ἐκ τοιαύτης μίξεως γινομένων συνδυαζόμενα φαίνετ μι πάλιν ἀλλήλοις καὶ
μιγνύμενα καὶ δυνάμενα τό τε θῆλυ καὶ τὸ ἄρρεν
15 γεννᾶν, οἱ δ' ὀρεῖς ἄγονοι μόνοι τῶν τοιούτων· οὔτε
γὰρ ἐξ ἀλλήλων οὔτ' ἄλ οις μιγνύμενοι γεννῶσιν.
ἔστι δὲ τὸ πρόβλημα κα)όλου μέν, διὰ τίν' αἰτίαν
ἄγονον ἢ ἄρρεν ἢ θῆλύ ἐστιν· εἰσὶ γὰρ καὶ γυναῖκες
καὶ ἄνδρες ἄγονοι, καὶ τῶν ἄλλων ζῴων ἐν τοῖς
γένεσιν ἑκάστοις, οἷον ἐν ἵπποις καὶ προβάτοις.
20 ἀλλὰ τοῦτο τὸ γένος ὅλον ἄγονόν ἐστι, τὸ τῶν
ἡμιόνων. τὰ δ' αἴτια τῆς ἀγονίας ἐπὶ μὲν τῶν
ἄλλων πλείω συμβαίνει· καὶ γὰρ ἐκ γενετῆς, ὅταν
πηρωθῶσι τοὺς τόπους τοὺς πρὸς τὴν μίξιν χρησίμους, ἄγονοι γίνονται καὶ γυναῖκες καὶ ἄνδρες,
ὥστε τὰς μὲν μὴ ἡβᾶν τοὺς δὲ μὴ γενειᾶν, ἀλλ'
25 εὐνουχίας διατελεῖν ὄντας· τοῖς δὲ προϊούσης τῆς
ἡλικίας ταὐτὸν συμβαίνει πάσχειν, ὁτὲ μὲν δι'
εὐτροφίαν τῶν σωμάτων (ταῖς μὲν γὰρ πιοτέραις

---

a The *batos* is a flat-fish (*P.A.* 695 b 27, 696 a 26), called
by Thompson (translation of *H.A.* 566 a 27) the " skate," by
Platt, a " ray." The *rhine* is called by Thompson the " angel-
fish " (note on *H.A.* 540 b 11), by Platt, a " shark." At
H.A. 566 a 27 ff. Aristotle again refers to the *rhinobates* as
a cross between these two fishes, and says that it has the head
and foreparts of the *batos* and the hindparts of the *rhine*.

lation of the *rhine* and the *batos*.[a] Also, the origin of the proverb about Libya, to the effect that " Libya is always bringing forth something new," [b] is said to be that there animals of different species unite, owing to the fact that as there is very little water they all meet together at the few places where springs are to be found, and so animals of different species unite.

It is known that with one exception all the animals which are produced as a result of such unions copulate with each other and unite in their turn and are able to produce young of both sexes. Mules are the one exception. They are sterile and do not generate either by union with each other or with other animals. It is, of course, a general problem why any particular male or female is sterile : there are men and women who are sterile, and there are instances in the several kinds of animals, *e.g.*, horses and sheep. But with the mules we have a whole race which is sterile. Leaving this exception for the moment : elsewhere the causes of sterility are numerous. (*a*) Men and women alike are sterile from birth if they are deformed in the regions employed for copulation ; as a result, the men do not grow a beard but remain as eunuchs, while the women do not reach puberty ; (*b*) others become sterile as they advance in age, sometimes (i) because they have put on too much flesh : in men

Platt thinks the *rhinobates* is the angel-fish ; Thompson offers the opinion that it is " probably the modern genus *Rhinobatus* " ; Platt says " it certainly did not belong to the modern genus of that name."

[b] For this proverb and its explanation, *cf.* the similar passage *H.A.* 606 b 19 ff. Platt suggests that a mutilated passage in Hippocrates, π. ἀέρων ὑδάτων τόπων 12 *fin.*, contained a statement on this subject.

γινομέναις τοῖς δ' εὐεκτικωτέροις εἰς τὸ σῶμα
καταναλίσκεται τὸ περίττωμα τὸ σπερματικόν, καὶ
ταῖς μὲν οὐ γίνεται καταμήνια τοῖς δὲ γονή), ὁτὲ
30 δὲ διὰ νόσον οἱ μὲν ὑγρὸν καὶ ψυχρὸν προΐενται,
ταῖς δὲ γυναιξὶν αἱ καθάρσεις φαῦλαι καὶ πλήρεις
νοσηματικῶν περιττωμάτων. πολλοῖς δὲ καὶ πολλαῖς καὶ διὰ πηρώματα τοῦτο συμβαίνει τὸ πάθος
περὶ τὰ μόρια καὶ τοὺς τόπους τοὺς περὶ τὴν
ὁμιλίαν χρησίμους. γίνεται δὲ τὰ μὲν ἰατὰ τὰ δ'
ἀνίατα τῶν τοιούτων, μάλιστα δὲ διατελοῦσιν
35 ἄγονα ⟨τὰ⟩[1] κατὰ τὴν πρώτην σύστασιν τοιαῦτα

γενόμενα· γίνονται γὰρ γυναῖκές τε ἀρρενωποὶ καὶ
ἄνδρες θηλυκοί, καὶ ταῖς μὲν οὐ γίνεται τὰ καταμήνια, τοῖς δὲ τὸ σπέρμα λεπτὸν καὶ ψυχρόν.
διόπερ εὐλόγως βασανίζεται ταῖς πείραις τό γε
τῶν ἀνδρῶν, εἰ ἄγονον, ἐν τῷ ὕδατι· ταχὺ γὰρ
5 διαχεῖται τὸ λεπτὸν καὶ ψυχρὸν ἐπιπολῆς, τὸ δὲ
γόνιμον εἰς βυθὸν χωρεῖ· θερμὸν μὲν γὰρ τὸ πεπεμμένον ἐστί, πέπεπται δὲ τὸ συνεστηκὸς καὶ
πάχος ἔχον. τὰς δὲ γυναῖκας βασανίζουσι τοῖς τε
προσθέτοις, ἐὰν διικνῶνται αἱ ὀσμαὶ πρὸς τὸ
πνεῦμα τὸ θύραζε κάτωθεν ἄνω, καὶ τοῖς ἐγχρί-
10 στοις εἰς τοὺς ὀφθαλμοὺς χρώμασιν, ἂν χρωματί-
ζωσι τὸ ἐν τῷ στόματι πτύελον. ταῦτα γὰρ οὐ
συμβαίνοντα δηλοῖ τὸ σῶμα τοὺς πόρους δι' ὧν
ἀποκρίνεται τὸ περίττωμα συγκεχυμένους ἔχειν
καὶ συμπεφυκότας. ὅ τε γὰρ περὶ τοὺς ὀφθαλμοὺς
τόπος τῶν περὶ τὴν κεφαλὴν σπερματικώτατός

---

[1] τὰ supplevi : post σύστασιν P.

---

[a] And therefore might be expected to rise.

who are too well fed and in women who are too fat
the seminal residue is used up for the benefit of the
bodily system, so that no semen is formed in the men
and no menstrual discharge in the women ; sometimes (ii) because of disease ; the semen which the
men emit is fluid and cold, and the discharges of
the women are poor and full of morbid residues. But
in very many cases, in both sexes, this drawback is
due to deformities in the parts and regions employed
for intercourse. Some of these deformities are cur-
able, some are not ; those, however, who have become
deformed during the original constitution of the
embryo, have a special tendency to remain infertile
throughout ; thus, masculine-looking women are
produced in whom the menstrual discharges do not
occur, and effeminate men whose semen is thin and
cold. On this account the water-test is quite a fair
one for infertility in the male semen, because the
thin, cold semen quickly diffuses itself on the surface,
whereas the fertile semen sinks to the bottom ; for
though it is true that a substance which has been
concocted is hot,[a] yet that which has been set and
compacted and possesses thickness [b] has certainly
undergone concoction. Women are tested (a) by
means of pessaries : the test is whether the scent of
the pessary penetrates upwards from below to the
breath which is exhaled from the mouth ; (b) by
means of colours rubbed on to the eyes, the test being
whether they colour the saliva. If the required
result is not forthcoming, it is proved that the passages
of the body through which the residue is secreted
have got obstructed and have closed up, for of all the
regions in the head the eyes are the most seminal,

[b] As is shown by its sinking. *Cf.* 765 b 2.

747 a

15 ἐστιν. δηλοῖ δ' ἐν¹ ταῖς ὁμιλίαις μετασχηματιζό-
μενος ἐπιδήλως μόνος, καὶ τοῖς χρωμένοις πλείο-
σιν ἀφροδισίοις ἐνδιδόασι τὰ ὄμματα φανερῶς.
αἴτιον δ' ὅτι ἡ τῆς γονῆς φύσις ὁμοίως ἔχει τῇ
τοῦ ἐγκεφάλου· ὑδατώδης γάρ ἐστιν ἡ ὕλη αὐτῆς,
ἡ δὲ θερμότης ἐπίκτητος. καὶ αἱ σπερματικαὶ
20 καθάρσεις ἀπὸ τοῦ ὑποζώματός εἰσιν, ἡ γὰρ ἀρχὴ
τῆς φύσεως ἐντεῦθεν, ὥστε δικνεῖσθαι πρὸς τὸν
θώρακα τὰς κινήσεις ἀπὸ τῶν ἄρθρων· αἱ δ' ἐκ
τοῦ θώρακος ὀσμαὶ ποιοῦσιν αἴσθησιν διὰ τῆς
ἀναπνοῆς.

Ἐν μὲν οὖν τοῖς ἀνθρώποις καὶ τοῖς ἄλλοις
γένεσιν, ὥσπερ εἴρηται πρότερον, κατὰ μέρος ἡ
VIII 25 τοιαύτη συμβαίνει πήρωσις, τὸ δὲ τῶν ἡμιόνων
γένος ὅλον ἄγονόν ἐστιν. περὶ δὲ τῆς αἰτίας, ὡς
μὲν λέγουσιν Ἐμπεδοκλῆς καὶ Δημόκριτος, λέγων
ὁ μὲν οὐ σαφῶς, Δημόκριτος δὲ γνωρίμως μᾶλλον,
οὐ καλῶς εἰρήκασιν. λέγουσι γὰρ ἐπὶ πάντων
ὁμοίως τὴν ἀπόδειξιν τῶν παρὰ τὴν συγγένειαν
30 συνδυαζομένων. Δημόκριτος μὲν γάρ φησι δι-
εφθάρθαι τοὺς πόρους² τῶν ἡμιόνων ἐν ταῖς ὑστέραις διὰ τὸ μὴ ἐκ συγγενῶν γίνεσθαι τὴν ἀρχὴν
τῶν ζῴων. συμβαίνει δ' ἐφ' ἑτέρων ζῴων τοῦτο
μὲν ὑπάρχειν, γεννᾶν δὲ μηδὲν ἧττον· καίτοι χρῆν,
εἴπερ αἴτιον τοῦτ' ἦν, ἄγονα καὶ τἆλλ' εἶναι τὰ
35 μιγνύμενα τὸν τρόπον τοῦτον. Ἐμπεδοκλῆς δ'
747 b αἰτιᾶται τὸ μῖγμα τὸ τῶν σπερμάτων γίνεσθαι
πυκνὸν ἐκ μαλακῆς τῆς γονῆς οὔσης ἑκατέρας·
συναρμόττειν γὰρ τὰ κοῖλα τοῖς πυκνοῖς ἀλλήλων,

---

¹ ἐν P : ἐν μὲν vulg.   ² σπόρους Y (πόρους Z).

ᵃ Cf. Plato, Timaeus 91 A, B.

as is proved by the fact that this is the only region which unmistakably changes its appearance during sexual intercourse, and those who overfrequently indulge in it have noticeably sunken eyes. The reason is that the nature of the semen is similar to that of the brain[a]; its matter is watery whereas its heat is a mere supplementary acquisition.[b] Also the seminal discharges come from the diaphragm, because the first principle of the natural organism is there,[c] so that the movements initiated in the genital organs penetrate to the chest, and the scents from the chest become perceptible through the breathing.

As I said earlier, this particular deformity occurs Mules. in man and in the other kinds of animals to some extent, but with mules it is the whole race that is VIII infertile. What Empedocles has to say about the reason for this is obscure ; Democritus is more intelligible ; but they are both wrong. They give one omnibus explanation, covering all cases of copulation between animals of different kinds. Democritus[d] says that in mules the genital passages are destroyed in the uterus, because the formation of these animals has its origin in parents of different species. But we find this same situation with other animals, and yet they generate notwithstanding ; whereas, if Democritus's explanation was right, all other animals which unite in this way ought to be infertile too. The cause alleged by Empedocles is this : He says[e] *the mixture of the seeds becomes dense as a result of the two component portions of semen being both soft ; because the hollows of one fit into the densities of the other, and in*

---

[b] See Introd. § 69.   [c] See 719 a 14.
[d] See Diels, *Vorsokr.*[5] 68 A 151.
[e] Diels, *Vorsokr.*[5] 31 B 92 ; *cf.* 91 ; and 31 A 82.

ἐκ δὲ τῶν τοιούτων γίνεσθαι ἐκ μαλακῶν σκληρόν,
ὥσπερ τῷ καττιτέρῳ μιχθέντα τὸν χαλκόν, λέγων
οὔτ' ἐπὶ τοῦ χαλκοῦ καὶ τοῦ καττιτέρου τὴν αἰτίαν
5 ὀρθῶς (εἴρηται δ' ἐν τοῖς προβλήμασι περὶ αὐτῶν)
οὔθ' ὅλως ἐκ γνωρίμων ποιούμενος τὰς ἀρχάς. τὰ
γὰρ κοῖλα καὶ τὰ στερεὰ ἁρμόττοντα ἀλλήλοις πῶς
ποιεῖ τὴν μίξιν οἷον οἴνου καὶ ὕδατος; τοῦτο γὰρ
ὑπὲρ ἡμᾶς ἐστι τὸ λεγόμενον· πῶς γὰρ δεῖ λαβεῖν
10 τὰ κοῖλα τοῦ οἴνου καὶ τοῦ ὕδατος, λίαν ἐστὶ παρὰ
τὴν αἴσθησιν. ἔτι δ' ἐπειδὴ συμβαίνει καὶ ἐξ
ἵππων γίνεσθαι ἵππον καὶ ἐξ ὄνων ὄνον καὶ ἐξ
ἵππου καὶ ὄνου ἡμίονον, ἀμφοτέρως ἄρρενος καὶ
θήλεος ὁποτερουοῦν ὄντος, διὰ τί ἐκ μὲν τούτων
γίνεται πυκνὸν οὕτως ὥστ' ἄγονον εἶναι τὸ γενόμενον, ἐκ δ' ἵππου θήλεος καὶ ἄρρενος ἢ ὄνου
15 θήλεος καὶ ἄρρενος οὐ γίνεται ἄγονον; καίτοι
μαλακὸν καὶ τὸ τοῦ ἄρρενος ἵππου ἐστὶ καὶ τὸ
τοῦ θήλεος, μίγνυνται δὲ καὶ ὁ θῆλυς ἵππος καὶ ὁ
ἄρρην τῷ ὄνῳ, καὶ τῷ ἄρρενι καὶ τῷ θήλει. καὶ
διὰ τοῦτο γίνονται ἄγονα ἐξ ἀμφοτέρων, ὡς φησίν,
ὅτι ἐξ ἀμφοῖν ἕν τι γίνεται ⟨πυκνόν⟩,[1] μαλακῶν
20 ὄντων τῶν σπερμάτων. ἔδει οὖν καὶ τὸ ἐξ ἵππου
ἄρρενος καὶ θήλεος γινόμενον. εἰ μὲν γὰρ θάτερον
ἐμίγνυτο μόνον, ἐνῆν ἂν λέγειν ὅτι θάτερον αἴτιον
τοῦ μὴ γεννᾶν ἀνόμοιον ὂν[2] τῇ τοῦ ὄνου γονῇ· νῦν δ'
οἷάπερ οὔσῃ ἐκείνῃ μίγνυται, τοιαύτη καὶ τῇ τοῦ

---

[1] πυκνόν supplevi (πυκνόν τι pro ἕν τι Platt) : ὅτι . . . σπερμάτων om. Σ.
[2] ἀνόμοιον ὂν Platt (non assimilatur Σ): ὅμοιον vulg.:
ὅμοιον ὂν P (γεννᾶν ἡμίονον coniecerunt A.-W.).

such circumstances two softs give rise to one hard, just as bronze mixed with tin does. In the first place, he has got the reason wrong in the case of bronze and tin (see what I have written about this in the *Problems*),[a] and further, to put the objection generally, the principles from which he starts his argument are not intelligible.[b] How do the hollows and solids by " fitting on to one another " produce " the mixture as of wine and water " ? This saying of his is over our heads ; it is quite beyond our perception what we are to understand by the " hollows " of wine and water. Further, in point of fact, a horse is the off-spring of two horses, an ass of two asses, a mule of a horse and an ass—*i.e.*, its sire is a horse and its dam an ass or *vice versa*. Why is it then that a horse and an ass produce something so " dense " that the off-spring formed is infertile, whereas the offspring resulting from a male and female horse or from a male and female ass is not infertile ? After all, the secretion of both the male and of the female horse is " soft," and both sexes of the horse unite with asses of the opposite sex. The reason why in both these cases the offspring produced is infertile, according to Empedocles, is because the one product of the two soft " seeds " is something ⟨" dense "⟩. But then so it ought to be when the two seeds originate from two horses. If only one sex of the horse united with the ass, it would be open to Empedocles to say that the cause of the mule's infertility was the dissimilarity of that one sex to the semen of the ass. In fact, how-ever, there is no difference in quality between the seed of the ass with which it unites ⟨to form a mule⟩

---

[a] No such reference can be found.
[b] *Cf. Anal. Post.* 100 b 9.

συγγενοῦς. ἔτι δ' ἡ μὲν ἀπόδειξις κατ' ἀμφοτέρων
εἴρηται ὁμοίως καὶ[1] τοῦ θήλεος καὶ τοῦ ἄρρενος,
25 γεννᾷ δ' ὁ ἄρρην ἑπταέτης ὢν ἡμίονος,[2] ὥς φασίν·
ἀλλ' ἡ θήλεια ἄγονος ὅλως,[3] καὶ αὕτη τῷ μὴ ἐκτρέφειν εἰς τέλος, ἐπεὶ ἤδη κύημα ἔσχεν ἡμίονος.

Ἴσως δὲ μᾶλλον ἂν δόξειεν ἀπόδειξις εἶναι
πιθανὴ τῶν εἰρημένων λογική. λέγω δὲ λογικὴν
διὰ τοῦτο, ὅτι ὅσῳ καθόλου μᾶλλον, πορρωτέρω
30 τῶν οἰκείων ἐστὶν ἀρχῶν. ἔστι δὲ τοιαύτη τις.
εἰ γὰρ ἐξ ὁμοειδῶν ἄρρενος καὶ θήλεος ὁμοειδὲς
γίνεσθαι πέφυκε τοῖς γεννήσασιν ἄρρεν ἢ θῆλυ,
οἷον ἐκ κυνὸς ἄρρενος καὶ θήλεος κύων ἄρρην ἢ
θήλεια, καὶ ἐξ ἑτέρων τῷ εἴδει ἕτερον τῷ εἴδει,
οἷον εἰ κύων ἕτερον λέοντος, καὶ ἐκ κυνὸς ἄρρενος
35 καὶ λέοντος θήλεος ἕτερον καὶ ἐκ λέοντος ἄρρενος
καὶ κυνὸς θήλεος ἕτερον· ὥστ' ἐπειδὴ γίνεται
ἡμίονος ἄρρην καὶ θῆλυς ἀδιάφοροι ὄντες[4] τῷ
εἴδει ἀλλήλοις, γίνεται δ' ἐξ ἵππου καὶ ὄνου ἡμίονος, ἕτερα δ' ἐστὶ τῷ εἴδει ταῦτα καὶ οἱ ἡμίονοι,
ἀδύνατον γενέσθαι ἐξ ἡμιόνων· ἕτερον γὰρ γένος
5 οὐχ οἷόν τε διὰ τὸ ἐξ ἄρρενος καὶ θήλεος τῶν
ὁμοειδῶν ταὐτὸ γίνεσθαι τῷ εἴδει, ἡμίονος δ' ὅτι

---

[1] καὶ om. P, A.-W. : ὁμοίως hic om. A.-W., qui post ἄρρενος
inserunt, secuti cod. P, qui ibi ὁμοίως iterum, sed ὀρθῶς SYZ.
[2] ἡμίονος Peck : μόνος vulg. : Platt omisso (cum S) μόνος
scribit mox θήλεια ⟨μόνη⟩.
[3] ὅλως ἐκ παντός PYZ.
[4] correxi : ἀδιαφόρων ὄντων vulg.

---

[a] They are both " soft," according to Empedocles.
252

and the seed of an animal of its own species.[a] Further, Empedocles applies his argument equally to the male and the female. But, people say, the male mule does generate at the age of seven years ; it is the female which is totally infertile and that is simply because she fails to bring the nourishing of the fetation to its completion (as instances of fetations in mules have been known to occur).

Still, perhaps an abstract argument might be considered more convincing than those which we have already mentioned. I call it an abstract one, because in so far as it is a more general argument it is further removed from those principles which belong to this particular subject. It goes somewhat like this. In the normal course of nature the offspring which a male and a female of the same species produce is a male or female of that same species—for instance, the offspring of a male dog and a female dog is a male dog or a female dog. Two animals which differ in species produce offspring which differs in species ; for instance, a dog differs in species from a lion, and the offspring of a male dog and a female lion is different in species ; so is the offspring of a male lion and a female dog. This being so, it follows that as both male and female mules are produced, which of course do not differ in species, and as a mule is the offspring produced by a horse and an ass, both of which are different in species from the mule, it is impossible for any offspring to be produced by mules ; the reason being : (a) no offspring of a different species can be produced by them, because the offspring of two animals male and female of the same species belongs itself to that species, nor (b) can a mule be produced, because that is the offspring of a horse and an

748 a

ἐξ ἵππου καὶ ὄνου γίνεται ἑτέρων ὄντων τῷ εἴδει
[ἐκ δὲ τῶν ἑτέρων τῷ εἴδει ἕτερον ἐτέθη γίνεσθαι
ζῷον].¹ οὗτος μὲν οὖν ὁ λόγος καθόλου λίαν καὶ
κενός. οἱ γὰρ μὴ ἐκ τῶν οἰκείων ἀρχῶν λόγοι
κενοί, ἀλλὰ δοκοῦσιν εἶναι τῶν πραγμάτων οὐκ
10 ὄντες. οἱ γὰρ ἐκ τῶν ἀρχῶν τῶν γεωμετρικῶν
γεωμετρικοί, ὁμοίως δὲ καὶ ἐπὶ τῶν ἄλλων· τὸ δὲ
κενὸν δοκεῖ μὲν εἶναί τι, ἔστι δ' οὐθέν. οὐκ ἀληθὲς
δέ, ὅτι πολλὰ τῶν μὴ ⟨ἐξ⟩² ὁμοειδῶν γενομένων
γίνεται γόνιμα, καθάπερ ἐλέχθη πρότερον. τοῦτον
μὲν οὖν τὸν τρόπον οὔτε περὶ τῶν ἄλλων δεῖ ζητεῖν
15 οὔτε περὶ τῶν φυσικῶν· ἐκ δὲ τῶν ὑπαρχόντων τῷ
γένει τῷ τῶν ἵππων καὶ τῷ τῶν ὄνων θεωρῶν ἄν
τις μᾶλλον λάβοι τὴν αἰτίαν, ὅτι πρῶτον μὲν
ἑκάτερον αὐτῶν ἐστι μονοτόκον ἐκ τῶν συγγενῶν
ζῴων, ἔπειτ' οὐ συλληπτικὰ τὰ θήλεα ἐκ τῶν
ἀρρένων ἀεί, διόπερ τοὺς ἵππους διαλείποντες
20 ὀχεύουσι [διὰ τὸ μὴ δύνασθαι συνεχῶς φέρειν].³
ἀλλ' ἡ μὲν ἵππος οὐ καταμηνιώδης, ἀλλ' ἐλάχιστον
προΐεται τῶν τετραπόδων· ἡ δ' ὄνος οὐ δέχεται τὴν
ὀχείαν, ἀλλ' ἐξουρεῖ τὸν γόνον, διὸ μαστιγοῦσιν
ἀκολουθοῦντες. ἔτι δὲ ψυχρὸν τὸ ζῷον [ὁ ὄνος]⁴
ἐστί, διόπερ ἐν τοῖς χειμερινοῖς οὐ θέλει γίνεσθαι
25 τόποις διὰ τὸ δύσριγον εἶναι τὴν φύσιν, οἷον περὶ
Σκύθας καὶ τὴν ὅμορον χώραν, οὐδὲ περὶ Κελτοὺς
τοὺς ὑπὲρ τῆς Ἰβηρίας· ψυχρὰ γὰρ καὶ αὕτη ἡ

---

¹ ἐκ δὲ . . . ζῷον vulg. : eicit Platt.
² ἐξ supplevi (ἐκ post τῶν Ζ²).
³ seclusit Platt : habet vulg., Σ.   ⁴ seclusit Btf.

ᵃ Cf. H.A. 577 a 23.

254

ass, two animals which differ in species [and it was laid down that an animal of a different species is produced by two animals that differ in species]. Now this argument is too general ; there is nothing in it, because there is nothing in any argument which does not start from the first principles belonging to the particular subject. Such arguments may appear to be relevant, but in fact they are not. For a geometrical argument, you must start from geometrical principles, and the same applies elsewhere ; that which is empty, which has nothing in it, may appear to be somewhat but in fact is nothing at all. But also, this argument is false, because many of the animals that are produced from parents of differing species are fertile, as I have said earlier. No ; this method of inquiry is as wrong in natural science as it is elsewhere. We shall be more likely to discover the reason we are looking for if we consider the actual facts with regard to the two species, horse and ass. First, then, both horse and ass, when mated with their own kind, produce only one at a birth ; secondly, the females do not on every occasion conceive when covered by the male, and that is why breeders after an interval put the horse to the mare again [because the mare cannot bear it continuously]. Mares do not produce a large amount of menstrual discharge ; indeed they discharge less than any other quadruped ; she-asses too do not admit the impregnation, but pass the semen out with their urine ; and that is why people follow behind, flogging them.[a] Further, the animal is a cold subject ; and as it is by nature so sensitive to cold, it is not readily produced in wintry regions, such as Scythia and the neighbouring parts, or the Keltic country beyond Iberia, which is also a

748 a

χώρα. διὰ ταύτην δὲ τὴν αἰτίαν καὶ τὰ ὀχεῖα
ἐπιβάλλουσι τοῖς ὄνοις οὐχ ὥσπερ τοῖς ἵπποις κατ᾽
ἰσημερίαν, ἀλλὰ περὶ τροπὰς θερινάς, ὅπως ἐν
30 ἀλεεινῇ γίνηται ὥρα τὰ πωλία (ἐν τῇ αὐτῇ γὰρ
γίνεται ἐν ᾗ ἂν ὀχευθῇ· ἐνιαυτὸν γὰρ κύει καὶ
ἵππος καὶ ὄνος). ὄντος δ᾽ ὥσπερ εἴρηται ψυχροῦ
τὴν φύσιν, καὶ τὴν γονὴν ἀναγκαῖον εἶναι τοῦ
τοιούτου ψυχράν. (σημεῖον δὲ τούτου· διὰ τοῦτο
γάρ, ἐὰν μὲν ἵππος ἀναβῇ ἐπὶ ὠχευμένην ὑπὸ
ὄνου, οὐ διαφθείρει τὴν τοῦ ὄνου ὀχείαν, ὁ δ᾽ ὄνος
35 ἐὰν ἐπαναβῇ, διαφθείρει τὴν τοῦ ἵππου διὰ
748 b ψυχρότητα τὴν τοῦ σπέρματος.) ὅταν μὲν οὖν
ἀλλήλοις μιχθῶσι, σώζεται διὰ τὴν θατέρου θερμότητα, θερμότερον γὰρ τὸ ἀπὸ τοῦ ἵππου ἀπο-
κρινόμενον· ἡ μὲν γὰρ τοῦ ὄνου ψυχρὰ καὶ ἡ ὕλη
καὶ ἡ γονή, ἡ δὲ τοῦ ἵππου θερμοτέρα. ὅταν δὲ
5 μιχθῇ ἢ θερμὸν ἐπὶ ψυχρὸν ἢ ψυχρὸν ἐπὶ θερμόν,
συμβαίνει αὐτὸ μὲν τὸ ἐκ τούτων κύημα γενόμενον[1]
σώζεσθαι καὶ ταῦτ᾽ ἐξ ἀλλήλων εἶναι γόνιμα, τὸ
δ᾽ ἐκ τούτων μηκέτι γόνιμον ἀλλ᾽ ἄγονον εἰς
τελειογονίαν.

Ὅλως δ᾽ ὑπάρχοντος ἑκατέρου εὐφυοῦς πρὸς
ἀγονίαν, τῷ τε γὰρ ὄνῳ ὑπάρχει τὰ ἄλλα τὰ εἰρη-
10 μένα, καὶ ἐὰν μὴ μετὰ τὸν βόλον τὸν πρῶτον
ἄρξηται γεννᾶν, οὐκέτι γεννᾷ τὸ παράπαν· οὕτως
ἐπὶ[2] μικροῦ ἔχεται τοῦ[3] ἄγονον εἶναι τὸ σῶμα τῶν
ὄνων. ὁμοίως δὲ καὶ ὁ ἵππος· εὐφυὴς γὰρ πρὸς

---

[1] γεν- S*PYZ[1] : γιν- vulg.
[2] ἐπὶ om. Z[1].   [3] τοῦ PZ[2], Platt : τὸ vulg.

---

[a] i.e., a mare ; cf. H.A. 577 a 13, 28.
[b] According to H.A. 577 a 18, this happens at the age of
2½ years ; see also 545 b 20.

cold quarter. For this reason they do not put the jack-asses to the females at the equinox, as is done with horses, but at the time of the summer solstice, so that the asses' foals may be born when the weather is warm. (Since the period of gestation in both horse and ass is a year, the young are born at the same season as that when impregnation takes place.) As has been said, the ass is by nature cold ; and a cold animal's semen is, of necessity, cold like itself. (Here is a proof of it. If a horse mounts a female [a] which has been impregnated by an ass, he does not destroy the ass's impregnation ; but if an ass mounts her after a horse has done so, he does destroy the horse's impregnation—because of the coldness of his own semen.) Thus when they unite with each other, the impregnation remains intact by reason of the heat resident in one of the two, viz., that of the horse, whose secretion is the hotter. Both the semen from the male and the matter supplied by the female are hotter in the case of the horse ; with the ass, both are cold. So when they unite—either the hot one added to the cold, or the cold added to the hot—the result is (a) that the fetation which is formed by them continues intact, i.e., these two animals are fertile when crossed with each other, but (b) the animal formed by them is not itself fertile, and cannot pro-duce perfect offspring.

Besides, both horse and ass have a general natural disposition to be infertile. I have already mentioned several points about the ass, and another is that unless it begins to generate after the first shedding of teeth,[b] it never generates at all ; so close does the ass come to being infertile. It is the same with the horse ; it is naturally disposed to be infertile ; all

τὴν ἀγονίαν, καὶ τοσοῦτον λείπει τοῦ ἄγονος εἶναι
ὅσον τὸ γενέσθαι τὸ ἐκ τούτου ψυχρότερον· τοῦτο
δὲ γίνεται, ὅταν μιχθῇ τῇ τοῦ ὄνου ἀποκρίσει. καὶ
15 ὁ ὄνος δὲ ὡσαύτως μικροῦ δεῖν κατὰ τὸν οἰκεῖον
συνδυασμὸν ἄγονον γεννᾷ, ὥστε ὅταν προσγένηται
τὸ παρὰ φύσιν, εἰ τότε ἑνὸς μόλις γεννητικὸν ἐξ
ἀλλήλων ἦν, τὸ ἐκ τούτων ἔτι μᾶλλον ἄγονον καὶ
παρὰ φύσιν οὐθενὸς δεήσει τοῦ ἄγονον εἶναι, ἀλλ'
ἐξ ἀνάγκης ἔσται ἄγονον.
20 Συμβαίνει δὲ καὶ τὰ σώματα τὰ τῶν ἡμιόνων
μεγάλα γίνεσθαι διὰ τὸ τὴν ἀπόκρισιν τὴν εἰς τὰ
καταμήνια τρέπεσθαι εἰς τὴν αὔξησιν. ἐπεὶ δ'
ἐνιαύσιος ὁ τοκετὸς τῶν τοιούτων, οὐ μόνον συλλαβεῖν δεῖ τὴν ἡμίονον ἀλλὰ καὶ ἐκθρέψαι· τοῦτο
δ' ἀδύνατον μὴ γινομένων καταμηνίων. ταῖς δ'
25 ἡμιόνοις οὐ γίνεται, ἀλλὰ τὸ μὲν ἄχρηστον μετὰ
τοῦ περιττώματος τοῦ ἐκ τῆς κύστεως ἐκκρίνεται
(διόπερ οὐδὲ τῶν ἄρθρων οἱ ἡμίονοι οἱ ἄρρενες
ὀσφραίνονται τῶν θηλειῶν, ὥσπερ τἆλλα τὰ μώνυχα, ἀλλ' αὐτοῦ τοῦ περιττώματος), τὰ δ' ἄλλα
τρέπεται εἰς τὴν τοῦ σώματος[1] αὔξησιν καὶ τὸ
μέγεθος. ὥστε συλλαβεῖν μὲν ἐνδέχεταί ποτε τὴν
30 θήλειαν, ὅπερ ἤδη φαίνεται γεγονός, ἐκθρέψαι δὲ
καὶ ἐξενεγκεῖν εἰς τέλος ἀδύνατον. ὁ δ' ἄρρην
ποτὲ γεννήσειεν ἂν διά τε τὸ θερμότερον εἶναι τοῦ
θήλεος φύσει τὸ ἄρρεν, καὶ διὰ τὸ μὴ συμβάλ-

---

[1] τοῦ σώματος P, Platt : om. vulg.

---

that is wanting to make it such is that its secretion should be colder, and this occurs when it is united with that of the ass. In the same way the ass comes within an ace of generating infertile offspring even when it mates with its own kind ; so that when there is the additional factor of unnatural mating beside the difficulty it has in producing even a single young one in the normal way, the resultant offspring is still more infertile and unnatural ; in fact, it will lack nothing to make it completely infertile, and will be infertile of necessity.

Furthermore, female mules grow large in size. This is because the secretion intended for the menstrual flow is diverted to produce growth. And since the period of gestation in such animals lasts a year, the female mule not only has to conceive but has to nourish the embryo all that time ; and this is impossible unless menstrual flow is being produced. None is produced in mules : the unserviceable part of the nourishment is passed out together with the residue that comes from the bladder (which explains why male mules do not smell at the pudenda of the females as the other solid-hoofed animals do, but at the residue itself) ; the rest of the nourishment is diverted to growth of the body and to size. Hence although it is possible for the female to conceive occasionally—and indeed the fact is established that this has happened—it is impossible for her to nourish an embryo for the full period and bring it to the birth. The male may occasionally generate (a) because ᵃ the male is by nature hotter than the female, and (b) because the male does not contribute any corporeal

and are cited here to explain how the male mule may be able to generate.

748 b

749 a

λεσθαι πρὸς τὴν μίξιν σῶμα μηδὲν τὸ ἄρρεν. τὸ
δ᾽ ἀποτελεσθὲν γίνεται γίννος. τοῦτο δ᾽ ἐστὶν
35 ἡμίονος ἀνάπηρος· καὶ γὰρ ἐκ τοῦ ἵππου καὶ τοῦ
ὄνου γίνονται γίννοι, ὅταν νοσήσῃ τὸ κύημα ἐν τῇ
ὑστέρᾳ. ἔστι γὰρ ὁ γίννος ὥσπερ τὰ μετάχοιρα
ἐν τοῖς χοίροις· καὶ γὰρ ἐκεῖ τὸ πηρωθὲν ἐν τῇ
ὑστέρᾳ καλεῖται μετάχοιρον. γίνεται δὲ τοιοῦτος
ὃς ἂν τύχῃ τῶν χοίρων. ὁμοίως δὲ γίνονται καὶ
5 οἱ πυγμαῖοι· καὶ γὰρ οὗτοι πηροῦνται τὰ μέρη
καὶ τὸ μέγεθος ἐν τῇ κυήσει, καὶ εἰσὶν ὥσπερ
μετάχοιρα καὶ γίννοι.

ᵃ According to *H.A.* 577 b 21, a *ginnos* is the offspring of
a mule and a mare ; and there, as here, a *ginnos* is also said
to be the *diseased* offspring of a mare, and is compared with
dwarfs and *metachoira*. Aristotle thus compares the product
of the union of mule and mare with the *diseased* or *deformed*

ingredient to the mixture. The final result which is produced is a *ginnos.*[a] This is a deformed mule, for *ginnoi* are produced also from the horse and the ass when the fetation gets diseased in the uterus, the *ginnos* being comparable to the *metachoiron* which occurs among swine, since in that case too it is the offspring which has been deformed in the uterus that is called a *metachoiron* : any pig may happen to be born thus deformed. Human dwarfs too are formed in a similar way : they too become deformed in their parts and stunted in size during the time of gestation, and thus are comparable with *metachoira* and *ginnoi*.

offspring which sometimes result from the union of male and female of one and the same species. For *metachoira* see also 770 b 7.

Γ

I Περὶ μὲν οὖν τῆς τῶν ἡμιόνων ἀτεκνίας εἴρηται,
καὶ περὶ τῶν ζῳοτοκούντων καὶ θύραζε καὶ ἐν
αὑτοῖς· ἐν δὲ τοῖς ᾠοτοκοῦσι τῶν ἐναίμων τῇ μὲν
παραπλησίως ἔχει τὰ περὶ τὰς γενέσεις αὑτοῖς τε
καὶ τοῖς πεζοῖς καὶ ταὐτόν τι λαβεῖν ἔστι περὶ
πάντων, τῇ δ' ἔχει διαφορὰς καὶ πρὸς ἄλληλα καὶ
15 πρὸς τὰ πεζὰ τῶν ζῴων. γίνεται μὲν οὖν ἀπὸ
συνδυασμοῦ πάντα ὅλως, καὶ προϊεμένου γονὴν εἰς
τὸ θῆλυ τοῦ ἄρρενος· τῶν δ' ᾠοτοκούντων αἱ μὲν
ὄρνιθες προΐενται τέλειον ᾠὸν καὶ σκληρόδερμον,
ἐὰν μή τι πηρωθῇ διὰ νόσον, καὶ πάντα δίχροα τὰ
τῶν ὀρνίθων ἐστίν, τῶν δ' ἰχθύων οἱ μὲν σελαχώ-
20 δεις, ὥσπερ εἴρηται πολλάκις, ἐν αὑτοῖς ᾠοτοκή-
σαντες ζῳοτοκοῦσι, μεταστάντος τοῦ ᾠοῦ ἐξ ἄλλου
τόπου τῆς ὑστέρας εἰς ἄλλον, μαλακόδερμον δὲ
τὸ ᾠὸν καὶ ὁμόχρων ἐστὶν αὐτῶν. εἷς δὲ μόνος
οὐ ζῳοτοκεῖ τῶν τοιούτων ἐν αὑτῷ, ὁ καλούμενος
βάτραχος· περὶ οὗ τὴν αἰτίαν ὕστερον λεκτέον. οἱ
25 δὲ ἄλλοι ὅσοιπερ ᾠοτοκοῦσι τῶν ἰχθύων, μονόχρων

---

[a] Although most Ovipara are flying or swimming animals,
some of course are πεζά, but by πεζά Aristotle here means
viviparous animals only.

[b] i.e., an egg which does not increase in size after deposition ; see below, l. 25.

[c] i.e., there is no difference of yolk and white.

262

BOOK III

I. Blooded animals (continued). Ovipara :

WE have spoken about the sterility of mules, and about the animals which are viviparous both externally and internally. We now pass on to those blooded animals which are oviparous. The phenomena of generation here are on the one hand similar to those which obtain in the animals that walk,[a] so that the same statement will serve for all of them ; on the other hand, these animals exhibit certain differences not only as between themselves, but also when compared with the animals that walk. Their generation is the result of copulation, *i.e.*, of the emission of semen into the female by the male : this applies to all of them, of course. But beyond that there are variations : (*a*) Birds produce a perfect [b] egg with a hard shell (unless it be deformed by disease). All birds' eggs are of two colours. (*b*) The Selachian fishes, as I have often repeated, are internally oviparous but bring forth their young alive, after the egg has moved from one position in the uterus to another. Their egg is soft-shelled and of one colour only.[c] The fish known as the fishing-frog [d] is the only one in this class that is not internally viviparous. The cause of this will have to be stated later.[e] (*c*) All other fishes that are oviparous pro-

[d] Probably *Lophius piscatorius* ; see 754 a 26, n.
[e] At 754 a 25-31.

μὲν προΐενται τὸ ᾠόν, ἀτελὲς δὲ τοῦτο· λαμβάνει
γὰρ ἔξω τὴν αὔξησιν, διὰ τὴν αὐτὴν αἰτίαν δι’
ἥνπερ καὶ τὰ ἔσω τελειούμενα τῶν ᾠῶν.

Περὶ μὲν οὖν τῶν ὑστερῶν, τίνας ἔχουσι διαφορὰς καὶ διὰ τίνας αἰτίας, εἴρηται πρότερον.
καὶ γὰρ τῶν ζῳοτοκούντων τὰ μὲν ἄνω πρὸς τῷ
30 ὑποζώματι ἔχει τὰς ὑστέρας, τὰ δὲ κάτω πρὸς
τοῖς ἄρθροις, ἄνω μὲν τὰ σελαχώδη, κάτω δὲ τὰ
καὶ ἐν αὑτοῖς ζῳοτοκοῦντα καὶ θύραζε, οἷον ἄνθρωπος καὶ ἵππος καὶ τῶν ἄλλων ἕκαστον τῶν
τοιούτων. καὶ τῶν ᾠοτοκούντων τὰ μὲν κάτω,
καθάπερ τῶν ἰχθύων οἱ ᾠοτοκοῦντες, τὰ δ’ ἄνω,
καθάπερ οἱ ὄρνιθες.

35 Συνίσταται μὲν οὖν κυήματα τοῖς ὄρνισι καὶ
αὐτόματα, ἃ καλοῦσιν ὑπηνέμια καὶ ζεφύριά τινες,
γίνεται δὲ ταῦτα τοῖς μὴ πτητικοῖς μηδὲ γαμψώ-
νυξι τῶν ὀρνίθων, ἀλλὰ τοῖς πολυγόνοις, διὰ τὸ
πολὺ περίττωμα ταῦτ’ ἔχειν (τοῖς δὲ γαμψώνυξιν
εἰς τὰς πτέρυγας καὶ τὰ πτερὰ τρέπεσθαι τὴν
5 τοιαύτην ἀπόκρισιν, τὸ δὲ σῶμα μικρὸν ἔχειν καὶ
ξηρόν τε καὶ θερμόν[1]), τὴν δ’ ἀπόκρισιν τὴν κατα-
μηνιώδη καὶ τὴν γονὴν περίττωμα εἶναι· ἐπεὶ οὖν
καὶ ἡ τῶν πτερῶν φύσις καὶ ἡ τοῦ σπέρματος
γίνεται ἐκ περιττώσεως, οὐ δύναται ἡ φύσις ἐπ’
ἀμφότερα πολυχοεῖν. διὰ τὴν αὐτὴν δὲ ταύτην
10 αἰτίαν[2] τὰ μὲν γαμψώνυχα οὔτ’ ὀχευτικά ἐστιν

---

[1] acutum Σ.   [2] καὶ post αἰτίαν codd. : del. Platt.

---

[a] i.e., the cause which controls the growth of the egg to
perfection.

duce an egg of one colour only, but this egg is imperfect—its growth takes place away from the parent, and the Cause concerned [a] is just the same as for those eggs which are perfected within the parent.

I have already spoken about the uterus of these animals ; I have said what are the differences they show, and what are the Causes. Thus, some of the viviparous animals (the Selachian fishes) have the uterus high up towards the diaphragm,[b] others (the animals which are both internally and externally viviparous, such as man, horse, and all such animals) have it down by the pudenda. And of the oviparous animals some (such as the oviparous fishes) have it low down, others (such as the birds) have it high up.

Fetations arise in birds spontaneously as well ⟨as (i.) Birds. in the normal way⟩ ; some people call them wind-eggs or *zephyria*.[c] They occur in those birds [d] which are neither good fliers nor crook-taloned but which are prolific.[e] The reason is : (*a*) these have a great deal of residue, whereas in the crook-taloned birds this secretion is diverted to produce wings and wing feathers and their body is small [f] and solid and hot ; and (*b*) the menstrual secretion and the male semen are residue ; therefore, as both feathers and semen alike are formed out of residue, Nature cannot provide a large supply for both purposes. And it is for this same cause that the crook-taloned birds do not indulge much in copulation and are not very prolific,

[b] See note on 717 a 2.
[c] See note on 753 a 22.
[d] See table of birds, p. 368.
[e] *i.e.*, produce a large number of eggs (or young). I use " prolific " throughout to translate πολύγονος and πολυτόκος.
[f] For the smallness of the body of crook-taloned birds (apart from their wings), *cf. P.A.* 694 a 8 f.

οὔτε πολύγονα, τὰ δὲ βαρέα καὶ τῶν πτητικῶν
ὅσων τὰ σώματα ὀγκώδη, καθάπερ περιστερᾶς
καὶ τῶν τοιούτων. τοῖς μὲν γὰρ βαρέσι καὶ μὴ
πτητικοῖς, οἷον ἀλεκτορίσι καὶ πέρδιξι καὶ τοῖς
ἄλλοις τοῖς τοιούτοις, πολὺ γίνεται περίττωμα
15 τοιοῦτον· διὸ τά τε ἄρρενα αὐτῶν ὀχευτικὰ καὶ τὰ
θήλεα προΐεται πολλὴν ὕλην, καὶ τίκτει τῶν τοιούτων τὰ μὲν πολλὰ τὰ δὲ πολλάκις, πολλὰ μὲν οἷον
ἀλεκτορὶς καὶ πέρδιξ καὶ στρουθὸς ὁ Λιβυκός, τὰ
δὲ περιστερώδη πολλὰ μὲν οὔ, πολλάκις δέ· μεταξὺ
γάρ ἐστι ταῦτα τῶν γαμψωνύχων καὶ τῶν βαρέων·
20 πτητικὰ μὲν γάρ ἐστιν ὥσπερ τὰ γαμψώνυχα,
πλήθη δ' ἔχει τοῦ σώματος ὥσπερ τὰ βαρέα, ὥστε
διὰ μὲν τὸ πτητικὰ εἶναι καὶ ἐνταῦθα τρέπεσθαι τὸ
περίττωμα ὀλίγα τίκτουσι, διὰ δὲ τὸ πλῆθος τοῦ
σώματος καὶ διὰ τὸ θερμὴν ἔχειν τὴν κοιλίαν καὶ
πεπτικωτάτην, πρὸς δὲ τούτοις καὶ διὰ τὸ ῥᾳδίως
25 πορίζεσθαι τὴν τροφήν, τὰ δὲ γαμψώνυχα χαλεπῶς,
πολλάκις.

Ὀχευτικὰ δὲ καὶ πολύγονα καὶ τὰ μικρὰ τῶν
ὀρνέων ἐστί, καθάπερ ἐνίοτε καὶ τῶν φυτῶν· ἡ
γὰρ εἰς τὸ σῶμα αὔξησις γίνεται περίττωμα σπερματικόν. διὸ καὶ τῶν ἀλεκτορίδων αἱ Ἀδριανικαὶ
πολυτοκώταταί εἰσιν· διὰ γὰρ μικρότητα τοῦ σώ-
30 ματος εἰς τὴν τέκνωσιν καταναλίσκεται ἡ τροφή.
καὶ αἱ ἀγεννεῖς τῶν γενναίων πολυτοκώτεραι·
ὑγρότερα γὰρ τὰ σώματα τῶνδε καὶ[1] ὀγκωδέσ-

---

[1] τῶνδε καὶ vulg.: τῶν δὲ Y: αὐτῶν τῶν δὲ PZ: αὐτῶν
καὶ A.-W.

---

[a] Mentioned also at H.A. 558 b 17. Thompson (Glossary[2],
ἀλεκτρύων) considers them as a kind of bantam.

266

whereas the heavy birds and those fliers which have bulky bodies (such as pigeons and the like) do so. In those birds which are heavy and are not fliers, such as common fowls, partridges, and the like, a great deal of this residue is formed, and that is why their males copulate frequently and their females emit a great deal of matter; also, some birds of this sort lay many eggs, some lay many times; thus the common fowl, the partridge and the ostrich lay a large number; whereas the pigeon family do not lay a large number, but lay many times, the reason being that the last-named stand midway between the crook-taloned birds and the heavy birds; they are fliers, like the former, and have a bulky body, like the latter. The result is: (1) As they are fliers, the residue is diverted to their wings; hence they lay but few eggs; (2) they are bulky in build, their stomach is hot and very good at concoction, and, in addition, they can easily get their food, whereas the crook-taloned birds have difficulty in getting it; hence they lay often.

Small birds, too, copulate frequently and are very prolific, just as some small plants are: the material which might produce increase of bulk turns into seminal residue. On this account the Adrianic fowls [a] are extremely prolific; as they are small in size, the nourishment is used up for the production of offspring. Also, low-bred birds are more prolific than high-bred ones,[b] because their bodies are more

---

[b] Thompson's terms (*loc. cit.*). The definition of γενναῖος is given at *H.A.* 488 b 18 ff.: εὐγενὲς μὲν γάρ ἐστι τὸ ἐξ ἀγαθοῦ γένους, γενναῖον δὲ τὸ μὴ ἐξιστάμενον ἐκ τῆς αὐτοῦ φύσεως, whence it appears that γενναῖος = " thoroughbred," as Thompson there translates it.

τερα, τῶν δὲ ἰσχνότερα καὶ ξηρότερα· ὁ γὰρ θυμὸς
ὁ γενναῖος ἐν τοῖς τοιούτοις γίνεται σώμασι μᾶλλον.
35 ἔτι δὲ καὶ ἡ τῶν σκελῶν λεπτότης καὶ ἀσθένεια
συμβάλλεται πρὸς τὸ τὴν φύσιν τῶν τοιούτων

ὀχευτικὴν εἶναι καὶ πολύγονον, καθάπερ καὶ ἐπὶ
τῶν ἀνθρώπων· ἡ γὰρ εἰς τὰ κῶλα τροφὴ τρέπεται
τοῖς τοιούτοις εἰς περίττωμα σπερματικόν· ὃ γὰρ
ἐκεῖθεν ἀφαιρεῖ ἡ φύσις, προστίθησιν ἐνταῦθα. τὰ
5 δὲ γαμψώνυχα τὴν βάσιν ἰσχυρὰν ἔχει καὶ τὰ
σκέλη πάχος ἔχοντα διὰ τὸν βίον· ὥστε διὰ πάσας
ταύτας τὰς αἰτίας οὔτ᾽ ὀχευτικά ἐστιν οὔτε πολύγονα. μάλιστα δὲ ἡ κεγχρὶς πολύγονον· μόνον
γὰρ σχεδὸν τοῦτο καὶ πίνει τῶν γαμψωνύχων, ἡ
δ᾽ ὑγρότης καὶ ἡ σύμφυτος καὶ ἡ ἐπακτὸς σπερ-
10 ματικὸν μετὰ τῆς ὑπαρχούσης αὐτῇ θερμότητος.
τίκτει δ᾽ οὐδ᾽ αὕτη[1] πολλὰ λίαν, ἀλλὰ τέτταρα τὸ
πλεῖστον.

Ὁ δὲ κόκκυξ ὀλιγοτόκον ἐστὶν οὐκ ὢν γαμψώνυ-
χος, ὅτι ψυχρὸς τὴν φύσιν ἐστίν (δηλοῖ δ᾽ ἡ δειλία
τοῦ ὀρνέου), τὸ δὲ σπερματικὸν ζῷον δεῖ θερμὸν
καὶ ὑγρὸν εἶναι. ὅτι δὲ δειλόν, φανερόν· ὑπό τε
15 γὰρ τῶν ὀρνέων διώκεται πάντων καὶ ἐν ἀλλοτρίαις
τίκτει νεοττιαῖς.

Τὰ δὲ περιστερώδη δύο ὡς τὰ πολλὰ τίκτειν
εἴωθεν· οὔτε γὰρ μονοτόκοι εἰσὶν (οὐθεὶς γὰρ
μονοτόκος ὄρνις πλὴν ὁ κόκκυξ, καὶ οὗτος ἐνίοτε
διτοκεῖ) οὔτε πολλὰ τίκτουσιν, ἀλλὰ πολλάκις δύο

---

[1] αὕτη Peck : αὐτὴ vulg.

---

[a] For " solid " and " fluid " see Introd. § 38.
[b] Cf. the remarks on the chameleon at P.A. 692 a 22 ff.;

fluid and more bulky, whereas those of the high-bred birds are leaner and more solid,[a] this being the kind of body in which a thoroughbred and high-spirited temper tends rather to make its appearance ; also the thinness and weakness of their legs contribute towards making these birds prone to copulation and prolific—and this applies also to human beings : the nourishment which was intended for the legs is in such cases diverted to the seminal residue : what Nature takes away from one place she puts on at the other. The crook-taloned birds, on the other hand, have strong feet, and their legs are thick : this is due to their manner of life ; thus on account of all these causes they do not copulate much nor are they very prolific. The kestrel is the most prolific of them, for this is practically the only one of the crook-taloned birds which drinks, and the fluid, both that which is innate and that which it gets from without, is productive of semen when combined with the heat which is present in it. Even this bird does not lay many eggs ; four at the most.

The cuckoo lays but few eggs although it is not a crook-taloned bird, because it is cold by nature (as its cowardice [b] clearly shows), whereas an animal that is abundant in semen must be hot and fluid. That it is cowardly is shown by the fact that all other birds chase it and that it lays its eggs in other birds' nests.

Most birds of the pigeon kind usually lay a couple of eggs. They are neither one-egg birds (there is no one-egg bird beside the cuckoo, and this sometimes lays two), nor do they lay a large number ; but they

also 650 b 28 (ὁ γὰρ φόβος καταψύχει) and 667 a 17 ff., where a large heart is said to produce cowardice because the heart is so large that the heat is lost in so large a space.

ἢ τρία τὰ πλεῖστα γεννῶσι, τὰ δὲ πολλὰ δύο·
20 οὗτοι γὰρ οἱ ἀριθμοὶ μεταξὺ τοῦ ἑνὸς καὶ πολλῶν.

Ὅτι δὲ τοῖς πολυγόνοις τρέπεται εἰς τὸ σπέρμα
ἡ τροφή, φανερὸν ἐκ τῶν συμβαινόντων. τῶν τε
γὰρ δένδρων τὰ πολλὰ πολυκαρπήσαντα λίαν ἐξ-
αυαίνεται μετὰ τὴν φοράν, ὅταν μὴ ὑπολειφθῇ τῷ
σώματι τροφή, καὶ τὰ ἐπέτεια ταῦτὸ πάσχειν
25 ἔοικεν, οἷον τά τε χεδροπὰ καὶ ὁ σῖτος καὶ τἆλλα
τὰ τοιαῦτα· τὴν γὰρ τροφὴν ἀναλίσκουσιν εἰς τὸ
σπέρμα πᾶσαν· ἔστι γὰρ πολύσπερμον τὸ γένος
αὐτῶν. καὶ τῶν ἀλεκτορίδων ἔνιαι πολυτοκήσασαι
λίαν οὕτως ὥστε καὶ δύο τεκεῖν ἐν ἡμέρᾳ, μετὰ
τὴν πολυτοκίαν ἀπέθανον. ὑπέρινοι γὰρ γίνονται
30 καὶ οἱ ὄρνιθες καὶ τὰ φυτά· τοῦτο δ' ἐστὶ τὸ πάθος
ὑπερβολὴ περιττώματος ἐκκρίσεως. αἴτιον δὲ τὸ
τοιοῦτον πάθος καὶ τῷ λέοντι τῆς ἀγονίας τῆς
ὕστερον· τὸ μὲν γὰρ πρότερον τίκτει πέντε ἢ ἕξ,
εἶτα τῷ ὑστέρῳ ἔτει τέτταρας, πάλιν δὲ τρεῖς
σκύμνους, εἶτα τὸν ἐχόμενον ἀριθμὸν ἕως ἑνός, εἶτ'
35 οὐθέν, ὡς ἐξαναλισκομένου τοῦ περιττώματος καὶ
ἅμα τῆς ἡλικίας ληγούσης φθίνοντος τοῦ σπέρματος.

Τίσι μὲν οὖν γίνεται τὰ ὑπηνέμια τῶν ὀρνίθων,
ἔτι δὲ ποῖοι πολύγονοι καὶ ὀλιγόγονοι αὐτῶν, καὶ
διὰ τίνας αἰτίας, εἴρηται.

Γίνεται δὲ τὰ ὑπηνέμια, καθάπερ εἴρηται καὶ
πρότερον, διὰ τὸ ὑπάρχειν ἐν τῷ θήλει τὴν ὕλην
5 τὴν σπερματικήν, τοῖς δ' ὀρνέοις μὴ γίνεσθαι τὴν
τῶν καταμηνίων ἀπόκρισιν ὥσπερ τοῖς ζῳοτόκοις
τοῖς ἐναίμοις· πᾶσι γὰρ τούτοις γίνεται, τοῖς μὲν

lay often, producing two, or three at the most, generally two, as these numbers are intermediate between one and many.

The actual facts make it clear that in the prolific birds the nourishment is diverted to the semen. Most trees, if they have borne an excessive amount of fruit, wither away when the crop is over, when no nourishment is left over for themselves; annual plants, as it seems, have the same experience, *e.g.*, leguminous plants, corn, and the rest of that sort. The reason is that, as they belong to a kind which produces a great deal of seed, they use up all their nourishment for semen (seed). Some fowls, too, after having laid excessively—as many as two eggs in a day—have died after performing the feat. The birds and plants alike become completely exhausted, and this condition is simply one of excessive evacuation of residue. It is responsible for the sterility which besets the lion in the latter part of its life. To begin with, the lion [a] will produce five or six cubs in a litter, then four the next year, next time three, then two, after that one, and then none at all, which suggests that the residue is being used up and that the semen is diminishing as the prime of life abates.

We have now said which are the birds that produce wind-eggs, and what sorts of birds are prolific and not prolific, together with the causes thereof.

Why are wind-eggs formed? As has been said Wind-eggs. earlier, their formation is due to the fact that though seminal matter is present in the female, with birds no discharge of the menstrual fluid take place as it does with the blooded Vivipara; in all of the last-named it does take place, and it is greater in some, smaller

[a] *Cf.* 760 b 23.

πλείων, τοῖς δ' ἐλάττων, τοῖς δὲ τοσαύτη τὸ πλῆθος
ὥστε ὅσον γε ἐπισημαίνειν. ὁμοίως δ' οὐδὲ τοῖς
ἰχθύσι, καθάπερ[1] τοῖς ὄρνισιν· διὸ καὶ τούτοις
10 γίνεται μὲν ἄνευ ὀχείας σύστασις κυημάτων,
[ὁμοίως καὶ τοῖς ὄρνισιν,][2] ἧττον δ' ἐπιδήλως·
ψυχροτέρα γὰρ ἡ φύσις αὐτῶν. ἡ δὲ γινομένη τοῖς
ζῳοτόκοις ἀπόκρισις τῶν καταμηνίων συνίσταται
τοῖς ὄρνισι κατὰ τοὺς ἱκνουμένους χρόνους τοῦ
περιττώματος, καὶ διὰ τὸ τὸν τόπον εἶναι θερμὸν
15 τὸν πρὸς τῷ διαζώματι τελειοῦται τοῖς μεγέθεσιν,
πρὸς δὲ τὴν γένεσιν ἀτελῆ καὶ ταῦτα καὶ τὰ τῶν
ἰχθύων ὁμοίως ἄνευ τῆς τοῦ ἄρρενος γονῆς· ἡ δ'
αἰτία τούτων εἴρηται πρότερον. οὐ γίνεται δὲ τὰ
ὑπηνέμια τοῖς πτητικοῖς τῶν ὀρνίθων διὰ τὴν
αὐτὴν αἰτίαν δι' ἥνπερ οὐδὲ πολυτοκεῖ τὰ τοιαῦτα[3]·
τοῖς γὰρ γαμψώνυξιν ὀλίγον τὸ περίττωμα, καὶ
20 προσδέονται τοῦ ἄρρενος πρὸς τὴν ὁρμὴν τῆς τοῦ
περιττώματος[4] ἐκκρίσεως. πλείω δὲ τὰ ὑπηνέμια
γίνεται τῶν γονίμων ᾠῶν,[5] ἐλάττω δὲ τὸ μέγεθος
διὰ μίαν αἰτίαν καὶ τὴν αὐτήν· διὰ μὲν γὰρ τὸ
ἀτελῆ εἶναι ἐλάττω τὸ μέγεθος, διὰ δὲ τὸ τὸ μέγε-
25 θος ἔλαττον πλείω τὸν ἀριθμόν. καὶ ἧττον δὲ ἡδέα
διὰ τὸ ἀπεπτότερα εἶναι· ἐν πᾶσι γὰρ τὸ πεπεμμένον γλυκύτερον.

Ὅτι μὲν οὖν οὔτε τὰ τῶν ὀρνίθων οὔτε τὰ τῶν

---

[1] fort. ⟨οὐδὲ⟩ supplendum.
[2] secl. A.-W. : ὁμοίως om. S, ὄρνισιν om. Z¹.
[3] hic lacunam statuit Platt.
[4] περιττώματος PSYZ²Σ, A.-W., Platt : σπέρματος vulg.
[5] γονίμων ᾠῶν A.-W., ovis convenientibus generationi Σ : γόνῳ γιγνομένων Z, vulg. : γονῶν γιγ. PSY.

---

[a] i.e., to mark that it belongs to a class which exhibits the

in others, and in some just enough to serve as an indication.[a] Similarly, there is no discharge in fishes, any more than in birds : and therefore in fishes too, [just as in birds,] fetations arise without previous copulation, though they are less obvious ; that is because their nature is colder. What corresponds to the secretion of the menstrual fluid which occurs in viviparous animals arises in birds at the times proper for that residue, and as the region by the diaphragm is hot these fetations reach perfection in respect of size, though for the purpose of generation they are imperfect, both in birds and fishes, without the semen of the male. The cause of these things has been given earlier. Wind-eggs are not formed in the birds that are fliers ; the reason why this is so and why birds of this sort are not very prolific layers is one and the same [b] : in the crook-taloned birds the residue is scanty, and they need the male to give the impulse for the discharge of the residue. The wind-eggs are formed in larger numbers than the ones which are fertile but they are smaller in size ; both facts are due to one and the same cause : they are smaller in size because they are imperfect, and they are more in number because their size is smaller. They are less pleasant to eat because they are more unconcocted, for that which has been concocted [c] always makes the more tasty morsel.

Now it has been sufficiently established by ob-

phenomenon. A similar remark is made at *P.A.* 689 b 5 about the stumpy tail of certain animals.

[b] Platt's assumption of a lacuna here is unnecessary. Although πτητικά and γαμψώνυχα are not simply convertible, all γαμψώνυχα are πτητικά, and clearly Aristotle is here thinking of them as especially good examples of fliers.

[c] The Greek word also connotes " matured," " ripened."

750 b

ἰχθύων¹ τελειοῦται πρὸς τὴν γένεσιν ἄνευ τῶν
ἀρρένων, ἱκανῶς ὦπται, περὶ δὲ τοῦ γίνεσθαι
καὶ ἐν τοῖς ἰχθύσι κυήματα ἄνευ τῶν ἀρρένων,
30 οὐχ ὁμοίως, μάλιστα δ᾽ ἐπὶ τῶν ποταμίων ἑώραται
[περὶ τοὺς ἐρυθρίνους] ⟨τοῦ⟩το συμβαῖνον²· ἔνιοι
γὰρ εὐθὺς ἔχοντες ᾠὰ φαίνονται, καθάπερ ἐν ταῖς
ἱστορίαις γέγραπται περὶ αὐτῶν. ὅλως δ᾽ ἔν γε
τοῖς ὄρνισιν οὐδὲ τὰ γινόμενα διὰ τῆς ὀχείας ᾠὰ
θέλει ὡς ἐπὶ τὸ πολὺ λαμβάνειν αὔξησιν, ἐὰν μὴ
ὀχεύηται ἡ ὄρνις συνεχῶς. τούτου δ᾽ αἴτιον ὅτι
35 καθάπερ ἐπὶ τῶν γυναικῶν τὸ πλησιάζειν τοῖς
751 a ἄρρεσι κατασπᾷ τὴν τῶν γυναικείων ἀπόκρισιν
(ἕλκει γὰρ τὸ ὑγρὸν ἡ ὑστέρα θερμανθεῖσα, καὶ οἱ
πόροι ἀναστομοῦνται), τοῦτο συμβαίνει καὶ ἐπὶ
τῶν ὀρνίθων ἐπιόντος κατὰ μικρὸν τοῦ κατα-
μηνιώδους περιττώματος, ὃ θύραζε μὲν οὐκ ἀπο-
5 κρίνεται διὰ τὸ ὀλίγον εἶναι καὶ πρὸς τῷ διαζώματι
ἄνω τὰς ὑστέρας, συλλείβεται δ᾽ εἰς αὐτὴν τὴν
ὑστέραν. τοῦτο γὰρ αὔξει τὸ ᾠόν, ὥσπερ τὰ
ἔμβρυα τὰ τῶν ζῳοτόκων ⟨τὸ⟩³ διὰ τοῦ ὀμφαλοῦ,
τὸ ἐπιρρέον διὰ τῆς ὑστέρας, ἐπεὶ ὅταν ἅπαξ
ὀχευθῇ τὰ ὄρνεα, πάντα σχεδὸν ἀεὶ διατελεῖ ᾠὰ
10 ἔχοντα, μικρὰ δὲ πάμπαν. διὸ καὶ περὶ τῶν ὑπ-
ηνεμίων τινὲς εἰώθασι λέγειν ὡς οὐ γιγνομένων
ἀλλ᾽ ὡς ὑπολειμμάτων ἐκ προτέρας ὀχείας ὄντων.
τοῦτο δ᾽ ἐστὶ ψεῦδος· ὦπται γὰρ ἱκανῶς καὶ ἐπὶ

---

¹ οὔτε τὰ τῶν ἰχθύων om. Y., ova piscium non complen-
tur Σ.
² π. τοὺς ἐ. συμ. vulg.: τὸ συμ. Z¹, add. π. τοὺς ἐ. Z²: haec
aliena esse monuerat Buss., secl. A.-W.; τοῦτο Peck. συμβαῖνον post ἑώραται SY.   ³ ⟨τὸ⟩ Peck.

servation that neither in birds nor in fishes do the
fetations attain perfection for the purpose of generation apart from the males ; with regard to fetations
being formed apart from the males in fishes as well,
this has been observed, though to a less extent, to
occur, but it has been noticed most in the fresh-
water fishes.[a] Some of them, as we can see, have
eggs from the very outset, as is recorded in the
Researches.[b] Speaking generally, in birds at any
rate even the impregnated eggs usually do not grow
unless the hen is trodden continually. The reason
for this is, that, just as in the case of women intercourse with the males draws down the discharge of
the menstrual flow (since when the uterus has been
heated it draws [c] the liquid and the mouths of the
passages are opened), so with birds : the same thing
occurs ; the menstrual residue advances little by
little. It is not discharged externally because there
is not much of it and the uterus is high up towards
the diaphragm, but it runs down and collects in the
uterus itself. This liquid, of course, which percolates
through the uterus, makes the egg grow, just as that
which passes through the umbilical cord makes the
embryos of Vivipara grow, for when once the birds
have been trodden, they all continue almost always
to have eggs, albeit quite small ones. In view of this,
some people are in the habit of saying that wind-
eggs are not formed ⟨independently⟩ either, but are
merely relics of an earlier impregnation. This however is untrue. It has been sufficiently established by

---

[a] The reference to the *erythrinus* which several MSS. have
at this point is out of place ; *cf. H.A.* 567 a 27.

[b] At *H.A.* 567 a 30.

[c] See above, 739 a 35 ff., esp. b 11 ff.

νεοττῶν ἀλεκτορίδος καὶ χηνὸς γενόμενα ἄνευ
ὀχείας. ἔτι δὲ αἱ πέρδικες αἱ θήλειαι, αἵ τ᾽ ἀν-
15 όχευτοι καὶ αἱ ὠχευμέναι τῶν εἰς τὰς θήρας ἀγο-
μένων, ὀσφραινόμεναι[1] τοῦ ἄρρενος καὶ ἀκούουσαι
τῆς φωνῆς αἱ μὲν πληροῦνται αἱ δὲ τίκτουσι παρα-
χρῆμα. τοῦ δὲ πάθους[2] αἴτιον ταὐτὸν ὅπερ ἐπὶ τῶν
ἀνθρώπων καὶ τῶν τετραπόδων· ἐὰν γὰρ ὀργῶντα
τύχῃ τὰ σώματα πρὸς τὴν ὁμιλίαν, τὰ μὲν ἰδόντα
τὰ δὲ μικρᾶς γενομένης θίξεως προΐεται σπέρμα.
20 τὰ δὲ τοιαῦτα τῶν ὀρνέων ὀχευτικὰ καὶ πολύσπερμα τὴν φύσιν ἐστίν, ὥστε μικρᾶς δεῖσθαι τῆς
κινήσεως, ὅταν ὀργῶντα τύχῃ, καὶ γίνεσθαι ταχὺ
τὴν ἔκκρισιν αὐτοῖς, ὥστε τοῖς μὲν ἀνοχεύτοις
ὑπηνέμια συνίστασθαι, τοῖς δ᾽ ὠχευμένοις αὐξάνεσθαι καὶ τελειοῦσθαι ταχέως.

25   Τῶν δὲ θύραζε ὠοτοκούντων οἱ μὲν ὄρνιθες προΐενται τὸ ᾠὸν τέλειον, οἱ δ᾽ ἰχθύες ἀτελές, ἀλλ᾽
ἔξω λαμβάνει τὴν αὔξησιν, καθάπερ εἴρηται καὶ
πρότερον. αἴτιον δ᾽ ὅτι πολύγονόν ἐστι τὸ τῶν
ἰχθύων γένος· ἀδύνατον οὖν ἔσω πολλὰ λαμβάνειν
τέλος, διόπερ ἀποτίκτουσιν ἔξω. ταχεῖα δ᾽ ἡ
30 πρόεσις· αἱ γὰρ ὑστέραι πρὸς τοῖς ἄρθροις τῶν
θύραζε ὠοτοκούντων ἰχθύων.

   Ἔστι δὲ τὰ μὲν τῶν ὀρνίθων δίχροα, τὰ δὲ τῶν
ἰχθύων μονόχροα πάντων. τῆς δὲ διχροίας τὴν
αἰτίαν ἴδοι τις ἂν ἐκ τῆς δυνάμεως ἑκατέρου τῶν
μορίων, τοῦ τε λευκοῦ καὶ τοῦ ὠχροῦ. γίνεται μὲν
γὰρ ἡ ἀπόκρισις ἐκ τοῦ αἵματος [(οὐθὲν γὰρ ἄναιμον

---

[1] ὀσφραινόμεναι P : ὀσφρώμεναι SY : ὀσμώμεναι Z, vulg.
[2] πάθους] τάχους Z[1] ut vid.

---

[a] Cf. H.A. 560 b 10 ff.

observation that they have been formed in chickens and goslings without impregnation. Again, when the female partridges [a] which are taken out to act as decoy-birds smell the male and hear his note, those which have not been trodden by a male become full of eggs and those which have already been trodden at once lay their eggs. The reason why this happens is the same as in the case of human beings and quadrupeds : if they are in heat, some emit the semen at the mere sight of a female, others at a slight touch. Birds of this sort are by nature inclined to frequent intercourse and have abundance of semen, so that when they are in heat the impulse they need to set them off is small, and emission quickly takes place ; the result is that in those which have not been impregnated wind-eggs take shape, and in those which have been impregnated the eggs quickly grow and reach perfection.

In the group of animals which lay their eggs externally, birds produce their eggs in a perfected state, fish in an imperfect state ; but fishes' eggs continue and finish their growth apart from the parent, as indeed I have said earlier. The reason for this is that the fish tribe is very prolific; therefore it is impossible for a large number of eggs to reach perfection within the animal ; hence they are laid externally. Their discharge is quickly effected, for in the externally oviparous fishes the uterus is near the genital parts.

Birds' eggs are double-coloured, but all fishes' eggs are single-coloured. The cause of the two colours in birds' eggs can be seen from the specific character [b] of each of the two parts, the white and the yolk. The secretion ⟨for the egg⟩ is formed out of the blood

Difference between yolk and white.

[b] *Dynamis* ; see Introd. § 26.

751 b ὠοτοκεῖ ζῷον)],¹ τὸ δ' αἷμα ὅτι ἐστὶν ὕλη τοῖς
σώμασιν, εἴρηται πολλάκις. τὸ μὲν οὖν ἐστιν ἐγγύτερον αὐτοῦ τῆς μορφῆς τῶν [μορίων]² γινομένων,
τὸ θερμόν· τὸ δὲ γεωδέστερον τὴν τοῦ σώματος
παρέχεται σύστασιν καὶ πορρώτερόν ἐστιν. διόπερ
5 ὅσα δίχροά ἐστι τῶν ὠῶν, τὴν μὲν ἀρχὴν τὸ ζῷον
λαμβάνει ἐκ τοῦ λευκοῦ τῆς γενέσεως (ἐν γὰρ τῷ
θερμῷ ἡ ψυχικὴ ἀρχή), τὴν δὲ τροφὴν ἐκ τοῦ
ὠχροῦ. τοῖς μὲν οὖν τὴν φύσιν θερμοτέροις τῶν
ζῴων διακέκριται χωρὶς ἐξ οὗ τε ἡ ἀρχὴ γίνεται
καὶ ἐξ οὗ τρέφεται, καὶ τὸ μὲν λευκόν ἐστι τὸ δ'
10 ὠχρόν, καὶ πλέον ἀεὶ τὸ λευκὸν καὶ καθαρὸν τοῦ
ὠχροῦ καὶ γεώδους· τοῖς δ' ἧττον θερμοῖς καὶ
ὑγροτέροις τὸ ὠχρὸν πλέον καὶ ὑγρότερον. ὅπερ
συμβαίνει ἐπὶ τῶν λιμναίων ὀρνέων· ὑγρότεροι γὰρ
τὴν φύσιν καὶ ψυχρότεροι τῶν πεζευόντων εἰσὶν
ὀρνέων, ὥστε καὶ τὰ ὠὰ τῶν τοιούτων πολλὴν ἔχει
15 τὴν καλουμένην λέκιθον καὶ ἧττον ὠχρὰν διὰ τὸ
ἧττον ἀποκεκρίσθαι τὸ λευκόν. τὰ δ' ἤδη καὶ
ψυχρὰ τὴν φύσιν τῶν ὠοτοκούντων καὶ ἔτι ὑγρὰ
μᾶλλον (τοιοῦτον δ' ἐστὶ τὸ τῶν ἰχθύων γένος)
οὐδ' ἀποκεκριμένον ἔχει τὸ λευκὸν διά τε μικρότητα καὶ διὰ τὸ πλῆθος τοῦ ψυχροῦ καὶ γεώδους·
20 διόπερ γίνεται μονόχροα πάντα τὰ τῶν ἰχθύων,

---

¹ secl. A.-W., Platt. ² om. Z¹.

ᵃ The white; because hot substance has to do with Soul;
see immediately below, and 762 a 18 ff. and *P.A.* 652 b 7 ff.
ᵇ See 744 b 32 ff. and note.
ᶜ For the two sorts of τροφή see 744 b 32 ff. Both yolk
and white are now known to be nourishment; Harvey
demonstrated the unreality of the distinction here made.—
Aristotle of course knew nothing of the germinal area on the

[(no bloodless animal lays eggs)], the blood, as I have often stated, being the matter for animal organisms. One part of the egg, the hot part,[a] is closer to the form of the developing creatures; the other, the more earthy part, supplies the wherewithal for building up the bodily frame and is further removed from the form.[b] That is why in the case of all double-coloured eggs the young animal gets its " principle " of generation from the white, because hot substance is the place where the soul-principle is to be found, while it gets its nourishment from the yolk.[c] With those animals, therefore, whose nature tends to be hotter than others we find there is a clear distinction between the part from which the " principle " is formed and the part from which the nourishment is derived: the one is white, the other yellow, and there is always more of the pure, white part than there is of the earthy, yellow part. With the animals that are less hot and more fluid, there is more yolk in the egg and it is more fluid. This occurs in the case of the marsh-birds, since they are more fluid and colder in their nature than the land-birds, so that the eggs of such birds contain a great deal of what is called yelk (*lēkithos*) and it is less yellow, because the white is less distinctly separated from it. Pass on a further stage to those oviparous animals which are cold in their nature and also still more fluid (the fish tribe answers to this description), and in their eggs the white is not distinct at all; this is due to their small size and to the abundance of the cold and earthy matter. And that is why all fishes' eggs are single-

yolk; and it was again Harvey who demonstrated that the " cicatricula " was the point of origin of the embryo, " the first Principle of the Egge."

751 b

καὶ ὡς μὲν ὠχρὰ λευκά, ὡς δὲ λευκὰ ὠχρά. τὰ
δὲ τῶν ὀρνέων καὶ τὰ ὑπηνέμια ἔχει ταύτην τὴν
δίχροιαν· ἔχει γὰρ ἐξ οὗ ἑκάτερον ἔσται τῶν
μορίων, καὶ ὅθεν ἡ ἀρχὴ καὶ ὅθεν ἡ τροφή, ἀλλὰ
ταῦτ' ἀτελῆ καὶ προσδεόμενα τοῦ ἄρρενος· γίνεται
25 γὰρ τὰ ὑπηνέμια γόνιμα, ἐὰν ἔν τινι καιρῷ ὀχευθῇ
ὑπὸ τοῦ ἄρρενος. οὐκ ἔστι δὲ τῆς διχροίας αἴτιον
τὸ ἄρρεν καὶ τὸ θῆλυ, ὡς τοῦ μὲν λευκοῦ ὄντος
ἀπὸ τοῦ ἄρρενος, τοῦ δ' ὠχροῦ ἀπὸ τοῦ θήλεος·
ἀλλ' ἄμφω γίνεται ἀπὸ τοῦ θήλεος, ἀλλὰ τὸ μὲν
ψυχρὸν τὸ δὲ θερμόν. ἐν ὅσοις μὲν οὖν ἐστι πολὺ
30 τὸ θερμόν, ἀποκρίνεται, ἐν ὅσοις δ' ὀλίγον, οὐ
δύναται· διὸ μονόχροα τὰ κυήματα, καθάπερ εἴρηται, τὰ τῶν τοιούτων. ἡ δὲ γονὴ συνίστησι[1] μόνον·
καὶ διὰ τοῦτο τὸ μὲν πρῶτον φαίνεται λευκὸν καὶ
μικρὸν τὸ κύημα ἐν τοῖς ὄρνισι, προϊὸν δὲ ὠχρὸν
ἅπαν, συμμιγνυμένου ἀεὶ πλείονος αἱματώδους·
τέλος δ' ἀποκρινομένου τοῦ θερμοῦ κύκλῳ περι-

752 a

ίσταται τὸ λευκόν, ὥσπερ ὑγροῦ ζέοντος, ὁμοίως
πάντῃ· τὸ γὰρ λευκὸν φύσει μὲν ὑγρόν, ἔχει δ' ἐν
αὑτῷ τὴν θερμότητα τὴν ψυχικήν· διὸ κύκλῳ
ἀποκρίνεται, τὸ δ' ὠχρὸν καὶ γεῶδες ἐντός. κἂν
5 πολλὰ συνεράσας τις ᾠὰ εἰς κύστιν ἤ τι τοιοῦτον
ἕψῃ πυρὶ μὴ[2] θάττονα ποιοῦντι τὴν τοῦ θερμοῦ

---

[1] συνίστησι Peck : συνέστησε vulg. : συνέστη δὲ S.
[2] μὴ om. Z[1] sub fin. pag., add. Z[2].

---

[a] It is of course the hot substance which constitutes the
white.

coloured—they are white, judged by the colour of ordinary yolk ; yellow, judged by ordinary white. Not only the eggs but also the wind-eggs of birds have this double colouring, because they contain that out of which each of the two parts is to come (the part from which the " principle " arises and that from which the nourishment is derived), although they are imperfect, *i.e.*, they lack the male factor ; since, as we know, wind-eggs become fertile if they are impregnated by the male within a certain time. The cause of the double colouring is not the two different sexes (as if the white were derived from the male and the yolk from the female) ; both alike are derived from the female, and the real difference is that one is cold and the other hot. So then, in cases where a good deal of the hot constituent is present, the hot substance is separated from the cold ; but if there is not much of it this cannot occur ; and that is why the fetations of such animals are single-coloured, as I have said. All that the semen does is to " set " the fetations, and that is why in birds the fetation is small and white in appearance at first, but completely yellow as it advances and more bloodlike matter is continually being mixed in with it ; finally, as the hot substance separates off, the white takes up its position around on the outside *a* evenly in every direction, just as when a liquid boils. ⟨I make this comparison⟩, because the white (*a*) is in its nature liquid, and (*b*) contains in itself the soul-heat. Therefore it separates off ⟨and arranges itself⟩ all round ⟨on the outside⟩, while the yellow earthy part separates off within. Also, if anyone pours a number of eggs together into a bladder or some such receptacle and then boils them up by means of a fire which does not

κίνησιν ἢ τὴν ἐν τοῖς ᾠοῖς διάκρισιν, ὥσπερ ἐν ἑνὶ
ᾠῷ, ⟨οὕτω⟩¹ καὶ τὸ ἐκ πάντων τῶν ᾠῶν σύστημα
τὸ μὲν ὠχρὸν ἐν μέσῳ γίνεται, κύκλῳ δὲ τὸ λευκόν.

Διότι μὲν οὖν τὰ μὲν μονόχροα τὰ δὲ δίχροα τῶν
10 ᾠῶν, εἴρηται·

II Ἀποκρίνεται δ' ἐν τοῖς ᾠοῖς ἡ τοῦ ἄρρενος ἀρχὴ
καθ' ὃ προσπέφυκε τῇ ὑστέρᾳ τὸ ᾠόν, καὶ γίνεται
δὴ ἀνόμοιον τὸ τῶν διχρόων ᾠῶν, καὶ οὐ πάμπαν
στρογγύλον ἀλλ' ἐπὶ θάτερα ὀξύτερον, διὰ τὸ
διαφέρειν δεῖν² ⟨τὸ⟩³ τοῦ λευκοῦ ἐν ᾧ ἔχει τὴν
ἀρχήν. διόπερ σκληρότερον ταύτῃ τὸ ᾠόν ἢ
15 κάτωθεν· σκεπάζειν γὰρ δεῖ καὶ φυλάττειν τὴν
ἀρχήν. καὶ διὰ τοῦτο ἐξέρχεται ὕστερον τοῦ ᾠοῦ
τὸ ὀξύ· τὸ γὰρ προσπεφυκὸς ὕστερον ἐξέρχεται,
κατὰ τὴν ἀρχὴν δὲ προσπέφυκεν, ἐν τῷ ὀξεῖ δ' ἡ
ἀρχή. τὸν αὐτὸν δ' ἔχει τρόπον καὶ ἐν τοῖς τῶν
φυτῶν σπέρμασιν· προσπέφυκε γὰρ ἡ ἀρχὴ τοῦ
20 σπέρματος τὰ μὲν ἐν τοῖς κλάδοις, τὰ δ' ἐν τοῖς
κελύφεσι, τὰ δ' ἐν τοῖς περικαρπίοις. δῆλον δ' ἐπὶ
τῶν χεδροπῶν· ᾗ γὰρ συνῆπται τὸ δίθυρον τῶν
κυάμων καὶ τῶν τοιούτων σπερμάτων, ταύτῃ
προσπέφυκεν· ἡ δ' ἀρχὴ ἐνταῦθα τοῦ σπέρματος.

Ἀπορήσειε δ' ἄν τις περὶ τῆς αὐξήσεως τῶν
25 ᾠῶν, τίνα τρόπον ἐκ τῆς ὑστέρας συμβαίνει. τὰ
μὲν γὰρ ζῷα διὰ τοῦ ὀμφαλοῦ λαμβάνει τὴν τρο-

---

¹ ⟨οὕτω⟩ Rackham.   ² δεῖν ἀεὶ SYZ².   ³ ⟨τὸ⟩ Peck.

[a] Cf. H.A. 560 a 30 ff.
[b] Cf. 767 b 17 ff. et passim.
[c] That is, the "big" end, which is the first to leave the hen
when laid. Platt remarks that Aristotle must have been a
"little-endian," for the germ always floats up to the top
whichever way the egg is placed.

cause the movement of the heat to be faster than the separation in the eggs, the yolk settles in the middle and the white round the outside of it [a]; *i.e.*, the same happens with the conglomerated mass composed of all those eggs as with one single egg.

We have now stated why some eggs are single-coloured and others double-coloured.

In eggs the place where the " principle " derived from the male [b] becomes separate and distinct is the point where the egg is attached to the uterus, and that gives us the reason why the shape of double-coloured eggs is unsymmetrical, *i.e.*, not perfectly round but more pointed at one end ; the reason is that that part of the white in which the principle is situated must be different. And that is why the egg-shell is harder at that place than it is at the bottom [c] : the " principle " has to be protected and safeguarded. That also is why the pointed end of the egg comes out last : for of course the part that comes out last is the part that is fastened, which is the part where the " principle " is, which is the pointed end. The same arrangement obtains in the seeds of plants. In some plants the " principle " of the seed is fastened on to the twig, in others on to the husk, in others on to the pericarp. This is clear in the leguminous plants. The seeds of beans and plants of that sort are fastened on at the point where the two cotyledons [d] are joined ; and that is where the " principle " of the seed is.

A puzzle may be raised about how eggs grow— how, it may be asked, do they derive their growth from the uterus ? Animals, of course, obtain their nourishment through the umbilical cord ; but by

II
Shape of the egg.

Growth of the egg.

[d] The two halves of the pea or bean.

752 a

φήν, τὰ δ' ᾠὰ διὰ τίνος; ἐπειδήπερ οὐχ ὥσπερ οἱ
σκώληκες αὐτὰ δι' αὐτῶν λαμβάνει τὴν αὔξησιν.
εἰ δ' ἔστι τι δι' οὗ προσπέφυκε, τοῦτο ποῖ τρέπεται
τελεωθέντος;   οὐ γὰρ συνεξέρχεται, καθάπερ ὁ
30 ὀμφαλὸς τοῖς ζῴοις[1]· γίνεται γὰρ τὸ πέριξ ὄστρακον
τελεωθέντος. τὸ μὲν οὖν εἰρημένον ὀρθῶς ζητεῖται·
λανθάνει δ' ὅτι τὸ γινόμενον ὄστρακον τὸ πρῶτον
μαλακὸς ὑμήν ἐστιν, ἀλλὰ τελεωθέντος γίνεται
σκληρὸν καὶ κραῦρον, οὕτω συμμέτρως ὥστ' ἐξέρχεται μὲν ἔτι μαλακόν (πόνον γὰρ ἂν παρεῖχε
35 τικτόμενον), ἐξελθὸν δ' εὐθὺς πήγνυται ψυχθέν,
752 b   συνεξατμίζοντος τοῦ ὑγροῦ ταχὺ δι' ὀλιγότητα,
λειπομένου δὲ τοῦ γεώδους. τούτου δή τι τοῦ
ὑμένος κατ' ἀρχὰς ὀμφαλῶδές ἐστι κατὰ τὸ ὀξύ,
καὶ ἀπέχει ἔτι μικρῶν ὄντων οἷον αὐλός. φανερὸν
δ' ἐστὶν ἐν τοῖς ἐκβολίμοις τῶν μικρῶν ᾠῶν· ἐὰν
5 γὰρ βρεχθῇ ἢ ἄλλως πως ῥιγώσασα ἐκβάλῃ ἡ
ὄρνις, ἔτι αἱματῶδές τε φαίνεται τὸ κύημα καὶ
ἔχον δι' ἑαυτοῦ στόλον μικρὸν ὀμφαλῶδε. μείζονος δὲ γινομένου περιτείνεται[2] μᾶλλον οὗτος καὶ
ἐλάττων γίνεται. τελεωθέντος δὲ τὸ ὀξὺ τοῦ ᾠοῦ
τοῦτο συμβαίνει τὸ πέρας. ὑπὸ δὲ τούτῳ ὁ ἐντὸς
10 ὑμήν, ὃς ὁρίζει τὸ λευκὸν καὶ τὸ ὠχρὸν ἀπὸ τούτου. τελειωθέντος δ' ἀπολύεται ὅλον τὸ ᾠόν, καὶ

---

[1] fortasse ζῳοτοκουμένοις vel ζῳοτόκοις scribendum.
[2] ? περαίνεται Z[1]. pro καὶ ἐλάττων . . . πέρας et efficientur
ova citrina, et maxime apud complementum. et cum complen-
tur accidit ut sit emissio Σ.

---

[a] See 732 a 32 and note there.   Cf. also 758 b 13 ff.
[b] i.e., the young of viviparous animals. Perhaps we should
read " ⟨the young of viviparous⟩ animals."
[c] This is a reference to the chalazae, the function and
development of which are obscure.

what means do eggs get theirs ? (The possibility that they are themselves their own means of growth, as larvae are,[a] may be ruled out.) If there is something by means of which the egg is fastened on, what happens to it when the egg has reached its perfection ? It does not come out along with the egg, as the umbilical cord does in the case of animals,[b] because when the egg has reached perfection, the shell is formed which envelops it. Well, this is a question which it is quite right to ask ; but those who ask it fail to notice that the shell as it forms is at first a soft membrane, and that it is only when the egg has been perfected that it becomes hard and brittle ; and this adjustment is so well timed that it is still soft when it leaves the bird (otherwise it would be painful to lay), but as soon as it has left the bird it cools, and that makes it set hard, for the fluid part quickly evaporates, being very small in quantity, while the earthy part remains behind. Now at the outset a portion of this membrane, at the pointed end of an egg, is like an umbilical cord, and while the egg is still small, it sticks out like a pipe. It can be clearly seen in small, aborted eggs : if the hen is drenched ⟨with cold water⟩ or chilled in some other way and so drops ⟨the fetation⟩ before its time, the fetation still has a blood-like appearance and has a small tail,[c] like an umbilical cord, running through it ; as the fetation gets larger, this tail gets twisted round more and becomes smaller ; when ⟨the fetation⟩ has reached its complete development, this terminus finishes up as the pointed end of the egg. Underneath this is the inner membrane, which acts as a boundary between it on the one side and the white and the yolk on the other. When the development

οὐ φαίνεται εὐλόγως ὁ ὀμφαλός· αὐτοῦ γάρ ἐστι
τοῦ ἐσχάτου τὸ ἄκρον.

Ἡ δ' ἔξοδος τοὐναντίον γίνεται τοῖς ᾠοῖς ἢ τοῖς
ζῳοτοκουμένοις· τοῖς μὲν γὰρ ἐπὶ κεφαλὴν καὶ τὴν
ἀρχήν, τῷ δ' ᾠῷ γίνεται ἡ ἔξοδος οἷον ἐπὶ πόδας.
15 τούτου δ' αἴτιον τὸ εἰρημένον, ὅτι προσπέφυκε
κατὰ τὴν ἀρχήν.

Ἡ δὲ γένεσις ἐκ τοῦ ᾠοῦ συμβαίνει τοῖς ὄρνισιν
ἐπῳαζούσης καὶ συμπεττούσης τῆς ὄρνιθος, ἀποκρινομένου μὲν τοῦ ζῴου ἐκ μέρους τοῦ ᾠοῦ, τὴν
δ' αὔξησιν λαμβάνοντος καὶ τελειουμένου ἐκ τοῦ
λοιποῦ μέρους, ἡ γὰρ φύσις ἅμα τήν τε τοῦ ζῴου
20 ὕλην ἐν τῷ ᾠῷ τίθησι καὶ τὴν ἱκανὴν τροφὴν πρὸς
τὴν αὔξησιν· ἐπεὶ γὰρ οὐ δύναται τελεοῦν ἐν αὐτῇ
ἡ ὄρνις, συνεκτίκτει τὴν τροφὴν ἐν τῷ ᾠῷ. τοῖς
μὲν γὰρ ζῳοτοκουμένοις ἐν ἄλλῳ μορίῳ γίνεται ἡ
τροφή, τὸ καλούμενον γάλα, ἐν τοῖς μαστοῖς· τοῖς
δ' ὄρνισι τοῦτο ποιεῖ ἡ φύσις ἐν τοῖς ᾠοῖς, τοὐ-
25 ναντίον μέντοι ἢ οἵ τε ἄνθρωποι οἴονται καὶ Ἀλ-
κμαίων φησὶν ὁ Κροτωνιάτης. οὐ γὰρ τὸ λευκόν
ἐστι γάλα, ἀλλὰ τὸ ὠχρόν· τοῦτο γάρ ἐστιν ἡ
τροφὴ τοῖς νεοττοῖς· οἱ δ' οἴονται τὸ λευκὸν διὰ
τὴν ὁμοιότητα τοῦ χρώματος.

Γίνεται μὲν οὖν ἐπῳαζούσης, καθάπερ εἴρηται,
30 τῆς ὄρνιθος ὁ νεοττός· οὐ μὴν ἀλλὰ κἂν ἡ ὥρα

---

[a] The heart. [b] See 744 b 32 ff.
[c] See pp. xvii. f. [d] See 751 b 7, n.

of the fetation is complete, the whole egg is released, and, as we should expect, nothing is to be seen of the umbilical cord, because it is the tip of the extreme end of the egg.

Eggs and the young of viviparous animals come out facing opposite ways ; the latter come out with the head and the " principle " first [a]; the egg comes out as it were feet first.   And the reason I have stated : it is because the egg is fastened at the point where the " principle " is.

The formation of birds out of the egg is effected by the mother's sitting on the eggs and helping to concoct them.   One part of the egg yields the substance out of which the animal is constituted, the remaining part provides the substance whereby it grows and is perfected ; Nature puts both in the egg —the material for making the animal, and sufficient nourishment for its growth,[b] since the hen cannot bring the young to perfection within herself, and therefore when she lays an egg she lays the creature's nourishment in it as well.   The nourishment for the young of viviparous animals, what we call milk, is formed in the breasts, a different part of the body altogether ; but for birds Nature provides this inside their eggs.   The truth about it, however, is the reverse of what is commonly supposed and what is asserted by Alcmeon of Crotona.[c]   It is not the white of the egg that is the milk, but the yolk, because it is the yolk that is the nourishment for the chicks. These people suppose that the white is, owing to the similarity of colour.[d]

The formation of the chick, then, as I have said, is effected by the mother-bird's sitting upon the egg ; notwithstanding, if the climate is well-tempered or

Incubation.

287

ἢ εὔκρατος ἢ ὁ τόπος ἀλεεινὸς ἐν ᾧ ἂν κείμενα
τυγχάνωσιν, ἐκπέττεται καὶ τὰ τῶν ὀρνίθων καὶ
τὰ τῶν τετραπόδων καὶ ῳοτόκων (πάντα γὰρ εἰς
τὴν γῆν ἐκτίκτει, καὶ συμπέττονται ὑπὸ τῆς ἐν
τῇ γῇ θερμότητος· ὅσα δ' ἐπῳάζει φοιτῶντα τῶν
35 ῳοτόκων καὶ τετραπόδων, ταῦτα ποιεῖ μᾶλλον
φυλακῆς χάριν.

Τὸν αὐτὸν δὲ τρόπον γίνεται τά τε τῶν ὀρνίθων
ῳὰ καὶ τὰ τῶν ζῴων τῶν τετραπόδων· καὶ γὰρ
σκληρόδερμα καὶ δίχροα, καὶ πρὸς τῷ διαζώματι
συνίσταται καθάπερ καὶ τὰ τῶν ὀρνίθων, καὶ τἆλλα
ταὐτὰ πάντα συμβαίνει καὶ ἐντὸς καὶ ἐκτός, ὥστε
5 ἡ αὐτὴ θεωρία περὶ τῆς αἰτίας ἐστὶ πάντων. ἀλλὰ
τὰ μὲν τῶν τετραπόδων δι' ἰσχὺν ἐκπέττεται καὶ
ὑπὸ τῆς ὥρας, τὰ δὲ τῶν ὀρνέων ἐπικηρότερα, καὶ
δεῖται τῆς τεκούσης. ἔοικε δὲ καὶ ἡ φύσις βούλε-
σθαι τῶν[1] τέκνων αἴσθησιν ἐπιμελητικὴν παρα-
σκευάζειν· ἀλλὰ τοῖς μὲν χείροσι τοῦτ' ἐμποιεῖ
10 μέχρι τοῦ τεκεῖν μόνον, τοῖς δὲ καὶ περὶ τὴν τε-
λέωσιν, ὅσα δὲ φρονιμώτερα, καὶ περὶ τὴν ἐκτροφήν.
τοῖς δ' ἤδη[2] μάλιστα κοινωνοῦσι φρονήσεως καὶ
πρὸς τελεωθέντα γίνεται συνήθεια καὶ φιλία,
καθάπερ τοῖς τε ἀνθρώποις καὶ τῶν τετραπόδων
ἐνίοις, τοῖς δ' ὄρνισι μέχρι τοῦ γεννῆσαι καὶ ἐκ-
15 θρέψαι· διόπερ καὶ μὴ ἐπῳάζουσαι αἱ θήλειαι, ὅταν

---

[1] τῶν PZ : τὴν τῶν vulg. [2] δ' ἤδη Z : δὲ δὴ vulg.

---

[a] Cf. H.A. 559 a 1 ff., where " non-fliers " such as partridges and quails are said to " lay their eggs on the ground
and to cover them over." Another " non-flier," the ostrich,
was believed by the author of Job (xxxix. 14) to behave in a

288

the situation where they happen to be is sunny, the eggs of birds [a] as well as of oviparous quadrupeds get fully concocted without incubation (for all these quadrupeds lay their eggs on the ground, and they get concocted by the heat in the earth; any oviparous quadrupeds which visit their eggs and sit on them do so rather for the sake of protecting them than for any other reason).

The eggs of quadrupeds are formed in the same way as birds' eggs. They are hard-shelled, and double-coloured, take shape up towards the diaphragm (as birds' eggs do), and present the same features in every other respect both externally and internally; so that studying the cause of any of them is the same as studying the cause of them all. Only, whereas the eggs of quadrupeds, being so strong, get fully concocted by the agency of the climate, birds' eggs, being more fragile, need the mother-bird. It looks as though Nature herself desires to provide that there shall be a feeling of attention and care for the young offspring. In the inferior animals this feeling which she implants lasts only until the moment of birth; in others, until the offspring reaches its perfect development; and in those that have more intelligence, until its upbringing is completed. Those which are endowed with most intelligence show intimacy and attachment towards their offspring even after they have reached their perfect development (human beings and some of the quadrupeds are examples of this); birds show it until they have produced their chicks and brought them up; and on this account hen birds which have laid eggs but omit

similar way: "she leaveth her eggs on the earth, and warmeth them in the dust" (R.V.).

τέκωσι, διατίθενται χεῖρον, ὥσπερ ἑνός τινος στερισκόμεναι τῶν συμφύτων.

Τελεοῦται δ' ἐν τοῖς ᾠοῖς τὰ ζῷα θᾶττον ἐν ταῖς ἀλεειναῖς ἡμέραις· συνεργάζεται γὰρ ἡ ὥρα· καὶ γὰρ ἡ πέψις θερμότης τίς[1] ἐστιν. ἥ τε γὰρ γῆ
20 συμπέττει τῇ θερμότητι, καὶ ἡ ἐπῳάζουσα ταὐτὸ τοῦτο δρᾷ· προσεγχεῖ[2] γὰρ τὸ ἐν αὐτῇ θερμόν. καὶ διαφθείρεται δὲ τὰ ᾠὰ καὶ γίνεται τὰ καλούμενα οὔρια μᾶλλον κατὰ τὴν θερμὴν ὥραν εὐλόγως· ὥσπερ γὰρ καὶ οἱ οἶνοι ἐν ταῖς ἀλέαις ὀξύνονται ἀνατρεπομένης τῆς ἰλύος (τοῦτο γὰρ αἴτιον τῆς
25 διαφθορᾶς), καὶ ἐν τοῖς ᾠοῖς ἡ λέκιθος· τοῦτο γὰρ ἐν ἀμφοτέροις τὸ γεῶδες, διὸ καὶ ἀναθολοῦται ὁ οἶνος μιγνυμένης τῆς ἰλύος, καὶ τὰ διαφθειρόμενα ᾠὰ τῆς λεκίθου.

Τοῖς μὲν οὖν πολυτόκοις συμβαίνει τὸ τοιοῦτον εὐλόγως (οὐ γὰρ ῥᾴδιον τὴν ἁρμόττουσαν πᾶσιν ἀποδιδόναι θερμασίαν, ἀλλὰ τοῖς μὲν ἐλλείπειν τοῖς
30 δὲ πλεονάζειν, καὶ ἀναθολοῦν οἷον σήπουσαν), τοῖς δὲ γαμψώνυξιν ὀλιγοτόκοις οὖσιν οὐδὲν ἧττον συμβαίνει τοῦτο· πολλάκις μὲν γὰρ καὶ τοῖν δυοῖν θάτερον οὔριον γίνεται, τὸ δὲ τρίτον ὡς εἰπεῖν ἀεί· θερμὰ γὰρ ὄντα τὴν φύσιν οἷον ὑπερζεῖν ποιεῖ τὴν
35 ὑγρότητα τὴν ἐν τοῖς ᾠοῖς. ἔχει γὰρ δὴ καὶ τὴν φύσιν ἐναντίαν τό τε ὠχρὸν καὶ τὸ λευκόν. τὸ

μὲν γὰρ ὠχρὸν ἐν τοῖς πάγοις πήγνυται, θερμαινόμενον δὲ ὑγραίνεται· διὸ καὶ συμπεττόμενον ἐν τῇ

---

[1] θερμότητος coni. A.-W.
[2] προσέχει SY, Z[2] in ras.

---

[a] According to *H.A.* 560 a 5 ff., *ouria* is a name given to wind-eggs produced chiefly in summer, *zephyria* (see 749 b 1) to those produced in spring.   [b] *Cf.* 735 a 34 ff.

to sit on them, deteriorate in their condition, as though they were being deprived of one of their natural endowments.

Animals reach their perfect development in the eggs quicker when the days are sunny, for then the climate takes a share in the work, concoction being a form of heat : the earth helps in concocting them with its heat, and the sitting bird does exactly the same—she infuses her own heat into them as well. Eggs get spoilt and *ouria*[a] (as they are called) are produced in the hot season more often than at any other, as is to be expected. In hot, sunny weather wines turn sour because the sediment gets stirred up —this is what is really responsible for their being spoilt—and the same happens with the yolk in eggs. Sediment and yolk are the earthy part in each respectively, and as a result of this earthiness wine becomes turbid when the sediment mixes up with it, and these spoilt eggs also become turbid when the yolk does the same.

It is only to be expected that this should happen in the case of prolific animals, because it is not easy to provide all the eggs with their proper amount of heat ; some will get too little, and some too much ; and too much heat will make them turbid, by causing them to putrefy, as it were. Nevertheless, the same thing occurs with the crook-taloned birds, although they lay but few eggs ; out of two eggs, one will often turn rotten (*ourion*), and pretty well always one out of three. They are hot in their nature, and they cause the fluid in the eggs as it were to boil over. The yolk and the white, of course, are of an opposite nature to each other. Yolk congeals in frosty weather,[b] and becomes fluid when heated ; hence it

Development during incubation.

291

753 b

γῇ ἢ ὑπὸ τοῦ ἐπῳάζειν ὑγραίνεται, καὶ τοιοῦτον
ὂν γίνεται τροφὴ τοῖς συνισταμένοις ζῴοις. πυρού-
5 μενον δὲ καὶ ὀπτώμενον οὐ γίνεται σκληρὸν διὰ τὸ
εἶναι τὴν φύσιν γεῶδες οὕτως ὥσπερ κηρός· καὶ
διὰ τοῦτο θερμαινόμενα μᾶλλον, [ἐὰν ᾖ μὴ ἐξ
ὑγροῦ περιττώματος,]¹ διοροῦται καὶ γίνεται οὖρια.
τὸ δὲ λευκὸν ὑπὸ μὲν τῶν πάγων οὐ πήγνυται,
ἀλλ' ὑγραίνεται μᾶλλον (τὸ δ' αἴτιον εἴρηται πρότερον), πυρούμενον δὲ γίνεται στερεόν· διὸ καὶ
10 πεττόμενον περὶ τὴν γένεσιν τῶν ζῴων παχύνεται.
ἐκ τούτου γὰρ συνίσταται τὸ ζῷον, τὸ δ' ὠχρὸν
τροφὴ γίνεται, καὶ τοῖς ἀεὶ συνισταμένοις τῶν
μορίων ἐντεῦθεν ἡ αὔξησις. διὸ καὶ διώρισται τό
τε ὠχρὸν καὶ τὸ λευκὸν χωρὶς ὑμέσιν ὡς ἔχοντα
τὴν φύσιν ἑτέραν. δι' ἀκριβείας μὲν οὖν, ὃν τρόπον
15 ἔχουσι ταῦτα πρὸς ἄλληλα κατ' ἀρχάς τε τῆς γενέσεως καὶ συνισταμένων τῶν ζῴων, ἔτι δὲ περί τε
ὑμένων καὶ περὶ² ὀμφαλῶν, ἐκ τῶν ἐν ταῖς ἱστορίαις
γεγραμμένων δεῖ θεωρεῖν· πρὸς δὲ τὴν παροῦσαν
σκέψιν ἱκανὸν φανερὸν εἶναι τοσοῦτον, ὅτι συστάσης πρώτης τῆς καρδίας, καὶ τῆς μεγάλης
20 φλεβὸς ἀπὸ ταύτης ἀφορισθείσης, δύο ὀμφαλοὶ ἀπὸ

---

¹ secl. Platt, sed fortasse sanandum : *et propter hoc fit
molle* (μαλακόν scribendum pro μᾶλλον?) *quando calefit.
cum ergo acciderit ei humiditas ex superfluitate humiditatum
corrumpetur* Σ.   ² περὶ codd.*: om. Bekker.

---

a Aristotle's observation that the yolk liquefies is quite
correct. The white loses water, partly by evaporation
through the shell, and partly to the growing embryo *via* the
yolk-sac and the yolk.

b Perhaps this should be emended to read " when it is

becomes fluid when it is concocted in the earth or by means of incubation,[a] and in that condition it becomes nourishment for the animals that are taking shape. When subjected to fire, or roasted, it does not become hard, because it is by its nature earthy in the same way that wax is ; and that is the reason why, when eggs are overheated, [unless they are from a liquid residue][b] they become serous, and turn rotten (*ouria*). The white, on the other hand, does not congeal as a result of frost, but tends rather to become fluid (I have given the reason earlier) ; and when subjected to fire, it becomes solid. This is why, when it is concocted in connexion with the generation of the young animals, it thickens ; for it is the white out of which the animal forms and develops, while the yolk becomes nourishment for it, and is the source from which the parts as they are formed at the various stages derive their growth. That, too, is why the yolk and the white are kept distinct and separate from each other by membranes, as having a different nature from each other. For an exact account of how these stand to one another both at the beginning of the process of generation and during the process of the young animals' formation, also for an account of the membranes and umbilical cords, what is written in the *Researches*[c] should be studied ; for our present inquiry it is sufficient that thus much should be clear, viz., that once the heart has been formed (this comes first of all) and the Great Blood-vessel has been marked off from it, two umbilical cords extend from

heated, it becomes soft ; and so when it is subjected to fluid, it turns rotten owing to the excess of fluidity " (*cf.* 753 a 34, above).

[c] *H.A.* 561 a 3—562 b 2 ; but the description there is no fuller.

τῆς φλεβὸς τείνουσιν, ὁ μὲν εἰς τὸν ὑμένα τὸν
περιέχοντα τὸ ὠχρόν, ὁ δ' ἕτερος εἰς τὸν ὑμένα
τὸν χοριοειδῆ,[1] ὃς κύκλῳ περιέχει τὸ ζῷον· ἔστι
δ' οὗτος περὶ τὸν ὑμένα τὸν τοῦ ὀστράκου. διὰ μὲν
οὖν θατέρου λαμβάνει τὴν ἐκ τοῦ ὠχροῦ τροφήν,
25 τὸ δ' ὠχρὸν γίνεται πλέον· ὑγρότερον γὰρ γίνεται
θερμαινόμενον, δεῖ γὰρ τὴν τροφὴν σωματώδη
οὖσαν ὑγρὰν εἶναι καθάπερ τοῖς φυτοῖς, ζῇ δὲ τὸ
πρῶτον καὶ τὰ ἐν τοῖς ᾠοῖς γιγνόμενα καὶ τὰ ἐν
τοῖς ζῴοις φυτοῦ βίον· τῷ πεφυκέναι γὰρ ἔκ τινος
λαμβάνει τὴν πρώτην αὔξησιν καὶ τροφήν. ὁ δ'
30 ἕτερος ὀμφαλὸς τείνει εἰς τὸ περιέχον χόριον.
δεῖ γὰρ ὑπολαβεῖν τὰ ᾠοτοκούμενα τῶν ζῴων
πρὸς μὲν τὸ ὠχρὸν οὕτως ἔχειν [τὸν νεοττὸν][2]
ὥσπερ πρὸς τὴν μητέρα τὰ ζῳοτοκούμενα ἔμβρυα,
ὅταν ἐν τῇ μητρὶ ᾖ, ἐπεὶ γὰρ οὐκ ἐκτρέφονταί γε
ἐν τῇ μητρὶ τὰ ᾠοτοκούμενα, ἐκλαμβάνει τι μέρος
35 αὐτῆς· πρὸς δὲ τὸν ἐξωτάτω ὑμένα τὸν αἱματώδη
754 a ὡς πρὸς τὴν ὑστέραν. ἅμα δὲ περί τε τὸ ὠχρὸν
καὶ τὸ χόριον τὸ ἀνάλογον τῇ ὑστέρᾳ τὸ ὄστρακον
τοῦ ᾠοῦ περιπέφυκεν, ὥσπερ ἂν εἴ τις περιθείη
περί τε τὸ ἔμβρυον αὐτὸ καὶ περὶ τὴν μητέρα ὅλην.
ἔχει δ' οὕτως, διότι δεῖ τὸ ἔμβρυον ἐν τῇ ὑστέρᾳ
5 εἶναι καὶ πρὸς τῇ μητρί. ἐν μὲν οὖν τοῖς ζῳοτοκουμένοις ἡ ὑστέρα ἐν τῇ μητρί ἐστιν, ἐν δὲ τοῖς

---

[1] χοροειδῆ vulg.   [2] seclusit Sus.

[a] Aristotle's two umbilical cords here are (1) the yolk-sac
stalk and (2) the allantois. See figure, p. 369.
[b] See above, 753 b 2, n.
[c] Cf. Harvey, " An egge is, as it were, an exposed womb;
wherein there is a substance concluded, as the Representative
and Substitute or Vicar of the breasts."

this blood-vessel, one to the membrane which surrounds the yolk, the other to the chorion-like membrane which surrounds the animal on all sides ; this
one goes round inside the membrane of the shell.[a]
Through one of these cords the embryo receives the
nourishment from the yolk ; and the yolk increases
in bulk, becoming more fluid as it is heated,[b] since
the nourishment, being corporeal, must be available in fluid form, just as it must for plants, and the
embryos that are in process of formation, either
within the egg or within the uterus, are to begin with
living the life of a plant, since their first growth and
nourishment they obtain through being fastened on
to something. The other umbilical cord extends to
the chorion which surrounds the embryo. In the
case of the animals that are produced oviparously, we should think of them (a) as having the
same relationship to the yolk as the viviparously
formed embryos have to the mother, so long as
they are within the mother ; for since the nourishment of the oviparously formed embryos is not
completed within the mother, when they leave her
they take a part of her out with them ; (b) as having
the same relationship to the outermost—the bloodlike—membrane as the other embryos have to the
uterus. Also, the eggshell which encloses the yolk
and the chorion gives the egg an envelope analogous to the uterus : it is as though you were to
envelop both a viviparously produced embryo itself
and its mother entire.[c] The reason why this is so is
that the embryo must be in the uterus, *i.e.*, in contact
with the mother. Very well then : in the case of the
viviparously produced animals, the uterus is in the
mother ; but with the oviparously produced ones

ὠοτοκουμένοις ἀνάπαλιν, ὥσπερ ἂν εἴ τις εἴποι τὴν
μητέρα ἐν τῇ ὑστέρᾳ εἶναι· τὸ γὰρ ἀπὸ τῆς μητρὸς
γινόμενον [ἡ τροφὴ]¹ τὸ ὠχρόν ἐστιν. αἴτιον δ' ὅτι
ἡ ἐκτροφὴ οὐκ ἐν τῇ μητρί ἐστιν.

10 Αὐξανομένων δὲ πρότερον ὁ ὀμφαλὸς συμπίπτει
ὁ πρὸς τὸ χόριον, διότι ταύτῃ δεῖ τὸ ζῷον ἐξελθεῖν,
τὸ δὲ λοιπὸν τοῦ ὠχροῦ καὶ ὁ ὀμφαλὸς ὁ εἰς τὸ
ὠχρὸν ὕστερον· δεῖ γὰρ ἔχειν τροφὴν εὐθὺς τὸ
γενόμενον· οὔτε γὰρ ἀπὸ τῆς μητρὸς τιτθεύεται,
δι' αὑτοῦ τε οὐκ εὐθὺς δύναται πορίζεσθαι τὴν
15 τροφήν· διόπερ ἐντὸς εἰσέρχεται τὸ ὠχρὸν μετὰ
τοῦ ὀμφαλοῦ, καὶ περιφύεται ἡ σάρξ.

Τὰ μὲν οὖν ἐκ τῶν τελείων ᾠῶν γινόμενα θύραζε
τοῦτον γίγνεται τὸν τρόπον ἐπί τε τῶν ὀρνίθων
καὶ τῶν τετραπόδων, ὅσα ᾠοτοκεῖ τὸ ᾠὸν τὸ
σκληρόδερμον. διάδηλα δὲ ταῦτα μᾶλλον ἐπὶ τῶν
μειζόνων· ἐν γὰρ τοῖς ἐλάττοσιν ἀφανῆ διὰ μι-
20 κρότητα τῶν ὄγκων ἐστίν.

III Ἔτι δ' ἐστὶν ᾠοτόκον τὸ τῶν ἰχθύων γένος.

Τούτων δὲ τὰ μὲν ἔχοντα κάτω τὴν ὑστέραν
ἀτελὲς ᾠὸν τίκτει διὰ τὴν πρότερον εἰρημένην
αἰτίαν, τὰ δὲ καλούμενα σελάχη τῶν ἰχθύων ἐν
25 αὑτοῖς μὲν ᾠοτοκεῖ τέλειον ᾠὸν ἔξω δὲ ζῳοτοκεῖ,
πλὴν ἑνὸς ὃν καλοῦσι βάτραχον· οὗτος δ' ᾠοτοκεῖ
θύραζε τέλειον ᾠὸν μόνος. αἰτία δ' ἡ τοῦ σώματος
φύσις· τήν τε γὰρ κεφαλὴν πολλαπλασίαν ἔχει τοῦ
λοιποῦ σώματος, καὶ ταύτην ἀκανθώδη καὶ σφόδρα

---

¹ seclusit Sus.

ᵃ See 718 b 23.
ᵇ *Lophius piscatorius* does not conform to the habits of
the Selachians because it is not in fact a Selachian ; Aristotle
wrongly includes it among them.

it is the other way round—the mother is in the uterus, as you might say, because in this case that which comes from the mother [the nourishment] is the yolk. The reason is that the embryo's period of nourishment does not reach completion within the mother.

As the embryos grow, the first of the umbilical cords to collapse is the one which connects to the chorion, because that is the point at which the young animal will have to make its way out ; the rest of the yolk and the cord which connects to it collapse later, because the young animal must have nourishment immediately it is hatched, as it is neither nursed by its mother nor able immediately to get nourishment by means of itself. That is why the yolk goes inside it together with the umbilical cord and the flesh grows round it.

Such is the manner in which animals which are brought to birth out of perfect eggs are produced in the case of those birds and fishes which lay a hard-shelled egg. The points mentioned are to be seen more clearly in the larger animals ; in the smaller ones they are not so obvious owing to the small bulk of the animals.

Another member of the Ovipara is the tribe of fishes. ^{III (ii.) Fishes :}

Those fishes whose uterus is low down lay an imperfect egg. The cause of this I have stated previously.[a] The Selachian fishes as they are called produce a perfect egg internally though they are externally viviparous, except for one which they call the fishing-frog[b] ; this is the only one that lays a perfect egg externally. The cause of this is the nature of its body. Its head is several times as large as the rest of its body, and, besides that, spiny and extremely ^{(a) Selachia.}

τραχεῖαν· ὥστε[1] διόπερ οὐδ' ὕστερον εἰσδέχεται
30 τοὺς νεοττούς, οὐδ' ἐξ ἀρχῆς ζῳοτοκεῖ· τὸ γὰρ
μέγεθος καὶ ἡ τραχύτης τῆς κεφαλῆς ὥσπερ καὶ
εἰσελθεῖν κωλύει, οὕτω καὶ ἐξελθεῖν. ἐπεὶ δὲ
μαλακόδερμόν ἐστι τὸ ᾠὸν τὸ τῶν σελαχῶν (οὐ γὰρ
δύνανται σκληρύνειν καὶ ξηραίνειν[2] τὸ περίξ· ψυχρότεροι γὰρ τῶν ὀρνίθων εἰσίν), τὸ τῶν βατράχων
ᾠὸν μόνον στερεόν ἐστι καὶ στιφρὸν πρὸς τὴν ἔξω
35 σωτηρίαν, τὰ δὲ τῶν ἄλλων ὑγρὰ καὶ μαλακὰ τὴν

φύσιν· σκεπάζεται γὰρ ἐντὸς τῷ σώματι τῷ τῆς
ἐχούσης.

Ἡ δὲ γένεσις ἐκ τοῦ ᾠοῦ τοῖς τε βατράχοις ἔξω
τελειουμένοις καὶ τοῖς ἐντὸς ἡ αὐτή, τούτοις δὲ καὶ
τοῖς τῶν ὀρνίθων τῇ μὲν ὁμοία τῇ δὲ διάφορός
5 ἐστιν. πρῶτον μὲν γὰρ οὐκ ἔχουσι τὸν ἕτερον
ὀμφαλὸν τὸν εἰς τὸ χόριον τείνοντα, ὅ ἐστιν ὑπὸ τὸ
περιέχον ὄστρακον, τούτου δ' αἴτιον ὅτι τὸ περίξ
ὄστρακον οὐκ ἔχουσιν· οὐδὲν γὰρ αὐτοῖς χρήσιμον·
σκεπάζει γὰρ ἡ μήτηρ, τὸ δ' ὄστρακόν ἐστι τοῖς
ἐκτικτομένοις ᾠοῖς ἀλεωρὰ πρὸς τὰς θύραθεν
βλάβας. ἔπειθ' ἡ γένεσις ἐξ ἄκρου μέν ἐστι τοῦ
10 ᾠοῦ καὶ τούτοις, ἀλλ' οὐχ ᾗ προσπέφυκε πρὸς τὴν
ὑστέραν· οἱ γὰρ ὄρνιθες ἐκ τοῦ ὀξέος γίνονται,
ταύτῃ δ' ἦν ἡ τοῦ ᾠοῦ πρόσφυσις. αἴτιον δ' ὅτι
τὸ μὲν τῶν ὀρνίθων χωρίζεται τῆς ὑστέρας, τῶν δὲ
τοιούτων οὐ πάντων ἀλλὰ τῶν πλείστων πρὸς τῇ

---

[1] ὥστε PZ[1] : om. vulg.
[2] καὶ ξηραίνειν PZΣ : om. vulg.

---

[a] In several of the Selachia the young have the habit of
swimming into the mouth of the parent for shelter. This

rough ; so that the reason why it does not take its young ones in afterwards [a] is also the reason why it does not produce them alive at the outset : just as the size and roughness of its head prevents them from going in, so also it prevents them from coming out. Since, then, the egg of the Selachia has a soft shell (because they cannot make the envelope hard and solid, being colder creatures than birds are), the egg of the fishing-frog is the only one that is hard and stout, so as to keep it safe in the outside world ; the others' eggs are liquid and soft in nature, because they are inside the mother and get their shelter from her body.

The process of generation out of the egg is the same both for the fishing-frogs, which are perfected externally, and for those Selachia which are perfected internally ; and as between the latter and the birds, it is partly similar, partly dissimilar. First of all, they lack the second umbilical cord which extends to the chorion under the surrounding shell, and the reason for this is that they have not got this shell round them, as it is no use to them, their shelter being provided by the mother ; whereas for eggs that are laid externally the shell is there to act as a protection against injury from without. Secondly, with these, as with birds, the process of generation originates from the extremity of the egg, though not at the place where it is attached to the uterus. A bird's development begins from the pointed end, which is the place where the egg was attached, the reason being that a bird's egg becomes separated from the uterus, whereas the eggs of most, though not all,

Development of the embryo.

may be the foundation of this remark ; *cf.* also *H.A.* 565 b 24 ff.

ὑστέρᾳ προσπέφυκε τὸ ᾠὸν τέλειον ⟨ὄν⟩.[1] ἐπ'
15 ἄκρῳ δὲ γιγνομένου τοῦ ζῴου καταναλίσκεται τὸ
ᾠόν, ὥσπερ καὶ ἐπὶ τῶν ὀρνίθων καὶ τῶν ἄλλων
⟨ᾠῶν⟩[2] τῶν ἀπολελυμένων, καὶ τέλος πρὸς τῇ
ὑστέρᾳ ὁ ὀμφαλὸς προσπέφυκε τῶν ἤδη τελείων.
ὁμοίως δ' ἔχει καὶ ὅσων ἀπολέλυται τὰ ᾠὰ τῆς
ὑστέρας· ἐνίοις γὰρ αὐτῶν, ὅταν τέλειον γένηται τὸ
ᾠόν, ἀπολύεται.

20 Ἀπορήσειεν ἂν οὖν τις διὰ τί διαφέρουσιν αἱ
γενέσεις τοῖς ὄρνισι κατὰ τοῦτο καὶ τοῖς ἰχθύσιν.
αἴτιον δ' ὅτι τὰ μὲν τῶν ὀρνίθων κεχωρισμένον
ἔχει τὸ λευκὸν καὶ τὸ ὠχρόν, τὰ δὲ τῶν ἰχθύων
μονόχροα, καὶ πάντῃ μεμιγμένον τὸ τοιοῦτον, ὥστ'
οὐθὲν κωλύει ἐξ ἐναντίας ἔχειν τὴν ἀρχήν· οὐ γὰρ
25 μόνον κατὰ τὴν πρόσφυσίν ἐστι τοιοῦτον ἀλλὰ καὶ
καταντικρύ, τὴν δὲ τροφὴν ῥᾴδιον[3] ἕλκειν ἐκ τῆς
ὑστέρας πόροις τισὶν ἀπὸ ταύτης τῆς ἀρχῆς.
δῆλον δ' ἐπὶ τῶν μὴ ἀπολυομένων ᾠῶν· ἐν[4] ἐνίοις
γὰρ τῶν σελαχῶν οὐκ ἀπολύεται τῆς ὑστέρας τὸ
ᾠόν, ἀλλ' ἐχόμενον μεταχωρεῖ κάτω πρὸς τὴν
30 ζῳοτοκίαν, ἐν οἷς τελεωθὲν τὸ ζῷον ἔχει τὸν
ὀμφαλὸν ἐκ τῆς ὑστέρας ἀνηλωμένου τοῦ ᾠοῦ.
φανερὸν οὖν ὅτι καὶ πρότερον ἔτεινον οἱ πόροι

---

[1] ⟨ὄν⟩ coni. Platt.
[3] ⟨ᾠῶν⟩ Peck ; cf. infra v. 27 ubi ᾠῶν om. Z[1].
[3] ῥᾴδιον Y, leviter Σ : ῥᾷον vulg.
[4] et ᾠῶν et ἐν om. Z[1] ut vid., suppl. Z[2].

---

[a] As in the " smooth dogfish " ; see note on 754 b 34,
below.
[b] Excluding, of course, the statement immediately pre-

fishes of this class remain attached to the uterus
even when they are perfect. As the young animal
develops at the extremity, the egg gets used up
(just as in the case of birds and the other eggs that
have been released from the uterus), and at the
final stage, by which the animal has reached its
perfect development, the umbilical cord remains
attached to the uterus.[a] The like [b] holds good in
the case of those Selachia whose eggs have been released from the uterus, there being some whose egg
is released as soon as it is perfected.[c]

In view of what has been said, the puzzle may be Differences
as between
Birds and
Selachia.
raised why the processes of generation in birds and
fishes differ in this respect. The reason is that in
birds' eggs the white and the yolk are separate,
whereas fishes' eggs are single-coloured, the contents
being mixed up together throughout, so that there
is nothing to prevent the " principle " in them being
at the opposite end ; the egg is of similar composition both at the end where it is fastened and at the
opposite end, and it is easy for it to draw the nourishment out of the uterus by means of passages which
lead from this principle. This can clearly be seen in
those eggs which do not get released, for in the case
of some of the Selachia the egg does not get released
from the uterus, but remains connected as it proceeds
downwards to produce the young alive. In these
cases, the young animal, after it has reached its
perfect development, retains its umbilical cord joined
to the uterus when the egg has been consumed. Thus
it is plain that during the earlier stages also, while

ceding. He means the embryo develops at the extremity.
The process is *similar* (" like "), not identical.
  [c] That of the " fishing-frog " ; but see 754 a 26, n.

τοῦ ᾠοῦ ἔτι ὄντος περὶ ἐκεῖνο πρὸς τὴν ὑστέραν. τοῦτο δὲ συμβαίνει, καθάπερ εἴπομεν, ἐν τοῖς γαλεοῖς τοῖς λείοις.

Διαφέρει μὲν οὖν ἡ γένεσις κατὰ ταῦτα τῶν
35 ἰχθύων τοῖς ὄρνισι, καὶ διὰ τὰς εἰρημένας αἰτίας·

τὰ δ' ἄλλα συμβαίνει τὸν αὐτὸν τρόπον. τόν τε γὰρ ὀμφαλὸν ἔχουσι τὸν ἕτερον ὡσαύτως, ὥσπερ οἱ ὄρνιθες πρὸς τὸ ᾠχρόν, οὕτως οἱ ἰχθύες πρὸς τὸ ὅλον ᾠόν (οὐ γάρ ἐστιν αὐτοῦ τὸ μὲν λευκὸν τὸ δ' ᾠχρόν, ἀλλὰ μονόχρων πᾶν), καὶ τρέφονται
5 ἐκ τούτου, καταναλισκομένου τε ἐπέρχεται καὶ περιφύεται ἡ σὰρξ ὁμοίως.

Περὶ μὲν οὖν τῶν ἐν αὑτοῖς μὲν ᾠοτοκούντων τέλειον ᾠὸν θύραζε δὲ ζῳοτοκούντων τοῦτον ἔχει
IV τὸν τρόπον ἡ γένεσις, οἱ δὲ πλεῖστοι τῶν ἄλλων ἰχθύων ἐκτὸς ᾠοτοκοῦσιν, ἀτελὲς δ' ᾠὸν πάντες πλὴν βατράχου· περὶ δὲ τούτου τὸ αἴτιον εἴρηται
10 πρότερον. εἴρηται δὲ καὶ περὶ τῶν ἀτελῆ τικτόν-των τὸ αἴτιον.

Ἡ δὲ γένεσις καὶ τούτων ἡ μὲν ἐκ τοῦ ᾠοῦ τὸν αὐτὸν ἔχει τρόπον ὅνπερ καὶ τῶν σελαχῶν τῶν ἐντὸς ᾠοτοκούντων, πλὴν ἥ γ' αὔξησις ταχεῖα καὶ ἐκ μικρῶν, καὶ τὸ ἔσχατον τοῦ ᾠοῦ σκληρότερον.
15 ἡ δὲ τοῦ ᾠοῦ αὔξησις ὁμοία τοῖς σκώληξίν ἐστιν· καὶ γὰρ τὰ σκωληκοτοκοῦντα τῶν ζῴων μικρὸν ἀποτίκτει τὸ πρῶτον, τοῦτο δ' αὐξάνεται δι' αὑτοῦ

---

^a The *Mustelus laevis.* The remarkable description of the placentoid structure in the embryo of this species will be found in *H.A.* 565 b 2 ff. The structure is similar both in form and function to the placenta of a mammal, although its origin is not the same. It was rediscovered by Johannes

302

the creature was still enveloped in the egg, the passages extended to the uterus. This occurs, as we have said, in the smooth dogfish.[a]

I have now mentioned the respects in which the process of generation of fishes differs from that of birds, and also the causes thereof. Otherwise, they both follow the same course. The fishes have one of the two umbilical cords, just as the birds have (in birds it connects with the yolk, in fishes with the entire egg, because the fish's egg is all single-coloured and lacks the distinction into white and yolk), and they obtain their nourishment by means of this; as it gets consumed the flesh in like manner encroaches upon it and grows round it.

I have now described the manner of formation of those fishes which produce a perfect egg internally and are viviparous externally.

The majority of the remaining fishes are externally oviparous; and all of them except the fishing-frog produce an imperfect egg. The reason for this exception I have given earlier.[b] I have also given the reason why the others produce imperfect eggs.[c]

So far as the process of formation is concerned, the development from the egg follows the same lines as the internally oviparous Selachia, except that they start very small and grow very quickly, and the outside of the egg is harder. The growth of the egg is like ⟨that of⟩ larvae, for those animals which produce larvae produce something small to start with, which

IV ⟨b⟩ Other fishes.

Müller in the 19th century (see J. Müller, *Über den glatten Hai des Aristoteles*, Berlin, 1842; paper read Apr. 1839 and Aug. 1840). An account of the discovery, with Müller's letters, is given by W. Haberling, *Archiv f. Gesch. der Math., der Naturw. und der Technik*, 10 (1927), 166-184. *Cf.* p. ix, n.

[b] At 754 a 26.   [c] At 718 b 8.

755 a

καὶ οὐ διὰ πρόσφυσιν οὐδεμίαν. τὸ δ' αἴτιον
παραπλήσιον ὅπερ ἐπὶ τῆς ζύμης· καὶ γὰρ ἡ ζύμη
ἐκ μικρᾶς μεγάλη γίνεται τοῦ μὲν στερεωτέρου
ὑγραινομένου, τοῦ δ' ὑγροῦ πνευματουμένου. δη-
20 μιουργεῖ δὲ τοῦτο ἡ τοῦ ψυχικοῦ θερμοῦ φύσις ἐν
τοῖς ζῴοις,[1] ἐν δὲ ταῖς ζύμαις ἡ τοῦ χυμοῦ τοῦ
συγκραθέντος θερμότης. αὐξάνεται μὲν οὖν τὰ
ᾠὰ ἐξ ἀνάγκης μὲν διὰ ταύτην τὴν αἰτίαν (ἔχει
γὰρ περίττωμα ζυμῶδες), χάριν δὲ τοῦ βελτίονος·
ἐν ταῖς ὑστέραις γὰρ ἀδύνατον αὐτοῖς λαμβάνειν
25 ὅλην τὴν αὔξησιν διὰ τὴν τῶν ζῴων πολυτοκίαν
τούτων. διὰ τοῦτο γὰρ καὶ μικρὰ πάμπαν ἀποκρίνεται καὶ ταχεῖαν λαμβάνει τὴν αὔξησιν, μικρὰ μὲν
διὰ τὸ στενοχωρῇ τὴν ὑστέραν εἶναι πρὸς τὸ πλῆθος
τῶν ᾠῶν, ταχὺ δ' ὅπως μὴ χρονιζόντων ἐν τῇ
γενέσει περὶ τὴν αὔξησιν φθείρηται τὸ γένος, ἐπεὶ
30 καὶ νῦν τὰ πολλὰ φθείρεται τῶν ἐκτικτομένων
κυημάτων. διόπερ πολύγονόν ἐστι τὸ γένος τὸ
τῶν ἰχθύων· ἀναμάχεται γὰρ ἡ φύσις τῷ πλήθει
τὴν φθοράν. εἰσὶ δέ τινες οἳ διαρρήγνυνται τῶν
ἰχθύων, οἷον ἡ καλουμένη βελόνη, διὰ τὸ μέγεθος
τῶν ᾠῶν· αὕτη γὰρ ἀντὶ τοῦ πολλὰ μεγάλα τὰ
35 κυήματα ἴσχει· τοῦ γὰρ πλήθους ἡ φύσις ἀφελοῦσα
προσέθηκε πρὸς τὸ μέγεθος.

Ὅτι μὲν οὖν αὐξάνεταί τε καὶ δι' ἣν αἰτίαν τὰ
755 b τοιαῦτα τῶν ᾠῶν, εἴρηται.

---

[1] ᾠοῖς coni. Platt.

---

[a] Such as an umbilical cord.
[b] Or " becoming inflated with *pneuma*." *Cf.* 762 a 19.
[c] Lit., "of the natural substance of the soul-heat " (a
periphrasis).　　　　　　　　　　[d] *Cf.* 739 b 23, n.
[e] *Cf. H.A.* 567 b 23.　One of the " pipe-fishes," perhaps

304

grows by its own means and not in virtue of any attachment.[a] The reason for this is on a par with the reason why yeast grows. Yeast, like these, is small in bulk to start with and gets larger : this growth is due to its more solid portion turning fluid, and the fluid turning into *pneuma*.[b] This is the handiwork of the soul-heat [c] in the case of animals, of the heat of the humour blent with it in the case of the yeast. Eggs thus grow *of necessity* on account of this cause (*i.e.*, they contain a yeast-like residue [d]), but also they grow *for the sake of what is better*, since it is impossible for them to obtain all their growth in the uterus owing to the prolific habit of these animals. That is why the eggs are quite small when they are discharged and why they grow quickly : they are small because the uterus is not roomy enough to hold so large a number of eggs, and they grow quickly to prevent the destruction of their kind which would occur as a result of their spending a long time over the growing period of their formation. Even as it is, the majority of the fetations that are laid externally get destroyed. That is why the fish tribe is prolific : Nature makes good the destruction by sheer weight of numbers. There are also some fishes, such as the one known as *belonē*,[e] which burst asunder owing to the *size* of the eggs, the fetations of this fish being large instead of numerous ; here Nature has taken away from their number and added to their size.

I have now described the growth of eggs of this sort and have stated the Cause of it.

Syngnathus acus. In this group (of which the well-known " sea-horse " is another member) the male incubates the eggs in a brood-pouch formed by the pelvic fins. Aristotle correctly states at *H.A. loc. cit.* that the fish is none the worse for its " bursting asunder."

V Ὅτι δ' ᾠοτοκοῦσι καὶ οὗτοι[1] οἱ ἰχθύες, σημεῖον
τὸ καὶ τοὺς ζῳοτοκοῦντας τῶν ἰχθύων, οἷον τὰ
σελάχη, ᾠοτοκεῖν ἐν αὑτοῖς πρῶτον. δῆλον γὰρ
ὅτι τὸ γένος ὅλον ἐστὶν ᾠοτόκον τὸ τῶν ἰχθύων.
5 τέλος μέντοι οὐθὲν λαμβάνει τῶν τοιούτων ᾠῶν,
ὅσων ἐστὶ τὸ μὲν θῆλυ τὸ δ' ἄρρεν καὶ γίγνονται ἐξ
ὀχείας, ἐὰν μὴ ἐπιρράνῃ ὁ ἄρρην τὸν θορόν. εἰσὶ
δέ τινες οἵ φασι πάντας εἶναι τοὺς ἰχθύας[2] θήλεις
ἔξω τῶν σελαχῶν, οὐκ ὀρθῶς λέγοντες. οἴονται
γὰρ διαφέρειν τῶν νομιζομένων ἀρρένων τοὺς
10 θήλεις αὐτῶν ὥσπερ τῶν φυτῶν ἐν ὅσοις τὸ μὲν
καρποφορεῖ τὸ δ' ἄκαρπόν ἐστιν, οἷον ἐλαία καὶ
κότινος καὶ συκῆ καὶ ἐρινεός· ὁμοίως δὲ καὶ τοὺς
ἰχθῦς πλὴν τῶν σελαχῶν· τούτοις γὰρ οὐκ ἀμφισ-
βητοῦσιν. καίτοι ὡσαύτως τε διάκεινται οἱ ἄρρενες περὶ τὰ θορικὰ οἵ τε σελαχώδεις καὶ οἱ ἐν
15 τῷ γένει τῷ τῶν ᾠοτόκων, καὶ σπέρμα κατὰ τὴν

---

[1] καὶ οὗτοι om. Z¹.   [2] sic SYZ: ἰχθῦς vulg.

---

[a] The argument seems to be this. Aristotle is arguing
from the principle that the production of eggs, if a characteristic of any fishes, must be a characteristic of the *whole*
tribe of fishes (*cf.* his enunciation of a similar principle below,
755 b 36 : it would be fantastic, he says, if the distinction of
sexes were found in some fishes and not throughout the
whole tribe of them, just as it is found throughout the *whole*
tribe of Vivipara. *Cf.* also 759 b 14 and 34). Nobody,
however, disputes that the Selachia, which are fishes, are
oviparous (internally), nor that they have the distinction of
sexes. Hence, *ex hypothesi*, the *whole* tribe of fishes is oviparous (though of course the eggs are " imperfect " ones),
and has the sexes distinct. Thus the argument will be
against those who hold that fish produce not eggs but larvae
(see 757 a 29 ff.), and do not have the sexes distinct. No

A proof that these fishes as well as the others V
produce eggs is that even the viviparous fishes, such Erroneous
theories :
as the Selachia, produce eggs internally at the first (1) Fish are
stage. Why is this a proof? Because *a* it is plain that not ovipar-
ous and do
the *whole* of the tribe of fishes is oviparous. At the not have
sexes.
same time, no eggs of this sort reach perfection,—*i.e.*,
eggs of species where both males and females exist,
and which are formed as the result of copulation *b*—
unless the male sprinkles his genital fluid (milt) upon
them ; though there are some people who hold—
incorrectly—that all fish are female apart from the
Selachia. Their view is that the females differ from
what are reputed to be males in the same way as
those species of plants in which one tree will bear
fruit and another will bear none (*e.g.*, the olive and
oleaster, the fig and caprifig).*c* They say it is just
the same with fish, except in the case of the Selachia,
where they do not dispute the point. But as a matter
of fact there is no difference as regards their seminal
parts between males of the Selachian fishes and males
which belong to the oviparous group, and semen can

doubt there were some who maintained that the eggs of
fishes, which Aristotle holds to be true, though " imperfect,"
eggs, were on a par with the " eggs " out of which caterpillars
and the like developed ; the latter, however, Aristotle holds
to be " larvae " and not true eggs (see 758 b 9 ff.) ; and
larvae, of course, are often found in connexion with creatures
in which (according to Aristotle) the sexes are not distinct and
are formed without copulation. Thus, the two points on
which Aristotle insists, (1) that fishes have sexes and copulate,
and (2) that they produce eggs, not larvae, are mutually
corroborative.

b The exception is the *erythrinus* ; see 741 a 36, n.

c See above, 715 b 25 ; also *H.A.* 557 b 31. There seems
to be no similar phenomenon in the case of the olive, but it
was a common practice to call some trees male and others female : see Theophr. *Hist. plant.* I. 8. 2, and *cf.* Soph. *Tr.* 1196.

755 b

ὥραν φαίνεται ἀμφοῖν ἐκθλιβόμενον. ἔχουσι δὲ
καὶ ὑστέρας αἱ θήλειαι· ἔδει δ' οὐ μόνον τοὺς
ᾠοτοκοῦντας ἀλλὰ καὶ τοὺς ἄλλους ἔχειν μέν, ἀλλὰ
διαφερούσας τῶν ᾠοτοκούντων, [ὥσπερ αἱ ἡμίονοι
ἐν τῷ γένει τῶν λοφούρων,]¹ εἴπερ ἦν θῆλυ τὸ
20 γένος πᾶν, ἀλλ' ἄτεκνοί τινες αὐτῶν. νῦν δ' οἱ
μὲν ἔχουσι θορικὰ οἱ δ' ὑστέρας, καὶ ἐν ἅπασιν
ἔξω δυοῖν, ἐρυθρίνου καὶ χάννης, αὕτη ἐστὶν ἡ
διαφορά· οἱ μὲν γὰρ θορικὰ ἔχουσιν, οἱ δ' ὑστέρας.²
ἡ δ' ἀπορία δι' ἣν οὕτως ὑπολαμβάνουσιν, εὔλυτος
τὸ συμβαῖνον ἀκούσασιν. οὐθὲν γὰρ τῶν ὀχευο-
μένων πολλά φασι τίκτειν, λέγοντες ὀρθῶς· ὅσα
25 γὰρ ἐξ αὑτῶν γεννᾷ τέλεια ἢ ζῷα ἢ ᾠά, οὐ πολυτοκεῖ οὕτως ὥσπερ οἱ ᾠοτοκοῦντες τῶν ἰχθύων·
ἄπλετον γάρ τι τὸ³ τούτων πλῆθος τῶν ᾠῶν ἐστιν.
ἀλλὰ τοῦτο οὐχὶ συνεωράκεσαν, ὅτι οὐχ ὁμοιο-
τρόπως τοῖς τῶν ὀρνίθων ἔχει τὰ περὶ τὰ ᾠὰ τῶν
ἰχθύων. οἱ μὲν γὰρ ὄρνιθες καὶ τῶν τετραπόδων
30 ὅσα ᾠοτοκεῖ, καὶ εἴ τινα τῶν σελαχωδῶν, τέλειον
ᾠὸν γεννῶσι, καὶ οὐ λαμβάνει ἐξελθὸν αὔξησιν· οἱ
δ' ἰχθύες ἀτελῆ, καὶ λαμβάνει θύραζε τὰ ᾠὰ τὴν
αὔξησιν. ἔτι καὶ ἐπὶ τῶν μαλακίων τὸν αὐτὸν
ἔχει τρόπον καὶ ἐπὶ τῶν μαλακοστράκων, ἃ καὶ

---

¹ haec verba post τινες αὐτῶν transtulit Platt ; ego seclusi.
fortasse plura corrupta.
² οἱ μὲν . . . ὑστέρας secl. A.-W.
³ τι τὸ Z : τι vulg.

---

[a] i.e., those which are in fact males.
[b] See note on 777 b 5.
[c] Platt transposes these words to follow " of young " a few
lines above ; no doubt they were part of a marginal note on

clearly be seen oozing out from males of both groups at the proper season. Also, the females have a uterus ; but if the whole tribe of fishes really were female, some of them being unproductive of young,[a] then not only those fishes which lay eggs but all the others as well ought to have a uterus, though no doubt different in form from that of the ones which lay the eggs [like female mules in the class of bushy-tailed [b] animals].[c] In fact, however, while some fish have a uterus, others have seminal parts, and this distinction is found in all species except two, the *erythrinus* and the *channa*[d] : some have seminal parts, others have an uterus. The puzzle which makes people put forward this theory is easily solved when we hear what the facts are. These people allege—and here they are quite correct—that none of the animals which copulate produces many young, for of all the animals which generate out of themselves either perfect animals or perfect eggs, none is so prolific as the oviparous fishes, the number of their eggs of course being something enormous. But this point they have overlooked : eggs of fishes do not behave in precisely the same way as those of birds. Birds, oviparous quadrupeds, and any oviparous Selachians there may be,[e] produce a perfect egg, and once it has left the parent it grows no further ; fish on the other hand produce imperfect eggs, which do grow after they have left the parent. Furthermore, the same occurs in the case of the Cephalopods and Crustacea ; and these creatures can actually be seen

the word ἄτεκνοι, but they are meaningless and irrelevant anywhere in the text.
    [d] For *erythrinus* see note, 741 a 36 ; the *channa* is another species of *Serranus*, probably *S. scriba*.
    The fishing-frog ; but see 754 a 26, n.

755 b

συνδυαζόμενα ὁρᾶται διὰ τὸ χρόνιον εἶναι τὸν
35 συνδυασμὸν αὐτῶν· καὶ τούτων φανερόν ἐστι τὸ
μὲν ἄρρεν ὄν, τὸ δ' ἔχον ὑστέραν. ἄτοπον δὲ καὶ
756 a τὸ μὴ ἐν παντὶ ⟨τῷ⟩[1] γένει ταύτῃ εἶναι τὴν δύναμιν, ὥσπερ ἐν τοῖς ζῳοτόκοις τὸ μὲν ἄρρεν τὸ
δὲ θῆλυ. αἴτιον δὲ τοῖς ἐκείνως λέγουσι τῆς
ἀγνοίας τὸ τὰς διαφορὰς μὴ δήλας εἶναι παντο-
δαπὰς οὔσας περί τε τὰς ὀχείας τῶν ζῴων καὶ τὰς
5 γενέσεις, ἀλλ' ἐξ ὀλίγων[2] θεωροῦντας οἴεσθαι δεῖν
ἔχειν ὁμοίως ἐπὶ πάντων.

Διὸ καὶ οἱ λέγοντες τὰς κυήσεις εἶναι ἐκ τοῦ
ἀνακάπτειν τὸ σπέρμα τοὺς θήλεις τῶν ἰχθύων, οὐ
κατανενοηκότες ἔνια λέγουσιν οὕτως. ὑπὸ τὸν
αὐτὸν γὰρ καιρὸν οἵ τ' ἄρρενες τὸν θορὸν καὶ αἱ
θήλειαι τὰ ᾠὰ ἔχουσι, καὶ ὅσῳ ἂν ᾖ ἐγγυτέρω ἡ
10 θήλεια τοῦ τίκτειν, τότε πλείων καὶ ὑγρότερος ὁ
θορὸς ἐν τῷ ἄρρενι ἐγγίνεται. καὶ ὥσπερ ἡ αὔξησις
κατὰ τὸν αὐτὸν χρόνον τοῦ θοροῦ ἐν τῷ ἄρρενι καὶ
τοῦ ᾠοῦ ἐν τῇ θηλείᾳ, οὕτω καὶ ἡ ἄφεσις συμβαίνει·
οὔτε γὰρ αἱ θήλειαι ἀθρόα ἐκτίκτουσιν, ἀλλὰ κατὰ
μικρόν, οὔθ' οἱ ἄρρενες ἀθρόον ἀφιᾶσι τὸν θορόν.
15 καὶ ταῦτα πάντα συμβαίνει κατὰ λόγον. ὥσπερ
γὰρ καὶ τὸ τῶν ὀρνέων γένος ἐν ἐνίοις ἴσχει μὲν
ᾠὰ ἄνευ ὀχείας,[3] ὀλίγα δὲ καὶ ὀλιγάκις, ἀλλ' ἐξ
ὀχείας τὰ πολλά, τοῦτ' αὐτὸ συμβαίνει καὶ ἐπὶ τῶν
ἰχθύων, ἧττον δέ. ἄγονα δὲ καὶ ἀμφοτέροις γί-
20 νεται τὰ αὐτόματα ἐὰν μὴ ἐπιρράνῃ τὸ ἄρρεν, ἐν
ὅσοις γένεσιν αὐτῶν καὶ τὸ ἄρρεν ἐστίν. τοῖς μὲν
οὖν ὄρνισι, διὰ τὸ τέλεια ἐξιέναι τὰ ᾠά, ἔτι ἐντὸς

[1] τῷ supplevit Platt.
[2] sic P*SYZ : ὀλίγου vulg.
[3] ὀχείας Peck : κυήσεως vulg.

copulating, for with them copulation goes on for quite a long time, and it is plain here that one is male and the other has a uterus. Also, it would be odd if this characteristic [a] were present in a portion of the group and not in the whole of it, just as male and female are found in all the Vivipara. The reason for the ignorance of those who make the statement mentioned is that the differences in the copulation and generation of the various animals are manifold, but they are not obvious, and our friends base their study on a few instances and think the same holds good for all.

So too those who assert that female fishes conceive as a result of swallowing the male's semen have failed to notice certain points. Thus in fact milt is present in the male and eggs in the female at about the same time, and the closer the female is to laying the eggs the more abundant and the more fluid becomes the milt in the male. And just as the growth of the milt in the male and that of the egg in the female is simultaneous, so also the emission of them both is simultaneous : the females do not lay all their eggs at once, but a few at a time, and the males do not emit all their milt at once. All this is as we should expect. In the bird tribe, eggs are in some instances present without impregnation, though such eggs are not numerous and they occur but seldom, most eggs being the result of impregnation. Exactly the same occurs in fish, though to a smaller extent. These spontaneous eggs, both in birds and fish, are infertile unless (in those species where there are males as well) the male sprinkles them. With birds, owing to the fact that the eggs have reached

(2) The function of milt.

---

[a] *Dynamis, i.e.,* the existence of the two sexes. *Cf.* the beginning of ch. 5.

756 a

ὄντων ἀνάγκη τοῦτο συμβῆναι· τοῖς δ' ἰχθύσι διὰ
τὸ ἀτελῆ καὶ ἔξω λαμβάνειν τὴν αὔξησιν πᾶσιν,[1]
κἂν ἐξ ὀχείας γένηται τὸ ᾠόν, ὅμως τὰ ἔξω ἐπιρ-
25 ραινόμενα[2] σώζεται, καὶ ἐνταῦθα ἀναλίσκεται ὁ
θορὸς τοῖς ἄρρεσιν. διὸ καὶ συγκαταβαίνει ἐλατ-
τούμενος ἅμα τοῖς ᾠοῖς τοῖς ἐν τοῖς θήλεσιν· ἀεὶ γὰρ
τοῖς ἐκτικτομένοις ἐπιρραίνουσι παρακολουθοῦντες.
Ὥστε ἄρρενες μὲν καὶ θήλεις εἰσὶ καὶ ὀχεύονται
πάντες, εἰ μὴ ἔν τινι γένει ἀδιόριστόν ἐστι τὸ θῆλυ
30 καὶ τὸ ἄρρεν, καὶ ἄνευ τῆς τοῦ ἄρρενος γονῆς οὐ
γίνεται τῶν τοιούτων οὐθέν.
Συμβάλλεται δὲ πρὸς τὴν ἀπάτην αὐτοῖς καὶ τὸ
ταχὺν εἶναι τὸν συνδυασμὸν τῶν τοιούτων ἰχθύων,
ὥστε πολλοὺς λανθάνειν καὶ τῶν ἁλιέων· οὐθεὶς
γὰρ αὐτῶν οὐθὲν τηρεῖ τοιοῦτον τοῦ γνῶναι χάριν·
ἀλλ' ὅμως ὠμμένος ὁ συνδυασμός ἐστιν. τὸν

756 b

αὐτὸν γὰρ τρόπον οἵ τε δελφῖνες ὀχεύονται παραπίπτοντες καὶ οἱ ἰχθύες [ὅσοις ⟨μὴ⟩[3] ἐμποδίζει τὸ
οὐραῖον],[4] ἀλλὰ τῶν μὲν δελφίνων χρονιωτέρα ἡ
ἀπόλυσίς ἐστι, τῶν δὲ τοιούτων ἰχθύων ταχεῖα.
5 διόπερ ταύτην οὐχ ὁρῶντες, τὰς δ' ἀνακάψεις τοῦ
θοροῦ καὶ τῶν ᾠῶν, καὶ οἱ ἁλιεῖς περὶ τῆς κυήσεως
τῶν ἰχθύων τὸν εὐήθη λέγουσι λόγον καὶ τεθρυλη-

---

[1] locus fortasse corruptus. pro καὶ ἔξω λαμβάνειν habent
συμβαίνειν PSY; pro κἂν habent εἰ μὴ ἐντός Y, εἰ καὶ μὴ
ἐντός PS. fortasse scribendum διὰ τὸ ἀτελῆ ⟨ἐξιέναι⟩ καὶ ἔξω
(ἔστι τέλος) λαμβάνειν ⟨ὥσπερ καὶ ἔξω⟩ συμβαίνειν τὴν αὔξησιν
πᾶσιν, καὶ μὴ ἐντός· ⟨ὥστε⟩ κἂν κτλ.   cf. 757 a fin.

[2] ⟨μόνον⟩ A.-W., ⟨μόνα⟩ Sus., Btf.; pro κἂν ἐξ ὀχείας . . .
σώζεται et cum mas eiecerit sperma super ipsa recipiunt
virtutem suam et fiunt convenientia generationi (=γίνεται
γόνιμα) Σ.   [3] ⟨μὴ⟩ Platt, cf. H.A. 540 b 10 et 22.
[4] secl. Platt, coll. H.A. loc. cit.

a perfected state when they are discharged, this
must happen while they are as yet within the mother ;
but the eggs of fish, without exception, are imperfect
when discharged and continue their growth afterwards ; hence, even if the egg has come into being
as the result of impregnation, still, the ones which
persist safe and sound are those which get sprinkled
after they have been discharged ; that is where the
milt of the males is used up, and that is why it
comes down in smaller quantities at the same time
that the production of eggs by the females dimin-
ishes, for the males always follow up the eggs and
sprinkle them as they are laid.

Thus fish are male and female, and they copulate,
all of them (unless there be some species [a] where the
sexes are not distinct), and no fish at all of any sort
comes into being apart from the semen of the male.

Another point which helps to deceive these people
is this. Fish of this sort take only a very short
time over their copulation, with the result that many
fishermen even never see it happening, for of course
no fisherman ever watches this sort of thing for the
sake of pure knowledge. All the same, the copulation has been observed. The fish copulate in the
same way as dolphins do, by placing themselves
alongside of each other [that is, those which are
⟨not⟩ hampered by the tail]. Dolphins, however,
take longer to relieve [b] themselves, whereas fish of
this sort do so quickly. The fishermen do not notice
this, but they do notice the swallowing of the milt
and eggs by the female, and so they join the chorus
and repeat the same old stupid tale that we find told

[a] *Erythrinus* and *channa*.
[b] See note, 718 a 2.

μένον, ὅπερ καὶ Ἡρόδοτος ὁ μυθολόγος, ὡς
κυϊσκομένων τῶν ἰχθύων ἐκ τοῦ ἀνακάπτειν τὸν
θορόν, οὐ συνορῶντες ὅτι τοῦτ᾽ ἐστὶν ἀδύνατον. ὁ
γὰρ πόρος ὁ διὰ τοῦ στόματος εἰσιὼν εἰς τὴν
10 κοιλίαν φέρει, ἀλλ᾽ οὐκ εἰς τὰς ὑστέρας· καὶ τὸ
μὲν εἰς τὴν κοιλίαν ἐλθὸν ἀνάγκη τροφὴν γίνεσθαι
(καταπέττεται γάρ), αἱ δ᾽ ὑστέραι φαίνονται πλήρεις
ᾠῶν, ἃ πόθεν εἰσῆλθεν;[1]

VI   Ὁμοίως δὲ καὶ περὶ τὴν τῶν ὀρνίθων γένεσιν
ἔχει. εἰσὶ γάρ τινες οἳ λέγουσι κατὰ τὸ στόμα
15 μίγνυσθαι τούς τε κόρακας καὶ τὴν ἶβιν, καὶ τῶν
τετραπόδων τίκτειν κατὰ τὸ στόμα τὴν γαλῆν.
ταῦτα γὰρ καὶ Ἀναξαγόρας καὶ τῶν ἄλλων τινὲς
φυσικῶν λέγουσι, λίαν ἁπλῶς καὶ ἀσκέπτως λέγοντες. περὶ μὲν οὖν τῶν ὀρνίθων ἐκ συλλογισμοῦ
διαψευδόμενοι τῷ τὴν μὲν ὀχείαν ὀλιγάκις ὁρᾶσθαι
20 τὴν τῶν κοράκων, τὴν δὲ τοῖς ῥύγχεσι πρὸς ἄλληλα
κοινωνίαν πολλάκις, ἣν πάντα ποιεῖται τὰ κορα-
κώδη τῶν ὀρνέων· δῆλον δὲ τοῦτο ἐπὶ τῶν τιθα-
σευομένων κολοιῶν. τὸ δ᾽ αὐτὸ τοῦτο ποιεῖ καὶ
τὸ τῶν περιστερῶν γένος· ἀλλὰ διὰ τὸ καὶ ὀχευόμενα φαίνεσθαι, διὰ τοῦτο ταύτης οὐ τετυχήκασι
25 τῆς φήμης.   τὸ δὲ κορακῶδες γένος οὐκ ἔστιν
ἀφροδισιαστικόν (ἔστι γὰρ τῶν ὀλιγογόνων), ἐπῶπται δ᾽ ἤδη[2] καὶ τοῦτο ὀχευόμενον.   τὸ δὲ δὴ μὴ
συλλογίζεσθαι πῶς εἰς τὰς ὑστέρας ἀφικνεῖται τὸ

---

[1] sic interpunx. A.-W.; εἰσῆλθεν. vulg.; fortasse ἃ πόθεν
εἰσῆλθεν. scribendum.
[2] ἐπῶπται δ᾽ ἤδη Z¹ : ἐπεὶ ὦπταί γ᾽ ἤδη vulg., Z² (γε δὴ SY).

by Herodotus [a] the fable-teller, to the effect that fish conceive by swallowing the milt. It never strikes them that this is impossible, but of course it is, because the passage whose entrance is through the mouth passes down into the stomach, not into the uterus, and whatever goes down into the stomach must of necessity be turned into nourishment, because it undergoes concoction. The uterus, however, as we can see is full of eggs; so we ask, how did they find their way there?

It is the same with the generation of birds. Thus there are those who say that ravens and ibises unite by the mouth, and that one of the quadrupeds, the weasel, brings forth its young by the mouth. This is, in fact, alleged by Anaxagoras and some of the other physiologers; but their verdict is based on insufficient evidence and inadequate consideration of the matter. (1) So far as the birds are concerned, they have reasoned themselves into an erroneous conclusion, since the copulation of ravens is seldom witnessed, whereas they are frequently observed uniting with each other by their beaks, which is something that all birds of the raven family do, as is plain for everyone to see in the case of domesticated jackdaws. Precisely the same thing is done by birds of the pigeon family; but as their copulation is plainly observable as well, they have not succeeded in having this tale told about them. Actually, birds of the raven group are not unduly sexual: it is one of the groups that produce but few young; still, like other birds, they have been observed in the act of copulation. It is odd, however, that our friends do not reason out how the

VI Erroneous theories about copulation of Birds, etc.

* Hdt. II. 93.

σπέρμα διὰ τῆς κοιλίας πεττούσης ἀεὶ τὸ ἐγγινόμενον, καθάπερ τὴν τροφήν, ἄτοπον. ὑστέρας δ'
30 ἔχουσι καὶ ταῦτα τὰ ὄρνεα, καὶ ᾠὰ φαίνεται πρὸς
τοῖς ὑποζώμασιν. καὶ ἡ γαλῆ, καθάπερ τἆλλα
τετράποδα, τὸν αὐτὸν τρόπον ἔχει ἐκείνοις τὰς
ὑστέρας· ἐξ ὧν εἰς τὸ στόμα πῇ βαδιεῖται τὸ
ἔμβρυον; ἀλλὰ διὰ τὸ τίκτειν πάμπαν μικρὰ τὴν
γαλῆν, καθάπερ καὶ τἆλλα τὰ σχιζόποδα, περὶ ὧν

ὕστερον ἐροῦμεν, τῷ δὲ στόματι πολλάκις μετα-
φέρειν τοὺς νεοττούς, ταύτην πεποίηκε τὴν δόξαν.

Εὐηθικῶς δὲ καὶ λίαν διεψευσμένοι καὶ οἱ περὶ
τρόχου καὶ ὑαίνης λέγοντες. φασὶ γὰρ τὴν μὲν
ὕαιναν πολλοί, τὸν δὲ τρόχον Ἡρόδωρος ὁ Ἡρα-
5 κλεώτης, δύο αἰδοῖα ἔχειν, ἄρρενος καὶ θήλεος, καὶ
τὸν μὲν τρόχον αὐτὸν αὑτὸν ὀχεύειν, τὴν δ' ὕαιναν
ὀχεύειν καὶ ὀχεύεσθαι παρ' ἔτος. ὦπται γὰρ ἡ
ὕαινα ἓν ἔχουσα αἰδοῖον· ἐν ἐνίοις γὰρ τόποις οὐ
σπάνις τῆς θεωρίας· ἀλλ' ἔχουσιν αἱ ὕαιναι ὑπὸ
τὴν κέρκον ὁμοίαν γραμμὴν τῷ τοῦ θήλεος αἰδοίῳ.
10 ἔχουσι μὲν οὖν καὶ οἱ ἄρρενες καὶ αἱ θήλειαι τὸ
τοιοῦτον σημεῖον, ἀλλ' ἁλίσκονται οἱ ἄρρενες μᾶλ-
λον· διὸ τοῖς ἐκ παρόδου θεωροῦσι ταύτην ἐποίησε
τὴν δόξαν.

---

[a] In Bk. IV.
[b] This animal cannot be identified. It must be distinguished from the genus now called *Trochus*, which is shellfish. No species of mammal is normally hermaphrodite.
[c] See also *H.A.* 579 b 15 ff.

semen manages to pass through the stomach and arrive in the uterus, in view of the fact that the stomach concocts everything that gets into it, as it does the nourishment. Besides, these birds have a uterus, just like other birds, and eggs can plainly be seen up towards the diaphragm. (2) The weasel, too, like other quadrupeds, has a uterus of exactly the same sort as theirs ; and how is the embryo going to make its way from that uterus into the mouth ? This notion is really due to the fact that the weasel produces very tiny young ones (as do the rest of the fissipede animals, of which we shall speak later),[a] and that it often carries them about in its mouth.

(3) There is another silly and extremely wrong-headed story which is told about the *trochos* [b] and the hyena,[c] to the effect that they have two pudenda, male and female (there are many who assert this of the hyena ; Herodorus of Heraclea [d] asserts it of the *trochos*), and that whereas the *trochos* impregnates itself, the hyena mounts and is mounted in alternate years. In some localities, however, there is ample opportunity for inspection, and the hyena has been observed to possess one pudendum only ; but hyenas have under the tail a line similar to the female pudendum. Both male and female ones have this mark, but as the males are captured more frequently, casual inspection has given rise to this erroneous idea.[e]

[d] Heraclea Pontica, a colony of Megara, on the south shore of the Black Sea, about 100 miles east of the Bosporus. Herodorus (fl. *c.* 400 B.C.) was the father of the sophist Bryson (both are mentioned at *H.A.* 563 a 7 and 615 a 9). He wrote a *History of Heracles*, which seems to have contained a great variety of matter.   [e] See add. note, p. 565.

'Αλλὰ περὶ μὲν τούτων ἅλις τὰ εἰρημένα.

VII Περὶ δὲ τῆς τῶν ἰχθύων γενέσεως ἀπορήσειεν ἄν
15 τις διὰ τίνα ποτὲ αἰτίαν τῶν μὲν σελαχωδῶν οὔθ'
αἱ θήλειαι τὰ κυήματα οὔθ' οἱ ἄρρενες ἀπορραί-
νοντες ὁρῶνται τὸν θορόν, τῶν δὲ μὴ ζῳοτόκων
καὶ αἱ θήλειαι τὰ ᾠὰ καὶ οἱ ἄρρενες τὸν θορόν.
αἴτιον δ' ὅτι τὸ γένος οὐ πολύσπερμον ὅλως τὸ
τῶν σελαχωδῶν· καὶ ἔτι[1] αἵ γε θήλειαι πρὸς τῷ
20 διαζώματι τὰς ὑστέρας ἔχουσιν. τὰ γὰρ ἄρρενα
τῶν ἀρρένων καὶ τὰ θήλεα τῶν θήλεων[2] ὁμοίως
διαφέρουσιν· ὀλιγοχούστεροι γὰρ πρὸς τὴν γονὴν
οἱ σελαχώδεις εἰσίν. τὸ δ' ἄρρεν γένος ἐν τοῖς
ᾠοτόκοις, καθάπερ αἱ θήλειαι τὰ ᾠὰ διὰ πλῆθος
ἀποτίκτουσιν, οὕτως ἐκεῖνοι ἀπορραίνουσιν· πλείω
25 γὰρ ἔχουσι θορὸν ἢ ὅσον πρὸς τὴν ὀχείαν ἱκανόν·
μᾶλλον γὰρ βούλεται ἡ φύσις δαπανᾶν τὸν θορὸν
πρὸς τὸ συναύξειν τὰ ᾠά, ὅταν ἀποτέκῃ ἡ θήλεια,
ἢ πρὸς τὴν ἐξ ἀρχῆς σύστασιν. καθάπερ γὰρ ἔν
τε τοῖς ἄνω καὶ τοῖς ὑπογύοις εἴρηται λόγοις, τὰ
μὲν τῶν ὀρνέων ᾠὰ τελεοῦται ἐντός, τὰ δὲ τῶν
30 ἰχθύων ἐκτός. τρόπον γάρ τινα ἔοικε τοῖς σκωληκο-
τοκοῦσιν· ἔτι γὰρ ἀτελέστερον προΐεται τὸ κύημα
τὰ σκωληκοτόκα τῶν ζῴων. ἀμφοτέροις δὲ τὴν
τελείωσιν καὶ τοῖς τῶν ὀρνίθων ᾠοῖς καὶ τοῖς τῶν
ἰχθύων ποιεῖ τὸ ἄρρεν, ἀλλὰ τοῖς μὲν τῶν ὀρνίθων
ἐντός (τελεοῦται γὰρ ἐντός), τοῖς δὲ τῶν ἰχθύων
35 ἐκτὸς διὰ τὸ ἔξω προΐεσθαι ἀτελές, ἐπεὶ συμβαίνει
γε ἐπ' ἀμφοτέρων ταὐτόν.

---

[1] ὅτι Y.   [2] θήλεων E, Btf. : θηλειῶν vulg.

I have now said enough on these subjects.

With regard to the generation of fish, the puzzle VII
may be raised, what the Cause can possibly be why _{Various points.}
neither the females of Selachian fishes are seen
shedding their fetations nor the males their milt,
whereas the males and females are observed so doing
in the case of non-viviparous fishes. The reason is
that in general the class of the Selachians is not rich
in semen ; and also in the females the uterus is up
towards the diaphragm.[a] Of course males of one class
differ from males of another, and females similarly ;
and the fact is that the Selachians yield less semen
than most. With the oviparous fishes, the males
shed their milt, just as the females lay their eggs,
because there is such an abundance of both ; the
males have more milt than the amount which suffices
for copulation, because Nature prefers to expend the
milt in helping to enlarge the eggs after the female
has laid them, rather than in constituting the eggs
at the outset. This remark is explained by what has
been said both in our earlier discussion and also not
long ago, viz., the eggs of birds are perfected inside
the parent, but the eggs of fish outside. In a way,
fish resemble the larva-producing animals, for the
latter deposit a fetation which is even more imperfect
still. The perfecting in both cases, birds' eggs and
fishes', is accomplished by the male. With birds this
is done within the parent animal, because a bird's
egg is perfected inside ; with fishes, outside, because
the egg is in an imperfect state when it is deposited
outside. The upshot however is the same in both
cases.

---

[a] And therefore the eggs are brought to perfection inside
the parent.

Τῶν μὲν οὖν ὀρνίθων τά τε ὑπηνέμια γίνεται
γόνιμα,[1] καὶ τὰ προωχευμένα ὑφ' ἑτέρου γένους τῶν
ἀρρένων μεταβάλλει τὴν φύσιν εἰς τὸν ὕστερον
ὀχεύοντα· καὶ τὰ οἰκεῖα δέ,[2] ἀναύξητα[3] ὄντα ἂν
5 διαλίπῃ[4] τὴν ὀχείαν, ὅταν ὀχεύσῃ[5] πάλιν, ποιεῖ
ταχεῖαν λαμβάνειν τὴν αὔξησιν· οὐ μέντοι κατὰ
πάντα τὸν χρόνον, ἀλλ' ἐάνπερ πρότερον γένηται ἡ
ὀχεία πρὶν μεταβαλεῖν[6] εἰς τὴν τοῦ λευκοῦ ἀπόκρισιν. τοῖς δὲ τῶν ἰχθύων οὐθὲν ὥρισται[7] τοιοῦτον, ἀλλὰ πρὸς τὸ σῴζεσθαι ταχέως ἐπιρραίνουσιν[8]
10 οἱ ἄρρενες. αἴτιον δ' ὅτι οὐ δίχροα ταῦτα· διόπερ
οὐχ ὥρισται τοιοῦτος καιρὸς τούτοις οἷος ἐπὶ τῶν
ὀρνίθων. τοῦτο δὲ συμβέβηκεν εὐλόγως· ὅταν γὰρ
τὸ λευκὸν ἀφωρισμένον ᾖ καὶ τὸ ὠχρὸν ἀπ' ἀλλήλων, ἔχει ἤδη τὴν ἀπὸ τοῦ ἄρρενος ἀρχήν[9]· [εἰς][10]
15 ταύτην γὰρ συμβάλλεται τὸ ἄρρεν. τὰ μὲν οὖν
ὑπηνέμια λαμβάνει τὴν γένεσιν μέχρι τοῦ ἐνδεχο-
μένου αὐτοῖς. τελεωθῆναι μὲν γὰρ εἰς ζῷον
ἀδύνατον (δεῖ γὰρ αἰσθήσεως), τὴν δὲ θρεπτικὴν
δύναμιν τῆς ψυχῆς ἔχει καὶ τὰ θήλεα καὶ τὰ ἄρρενα

---

[1] hic addit Σ *quando femina coierit existentibus illis ovis
in matrice.*
[2] δή PSY. fort. ὠχευμένα δέ scribendum, vel potius καὶ
τὰ ὀχεῖα δέ, ἂν ἀναυξῇ ᾖ τὰ ᾠὰ διὰ τὸ διαλείπειν κτλ.
[3] αναυξη Z[1] : ἀναυξῆσθαι S et om. ὄντα.
[4] διαλίπῃ Platt : διαλείπῃ vulg.
[5] ὀχεύσῃ Platt : ὀχευθῇ vulg. : δ' ὀχευθῇ PSY.
[6] μεταβαλεῖν P : μεταβάλλειν vulg.   [7] συνίσταται Z[1].
[8] ἐπιρραίνουσιν Z : ἀπορραίνουσιν vulg.
[9] lacunam hic statuit Platt.
[10] εἰς om. S ; seclusi : εἰς τοῦτο coni. A.-W. *et per hunc
modum erit conveniencia spermatis maris* Σ. **fortasse**
αἴσθησιν scribendum, vel εἰς . . . ἄρρεν secludenda.

In birds, wind-eggs become fertile,[a] and eggs Wind-eggs. previously impregnated by the treading of one sort of cock change their nature to that of the cock which treads the hen later [b] ; and also, where one and the same cock is concerned,[c] if he has left off treading the hen and the eggs are not growing on that account, he makes them grow quickly when he resumes the treading. This however cannot happen at any and every period : the treading must take place before the change occurs when the white of the egg becomes separate. In the case of fishes' eggs there is no such point fixed, but the males sprinkle them without delay to keep them in sound condition. The reason is that fishes' eggs are not double-coloured: that is why in their case there is no such fixed time as there is for birds' eggs. This situation is what we should expect, for once the white and the yolk have been distinctly separated from each other, they already [d] possess the principle that comes from the male,[e] since the male contributes [towards] this. Thus wind-eggs attain to generation in so far as it is possible for them to do so. It is impossible for them to be perfected to the point of producing an animal, because sense-perception [f] is required for that ; the nutritive faculty of the Soul, however, is possessed by females as well as by males and by all

[a] Probably there should here be inserted " if the hen is trodden by the male while they are in the uterus."

[b] This is qualified below, 757 b 27 ff.

[c] The force of οἰκεῖα seems to be that the eggs are the cock's " own " in the sense that he and not some other cock originally impregnated them. But see critical note.

[d] And therefore cannot be altered by another cock.

[e] See 767 b 17 ff., and references there given in note.

[f] Which is supplied by the male.

757 b

καὶ πάντα τὰ ζῶντα, καθάπερ εἴρηται πολλάκις·
διόπερ αὐτὸ¹ τὸ ᾠὸν ὡς μὲν φυτοῦ κύημα τέλειόν
20 ἐστιν, ὡς δὲ ζῴου ἀτελές. εἰ μὲν οὖν μὴ ἐνῆν ἄρρεν
ἐν τῷ γένει αὐτῶν, ἐγίγνετ᾽ ἂν ὥσπερ καὶ ἐπὶ τῶν
ἰχθύων, εἴπερ ἔστι τι τοιοῦτον γένος οἷον ἄνευ
ἄρρενος γεννᾶν· εἴρηται δὲ περὶ αὐτῶν καὶ πρότερον, ὅτι οὔ πω ὦπται ἱκανῶς. νῦν δ᾽ ἐστὶν ἐν
25 πᾶσι τοῖς ὄρνισι τὸ μὲν θῆλυ τὸ δ᾽ ἄρρεν, ὥσθ᾽ ᾗ
μὲν φυτόν, τετελέωκεν (διόπερ οὐ μεταβάλλει πάλιν
μετὰ τὴν ὀχείαν),² ᾗ δ᾽ οὐ φυτόν,³ οὐ τετελέωκεν,
οὐδ᾽ ἀποβαίνει ἐξ αὐτοῦ ἕτερον οὐθέν· οὔτε γὰρ
ὡς φυτὸν ἁπλῶς οὔθ᾽ ὡς ζῷον⁴ ἐκ συνδυασμοῦ
γέγονεν. τὰ δ᾽ ἐξ ὀχείας μὲν γενόμενα ᾠά, δια-
κεκριμένα δ᾽ εἰς τὸ λευκόν, γίνεται κατὰ τὸ πρῶτον
30 ὀχεῦσαν· ἔχει γὰρ ἀμφοτέρας ἤδη τὰς ἀρχάς.

VIII Τὸν αὐτὸν δὲ τρόπον καὶ τὰ μαλάκια ποιεῖται
τὸν τόκον, οἷον σηπίαι καὶ τὰ τοιαῦτα, καὶ τὰ
μαλακόστρακα, οἷον κάραβοι καὶ τὰ συγγενῆ τού-
τοις· τίκτει γὰρ ἐξ ὀχείας καὶ ταῦτα, καὶ συνδυαζό-
35 μενον τὸ ἄρρεν τῷ θήλει πολλάκις ὦπται. διόπερ
οὐδ᾽ ἱστορικῶς οὐδὲ ταύτῃ φαίνονται λέγοντες οἱ
758 a   φάσκοντες τοὺς ἰχθῦς πάντας εἶναι θήλεις καὶ

---

¹ αὐτὸ Platt : αὐτοῦ vulg.
² haec verba ad finem cap. transtulit Platt, recte, nisi
omnino omittenda.
³ quia non sunt animalia Σ.
⁴ φυτὸν . . . ζῷον Platt : φυτοῦ . . . ζῴου vulg.

ᵃ At 741 a 34 ff.
ᵇ Platt transposes these words to the end of the chapter.
ᶜ See 731 a 2, 3.
ᵈ Nutritive soul and sensitive soul, the latter being supplied
by " the principle of the male."

living things, as has been said repeatedly ; hence the egg itself, regarded as the fetation of a plant, is perfect, but regarded as the fetation of an animal it is imperfect. If there were no such thing as a male in the class of birds, the egg would have been formed as it is in fishes, supposing there really is some species which generates without a male ; though I mentioned earlier [a] in this connexion that this has not yet been sufficiently observed. Actually, however, both sexes exist in all species of birds ; so that, *qua* plant, the wind-egg has reached perfection (and that is why it does not change any more after impregnation),[b] *qua* non-plant, on the other hand, it has not reached perfection, and nothing else results from it, since it has been formed neither as a plant simply and directly [c] nor as an animal by means of copulation. As for eggs which are the result of copulation, however, but which have been distinguished into white and yolk, these are formed according to the male which impregnated them first, since by that time they possess both the required principles.[d]

The production of their young is accomplished in the same manner by the Cephalopods—sepias and the like—and by the Crustacea—*caraboi* [e] and the creatures akin to them. They too lay eggs as a result of copulation ; many instances have been observed of the male uniting with the female. So here we have another score on which we can convict of a lack of scientific accuracy those who allege that all fish are female and produce eggs without copula-

VIII

II. Blood-lessanimals. (i.) Reproduction of Cephalopods and Crustacea.

---

[e] At *P.A.* 683 b 25 Aristotle makes four main groups of of Crustacea : (1) *caraboi*, (2) *astacoi*, (3) *carides*, (4) *carcinoi*, corresponding roughly to (1) lobsters, (2) crayfish, (3) prawns and shrimps, (4) crabs.

τίκτειν οὐκ ἐξ ὀχείας· τὸ γὰρ ταῦτα μὲν ἐξ ὀχείας
οἴεσθαι, ἐκεῖνα δὲ μή, θαυμαστόν· εἴ τε τοῦτ᾽
ἐλελήθει, σημεῖον ἀπειρίας. γίνεται δὲ ὁ συνδυα-
5 σμὸς τούτων χρονιώτερος πάντων, ὥσπερ τῶν
ἐντόμων, εὐλόγως· ἄναιμα γάρ ἐστι, διόπερ ψυχρὰ
τὴν φύσιν.

Ταῖς μὲν οὖν σηπίαις καὶ ταῖς τευθίσι δύο τὰ
ᾠὰ φαίνεται διὰ τὸ διηρθρῶσθαι τὴν ὑστέραν καὶ
φαίνεσθαι δικρόαν· τὸ δὲ τῶν πολυπόδων ἓν ᾠόν,
αἴτιον δ᾽ ἡ μορφὴ στρογγύλη τὴν ἰδέαν οὖσα καὶ
10 σφαιροειδής· ἡ γὰρ σχίσις ἄδηλος πληρωθείσης
ἐστίν. δικρόα δὲ καὶ ἡ τῶν καράβων ἐστὶν ὑστέρα.
ἀποτίκτουσι δὲ τὸ κύημα ἀτελὲς καὶ ταῦτα πάντα
διὰ τὴν αὐτὴν αἰτίαν. τὰ μὲν οὖν καραβώδη τὰ
θήλεα πρὸς αὑτὰ ποιεῖται τὸν τόκον (διόπερ μείζους
ἔχει τὰς πλάκας τὰ θήλεα αὐτῶν ἢ τὰ ἄρρενα,
15 φυλακῆς χάριν τῶν ᾠῶν), τὰ δὲ μαλάκια ἔξω.
καὶ τοῖς μὲν θήλεσι τῶν μαλακίων ἐπιρραίνει ὁ
ἄρρην, καθάπερ οἱ ἄρρενες ἰχθύες τοῖς ᾠοῖς, καὶ
γίγνεται συνεχὲς καὶ κολλῶδες· τοῖς δὲ καραβώδε-
σιν οὔτ᾽ ὦπται τοιοῦτον οὔτ᾽ εὔλογον· ὑπό τε γὰρ
τῇ θηλείᾳ τὸ κύημα καὶ σκληρόδερμόν ἐστι, καὶ
20 λαμβάνει αὔξησιν καὶ ταῦτα καὶ τὰ τῶν μαλακίων
ἔξω, καθάπερ καὶ τὰ τῶν ἰχθύων.

Προσπέφυκε δ᾽ ἡ γιγνομένη σηπία τοῖς ᾠοῖς
κατὰ τὸ πρόσθιον· ταύτῃ γὰρ ἐνδέχεται μόνον· ἔχει
γὰρ μόνον ἐπὶ ταὐτὸ τὸ ὀπίσθιον μέρος καὶ τὸ

---

tion. What an extraordinary thing, to hold that Cephalopods and Crustacea lay eggs as a result of copulation, but fish without copulation! Or alternatively, if they were not already aware that the other creatures copulate, then it just shows how ignorant they are. The copulation of all these creatures takes quite a long time, just as that of insects does, which is not surprising, because they are bloodless, and therefore cold in their nature.

In the sepias and calamaries the eggs appear to be two in number, because the uterus is divided and appears to be double. The octopuses appear to have a single egg; the reason is that the shape of the uterus is round and spherical in form, and when it is full the cleavage is not obvious.[a] The *caraboi* also have a double uterus. All these animals as well deposit the fetation in an imperfect condition, and for the same cause. Females of the caraboid group deposit their eggs on to themselves; that is why they have larger flaps than the males—in order to protect the eggs; the Cephalopods lay their eggs clear of themselves. The male Cephalopods sprinkle their milt over the females, just as male fishes do over the eggs, and it becomes a glutinous mass. Nothing of the kind has been observed to occur with the caraboids, nor should we expect it, because the fetation is situated under the female and is hardskinned, and both these eggs and those of the Cephalopods pursue their growth after they have left the parent, just as the eggs of fishes do.

The sepia while in process of formation is fastened to the egg by its front part, which is the only possible place, because its front and back parts face in the same direction [b] (in this respect it is unique). For a

758 a

πρόσθιον. τὸ δὲ σχῆμα τῆς θέσεως ὃν ἔχει γιγνό-
25 μενα τρόπον, δεῖ θεωρεῖν ἐκ τῶν ἱστοριῶν.

Περὶ μὲν οὖν τῶν ἄλλων ζῴων τῆς γενέσεως
IX εἴρηται, καὶ πεζῶν καὶ πτηνῶν καὶ πλωτῶν· περὶ
δὲ τῶν ἐντόμων καὶ τῶν ὀστρακοδέρμων λεκτέον
κατὰ τὴν ὑφηγημένην μέθοδον. εἴπωμεν δὲ πρῶτον
περὶ τῶν ἐντόμων.

30   Ὅτι μὲν οὖν τὰ μὲν ἐξ ὀχείας γίνεται τῶν τοιούτων τὰ δ᾽ αὐτόματα, πρότερον ἐλέχθη, πρὸς δὲ
τούτοις ὅτι σκωληκοτοκεῖ καὶ διὰ τίν᾽ αἰτίαν
σκωληκοτοκεῖ. σχεδὸν γὰρ ἔοικε πάντα τρόπον
τινὰ σκωληκοτοκεῖν τὸ[1] πρῶτον· τὸ γὰρ ἀτελέστα-
τον κύημα τοιοῦτόν ἐστιν, ἐν πᾶσι δὲ καὶ τοῖς ζῳο-
35 τοκοῦσι καὶ τοῖς ᾠοτοκοῦσι τέλειον ᾠὸν τὸ κύημα
τὸ πρῶτον ἀδιόριστον ὂν λαμβάνει τὴν αὔξησιν·
τοιαύτη δ᾽ ἐστὶν ἡ τοῦ σκώληκος φύσις. μετὰ δὲ
τοῦτο τὰ μὲν ᾠοτοκεῖ τὸ κύημα τέλειον, τὰ δ᾽
758 b ἀτελές, ἔξω δὲ γίγνεται τέλειον, καθάπερ ἐπὶ τῶν
ἰχθύων εἴρηται πολλάκις. τὰ δ᾽ ἐν αὑτοῖς ζῳοτοκοῦντα τρόπον τινὰ μετὰ τὸ σύστημα τὸ ἐξ ἀρχῆς
ᾠοειδὲς γίνεται· περιέχεται γὰρ τὸ ὑγρὸν ὑμένι
λεπτῷ, καθάπερ ἂν εἴ τις ἀφέλοι τὸ τῶν ᾠῶν
5 ὄστρακον· διὸ καὶ καλοῦσι τὰς τότε γιγνομένας
τῶν κυημάτων φθορὰς ἐκρύσεις.

Τὰ δ᾽ ἔντομα καὶ γεννᾷ τὰ γεννῶντα σκώληκας,
καὶ τὰ γιγνόμενα μὴ δι᾽ ὀχείας ἀλλ᾽ αὐτόματα ἐκ
τοιαύτης γίνεται πρῶτον συστάσεως. δεῖ γὰρ

————
[1] τὸ PZ : om. vulg.

326

figure showing the way in which it is situated during the process of formation, the *Researches* [a] should be consulted.

We have now spoken about the generation of the animals that walk, fly and swim. Following the IX plan we have laid down, there remain the Insects and the Testacea to be discussed. We will deal with the Insects first.

I said earlier that some Insects are formed by (ii.) Reproduction of means of copulation, others spontaneously; further, Insects: that they produce a larva, and I stated the cause of their so doing. In a way, it looks as though practically all animals produce a larva to begin with, for the fetation in its most imperfect state is something of this sort; and in all the Vivipara and all the Ovipara that produce a perfect egg, the fetation in its earliest stage is still undifferentiated and is growing, and this is just the sort of thing a larva is. At the next step, some of the Ovipara produce their fetation as a perfect egg, some as an imperfect one which reaches its perfection after it has left the parent, as I have often stated with regard to fish. In the case of the internally viviparous animals, the fetation, after it has been constituted at the outset, in a way becomes egglike: its fluid content becomes enclosed in a fine membrane—like an egg with its shell taken off—and that is why a fetation aborted at this stage is known as an " efflux." [b]

Those Insects which generate, generate larvae; (α) Larvae. and those Insects also which are formed spontaneously and not by means of copulation are, to begin with, formed from an organism of this sort. This is

[a] See *H.A.* 550 a 10 ff.
[b] *Cf. H.A.* 583 b 12.

καὶ τὰς κάμπας εἶδός τι[1] τιθέναι σκώληκος, καὶ
10 τὰ τῶν ἀραχνίων. καίτοι δόξειεν ἂν ᾠοῖς ἐοικέναι
διὰ τὴν τοῦ σχήματος περιφέρειαν καὶ τούτων ἔνια
καὶ πολλὰ τῶν ἄλλων· ἀλλ' οὐ τῷ σχήματι λεκτέον
οὐδὲ τῇ μαλακότητι καὶ σκληρότητι (καὶ γὰρ
σκληρὰ τὰ κυήματα γίγνεται ἐνίων) ἀλλὰ τῷ ὅλον
μεταβάλλειν καὶ μὴ ἐκ μορίου τινὸς γίνεσθαι τὸ
15 ζῷον. προελθόντα δὲ πάντα τὰ σκωληκώδη καὶ
τοῦ μεγέθους λαβόντα τέλος οἷον ᾠὸν γίγνεται·
σκληρύνεταί τε γὰρ περὶ αὐτὰ τὸ κέλυφος, καὶ
ἀκινητίζουσι κατὰ τοῦτον τὸν καιρόν. δῆλον δὲ
τοῦτο ἐν τοῖς σκώληξι τοῖς τῶν μελιττῶν καὶ
σφηκῶν καὶ ταῖς κάμπαις. τούτου δ' αἴτιον ὅτι ἡ
20 φύσις ὡσπερανεὶ πρὸ ὥρας ᾠοτοκεῖ διὰ τὴν
ἀτέλειαν τὴν αὑτῆς, ὡς ὄντος τοῦ σκώληκος ἔτι ἐν
αὐξήσει ᾠοῦ μαλακοῦ. τὸν αὐτὸν δὲ τρόπον καὶ
ἐπὶ τῶν ἄλλων συμβαίνει πάντων τῶν μὴ[2] ἐξ ὀχείας
γιγνομένων ἐν ἐρίοις ἤ τισιν ἄλλοις τοιούτοις, καὶ
τῶν ἐν τοῖς ὕδασιν. πάντα γὰρ μετὰ τὴν τοῦ
25 σκώληκος φύσιν ἀκινητίσαντα, καὶ τοῦ κελύφους
περιξηρανθέντος, μετὰ ταῦτα τούτου ῥαγέντος
ἐξέρχεται καθάπερ ἐξ ᾠοῦ ζῷον ἐπιτελεσθὲν ἐπὶ

---

[1] τι P : om. vulg.   [2] μὴ om. PSZ (καὶ τῶν Z).

---

[a] This apparently means the eggs from which they are

correct, for we are bound to reckon caterpillars [a] and the product of spiders as a form of larva. True, some of these, and many belonging to other Insects, would appear to resemble eggs on account of their circular shape ; but our decision must not be determined by their shape nor yet by their softness or hardness (the fetations of some of these creatures are hard), but by the fact that the *whole* of the object undergoes change —the animal is formed out of the *whole* of it and not some *part* of it.[b] All these larva-like objects, when they have advanced and reached their full size, become as it were an egg : the shell around them gets hard, and they remain motionless during this period. This is clearly to be seen with the larvae of bees and wasps, and with caterpillars. The reason for this is that their Nature, owing to its own imperfection, deposits the eggs as it were before their time, which suggests that the larva, while it is yet in growth, is a soft egg. A comparable thing occurs in the case of all other creatures which are formed independently of copulation in wool [c] and other such material and in water. All of these first have the nature of a larva, then they remain motionless once the covering has solidified round them ; after that the covering bursts and there emerges, as from an egg, an animal which, at this its third genesis,[d] is at last

produced. Aristotle however calls them larvae, and not eggs, at this stage, because according to him the stage which really corresponds to the egg-stage is not reached until later, when the creature becomes immobilized as a " pupa."

[b] The distinction which Aristotle makes here is an important one. See note on 732 a 32.

[c] See *H.A.* 557 b 2 ; the dustier your clothes are, the more moths are produced.

[d] The stages are : larva, pupa, imago.

τῆς τρίτης γενέσεως· ὧν τὰ [πλεῖστα] πτερωτὰ
τῶν πεζῶν ⟨μείζω⟩ ἐστίν.[1]

Κατὰ λόγον δὲ συμβαίνει καὶ τὸ θαυμασθὲν ἂν
δικαίως ὑπὸ πολλῶν, αἵ τε γὰρ κάμπαι λαμβά-
30 νουσαι τὸ πρῶτον τροφὴν μετὰ ταῦτα οὐκέτι
λαμβάνουσιν, ἀλλ' ἀκινητίζουσιν αἱ καλούμεναι
ὑπό τινων χρυσαλλίδες, καὶ τῶν σφηκῶν οἱ σκώληκες καὶ τῶν μελιττῶν μετὰ ταῦτα αἱ καλούμεναι
νύμφαι γίνονται, [καὶ τοιοῦτον οὐδὲν ἔχουσιν·][2] καὶ
γὰρ ἡ τῶν ᾠῶν φύσις ὅταν λάβῃ τέλος, ἀναυξής
35 ἐστι, τὸ δὲ πρῶτον αὐξάνεται καὶ λαμβάνει τροφήν,
ἕως ἂν διορισθῇ καὶ γένηται τέλειον ᾠόν. τῶν δὲ
σκωλήκων οἱ μὲν ἔχουσιν ἐν ἑαυτοῖς τὸ τοιοῦτον
759 a ὅθεν τρεφομένοις ἐπιγίγνεται [τοιοῦτον][3] περίττωμα, οἷον οἱ[4] τῶν μελιττῶν καὶ σφηκῶν· οἱ δὲ
λαμβάνουσι θύραθεν, ὥσπερ αἵ τε κάμπαι καὶ τῶν
ἄλλων τινὲς σκωλήκων.

Διότι μὲν οὖν τριγενῆ τε γίγνεται τὰ τοιαῦτα,
καὶ δι' ἣν αἰτίαν ἐκ κινουμένων ἀκινητίζει πάλιν,
5 εἴρηται· γίγνεται δὲ τὰ μὲν ἐξ ὀχείας αὐτῶν,
καθάπερ οἵ τε ὄρνιθες καὶ τὰ ζῳοτόκα καὶ τῶν
ἰχθύων οἱ πλεῖστοι, τὰ δ' αὐτόματα, καθάπερ ἔνια
τῶν φυομένων.

---

[1] correxi (cf. 763 a 23). Σ vertit *et volatilia ex eis sunt
maiora quam ambulantia.*
[2] ante haec verba lacunam plurimorum vv. statuit Platt
(τροφὴν pro τοιοῦτον coni. A.-W., cf. 759 a 1); ego seclusi;
fort. transferenda ad 759 a 1-2 et ita scribendum οἱ δὲ οὐδὲν
τοιοῦτον ἔχουσιν ⟨ἀλλὰ⟩ λαμβάνουσιν κτλ. cf. infra 763 a 12 sqq.

perfected. Of these creatures, the winged ones are larger than those that walk.

Another occurrence, which may well cause surprise to many people, is really quite regular and normal. Caterpillars at first take nourishment, but afterwards they cease doing so, the chrysalis (as some call it) being motionless ; so too the larvae of wasps and bees afterwards turn into pupae as they are called [and have nothing of the sort]. This is not abnormal, for an egg also, when it has reached the perfection of its nature, does not grow, whereas to begin with it does grow and takes nourishment, until its differentiation is effected and it has become a perfect egg. Some larvae contain in themselves material from which as they feed on it residue is produced,[a] e.g., those of bees and wasps ; others get the material from without, as caterpillars and some other larvae do.

I have now stated why it is that it takes a threefold generation [b] to produce creatures of this sort, and the cause which, after they have begun as mobile creatures, makes them become immobile again. Also, some of them are formed in consequence of copulation, just as birds, Vivipara and the majority of fishes are ; others are formed spontaneously, as certain plants [c] are.

[a] Cf. H.A. 551 a 29 ff. " the larvae of bees . . . and wasps, while they are young, take nourishment and are seen to have excrement " ; cf. also ibid. a 25.

[b] See above, 758 a 28 et praeced.

[c] e.g., the mistletoe, 715 b 28.

---

[3] om. Z[1] : τοιοῦτο ἡ τροφὴ S : habent in se id quo cibantur et eiciunt superfluitatem cibi Σ.

[4] οἷον οἱ Peck (sicut Σ) : οἷ τε vulg.

759 a

X Ἡ δὲ τῶν μελιττῶν γένεσις ἔχει πολλὴν ἀπορίαν.
εἴπερ¹ γάρ ἐστι καὶ περὶ τοὺς ἰχθῦς τοιαύτη τις
10 γένεσις ἐνίων ὥστ' ἄνευ ὀχείας γεννᾶν, τοῦτο
συμβαίνειν ἔοικε καὶ περὶ τὰς μελίττας ἐκ τῶν
φαινομένων. ἀνάγκη γὰρ ἤτοι φέρειν αὐτὰς
ἀλλόθεν τὸν γόνον, ὥσπερ τινές φασι, καὶ τοῦτον
ἢ φυόμενον αὐτόματον ἢ ἄλλου τινὸς ζῴου τίκτοντος, ἢ γεννᾶν αὐτάς, ἢ τὸν μὲν φέρειν τὸν δὲ
15 γεννᾶν (καὶ γὰρ τοῦτο λέγουσί τινες, ὡς τὸν τῶν
κηφήνων μόνων φέρουσι γόνον), καὶ γεννᾶν ἢ
ὀχευομένας ἢ ἀνοχεύτους, καὶ ὀχευομένας γεννᾶν
ἤτοι ἕκαστον γένος καθ' αὑτὸ ἢ ἕν τι αὐτῶν τἆλλα
ἢ συνδυαζόμενον ἄλλο γένος ἄλλῳ, λέγω δ' οἷον
μελίττας μὲν γίγνεσθαι ἐκ μελιττῶν συνδυαζο-
20 μένων, κηφῆνας δ' ἐκ κηφήνων καὶ τοὺς βασιλεῖς
ἐκ τῶν βασιλέων, ἢ πάντα τἆλλα ἐξ ἑνὸς οἷον ἐκ
τῶν καλουμένων βασιλέων καὶ ἡγεμόνων, ἢ ἐκ τῶν
κηφήνων καὶ τῶν μελιττῶν· φασὶ γάρ τινες τοὺς

---

¹ εἰπερ ut vid. Z¹ : ἐπεὶ vulg., Z² in ras.

---

[a] The facts about bees, so far as they are known, are these.
There are three sorts of bees : (1) the Queen, which is a fully
developed female ; (2) the worker, which is a partially
developed female ; and (3) the drone, which is a male.   Eggs
are laid by the Queen, and it is generally agreed that the
unfertilized eggs produce drones and the fertilized eggs
Queens or workers.   When a hive becomes over-populated,
" swarming " takes place, and after the colony has settled
down in its new home, the Queen takes the " marriage flight,"
followed by a number of males ; copulation takes place in
mid-air, and the Queen returns to the nest.   At the end of
the summer the drones are ejected by the workers.   Queens

The generation of bees is a great [a] puzzle. If it
is a fact that certain fishes are generated without
copulation, the same probably occurs among bees as
well—or so it seems from appearances. The possible methods are these : Bees must either (a) fetch
the offspring [b] from elsewhere (some hold this view) ;
in which case the offspring will either have sprung
into being spontaneously or have been produced by
some other animal ; or (b) generate the young themselves ; or (c) fetch some and generate some (this,
too, is a view held by certain people, who maintain
that the young of the drones only are fetched). If
they generate the young themselves, this must be
done either with or without copulation ; if with
copulation, then either (i) each kind generates its
own kind,[c] or (ii) one of the three kinds generates the
others, or (iii) one kind unites with another kind.
What I mean is, e.g., either (i) " bees " are formed
from the union of " bees," drones from the union
of drones, kings from the union of kings ; or (ii) all
the rest are generated by one kind only : e.g., by the
kings or leaders as they are called ; or (iii) by the
union of drones and " bees " (some people of course

and workers are produced from similar eggs, though the
queen-cells are larger ; but the larva of a Queen is fed on
" royal jelly " (a special food produced by the workers)
throughout its development, whereas those of workers are
fed on this for a short time (3 or 4 days) only, and for the
remainder of the time on honey and digested pollen. It is
thought that in rare cases the workers may produce Queens
and other workers from unfertilized eggs. A worker's
development is completed in 3 weeks ; a Queen's in 16 days
and a drone's in 24 days.

[b] The larvae.

[c] The three " kinds " are : " kings " or " leaders " (i.e.,
queens) ; " bees " (i.e., workers) ; and drones.

333

μὲν ἄρρενας εἶναι τοὺς δὲ θήλεις, οἱ δὲ[1] τὰς μὲν
μελίττας ἄρρενας τοὺς δὲ κηφῆνας θήλεας.

25 Ταῦτα δ᾽ ἐστὶ πάντα ἀδύνατα συλλογιζομένοις τὰ
μὲν ἐκ τῶν συμβαινόντων ἰδίᾳ περὶ τὰς μελίττας,
τὰ δ᾽ ἐκ τῶν κοινοτέρων τοῖς ἄλλοις ζώοις. εἴτε
γὰρ μὴ τίκτουσαι φέρουσιν ἄλλοθεν, ἔδει γίγνεσθαι
μελίττας καὶ μὴ φερουσῶν τῶν μελιττῶν ἐν τοῖς
τόποις ἐξ ὧν[2] τὸ σπέρμα φέρουσιν. διὰ τί γὰρ
30 μετενεχθέντος μὲν ἔσται, ἐκεῖ δ᾽ οὐκ ἔσται; προσήκει γὰρ οὐδὲν ἧττον, εἴτε φυόμενον ἐν τοῖς
ἄνθεσιν αὐτόματον εἴτε ζώου τινὸς τίκτοντος. κἂν
εἴ γε ζώου τινὸς ἑτέρου τὸ σπέρμα ἦν, ἐκεῖνο ἔδει
γίγνεσθαι ἐξ αὐτοῦ, ἀλλὰ μὴ μελίττας. ἔτι δὲ τὸ
μὲν μέλι κομίζειν εὔλογον (τροφὴ γάρ),[3] τὸ δὲ τὸν
35 γόνον ἀλλότριον ὄντα καὶ μὴ τροφὴν ἄτοπον.
τίνος γὰρ χάριν;[4] πάντα γὰρ ὅσα πραγματεύεται
περὶ τὰ τέκνα, περὶ τὸν φαινόμενον οἰκεῖον δια-
πονεῖται γόνον.

᾽Αλλὰ μὴν οὐδὲ τὰς μὲν μελίττας θηλείας εἶναι
τοὺς δὲ κηφῆνας ἄρρενας εὔλογον· οὐδενὶ γὰρ τὸ
πρὸς ἀλκὴν ὅπλον τῶν θηλειῶν ἀποδίδωσιν ἡ
φύσις, εἰσὶ δ᾽ οἱ μὲν κηφῆνες ἄκεντροι, αἱ δὲ
5 μέλιτται πᾶσαι κέντρον ἔχουσιν. οὐδὲ τοὐναντίον
εὔλογον, τὰς μὲν μελίττας ἄρρενας τοὺς δὲ κηφῆνας
θήλεις[5]· οὐδὲν γὰρ τῶν ἀρρένων εἴωθε διαπονεῖσθαι
περὶ τὰ τέκνα, νῦν δ᾽ αἱ μέλιτται τοῦτο ποιοῦσιν.
ὅλως δ᾽ ἐπειδὴ φαίνεται ὁ μὲν τῶν κηφήνων γόνος

---

[1] οἱ δὲ PSYZ : οἷον vulg.
[2] ἐν τοῖς τόποις ἐξ ὧν Z : ἐκ τοῦ τόπου ἐξ οὗ vulg.
[3] τροφὴ γάρ om. SY. [4] τίνος γὰρ χάριν om. SZ[1].
[5] θήλεις P : θηλείας SZ : θήλεας vulg.

say that drones are male and " bees " female ; others that " bees " are male and drones female).

We have only to bring before our minds the special and particular facts concerning bees, on the one side, and on the other the facts more generally applicable to other animals, to see that all of these theories are impossible. Suppose they do not generate offspring themselves but fetch them from elsewhere. In that case bees ought to be formed, even if the bees failed to fetch them away, in those places whence they fetch the seed (semen). For why should a bee be produced if the seed is fetched away, and not if it is left where it is ? Surely it ought to be produced none the less, no matter whether it springs spontaneously to life in the blossoms or whether some animal generates it. Also, if the seed were that of some other animal, then that animal ought to be formed out of it, and not bees. Further, it is reasonable enough that bees should collect honey, for honey is their food ; but it is absurd that they should collect offspring which (*a*) is produced by some animal other than themselves, and (*b*) is not food. After all, why should they ? All creatures which concern themselves about young ones take that trouble over what appears to them to be their *own proper* offspring.

Nor is it reasonable to hold that " bees " are female and drones male ; because Nature does not assign defensive weapons to any female creature ; yet while drones are without a sting, all " bees " have one. Nor is the converse view reasonable, that " bees " are male and drones female, because no male creatures make a habit of taking trouble over their young, whereas in fact " bees " do. But generally, since it is apparent that the brood of the drones is produced

ἐγγινόμενος καὶ μηθενὸς ὄντος κηφῆνος, ὁ δὲ τῶν
10 μελιττῶν οὐκ ἐγγινόμενος ἄνευ τῶν βασιλέων (διὸ
καὶ φασί τινες τὸν τῶν κηφήνων φέρεσθαι μόνον),
δῆλον ὡς οὐκ¹ ἐξ ὀχείας γίνονται, οὔτ' ἐξ ἑκατέρου
τοῦ γένους αὐτοῦ αὑτῷ συνδυαζομένου, οὔτ' ἐκ
μελιττῶν καὶ κηφήνων. τό τε τοῦτον φέρειν μόνον
διά τε τὰ εἰρημένα ἀδύνατον, καὶ οὐκ εὔλογον μὴ
15 περὶ πᾶν τὸ γένος αὐτῶν ὅμοιόν τι συμβαίνειν
πάθος. ἀλλὰ μὴν οὐδ' αὐτὰς τὰς μελίττας ἐνδέχεται τὰς μὲν ἄρρενας εἶναι τὰς δὲ θηλείας· ἐν
πᾶσι γὰρ διαφέρει τοῖς γένεσι τὸ θῆλυ καὶ τὸ
ἄρρεν. κἂν ἐγέννων αὐταὶ αὑτάς· νῦν δ' οὐ φαίνεται γιγνόμενος ὁ γόνος αὐτῶν, ἐὰν μὴ ἐνῶσιν οἱ
20 ἡγεμόνες, ὥς φασίν. κοινὸν δὲ καὶ πρὸς τὴν ἐξ
ἀλλήλων γένεσιν καὶ πρὸς τὴν ἐκ τῶν κηφήνων,
καὶ χωρὶς καὶ μετ' ἀλλήλων, τὸ μηδέποτε ὦφθαι
ὀχευόμενον μηθὲν αὐτῶν· εἰ δ' ἦν ἐν αὐτοῖς τὸ μὲν
θῆλυ τὸ δ' ἄρρεν, πολλάκις ἂν τοῦτο συνέβαινεν.
λείπεται δ', εἴπερ ἐξ ὀχείας γίγνεται, τοὺς βασιλεῖς
25 γεννᾶν συνδυαζομένους. ἀλλ' οἱ κηφῆνες φαίνονται
γιγνόμενοι καὶ μὴ ἐνόντων ἡγεμόνων, ὧν οὔτε
φέρειν οἷόν τε τὸν γόνον τὰς μελίττας οὔτε γεννᾶν
αὐτὰς ὀχευομένας. λείπεται δή, καθάπερ φαίνεται

---

¹ οὐκ Z : οὔτ' vulg.

ᵃ Cf. above, 755 b 3, n.

even when there is no drone present to start with,
whereas young "bees" are produced only if the kings
are present (and this is why some people say that the
brood of the drones are the only ones they fetch
from away), it is plain that they are not formed as a
result of copulation, either (1) of " bee " with " bee "
or drone with drone, or (2) of " bee " with drone.
And anyway, not only is it impossible that drones
are the only ones they fetch in, for the reasons stated,
but also it is unreasonable to suppose that a similar
thing does not happen in respect of the whole tribe
of them.[a] Again, it is impossible that some of the
" bees themselves " should be male and some female,
since in all kinds of animals the male and the female
are different. And besides, if it were so, " bees " by
themselves would generate " bees," but in actual fact
we see that the brood of "bees" is not formed unless,
as they say, " the kings are within." And here is a
point which strikes at either theory (that they are
produced (a) by the union of " bees " with one
another, and (b) by their union with the drones, i.e.,
by one kind apart from the other, or by the two kinds
together with one another) : none of them has ever
been seen in the act of copulation, whereas if there
had been male and female among them this would
often be occurring. The remaining possibility, assuming that they are generated by means of copulation at
all, is that the kings unite and so generate them.
But, as against this, the drones, as we see, are formed
even if no " leaders " are " within " ; and as it is impossible that the " bees " should either fetch in
the brood of drones from away or generate them by
copulation themselves,[b] plainly the only possibility

[b] Proved already.

337

ARISTOTLE

συμβαῖνον ἐπί τινων ἰχθύων, τὰς μελίττας ἄνευ
ὀχείας γεννᾶν τοὺς κηφῆνας, τῷ μὲν γεννᾶν οὔσας
30 θηλείας, ἐχούσας δ' ἐν αὑταῖς, ὥσπερ τὰ φυτά, καὶ
τὸ θῆλυ καὶ τὸ ἄρρεν, διὸ καὶ τὸ πρὸς τὴν ἀλκὴν
ἔχουσιν ὄργανον· οὐ γὰρ δεῖ θῆλυ καλεῖν ἐν ᾧ
ἄρρεν μή ἐστι κεχωρισμένον.

Εἰ δ' ἐπὶ τῶν κηφήνων τοῦτο φαίνεται συμβαῖνον
καὶ γιγνόμενοι μὴ ἐξ ὀχείας, ἤδη καὶ κατὰ τῶν
35 μελιττῶν καὶ τῶν βασιλέων τὸν αὐτὸν ἀναγκαῖον
εἶναι λόγον καὶ μὴ γεννᾶσθαι ἐξ ὀχείας. εἰ μὲν
οὖν ἄνευ τῶν βασιλέων ἐφαίνετ' ἐγγινόμενος ὁ
760 a γόνος τῶν μελιττῶν, κἂν τὰς μελίττας ἀναγκαῖον
ἦν ἐξ αὑτῶν ἄνευ ὀχείας γίγνεσθαι. νῦν δ' ἐπειδὴ
τοῦτ' οὔ φασιν οἱ περὶ τὴν θεραπείαν τούτων τῶν
ζῴων ὄντες, λείπεται τοὺς βασιλεῖς καὶ αὐτοὺς
γεννᾶν καὶ τὰς μελίττας.

5 Ὄντος δὴ[1] περιττοῦ τοῦ γένους καὶ ἰδίου τοῦ τῶν
μελιττῶν, καὶ ἡ γένεσις αὐτῶν ἴδιος εἶναι φαίνεται.
τὸ μὲν γὰρ γεννᾶν τὰς μελίττας ἄνευ ὀχείας εἴη ἂν
καὶ ἐπ' ἄλλων ζῴων συμβαῖνον, ἀλλὰ τὸ μὴ τὸ[2]
αὐτὸ γένος γεννᾶν ἴδιον· οἱ γὰρ ἐρυθρῖνοι γεννῶσιν
ἐρυθρίνους καὶ αἱ χάνναι χάννας. αἴτιον δ' ὅτι
10 καὶ αὐταὶ γεννῶνται αἱ μέλιτται οὐχ ὥσπερ αἱ
μυῖαι καὶ τὰ τοιαῦτα τῶν ζῴων, ἀλλ' ἐξ ἑτέρου

[1] δὴ Rackham : δὲ vulg.
[2] τὸ μὴ τὸ Z (i.e., τὸ μὴ Z² in ras., tunc τὸ intactum Z¹ ;
totum fuerat ut vid. μὴν τὸ Z¹) : τὸ μὴ P : μὴ τὸ vulg.

[a] e.g., erythrinus and channa (below, 760 a 9) ; see also
762 b 23, and H.A. 569 a 17, 570 a 2 (cestreus and eel).
[b] See above, 759 b 4. They are as much male as female ;
hence it is not irregular for them to possess a sting.

remaining is something parallel to what we find occurs in certain fishes [a] : the " bees " generate the drones without copulation, *i.e.*, although so far as generating is concerned they are female, yet they contain in themselves the male as well as the female ⟨factor⟩, just as plants do ; and this also is why they possess the organ for self-defence,[b] for of course it is wrong to apply the term " female " to creatures where no separate male exists.

We find then that this is what occurs in the case of the drones : they are formed independently of copulation. And if this is so, then surely the same argument must apply to the " bees " and the kings ; they too must be generated independently of copulation. Now if we were sure that the brood of the " bees " made their appearance without the kings being there, then it would follow of necessity that the " bees " as well as the drones are produced from " bees " without copulation. This however is denied by those whose business it is to look after these creatures. Hence the only possibility left is that the kings generate their own kind and the " bees " as well.

We see then that the manner in which bees are generated appears to be peculiar, in keeping with their extraordinary and peculiar character. Bees' generating without copulation might be paralleled by the behaviour of other animals, but their generating some different kind of creature is peculiar and unique, for even *erythrinoi* and *channae* generate creatures of the same kind as themselves. The reason is that the " bees themselves " are not generated in the same way as flies and other such creatures, but from a kind which though different is akin to

760 a

μὲν συγγενοῦς δὲ γένους· γίγνονται γὰρ ἐκ τῶν
ἡγεμόνων. διὸ καὶ ἔχει ἀνάλογόν πως ἡ γένεσις
αὐτῶν· [1][οἱ μὲν γὰρ ἡγεμόνες μεγέθει μὲν ὅμοιοί
εἰσι τοῖς κηφῆσι, τῷ δὲ κέντρον ἔχειν ταῖς μελίτ-
15 ταις· αἱ μὲν οὖν μέλιτται κατὰ τοῦτ' ἐοίκασιν
αὐτοῖς, οἱ δὲ κηφῆνες κατὰ τὸ μέγεθος·] ἀνάγκη
γάρ τι παραλλάττειν, εἰ μὴ δεῖ ἀεὶ τὸ αὐτὸ γένος
ἐξ ἑκάστου γίνεσθαι. τοῦτο δ' ἀδύνατον· πᾶν γὰρ
ἂν τὸ γένος ἡγεμόνες ἦσαν. αἱ μὲν οὖν μέλιτται
κατὰ τὴν δύναμιν αὐτοῖς ὡμοίωνται [καὶ τῷ[2]
20 τίκτειν], οἱ δὲ κηφῆνες κατὰ τὸ μέγεθος· [εἰ δ'
εἶχον καὶ κέντρον, ἡγεμόνες ἂν ἦσαν. νῦν δὲ τοῦτο
λείπεται[3] τῆς ἀπορίας[4]· οἱ γὰρ ἡγεμόνες ἀμφοτέροις
ἐοίκασιν ἐν τῷ αὐτῷ τοῖς γένεσι, τῷ μὲν κέντρον
ἔχειν ταῖς μελίτταις, τῷ δὲ μεγέθει τοῖς κηφῆσιν.][5]
ἀναγκαῖον δὲ καὶ τοὺς ἡγεμόνας γίνεσθαι ἔκ τινος.
25 ἐπεὶ οὖν οὔτ' ἐκ τῶν μελιττῶν οὔτ' ἐκ τῶν
κηφήνων, αὐτοῖς ἀναγκαῖον καὶ αὐτοὺς γεννᾶν.
[γίνονται δ' ἐπὶ τέλει οἱ κύτταροι αὐτῶν καὶ οὐ
πολλοὶ τὸν ἀριθμόν.][6] ὥστε συμβαίνει τοὺς μὲν

---

[1] in seqq. plurima irrepsisse videntur. αἱ μὲν οὖν . . .
μέγεθος om. Σ.
[2] τὸ Υ : τοῦ coni. A.-W. ; καὶ τῷ τίκτειν seclusi.
[3] λέλυται coni. Platt.
[4] hic addit Υ καὶ ἤδη λέλυται· τὰ προειρημένα γὰρ ἡ λύσις
τῆς ἀπορίας.
[5] secl. A.-W.   [6] haec verba hic aliena.

---

[a] The full explanation of this statement comes at ll. 27 ff.
below, but owing to a number of interpolations in the text the
clarity of the passage has become obscured. The ἀναλογία is:
    Kings can generate two kinds, their own and another
    (viz., kings and " bees ");

them—they are, of course, generated from the
" leaders." Hence their manner of generation is
in fact arranged in a sort of proportionate series [a];
[thus, the leaders are similar to the drones in size,
but similar to the " bees " in possessing a sting;
therefore the " bees " are similar to them in this
respect, but the drones are similar to them in size,]
for of course the three kinds must of necessity fail to
coincide in some respect, unless the same kind is
always going to be bound to be generated from each,
and this is impossible, because then the whole tribe
of them would be " leaders." Therefore the " bees "
have been made similar to them in respect of characteristic properties,[b] [i.e., in virtue of generating
young,] while the drones have been made similar to
them in respect of size [and if they had a sting as
well, they would be " leaders." As it is, this portion
of the puzzle remains, since the leaders resemble
both kinds at the same time, the bees in possessing
a sting, the drones in size.] [c] But the leaders too
must be generated from something; and since they
are generated neither from the bees nor from the
drones, they must of necessity generate their own
kind as well. [And their cells are the last to be
formed, and are not many in number.] [d] So it turns

" Bees " can generate one kind, i.e., a kind other than
their own (viz., drones);
Drones can generate no kind. This is the πέρας of the
ἀναλογία (see 760 a 33).
[b] Dynamis: referring to the special and distinctive characteristic, viz., ability to generate, as the gloss explains.
[c] I have tentatively bracketed the passages which seem to
have been interpolated. The main argument is about the
power to generate, not about size or sting.
[d] This sentence seems to have been misplaced; it is more
relevant if moved to 760 b 27 below.

ἡγεμόνας γεννᾶν μὲν καὶ αὐτούς, γεννᾶν δὲ καὶ
ἄλλο τι γένος (τοῦτο δ᾽ ἐστὶ τὸ τῶν μελιττῶν), τὰς
30 δὲ μελίττας ἄλλο μέν τι γεννᾶν, τοὺς κηφῆνας,
αὐτὰς δὲ μηκέτι γεννᾶν, ἀλλὰ τοῦτ᾽ ἀφῃρῆσθαι
αὐτῶν. ἐπεὶ δ᾽ ἀεὶ τὸ κατὰ φύσιν ἔχει τάξιν, διὰ
τοῦτο τῶν κηφήνων ἀναγκαῖον καὶ τὸ ἄλλο τι
γένος γεννᾶν ἀφῃρῆσθαι. ὅπερ καὶ φαίνεται συμ-
βαῖνον· αὐτοὶ μὲν γὰρ γίγνονται, ἄλλο δ᾽ οὐθὲν
35 γεννῶσιν, ἀλλ᾽ ἐν τῷ τρίτῳ ἀριθμῷ πέρας ἔσχεν
ἡ γένεσις. καὶ οὕτω δὴ συνέστηκε τῇ φύσει καλῶς
ὥστ᾽ αἰεὶ διαμένειν ὄντα τὰ γένη καὶ μηδὲν ἐλλείπειν, μὴ πάντων γεννώντων. [εὔλογον δὲ καὶ
τοῦτο συμβαίνειν, ἐν μὲν ταῖς εὐετηρίαις μέλι καὶ
κηφῆνας γίνεσθαι πολλούς, ἐν δὲ ταῖς ἐπομβρίαις
5 ὅλως γόνον πολύν. αἱ μὲν γὰρ ὑγρότητες περίττωμα ποιοῦσι πλεῖον ἐν τοῖς σώμασι τῶν ἡγεμόνων,
αἱ δ᾽ εὐετηρίαι ἐν τοῖς τῶν μελιττῶν· ἔλαττω γὰρ
ὄντα[1] τῷ μεγέθει δεῖται τῆς εὐετηρίας μᾶλλον.][2] εὖ
δὲ καὶ τὸ τοὺς βασιλεῖς ὥσπερ πεποιημένους ἐπὶ
10 τέκνωσιν ἔσω μένειν, ἀφειμένους τῶν ἀναγκαίων
ἔργων, καὶ μέγεθος δὲ ἔχειν, ὥσπερ ἐπὶ τεκνο-
ποιίαν συστάντος τοῦ σώματος αὐτῶν· τούς τε
κηφῆνας ἀργοὺς ἅτ᾽ οὐδὲν ἔχοντας ὅπλον πρὸς τὸ
διαμάχεσθαι περὶ τῆς τροφῆς, καὶ διὰ τὴν βραδυτῆτα τὴν τοῦ σώματος. [αἱ δὲ μέλιτται μέσαι[3] τὸ
μέγεθός εἰσιν ἀμφοῖν (χρήσιμαι[4] γὰρ οὕτω πρὸς τὴν

---

[1] ἐλάττω γὰρ ὄντα P : ἔλαττον γὰρ ὂν vulg.   [2] aliena hic.
[3] μείους coni. Btf. ; τὸ μέγεθος del. Sus.
[4] χρήσιμαι P : χρήσιμοι vulg.

out that the leaders generate their own kind, and another kind as well (viz., the " bees ") ; while the " bees " generate another kind (the drones), but *not* their own kind ; this they have been deprived of doing. And since any business of Nature's always has an orderly arrangement, on that account necessity requires that the drones shall have been deprived even of generating some other kind. And this is what is found to be the case in actual fact : they are generated themselves, but generate no other creature ; thus the progression of generation reaches its limit at the third term of the series. And this arrangement has been so well constituted by Nature that the three kinds continue ever in existence and none of them fails, though not all of them generate. [Another point about them, which is in accord with what we should expect, is this. In fine seasons, much honey and a large number of drones is produced, in rainy seasons a large number of offspring generally. The reason is that wet conditions produce more residue in the bodies of the leaders, whereas fine seasons do the same in those of the bees, for being smaller in size they have greater need of fine weather.] [a] Besides, it is well that the kings, who have, as it were, been made specially for the purpose of procreation, should stay within, released from the drudgery that has got to be done by somebody ; and that they should be large, since their body has been constituted as it were for procreation, and that the drones should be idle, as they have no weapon for engaging in combat to secure their food, and also on account of the slowness of their bodies. [The bees, however, are as regards size midway between the two, for thus they are serviceable for active work,

15 ἐργασίαν), καὶ ἐργάτιδες ὡς καὶ τέκνα τρέφουσαι
καὶ πατέρας.]¹ ὁμολογούμενον δ' ἐστὶ καὶ τὸ ἐπ-
ακολουθεῖν τοῖς βασιλεῦσι τῷ τὴν γένεσιν ἐκ τούτων εἶναι τὴν τῶν μελιττῶν (εἰ γὰρ μηθὲν τοιοῦτον
ὑπῆρχεν, οὐκ εἶχε λόγον τὰ συμβαίνοντα περὶ τὴν
ἡγεμονίαν αὐτῶν), καὶ τὸ τοὺς μὲν ἐᾶν μηθὲν ἐρ-
20 γαζομένους ὡς γονεῖς, τοὺς δὲ κηφῆνας κολάζειν ὡς
τέκνα· κάλλιον γὰρ τὰ τέκνα κολάζειν καὶ ὧν μηθέν
ἐστιν ἔργον. τὸ δὲ τὰς μελίττας γεννᾶν πολλὰς
αὐτοὺς ὄντας ὀλίγους τοὺς ἡγεμόνας παραπλήσιον
ἔοικε συμβαίνειν τῇ γενέσει τῇ τῶν λεόντων, οἳ
τὸ πρῶτον πέντε γεννήσαντες ὕστερον ἐλάττω γεν-
25 νῶσι καὶ τέλος ἕν, εἶτ' οὐδέν. οἱ δ' ἡγεμόνες τὸ
μὲν πρῶτον πλῆθος, ὕστερον δ' ὀλίγους αὐτούς,
κἀκείνων² μὲν ἐλάττω τὸν γόνον, αὐτῶν δ' ἐπεὶ
τοῦ πλήθους ἀφεῖλε, τὸ³ μέγεθος αὐτοῖς⁴ ἀπέδωκεν
ἡ φύσις.⁵

Ἐκ μὲν οὖν τοῦ λόγου τὰ περὶ τὴν γένεσιν τῶν
μελιττῶν τοῦτον ἔχειν φαίνεται τὸν τρόπον, καὶ
30 ἐκ τῶν συμβαίνειν δοκούντων περὶ αὐτάς· οὐ μὴν
εἴληπταί γε τὰ συμβαίνοντα ἱκανῶς, ἀλλ' ἐάν ποτε
ληφθῇ, τότε τῇ αἰσθήσει μᾶλλον τῶν λόγων πιστευ-

---

¹ monent corrupta esse A.-W.: pro πρὸς τὴν . . .
πατέρας creationi pullorum Σ (=πρὸς τὴν τέκνωσιν). unde
et credo leg. esse v. 17 ⟨τὰς μελίττας⟩ τοῖς βασιλεῦσι . . . [τὴν
τῶν μελιττῶν].   ² κἀκεῖνοι PS.
³ ἀφεῖλε τὸ] ἀφεῖλεν. τὸ Z¹ : ἀφείλετο YZ².
⁴ ὃ αὐτοῖς Y.
⁵ αὐτούς . . . φύσις] quoniam diminuuntur superfluitates
que sunt in corpore Σ.

and they are workers inasmuch as they support and feed their children and fathers alike.] [a] Other facts which fit in well are these : (a) the bees attend upon the kings—because the bees are generated from the kings ; since, if nothing of this kind were the case, the facts about their leadership would be lacking in reason ; (b) they allow the leaders to do no work, as being their parents, and they punish the drones, as being their children, because it is a finer thing to punish children and those who have no function to perform.[b] The fact that the leaders, though few themselves in number, generate a large number of bees looks like a parallel phenomenon to the generation of lions. Lions [c] to begin with generate five, then fewer, finally one, then none at all. The "leaders" generate a multitude to begin with, and later on a few—these are of their own kind,[d] and though the brood of these is smaller in number, Nature, because she has taken away from their numbers makes up for it by giving them more in the way of size.

This, then, appears to be the state of affairs with regard to the generation of bees, so far as theory can take us, supplemented by what are thought to be the facts about their behaviour. But the facts have not been sufficiently ascertained ; and if at any future time they are ascertained, then credence must be given to the direct evidence of the senses more than

---

[a] Part of this sentence is inconsistent with what has already been said about the comparative sizes of the three kinds, and part anticipates what is to be said in the next sentence.

[b] I suppose attention should be called to this statement.

[c] See 750 a 31 ff.

[d] The statement at 760 a 26 above seems relevant here.

τέον, καὶ τοῖς λόγοις, ἐὰν ὁμολογούμενα δεικνύωσι τοῖς φαινομένοις.

[Πρὸς δὲ τὸ μὴ ἐξ ὀχείας γίνεσθαι σημεῖον καὶ τὸ τὸν γόνον φαίνεσθαι μικρὸν ἐν τοῖς τοῦ κηρίου 35 κυτταρίοις· ὅσα δ' ἐξ ὀχείας τῶν ἐντόμων γεννᾶται,

συνδυάζεται μὲν πολὺν χρόνον, τίκτει δὲ ταχέως καὶ μέγεθος ἔχον σκωληκοειδές.][1]

Περὶ δὲ τὴν γένεσιν τὴν τῶν συγγενῶν ζῴων αὐταῖς, οἷον ἀνθρηνῶν τε καὶ σφηκῶν, τρόπον τιν' ἔχει παραπλησίως πᾶσιν, ἀφήρηται δὲ τὸ περιττὸν 5 εὐλόγως· οὐ γὰρ ἔχουσιν οὐθὲν θεῖον, ὥσπερ τὸ γένος τὸ τῶν μελιττῶν. γεννῶσι μὲν γὰρ αἱ μῆτραι καλούμεναι, καὶ τὰ πρῶτα συμπλάττουσι τῶν κηρίων, ὀχευόμεναι δὲ γεννῶσιν ὑπ' ἀλλήλων· ὦπται γὰρ πολλάκις ὁ συνδυασμὸς αὐτῶν. πόσας δ' ἔχουσι διαφορὰς ἢ πρὸς ἄλληλα τῶν τοιούτων 10 γενῶν ἕκαστον ἢ πρὸς τὰς μελίττας, ἐκ τῶν περὶ τὰς ἱστορίας ἀναγεγραμμένων δεῖ θεωρεῖν.

Καὶ περὶ μὲν τῶν ἐντόμων τῆς γενέσεως εἴρηται πάντων, περὶ δὲ τῶν ὀστρακοδέρμων λεκτέον.

XI Ἔχει δὲ καὶ τούτων τὰ περὶ τὴν γένεσιν τῇ μὲν 15 ὁμοίως τῇ δ' οὐχ ὁμοίως τοῖς ἄλλοις. καὶ τοῦτ' εὐλόγως συμβαίνει· πρὸς μὲν γὰρ τὰ ζῷα φυτοῖς ἐοίκασι, πρὸς δὲ τὰ φυτὰ ζῴοις, ὥστε τρόπον μέν τινα ἀπὸ σπέρματος φαίνεσθαι γινόμενα, τρόπον δ' ἄλλον οὐκ ἀπὸ σπέρματος, καὶ τῇ μὲν αὐτόματα

---

[1] haec non proprio loco sita.

[a] The most important principle announced in this paragraph deserves very special attention.
[b] This is another misplaced paragraph.

to theories,—and to theories too provided that the results which they show agree with what is observed.[a]

[Another piece of evidence which goes to show that bees are generated without copulation is that the brood appears to be quite small in the cells of the comb, whereas those insects which are generated by means of copulation (a) spend a long time in intercourse, and (b) quickly bring forth their offspring, which is of the nature of a larva and of considerable size.] [b]

With regard to the generation of the animals that are akin to bees, such as hornets and wasps,[c] the situation is in a way similar in all of them, but the extraordinary features are lacking, and this is what we should expect, because they contain no divine ingredient as the tribe of bees does. Although the " mother-wasps " as they are called do indeed generate, and mould the first of the cells, it is by copulation with one another that they generate, as their copulation has often been observed. To find out the various differences between each of these kinds of creatures, and between them and bees, the records given in the *Researches* [d] should be studied. _Hornets and wasps._

We have now described the generation of all the Insects, and we have next to describe the Testacea.

The circumstances of the generation of these animals also is to some extent similar, to some extent dissimilar, to those of the others. And this is what we should expect, for compared with animals, they resemble plants, compared with plants, they resemble animals, so that in a way it seems that they are generated from semen, but in another way not ; _XI (iii.) Reproduction of Testacea._

***

[c] See *H.A.* 627 b 23 ff., 628 b 32 ff.
[d] At *H.A., locc. cit.*

τῇ δ᾽ ἀφ᾽ αὑτῶν, ἢ τὰ μὲν οὕτως τὰ δ᾽ ἐκείνως.
20 διὰ δὲ τὸ τοῖς φυτοῖς ἀντίστροφον ἔχειν τὴν φύσιν,
διὰ τοῦτο ἐν μὲν τῇ γῇ τῶν ὀστρακοδέρμων οὐθὲν
ἢ μικρόν τι γίγνεται γένος, οἷον τὸ τῶν κοχλιῶν
κἂν ᾖ τι τοιοῦτον ἕτερον μὲν σπάνιον δέ, ἐν δὲ τῇ
θαλάττῃ καὶ τοῖς ὁμοίοις ὑγροῖς πολλὰ καὶ παντοδαπὴν ἔχοντα μορφήν. τὸ δὲ τῶν φυτῶν γένος ἐν
25 μὲν τῇ θαλάττῃ καὶ τοῖς τοιούτοις[1] μικρὸν καὶ
πάμπαν ὡς εἰπεῖν οὐθέν, ἐν δὲ τῇ γῇ τὰ τοιαῦτα
γίνεται πάντα· τὴν γὰρ φύσιν ἀνάλογον ἔχει, καὶ
διέστηκεν, ὅσῳ[2] ζωτικώτερον τὸ ὑγρὸν τοῦ ξηροῦ
καὶ γῆς ὕδωρ, τοσοῦτον ἡ τῶν ὀστρακοδέρμων
φύσις τῆς τῶν φυτῶν, ἐπεὶ βούλεταί γε ὡς τὰ
30 φυτὰ πρὸς τὴν γῆν, οὕτως ἔχειν τὰ ὀστρακόδερμα
πρὸς τὸ ὑγρόν, ὡς ὄντα τὰ μὲν φυτὰ ὡσπερανεὶ
ὄστρεα χερσαῖα, τὰ δὲ ὄστρεα ὡσπερανεὶ φυτὰ
ἔνυδρα.

Διὰ τοιαύτην δ᾽ αἰτίαν καὶ πολύμορφα τὰ ἐν τῷ
ὑγρῷ μᾶλλόν ἐστι τῶν ἐν τῇ γῇ· τό τε γὰρ ὑγρὸν
εὐπλαστοτέραν ἔχει τὴν φύσιν τῆς γῆς καὶ σωματι-
35 κὴν οὐ πολλῷ ἧττον, καὶ μάλιστα τὰ ἐν τῇ θαλάττῃ

τοιαῦτα· τὸ μὲν γὰρ πότιμον γλυκὺ μὲν καὶ

---

[1] ποταμοῖς Z.   [2] ὅσῳ δὲ PSY.

---

[a] The scheme which Aristotle has in mind is :

Place:	Earth	Water	Air
Creature:	Plants	Testacea	Land-animals.

(From the passage 761 b 16-23 (see n., p. 352) we may add
a fourth pair, Fire, and Moon-animals ; but it is not essential
to Aristotle's main argument, and Aristotle himself does not
seem too sure of the existence of such creatures.) Aristotle
holds that water supports life better than earth (l. 27) ; and
also that the more " perfect " animals are those which breathe,
i.e., which live in the air (see 732 b 28 ff.) ; hence the three

and in one sense that they are spontaneously generated, in another that they are generated from themselves, or some by the one method, some by the
other. In virtue of the Testacea being in their
nature the correlative of plants,[a] no part, or only a
small part, of this tribe comes into being in the earth
(examples are snails, and any such species there may
be besides, but there are not many), whereas many
species, of all kinds of shapes, live in the sea and
similar watery places. The plant tribe, on the other
hand, makes very little show—practically none at all,
in fact—in the sea and such places, but all members
of this tribe grow in the earth. The reason is that
in respect of their nature the two tribes stand in
a correlative position [b] : the nature of Testacea is
removed from that of plants by an interval corresponding to that by which water and fluid matter are
better able to support life than earth and solid matter,
since Testacea aim at being so related to the water
as plants are related to the earth : it is as though
plants were a sort of land-shellfish, and shellfish a sort
of water-plant.

And it is for some such cause as this that the things
which grow in the water are more various in shape
than those which grow in the earth. It is because
a fluid substance is in its nature more plastic than
earth, and not much less substantial ; and this is a
characteristic possessed to a marked degree by the
creatures in the sea, since fresh water, though sweet

Various animals proper to various Elements.

stages are in order of increasing " perfection." We thus
get the ἀναλογία (l. 27) :

> Testacea : Water : : Plants : Earth,   or
> Testacea : Plants : : Water : Earth.

[b] Or, " proportionate relationship."

τρόφιμον, ἧττον δὲ σωματῶδες καὶ ψυχρόν ἐστιν.
διόπερ ὅσα ἄναιμα καὶ μὴ θερμὰ τὴν φύσιν, οὐ
γίνεται ἐν ταῖς λίμναις οὐδὲ τῶν ἁλμυρῶν ἐν τοῖς
5 ποτιμωτέροις ἀλλ' ἧττον, οἷον τὰ ὀστρακόδερμα
καὶ τὰ μαλάκια καὶ τὰ μαλακόστρακα (πάντα γὰρ
ἄναιμα καὶ ψυχρὰ ταῦτα τὴν φύσιν ἐστίν), ἐν δὲ
ταῖς λιμνοθαλάτταις καὶ πρὸς ταῖς ἐκβολαῖς τῶν
ποταμῶν γίνονται· ζητοῦσι γὰρ ἅμα τήν τ' ἀλέαν
καὶ τὴν τροφήν, ἡ δὲ θάλαττα ὑγρά τε καὶ σω-
10 ματῶδης πολλῷ μᾶλλον τοῦ ποτίμου καὶ θερμὴ
τὴν φύσιν ἐστί, καὶ κεκοινώνηκε πάντων τῶν
μορίων, ὑγροῦ καὶ πνεύματος καὶ γῆς, ὥστε καὶ
πάντων μετέχειν τῶν καθ' ἕκαστον γινομένων [ἐν
τοῖς τόποις ζῴων].[1] τὰ μὲν γὰρ φυτὰ θείη τις ἂν
15 γῆς, ὕδατος δὲ τὰ ἔνυδρα, τὰ δὲ πεζὰ ἀέρος· τὸ
δὲ μᾶλλον καὶ ἧττον καὶ ἐγγύτερον καὶ πορρώτερον
πολλὴν ποιεῖ καὶ θαυμαστὴν διαφοράν.[2] τὸ δὲ
τέταρτον γένος οὐκ ἐπὶ τούτων τῶν τόπων δεῖ

---

[1] seclusit Platt, ὥστε καὶ . . . ζῴων om. Σ.
[2] haec sensu carere monet Platt. post πορρώτερον addit Z
δεῖ τιθέναι, et pro ποιεῖ PZ habent ποιεῖν.

---

[a] Aristotle apparently did much of his zoological work in
lakes and lagoons; he refers to the lake at Siphae, *P.A.*
696 a 6, *I.A.* 708 a 5, *H.A.* 504 b 32. The difference between
a lake and a lagoon, as distinguished in the present passage,
is that the former is fresh, the latter salt. For lagoons *cf.*
H.A. 598 a 20; the whole passage is apposite. *Cf.* also
763 a 29, 763 b 2.

[b] It is now known that the blood serum (the fluid part of
the blood remaining after the cellular portion has been
removed by clotting) of both sea- and land-vertebrates has

(palatable) and nutritious, is less substantial and is cold. Hence, those animals which are bloodless and not hot by nature are not produced in lakes nor in the fresher of brackish waters, except to a somewhat small extent—such as the Testacea, Cephalopods and Crustacea, all of which are bloodless and cold by nature—whereas in lagoons ^a and near the mouths of rivers they *are* produced.^b The reason is that they seek both warmth and food together; and sea-water is fluid and much more substantial than fresh water and it is hot by nature,^c and it contains a quota of all the parts ^d—of fluid, of *pneuma*, and of earth—so that it also contains a quota of all the creatures which grow in each of them, because we may say that plants belong to the earth, aquatic creatures to the water, and land-animals to the air, but the more and less and nearer and further make a surprisingly great difference.^e As for the fourth tribe, we must not look for

Fire-animals.

a composition closely approximating to that of sea-water, which suggests that all vertebrates originated in the sea; and this receives support from comparative anatomical and embryological studies. Anaximander had asserted that human beings originated in fishes; see Plut. *Symp.* viii. 8. 4, p. 730 ε ἐν ἰχθύσιν ἐγγενέσθαι τὸ πρῶτον ἀνθρώπους . . . ὥσπερ οἱ γαλεοί [παλαιοί mss.] (see note, 754 b 32).

^c The rest of the paragraph from this point is obscure, and other passages do not help much in its elucidation. For Aristotle's theory of the structure of the universe, see App. A §§ 2 ff.

^d As Platt says, the sea " shares " in all three, earth, water, and air: it *is* fluid; it is σωματικόν, and so contains earthy matter; and it has *pneuma* in it, being warm—for πνεῦμα is " hot air " (736 a 1), and also, as Aristotle says at 762 a 19 ff., the things which are produced spontaneously in water are produced mainly in virtue of the *pneuma* in it, which contains Soul.

^e It is difficult to attach any meaning to this statement.

ζητεῖν· καίτοι βούλεταί γέ τι κατὰ τὴν τοῦ πυρὸς
εἶναι τάξιν· τοῦτο γὰρ τέταρτον ἀριθμεῖται τῶν
σωμάτων. ἀλλὰ τὸ μὲν πῦρ ἀεὶ φαίνεται τὴν
20 μορφὴν οὐκ ἰδίαν ἔχον, ἀλλ' ἐν ἑτέρῳ τῶν σω-
μάτων· ἢ γὰρ ἀὴρ ἢ καπνὸς ἢ γῆ φαίνεται τὸ
πεπυρωμένον. ἀλλὰ δεῖ τὸ τοιοῦτον γένος ζητεῖν
ἐπὶ τῆς σελήνης· αὕτη γὰρ φαίνεται κοινωνοῦσα
τῆς τετάρτης ἀποστάσεως. ἀλλὰ περὶ μὲν τούτων
ἄλλος ἂν εἴη λόγος.

Ἡ δὲ τῶν ὀστρακοδέρμων συνίσταται φύσις τῶν
25 μὲν αὐτομάτως, ἐνίων δὲ προϊεμένων τινὰ δύναμιν
ἀφ' αὑτῶν, πολλάκις δὲ γινομένων καὶ τούτων ἀπὸ
συστάσεως αὐτομάτης. δεῖ δὴ¹ λαβεῖν τὰς γενέσεις
τὰς τῶν φυτῶν. τούτων γὰρ γίνεται τὰ μὲν ἀπὸ
σπέρματος, τὰ δ' ἀπὸ σπαραγμάτων ἀποφυτευο-
μένων, ἔνια δὲ τῷ παραβλαστάνειν, οἷον τὸ τῶν
30 κρομμύων γένος. τοῦτον μὲν οὖν οἱ μύες γίνονται
τὸν τρόπον· παραφύονται γὰρ ἐλάττους ἀεὶ παρὰ
¹ δὴ Peck, coll. 762 b 6 : δὲ vulg.

ᵃ According to Aristotle, the "heavens" and the heavenly
bodies were composed of the "fifth element," *aither*, whose
natural movement is circular (see 736 b 35 ff. and n.,
and App. A § 2). As fire is the outermost of the sublunary
elements and is therefore in contact with the "heaven"
which is nearest to the earth, and as this "heaven" carries
the moon, it follows that the moon can be said to "have a
share in the fourth degree of remove," viz., fire. *Aither*
must be clearly distinguished from fire; and, according to
G.A. 737 a 1 (*cf. Meteor.* 382 a 7), fire generates no animal,
whereas *aither*, the "element of the stars," is a form of
θερμόν which *can* produce living creatures (ποιεῖ γόνιμα
τὰ σπέρματα; see 736 b 30-35). But at *H.A.* 552 b 10
Aristotle speaks of a creature which is engendered in the
fire in places where ore is smelted; and also mentions

it in these regions, although there wants to be a kind corresponding to the position of fire [a] in the series, since fire is reckoned as the fourth of the corporeal substances. But always, as we see, the shape and appearance which fire has is not its own ; on the contrary, fire is always in some other one of the substances, for the object which is on fire appears either as air or smoke or earth.[b] No ; this fourth tribe must be looked for on the moon, since the moon, as it appears, has a share in the fourth degree of remove. However, these matters should form the subject of another treatise.

With regard to the Testacea,[c] then : some of them take shape spontaneously, others by means of the emission of some special substance from themselves, though these too are often formed from a spontaneous composition. We must here apprehend the ways in which plants are generated. Some plants are formed from seed, some from slips planted out, others by sideshoots (*e.g.*, the onion tribe). Now the last-named is the method by which mussels are formed ; small ones are always growing up by the

(a) Side-shoot-propagation.

the salamander, which cannot be destroyed by fire ; the *History of Animals* passage is, however, excised by A.-W. There is a long discussion in Jaeger, *Aristotle*, 144-148, in which the doctrine of fire-animals is involved. Jaeger tries to prove that the doctrine that there were animals that were *engendered* in fire must have come in one of Aristotle's dialogues (*On Philosophy*), and by a curious blunder states that it does not come in *History of Animals* (*loc. cit.*, to which he actually refers) ; but in fact Aristotle's words are γίνεται θηρία ἐν τῷ πυρί. Jaeger makes no reference at all to the present passage.

[b] *Cf. P.A.* II. 649 a 22 ff., *G. & C.* II. 331 b 25, *Meteor.* I, chh. 3, 4, etc.

[c] Lit., the nature, *i.e.*, the physical structure, of the Testacea. See Introd. §§ 26, 27.

τὴν ἀρχήν. κήρυκες δὲ καὶ πορφύραι καὶ τὰ
λεγόμενα κηριάζειν οἷον ἀπὸ σπερματικῆς φύσεως
προΐενται μυξώδεις ὑγρότητας (σπέρμα δ' οὐθὲν
τούτων δεῖ νομίζειν, ἀλλὰ κατὰ τὸν εἰρημένον
35 τρόπον μετέχειν τῆς ὁμοιότητος τοῖς φυτοῖς· διὸ
καὶ γίνεται πλῆθος τῶν τοιούτων ὅταν ἅπαξ

γένηταί τι, πάντα μὲν γὰρ ταῦτα καὶ αὐτόματα
συμβαίνει γίνεσθαι, κατὰ λόγον δὲ καὶ ὑπαρξάντων
συνίστασθαι μᾶλλον). περιγίγνεσθαι γάρ τι περίττωμα πρὸς ἑκάστῳ τῆς ἀρχῆς εὔλογον, ἀφ' ἧς[1]
5 παραβλαστάνει τῶν παραφυομένων ἕκαστον. ἐπεὶ
δὲ παραπλησίαν ἔχει τὴν δύναμιν ἡ τροφὴ καὶ τὸ
ταύτης περίττωμα, ⟨τὸ⟩[2] τῶν κηριαζόντων ὅμοιον[3]
εἰκός ἐστιν εἶναι τῇ ἐξ ἀρχῆς[4] συστάσει [οὐσίαν][5]·
διόπερ εὔλογον γίνεσθαι καὶ ἐκ ταύτης.[6]

Ὅσα δὲ μήτε παραβλαστάνει μήτε κηριάζει,
τούτων δὲ πάντων ἡ γένεσις αὐτόματός ἐστιν.
10 πάντα δὲ τὰ συνιστάμενα τὸν τρόπον τοῦτον καὶ
ἐν γῇ καὶ ἐν ὕδατι φαίνεται γινόμενα μετὰ σήψεως
καὶ μιγνυμένου τοῦ ὀμβρίου ὕδατος· ἀποκρινομένου γὰρ τοῦ γλυκέος εἰς τὴν συνισταμένην ἀρχὴν
τὸ περιττεῦον τοιαύτην λαμβάνει μορφήν. γίνεται
δ' οὐθὲν σηπόμενον ἀλλὰ πεττόμενον· ἡ δὲ σῆψις

---

[1] ἀφ' οὗ Platt. [2] ⟨τὸ⟩ Peck.
[3] ὅμοιον Peck : ὁμοίαν vulg. [4] fort. τῆς ἀρχῆς leg.
[5] οὐσίαν om. Z[1], secl. A.-W. : pro οὐσίαν coni. Platt τὴν
παράφυσιν.
[6] περιγίγνεσθαι (a 3) . . . ταύτης om. Σ.

---

[a] The " honeycombs " are really the eggs of these
Gastropods, and Aristotle rightly recognizes their nature, as
against later scientists who regarded them as distinct species
of animals.

[b] As against none, in the case of spontaneous generation.

side of the original one. The whelks and purpuras
and those which, as the phrase goes, are " honeycombers " [a] emit quantities of slimy fluid emanating
as it were from some seminal substance. (We must
not, however, consider any of these substances as
being semen proper ; instead, we should regard them
as sharing in the resemblance to plants in the way
already mentioned. And that is why a large number
of such creatures is produced when once one has been
produced, since, as all these creatures are in fact produced spontaneously as well, *pro rata* more of them
arise if there are actually some [b] present to start
with.) After all, it is reasonable to suppose that
there is a surplus portion of residue close by each of
the original stock, from which each of the sideshoots
springs up. And since the residue is a substance
possessing one and the same character as the nourishment of which it is the residue, it is probable that
the stuff produced by the " honeycombers " is similar
to the substance out of which they were originally
constituted ; hence it is reasonable to suppose that
it too [c] gives rise to young ones.

All which neither produce sideshoots nor make
" honeycombs " reproduce by spontaneous generation ; and all which arise in this manner whether on
land or in the water come to be formed, as can be
seen, to the accompaniment of putrefaction and admixture of rainwater : as the sweet ingredients are
separated off into the principle which is taking form,
that which remains over assumes a putrefying aspect.[d]
Nothing, however, is formed by a process of putrefaction, but by a process of concoction : the putrefaction

(b) Spon-
taneous
generation.

---

[e] *i.e.*, as well as residues such as semen.
[d] *i.e.*, putrefies.

15 καὶ τὸ σηπτὸν περίττωμα τοῦ πεφθέντος ἐστίν·
οὐθὲν γὰρ ἐκ παντὸς γίνεται, καθάπερ οὐδ' ἐν τοῖς
ὑπὸ τῆς τέχνης δημιουργουμένοις· οὐθὲν γὰρ ἂν
ἔδει ποιεῖν· νῦν δὲ τὸ μὲν ἡ τέχνη τῶν ἀχρήστων
ἀφαιρεῖ, τὸ δ' ἡ φύσις.

Γίνεται δ' ἐν γῇ καὶ ἐν ὑγρῷ τὰ ζῷα καὶ τὰ
20 φυτὰ διὰ τὸ ἐν γῇ μὲν ὕδωρ ὑπάρχειν, ἐν δ' ὕδατι
πνεῦμα, ἐν δὲ τούτῳ παντὶ θερμότητα ψυχικήν,
ὥστε τρόπον τινὰ πάντα ψυχῆς εἶναι πλήρη· διὸ
συνίσταται ταχέως, ὁπόταν ἐμπεριληφθῇ. ἐμπερι-
λαμβάνεται δὲ καὶ γίνεται θερμαινομένων τῶν
σωματικῶν ὑγρῶν οἷον ἀφρώδης πομφόλυξ. αἱ
25 μὲν οὖν διαφοραὶ τοῦ τιμιώτερον εἶναι τὸ γένος
καὶ ἀτιμότερον τὸ συνιστάμενον ἐν τῇ περιλήψει
τῆς ἀρχῆς τῆς ψυχικῆς εἰσιν.[1] τούτου[2] δὲ καὶ οἱ
τόποι αἴτιοι καὶ τὸ σῶμα τὸ περιλαμβανόμενον.
ἐν δὲ τῇ θαλάττῃ πολὺ τὸ γεῶδες ἔνεστιν· διόπερ
ἐκ τῆς τοιαύτης συστάσεως ἡ τῶν ὀστρακοδέρμων
30 γίνεται φύσις, κύκλῳ μὲν τοῦ γεώδους σκληρυνο-
μένου καὶ πηγνυμένου τὴν αὐτὴν πῆξιν τοῖς ὀστοῖς
καὶ τοῖς κέρασι (πυρὶ γὰρ ἄτηκτα ταῦτ' ἐστίν),
ἐντὸς δὲ περιλαμβανομένου τοῦ τὴν ζωὴν ἔχοντος
σώματος.

Μόνον δὲ τῶν τοιούτων συνδυαζόμενον ἑώραται
τὸ τῶν κοχλιῶν γένος. εἰ δ' ἐκ τοῦ συνδυασμοῦ

---

[1] εἰσιν Peck : ἐστιν vulg.   [2] τούτων P.

---

[a] This of course is not intended to cover the development
of a larva once it has been constituted.
[b] Cf. above, 736 b 35 ff., and App. B §§ 13-17.

and the putrefied matter are a residue of that which
has been concocted, for no creature's formation uses
up the *whole* of the material,[a] any more than in the
case of objects fashioned by the agency of art, otherwise there would be no need to make anything at all,
whereas what happens in actual fact is that the useless material is removed in the one case by art and
in the other by Nature.

Animals and plants are formed in the earth and in
the water because in earth water is present, and in
water *pneuma* is present, and in all *pneuma* soul-heat is
present,[b] so that in a way all things are full of Soul ;
and that is why they quickly take shape once it has
been enclosed. Now it gets enclosed as the liquids
containing corporeal matter[c] become heated, and
there is formed as it were a frothy bubble. The
object which thus takes shape may be more valuable
in kind or less valuable ; and the differences herein
depend upon the envelope which encloses the soul-
principle ; and the causes which determine this are
the situations where the process takes place and the
physical substance which is enclosed. Now in the
sea earthy substance is plentiful, and that is why
the Testacea[d] are formed and constructed out of a
composition which is earthy in character : the earthy
substance hardens all round and congeals in the same
way that bones and horns do (since these cannot be
melted by fire), while within it the physical substance
that contains the life becomes enclosed.

Of such creatures the only tribe which has been
observed to copulate is that of the snails ; but whether

[c] Sea-water is such a liquid ; see above, 761 b 9 and immediately below, l. 27. Also App. B §§ 13-17.
[d] Lit., the nature of the Testacea ; *cf.* above, 761 b 24.

35 ἡ γένεσις αὐτῶν ἐστιν ἢ μή, οὔπω συνῶπται
ἱκανῶς.

Ζητήσειε δ' ἄν τις βουλόμενος ὀρθῶς ζητεῖν, τί
762 b τὸ κατὰ τὴν ὑλικὴν ἀρχὴν συνιστάμενόν ἐστιν ἐν
τοῖς τοιούτοις. ἐν μὲν γὰρ τοῖς θήλεσι περίττωμά
τι τοῦ ζῴου τοῦτ' ἐστίν, ὃ ἡ παρὰ τοῦ ἄρρενος
ἀρχὴ κινοῦσα, δυνάμει τοιοῦτον ὂν οἷον ἀφ' οὗπερ
ἦλθεν, ἀποτελεῖ τὸ ζῷον. ἐνταῦθα δὲ τί δεῖ λέγειν
5 τὸ τοιοῦτον, καὶ πόθεν καὶ τίς ἡ κινοῦσα ἀρχὴ ἡ
κατὰ τὸ ἄρρεν; δεῖ δὴ λαβεῖν ὅτι καὶ ἐν τοῖς
ζῴοις τοῖς γεννῶσιν ἐκ τῆς εἰσιούσης τροφῆς ἡ ἐν
τῷ ζῴῳ θερμότης ἀποκρίνουσα καὶ συμπέττουσα
ποιεῖ τὸ περίττωμα, τὴν ἀρχὴν τοῦ κυήματος.
ὁμοίως δὲ καὶ ἐν φυτοῖς, πλὴν ἐν μὲν τούτοις καὶ
10 ἔν τισι τῶν ζῴων οὐθὲν προσδεῖται τῆς τοῦ ἄρρενος
ἀρχῆς (ἔχει γὰρ ἐν αὑτοῖς μεμιγμένην), τὸ δὲ τῶν
πλείστων ζῴων περίττωμα προσδεῖται. τροφὴ δ'
ἐστὶ τοῖς μὲν ὕδωρ καὶ γῆ, τοῖς δὲ τὰ ἐκ τούτων,
ὥσθ' ὅπερ ἡ ἐν τοῖς ζῴοις θερμότης ἐκ τῆς τροφῆς
ἀπεργάζεται, τοῦθ' ἡ τῆς ὥρας ἐν τῷ περιέχοντι
15 θερμότης ἐκ θαλάττης καὶ γῆς συγκρίνει πέττουσα
καὶ συνίστησιν. τὸ δ' ἐναπολαμβανόμενον ἢ ἀποκρινόμενον ἐν τῷ πνεύματι τῆς ψυχικῆς ἀρχῆς
κύημα ποιεῖ καὶ κίνησιν ἐντίθησιν. ἡ μὲν οὖν τῶν

---

or not their generation is the result of such copulation has not so far been adequately observed.

Anyone who wishes to follow the right line of inquiry might well inquire what it is which, as it takes shape, corresponds in the case of these creatures to the " material principle." In females of course this is a residue produced by the animal, a residue which *potentially* is such as the parent is from which it came, and which is perfected into an animal by the principle from the male [a] imparting movement to it. In the present case, however, what are we to describe as holding this sort of position ? and whence comes the principle that imparts movement, corresponding to the male, and what is it ? Now we must apprehend that, even in the case of those animals which generate, it is the incoming nourishment that is the material out of which the heat residing in the animal produces the residue—the " principle " of the fetation—by setting it apart and concocting it. Similarly with plants, except that with them and certain of the animals there is no need of the principle of the male over and above that, because they contain in themselves this principle mixed ⟨with the female⟩ ; in most animals, however, the residue does need this principle. Of the one set, the nourishment is water and earth ; of the other, it is the things that are formed out of these ; so that in their case the seasonal heat present in their environment causes to accumulate and to take shape by means of concoction out of sea-water and earth that which in the case of animals the heat present in them produces out of the nourishment. And that portion of the soul-principle which gets enclosed or separated off within the *pneuma* makes a fetation and implants movement in it. Now

762 b

φυτῶν τῶν ἀπὸ ταὐτομάτου γινομένων σύστασις
ὁμοειδής ἐστιν· ἔκ τινος γὰρ μορίου γίνεται, καὶ
20 τὸ μὲν ἀρχὴ τὸ δὲ τροφὴ γίνεται ἡ πρώτη τοῖς
ἐκφυομένοις.[1] τὰ δὲ τῶν ζῴων σκωληκοτοκεῖται
καὶ τῶν ἀναίμων ὅσα μὴ ἀπὸ ζῴων γίνεται καὶ
τῶν ἐναίμων, οἷον γένος τι κεστρέων καὶ ἄλλων
ποταμίων ἰχθύων, ἔτι δὲ τὸ τῶν ἐγχέλεων γένος·
25 ἅπαντα γὰρ ταῦτα, καίπερ ὀλίγαιμον ἔχοντα τὴν
φύσιν, ὅμως ἔναιμά ἐστι, καὶ καρδίαν ἔχουσι τὴν
ἀρχὴν τὴν τῶν μορίων αἱματικήν. τὰ δὲ καλούμενα γῆς ἔντερα σκώληκος ἔχει φύσιν, ἐν οἷς ἐγγίνεται τὸ σῶμα τὸ τῶν ἐγχέλεων. διὸ καὶ περὶ
τῆς τῶν ἀνθρώπων καὶ τετραπόδων γενέσεως
ὑπολάβοι τις ἄν, εἴπερ ἐγίγνοντό ποτε γηγενεῖς,
30 ὥσπερ φασί τινες, δύο τρόπων τούτων[2] γίνεσθαι
τὸν ἕτερον· ἢ γὰρ ὡς σκώληκος συνισταμένου τὸ
πρῶτον ἢ ἐξ ᾠῶν, ἀναγκαῖον γὰρ ἢ ἐν αὐτοῖς
ἔχειν τὴν τροφὴν εἰς τὴν αὔξησιν (τὸ δὲ τοιοῦτον
κύημα σκώληξ ἐστίν) ἢ λαμβάνειν ἄλλοθεν, τοῦτο

---

[1] hic lacunam statuit Platt.
[2] τούτων PZ, *istorum* Σ : om. vulg.

---

[a] *Cf.* 715 b 27 " they are formed when . . . certain *parts*
in plants become putrescent . . . as for instance the mistletoe."
[b] See above, 741 b 1.

as for plants, the manner in which those plants take shape which are generated spontaneously is uniform : they are formed from a part [a] of something, and some of it forms into the " principle," some into the first nourishment of the germinating plants. As for the animals, however, some of them are brought forth as larvae, both the bloodless ones that are not formed from living animals, and some blooded ones (examples are a kind of *cestreus* [b] and other river fishes, also the eel tribe) : all of these, although by nature they have but little blood, nevertheless are blooded animals and have a heart, which is the " principle " of the parts and bloodlike in constitution. The " earth's-guts " as they are called have the nature of a larva ; the body of the eels forms within them.[c] Hence, too, with regard to the genera- Traditional view of the origin of man and animals. tion of human beings and quadrupeds, if once upon a time they were " earthborn " as some allege,[d] one might assume them to be formed in one of these two ways—either it would be by a larva taking shape to begin with, or else they were formed out of eggs, since of necessity they must either contain the nourishment for their growth within themselves (and a fetation of this sort is a larva) or they must get it from elsewhere, and that means either from

---

[c] The " earth's-guts " are apparently the round-worm *Gordius*. *Cf. H.A.* 570 a 15 ff., where they are said to be " formed spontaneously in mud and humid ground . . . for it is by the water's edge that the heat of the sun is strong and causes putrefaction." See note on eels, p. 565.

[d] This was an old and traditional belief ; *cf.* Plato, *Politicus* 269 B; in Hdt. VIII. 55 there is a reference to "Erechtheus, who is said to have been γηγενής " : *cf.* also Empedocles, Diels, *Vorsokr.*[5] 31 B 62 " First whole-natured forms sprang up from the earth, having a portion both of water and fire "; and *ibid.* B 57 ; 96 ; 98. And above, *G.A.* 722 b 20 ff.

δ' ἢ ἐκ τῆς γεννώσης ἢ ἐκ μορίου τοῦ κυήματος·
35 ὥστ' εἰ θάτερον ἀδύνατον, ἐπιρρεῖν ἐκ τῆς γῆς
ὥσπερ ἐν τοῖς¹ ζῴοις ἐκ τῆς μητρός, ἀναγκαῖον ἐκ
μορίου λαμβάνειν τοῦ κυήματος· τὴν δὲ τοιαύτην
ἐξ ᾠοῦ λέγομεν εἶναι γένεσιν. ὅτι μὲν οὖν, εἴπερ
ἦν τις ἀρχὴ τῆς γενέσεως πᾶσι τοῖς ζῴοις, εὔλογον
τοῖν δυοῖν τούτοιν εἶναι τὴν ἑτέραν, φανερόν· ἧττον
5 δ' ἔχει λόγον ἐκ τῶν ᾠῶν· οὐθενὸς γὰρ τοιαύτην
ὁρῶμεν ζῴου γένεσιν, ἀλλὰ τὴν ἑτέραν, καὶ τῶν
ἐναίμων τῶν ῥηθέντων καὶ τῶν ἀναίμων. τοιαῦτα
δ' ἐστὶ τῶν τ' ἐντόμων ἔνια καὶ τὰ ὀστρακόδερμα
περὶ ὧν ὁ λόγος· οὐ γὰρ ἐκ μορίου γίνονταί τινος,
ὥσπερ τὰ ᾠοτοκούμενα, ποιοῦνται δὲ καὶ τὴν
10 αὔξησιν ὁμοίως τοῖς σκώληξιν· ἐπὶ τὰ ἄνω γὰρ
καὶ τὴν ἀρχὴν αὐξάνονται οἱ σκώληκες· ἐν τῷ
κάτω γὰρ ἡ τροφὴ τοῖς ἄνω. καὶ τοῦτό γε ὁμοίως
ἔχει τοῖς ἐκ τῶν ᾠῶν, πλὴν ἐκεῖνα μὲν καταναα-
λίσκει πᾶν, ἐν δὲ τοῖς σκωληκοτοκουμένοις, ὅταν
αὐξηθῇ ἐκ τῆς ἐν τῷ κάτω μορίῳ συστάσεως τὸ
15 ἄνω μόριον, οὕτως ἐκ τοῦ ὑπολοίπου διαρθροῦται
τὸ κάτωθεν. αἴτιον δ' ὅτι καὶ ὕστερον ἡ τροφὴ ἐν
τῷ μορίῳ τῷ ὑπὸ τὸ ὑπόζωμα γίνεται πᾶσιν. ὅτι
δὲ τοῦτον τὸν τρόπον ποιεῖται τὰ σκωληκώδη τὴν

---

¹ ἄλλοις post τοῖς vulg.: om. PZ.

ᵃ i.e., in the uterus.
ᵇ i.e., the egg. Thus the three possibilities are—production as larvae; viviparously; oviparously. It should not
be supposed that Aristotle seriously envisages the possibility
of this sort of " evolution "; but in view of the popular
nature of the belief he thinks fit to show by which of the three
modes of generation these " earthborn " men would have
been produced, if they had been produced.
ᶜ Spontaneous generation from eggs.

the female parent [a] or from part of the fetation [b] ;
so that if the former way is impossible (*i.e.*, if it
cannot flow to them out of the earth as it flows to
animals from the mother), of necessity they must get
it from part of the fetation, and generation of this
sort we call generation from an egg. Thus much,
therefore, is plain : if there were a " principle " of
their generation in the case of all animals, we should
reasonably expect it to be one or other of these two,
larva or egg. It is, however, less reasonable to hold
that their generation would take place out of eggs,
because in the case of no animal do we observe this
sort of generation [c] to occur, whereas we do see the
other, in the case both of the blooded animals I
mentioned [d] and the bloodless ones. Under this
latter heading come certain of the Insects, and also
the Testacea with which our discussion is concerned :
they are not formed out of a *part* of something as
are the creatures produced from eggs, and further,
they effect their growth in a similar way to larvae,
for larvae grow towards the upper part, towards the
" principle," the nourishment for the upper parts
being in the lower part. In this respect they resemble the creatures that are produced from eggs,
except that the latter use up the *whole* of the egg,
whereas, in the case of those produced from larvae,
when the upper part has grown by drawing on the
substance in the lower part, then the lower part
becomes articulated out of what remains. The
reason for this is that ⟨not only in the early stages
but⟩ afterwards as well [e] the nourishment is produced
in the part below the diaphragm in all animals.
That the larva-like creatures effect their growth in

[a] *Cestreus* and eels.   [e] When they are fully grown.

763 a

αὔξησιν, δῆλον ἐπὶ τῶν μελιττῶν καὶ τῶν τοιού-
20 των· κατ᾽ ἀρχὰς γὰρ τὸ μὲν κάτω μόριον μέγα
ἔχουσι, τὸ δ᾽ ἄνω ἔλαττον. καὶ ἐπὶ τῶν ὀστρακοδέρμων δὲ τὸν αὐτὸν τρόπον ἔχει τὰ περὶ τὴν
αὔξησιν. φανερὸν δὲ καὶ τοῦτ᾽ ἐπὶ τῶν στρομβω-
δῶν ⟨ἐν⟩¹ ταῖς ἕλιξιν· ἀεὶ γὰρ αὐξανομένων
γίνονται μείζους² ἐπὶ τὸ πρόσθιον καὶ τὴν καλουμένην κεφαλήν.

Ὃν μὲν οὖν τρόπον ἔχει ἡ γένεσις καὶ τούτων καὶ
25 τῶν ἄλλων τῶν αὐτομάτων, εἴρηται σχεδόν.

Ὅτι δὲ συνίσταται αὐτόματα πάντα τὰ ὀστρακόδερμα, φανερὸν ἐκ τῶν τοιούτων, ὅτι πρός τε
τοῖς πλοίοις γίνεται σηπομένης τῆς ἀφρώδους
ἰλύος, καὶ πολλαχοῦ, οὗ πρότερον οὐθὲν ὑπῆρχε
τοιοῦτον, ὕστερον δι᾽ ἔνδειαν ὑγροῦ τοῦ τόπου
30 βορβορωθέντος ἐγένετο τὰ καλούμενα λιμνόστρεα
τῶν ὀστρακηρῶν, οἷον περὶ Ῥόδον παραβαλόντος
ναυτικοῦ στόλου καὶ ἐκβληθέντων κεραμίων εἰς
τὴν θάλατταν, χρόνου γενομένου καὶ βορβόρου περὶ
αὐτὰ συναλισθέντος, ὄστρεα εὑρίσκοντ᾽ ἐν αὐτοῖς.
ὅτι δ᾽ οὐδ᾽ ἀφίησι τὰ τοιαῦτα οὐδὲν ἀφ᾽ αὑτῶν

763 b

γεννητικόν, τεκμήριον· ἐπεὶ γὰρ Χῖοί τινες ἐκ
Πύρρας τῆς ἐν Λέσβῳ τῶν ὀστρέων διεκόμισαν

---

¹ ⟨ἐν⟩ A.-W. : καὶ ταῖς S : ἐπὶ ταῖς PZ : ταῖς vulg.
² μείζους Platt : πλείους vulg., om. Y.

ᵃ This does not entirely square with what has been said,
although Aristotle seems to think that even those which are
generated otherwise are also spontaneously generated ; see
761 b 25 ff.   ᵇ Cf. 736 a 13 ff.
ᶜ i.e., when there is only mud and no water in the lagoon ;
cf. H.A. VI, ch. 15.

this manner is plain in the case of bees and insects of that sort, as their lower part is large to start with and the upper part smaller. The arrangements for growth in the Testacea are on the same lines. This is shown in the convolutions of the spiral-shelled creatures, which as they grow always become larger towards the front and the " head " as it is called.

This practically completes our description of the manner of generation of these animals and of the others that are generated spontaneously.

The fact that all [a] the Testacea take shape spontaneously is shown by considerations like the following : They form on the side of boats when the frothy slime [b] putrefies ; and also, in many places where nothing of the kind had been present previously, after a time when the place has become muddy owing to lack of water,[c] lagoon-oysters,[d] as they are called, a kind of testaceous animal, have been formed ; for example, on an occasion when a naval squadron cast anchor off Rhodes, some earthenware pots were thrown out into the sea, and as time went on and mud had collected round them, oysters were continually found inside them. Here is a piece of evidence to show that animals of this kind emit no generative substance : people from Chios transported some live oysters across from Pyrrha in Lesbos,[e]

[d] Cf. H.A. 547 b 11. Apparently barnacles, which are, however, Crustacea, not Testacea.

[e] The lagoon at Pyrrha seems, as D'Arcy Thompson (prefatory note to translation of H.A.) suggests, to have been one of the chief places where Aristotle carried on his researches. The strait leading to it is mentioned again at P.A. 680 b 1 (a passage where also the " eggs " of sea-urchins and oysters are discussed), and several times in H.A. Cf. 761 b 4.

763 b

ζῶντα καὶ εἰς τόπους τινὰς τῆς θαλάττης εὐρι-
πώδεις καὶ ὁμόρρους[1] ἀφεῖσαν, πλείω μὲν τῷ
χρόνῳ οὐδὲν ἐγένετο, τὸ δὲ μέγεθος εἰς αὔξησιν
5 ἐπέδωκε πολύ. τὰ δὲ λεγόμενα ᾠὰ οὐθὲν συμβάλλεται πρὸς τὴν γένεσιν, ἀλλ' ἐστὶν εὐτροφίας
σημεῖον, οἷον ἐν τοῖς ἐναίμοις ἡ πιότης· διὸ καὶ
πρὸς τὴν ἐδωδὴν γίνεται εὔχυμα κατὰ τοὺς καιροὺς
τούτους. σημεῖον δ' ὅτι τὰ τοιαῦτα ἀεὶ ἔχουσιν,
οἷον αἱ πῖναι καὶ οἱ κήρυκες καὶ αἱ πορφύραι, πλὴν
10 ὁτὲ μὲν μείζω ὁτὲ δ' ἐλάττω. ἔνια δ' οὐκ ἀεί,
ἀλλὰ τοῦ μὲν ἔαρος ἔχουσι, προβαινούσης δὲ φθίνει
τῆς ὥρας, καὶ τέλος ἀφανίζεται πάμπαν, οἷον οἵ
τε κτένες καὶ οἱ μύες καὶ τὰ καλούμενα λιμνόστρεα·
ἡ γὰρ ὥρα αὕτη συμφέρει τοῖς σώμασιν αὐτῶν.
τοῖς δὲ συμβαίνει τοιοῦτον οὐδὲν ἐπίδηλον, οἷον
15 τοῖς τηθύοις. τὰ δὲ καθ' ἕκαστα περὶ τούτων, καὶ
ἐν οἷς γίνονται τόποις, ἐκ τῆς ἱστορίας θεωρείσθω.

---

[1] ὁμόρρους Platt : ὁμόρους Z : ὁμοίους vulg.

---

[a] The characteristic of a εὔριπος is the force and violence
of the currents sweeping through it ; hence there is no opportunity for mud to collect and so for any Testacea to arise.
Platt's conjecture ὁμόρρους is also supported by the use of the

and deposited them in some sea-straits where the currents met.[a] As time passed the oysters did not increase at all in number, but they grew greatly in size. As for their " eggs," [b] as they are called, these contribute nothing to generation ; they are just a sign of good nourishment, like fat in blooded animals, and that too is why they are tasty to eat at these seasons. A proof of this is that these creatures —e.g., pinnae, whelks and purpurae—have such " eggs " as these always, only sometimes they are larger, sometimes smaller. Others—e.g., pectens, mussels and the lagoon-oysters as they are called— do not have them always, but only in the spring ; as the season advances they wane, and finally disappear altogether ; the reason being that the spring-season is favourable to their physical condition. In others —e.g., the seasquirts—nothing of the kind is to be detected. For an account dealing with these individually, and the places where they grow, the student should consult the *Researches*.

verb ῥέω elsewhere in connexion with εὔριπος, e.g. E.N. 1167 b 7 μένει τὰ βουλεύματα καὶ οὐ μεταρρεῖ ὥσπερ εὔριπος : cf. Prob. 940 b 16 οἱ εὔριποι ῥέουσιν, and De somno et vig. 456 b 21. Gaza's translation *luto similia* seems to imply the reading βορβορώδεις, which is entirely against the sense.

  [b] See note on 763 b 1.

TABLE OF BIRDS

(This table has been constructed solely as an aid to the reading of Aristotle's discussions of birds. It has no value as a scientific classification.)

A	C	D	B
Non-fliers	Fliers	Fliers (*P.A.*)	Fliers (Crook-taloned)
Heavy, bulky bodies	Bulky bodies	Small	Small (apart from wings) (*P.A.* 694 a 5 f.)
	Residue to wings	Much residue	Residue to wings and feathers
Prolific, many eggs	Prolific; lay few, but often	Prolific	Not prolific
E.g., fowls, partridges, ostrich	Pigeons, ringdoves, turtledoves	Adrianic fowls and small birds named at 774 b 25 ff.: crow, rook, jay, sparrow, swallow	
Water-birds		Migrants (*P.A.*)	

At 749 b 1-25 Aristotle seems to make a threefold classification of birds, but he immediately goes on to speak of a class of "small" birds (also mentioned at 774 b 25 ff.) which does not appear to be allowed for in the threefold classification, though it has some characteristics in common with class C and some with class B (those in B, as appears also from *P.A.* 694 a 5 f., have small bodies, apart from their wings). These "small" birds must therefore be inserted as a fourth class, D, between C and B.

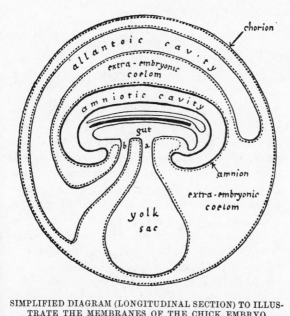

SIMPLIFIED DIAGRAM (LONGITUDINAL SECTION) TO ILLUS-
TRATE THE MEMBRANES OF THE CHICK EMBRYO

The dotted lines represent mesoblast.   The diagram shows the state
of development after about ten days.   The embryo itself is in the
central part of the diagram.   Immediately above the gut is the noto-
chord (shown in black), and immediately above that is the nerve-cord,
of which the right end is the brain.   The two " umbilical cords "
mentioned by Aristotle (III. 753 b 20 ff.) are shown : (a) the yolk-sac
stalk, (b) the stalk of the allantois.

To begin with, the embryo is a sort of thin plate on top of the yolk ;
and as time goes on, both the amniotic cavity (which encloses the
embryo) and the allantois (which acts as a respiratory organ and as a
receptacle for excreta) progressively encircle the yolk, which finally
becomes enclosed in the embryo (as Aristotle says).   The chorion and
allantois coalesce after a period, and the resulting chorio-allantois then
corresponds to the fetal placenta of mammals.   The chorion is really
the outer layer of the amnion.   The extra-embryonic coelom, which
is lined with mesoblast, is an extension of the coelom proper (the
main body-cavity), which is also lined with mesoblast.

369

Δ

763 b 20 I   Περὶ μὲν οὖν τῆς γενέσεως τῆς τῶν ζῴων εἴρηται
καὶ κοινῇ καὶ χωρὶς περὶ πάντων. ἐπεὶ δ' ἐν τοῖς
τελεωτάτοις αὐτῶν ἐστι τὸ θῆλυ καὶ τὸ ἄρρεν
κεχωρισμένον, καὶ ταύτας τὰς δυνάμεις ἀρχάς
φαμεν εἶναι πάντων καὶ ζῴων καὶ φυτῶν, ἀλλὰ τὰ
25 μὲν αὐτὰς ἀχωρίστους ἔχει τὰ δὲ κεχωρισμένας,
λεκτέον περὶ τῆς γενέσεως τῆς τούτων πρῶτον·
ἔτι γὰρ ἀτελῶν ὄντων ἐν τῷ γένει διορίζεται τὸ
θῆλυ καὶ τὸ ἄρρεν. πότερον δὲ καὶ πρὶν δήλην
τὴν διαφορὰν εἶναι πρὸς τὴν αἴσθησιν ἡμῶν τὸ μὲν
θῆλυ τὸ δ' ἄρρεν ἐστίν, ἐν τῇ μητρὶ λαβόντα τὴν
30 διαφορὰν ἢ πρότερον, ἀμφισβητεῖται. φασὶ γὰρ
οἱ μὲν ἐν τοῖς σπέρμασιν εἶναι ταύτην τὴν ἐναντίωσιν εὐθύς, οἷον Ἀναξαγόρας καὶ ἕτεροι τῶν
φυσιολόγων· γίνεσθαί τε γὰρ ἐκ τοῦ ἄρρενος τὸ

---

^a See Introd. § 76.
^b See Introd. § 26.   ^c See Introd. § 11.
^d The first microscopically visible signs of sex-differentia-
tion occur about the fifth day in the chick. Aristotle was
quite justified in his belief that sex-differentiation occurs
early. We know to-day that sex is determined genetically
from the moment of fertilization, since some animals have
two kinds of sperm and others have two kinds of egg.

370

BOOK IV

THE formation of animals, both in general and as concerns all of them separately, has now been dealt with. Since, however, in the most perfect [a] of them the male and the female are separate, and we hold that these characteristics [b] are "principles" [c] of all animals and all plants alike (the only difference being that in some these "principles" are inseparable while in others they are separate), we must deal with the formation of these first of all, for male and female become distinct while animals are still imperfect in kind.[d] It is however not agreed whether one is male and another female even before the difference is plain to our senses, the difference being acquired by them either within the mother or earlier. Thus, some people, such as Anaxagoras and certain other physiologers,[e] say that this opposition exists right back in the semens,[f] alleging that the semen comes

I

Origin of sex-differentiation.

Various theories:

Anaxagoras.

Aristotle's view will be found in the passage below, 766 a 30–b 3. The heart is the first thing to be formed in the embryo, because it is the seat of τὸ θρεπτικόν, the nutritive part of the Soul; and τὸ θρεπτικόν is also τὸ γεννητικόν (see 735 a 17 ff., 744 b 36, n.). Sex can be ultimately traced back to the heart, which, as also containing the principle of vital heat, is the source of concoction, upon which ability to produce semen, etc., depends.

[e] See pp. xvi f.

[f] This is an example of the view that the difference is acquired " earlier " than in the mother.

371

σπέρμα, τὸ δὲ θῆλυ παρέχειν τὸν τόπον, καὶ εἶναι
τὸ μὲν ἄρρεν ἐκ τῶν δεξιῶν τὸ δὲ θῆλυ ἐκ τῶν
ἀριστερῶν [καὶ τῆς ὑστέρας τὰ μὲν ἄρρενα ἐν τοῖς

δεξιοῖς εἶναι τὰ δὲ θήλεα ἐν τοῖς ἀριστεροῖς].[1] οἱ
δ' ἐν τῇ μήτρᾳ, καθάπερ Ἐμπεδοκλῆς· τὰ μὲν γὰρ
εἰς θερμὴν ἐλθόντα τὴν ὑστέραν ἄρρενα γίνεσθαί
φησι τὰ δ' εἰς ψυχρὰν θήλεα, τῆς δὲ θερμότητος
5 καὶ τῆς ψυχρότητος τὴν τῶν καταμηνίων αἰτίαν
εἶναι ῥύσιν, ἢ ψυχροτέραν οὖσαν ἢ θερμοτέραν, καὶ
ἢ παλαιοτέραν ἢ προσφατωτέραν. Δημόκριτος δὲ
ὁ Ἀβδηρίτης ἐν μὲν τῇ μητρὶ γίνεσθαί φησι τὴν
διαφορὰν τοῦ θήλεος καὶ τοῦ ἄρρενος, οὐ μέντοι
διὰ θερμότητά γε καὶ ψυχρότητα τὸ μὲν γίγνεσθαι
10 θῆλυ τὸ δ' ἄρρεν, ἀλλ' ὁποτέρου ἂν κρατήσῃ τὸ
σπέρμα τὸ ἀπὸ τοῦ μορίου ἐλθὸν ᾧ διαφέρουσιν
ἀλλήλων τὸ θῆλυ καὶ τὸ ἄρρεν. τοῦτο γὰρ ὡς
ἀληθῶς Ἐμπεδοκλῆς ῥαθυμότερον ὑπείληφεν, οἰό-
μενος ψυχρότητι καὶ θερμότητι διαφέρειν μόνον
ἀλλήλων, ὁρῶν ὅλα τὰ μόρια μεγάλην ἔχοντα
15 διαφορὰν τήν τε τῶν αἰδοίων καὶ τὴν τῆς ὑστέρας.
εἰ γὰρ πεπλασμένων τῶν ζῴων, τοῦ μὲν τὰ μόρια

---

[1] seclusi, nam argumento aliena ; cf. 765 a 22.

---

[a] This is a view put forward also in the *Eumenides* of
Aeschylus (658 ff.) by Apollo, who cites the apposite example
of Athena standing by his side :

οὐκ ἔστι μήτηρ τοῦ κεκλημένου τέκνου
τοκεύς, τροφὸς δὲ κύματος νεοσπόρου.
τίκτει δ' ὁ θρώσκων, etc.

(τοῦ in the first line is Headlam's emendation for ἡ.)   In his
commentary, ii. 293-294, G. Thomson gives references to a
similar belief among the Egyptians and primitive peoples in
Australia and South America ; the reference which he gives

into being from the male, while the female provides the space for it,[a] and that the male comes from the right side [b] and the female from the left [and, as regards the uterus, that the males are in the right side and the female in the left][c]. Others, like Empedocles, hold that the opposition begins in the womb; according to him, the semens which enter a hot womb become males, those which enter a cold one, females [d]; and that the cause of this heat and cold is the menstrual flow, according as it is hotter or colder, older or more recent.[e] Democritus of Abdera holds that the difference of male and female is produced in the womb, certainly, but denies that it is on account of heat and cold that one becomes male and another female; this is determined, he asserts, according to which of the two parents' semen prevails, the semen, that is to say, which has come from the part wherein male and female differ from one another.[f] After all, Empedocles was really rather slipshod in his assumption, in supposing that the two differ from each other merely in virtue of heat and cold, when he could see that the whole of the parts concerned— the male pudenda and the uterus—exhibit a great difference, for supposing that once the animals have been fashioned, and one has got all the parts of the

Empedocles and Democritus.

for the Pythagoreans is, however, to a different doctrine from this. See also G. Thomson, *Aeschylus and Athens* (1941), and for other references to such views and their social consequences, J. Needham, *History of Embryology*, 25 ff.

[b] *i.e.*, the right testis.

[c] These words must be an interpolation, as they are inconsistent with the view just described. *Cf.* 765 a 22.

[d] See quotation, 723 a 24.

[e] These terms, as Platt suggests, may echo Empedocles' own words. The hotter will of course be the more recent.

[f] See note on the theory of " pangenesis," 721 b 9.

ἔχοντος τὰ τοῦ θήλεος πάντα, τοῦ δὲ τὰ τοῦ ἄρρενος, καθάπερ εἰς κάμινον εἰς τὴν ὑστέραν τεθείη,
τὸ μὲν ἔχον ὑστέραν εἰς θερμήν, τὸ δὲ μὴ ἔχον εἰς
ψυχράν, ἔσται θῆλυ τὸ οὐκ ἔχον ὑστέραν καὶ ἄρρεν
20 τὸ ἔχον. τοῦτο δ' ἀδύνατον. ὥστε ταύτῃ γε
βέλτιον ἂν λέγοι Δημόκριτος· ζητεῖ γὰρ ταύτης τῆς
γενέσεως τὴν διαφορὰν[1] καὶ πειρᾶται λέγειν· εἰ δὲ
καλῶς ἢ μὴ καλῶς, ἕτερος λόγος. ἀλλὰ μὴν κἂν
εἰ[2] τῶν μορίων τῆς διαφορᾶς αἴτιον ἡ θερμότης καὶ
25 ἡ ψυχρότης, τοῦτο λεκτέον ἦν τοῖς ἐκείνως λέγου-
σιν· τοῦτο γάρ ἐστιν ὡς εἰπεῖν τὸ λέγειν περὶ
γενέσεως ἄρρενος καὶ θήλεος· τούτοις[3] γὰρ διαφέρει
φανερῶς. οὐ μικρὸν δὲ[4] ἔργον τὸ ἀπ' ἐκείνης τῆς
ἀρχῆς περὶ τῆς γενέσεως τούτων τῶν μορίων τὴν
αἰτίαν συναγαγεῖν, ὡς ἀναγκαῖον[5] ἀκολουθεῖν ψυχο-
μένῳ μὲν τῷ ζῴῳ γίνεσθαι τοῦτο τὸ μόριον ἣν
30 καλοῦσιν ὑστέραν, θερμαινομένῳ δὲ μὴ γίνεσθαι.
τὸν αὐτὸν δὲ τρόπον καὶ περὶ τῶν εἰς τὴν ὁμιλίαν
συντελούντων μορίων· καὶ γὰρ ταῦτα διαφέρει,
καθάπερ εἴρηται πρότερον.

Ἔτι δὲ γίνεται δίδυμα θῆλυ καὶ ἄρρεν ἅμα ἐν
τῷ αὐτῷ μορίῳ πολλάκις τῆς ὑστέρας, καὶ τοῦθ'

---

[1] διαφορᾶς τὴν γένεσιν coni. Platt.
[2] εἰ PSYZ : ᾖ vulg.
[3] τούτοις Peck : τοῦτο vulg.
[4] δὲ Platt : τε vulg.   [5] ⟨ὂν⟩ coni. Platt.

---

[a] Viz., primarily testes and uterus, not the parts employed
in intercourse; these are mentioned separately, ll. 30-32
below. See also 716 a 25-b 3.

[b] Empedocles'. Aristotle seems to assume all through
this discussion that according to Empedocles the fundamental difference between male and female was one of heat

male and the other all the parts of the female, they were to be put into the uterus as though it were into an oven, the one which has a uterus into a hot oven, and the one which has no uterus into a cold one, then it follows that the one that has no uterus will turn out a female and the one that has a uterus a male. And this is impossible. So that we may allow that in this respect Democritus's statement is the better of the two, because he is trying to find out what is the difference inherent in this process of formation of male and female, and endeavouring to state it, though whether he is right or not is another matter. Yet indeed, if heat and cold were the cause of the difference of the actual parts,[a] then those who hold the other view [b] ought to have stated this, because, one might say, this is tantamount to making a statement about the process of formation of male and female, since it is in these parts that the evident difference between the two lies. And also, if you start from this principle,[c] it is no light task to prove the cause of the process of formation of these parts, and to show that it necessarily follows that when the animal is cooled the part called the uterus is formed in it, but that when it is heated it is not formed. The same may be said about the parts which serve for intercourse, since these too differ, as has already been stated.

Further, male and female twins are often formed together in the same part of the uterus. This has

and cold (see above, l. 13), and that this had little or nothing to do with the difference of the sexual organs. But it seems impossible that Empedocles could have meant anything else than that heat and cold were the *cause* of the difference of the sexes, including that of the distinctive organs.

[c] *i.e.*, of heat and cold.

35 ἱκανῶς τεθεωρήκαμεν ἐκ τῶν ἀνατομῶν ἐν πᾶσι
τοῖς ζῳοτοκοῦσι, καὶ ἐν τοῖς πεζοῖς καὶ ἐν τοῖς
ἰχθύσιν· περὶ ὧν εἰ μὲν μὴ συνεωράκει, εὐλόγως

ἡμάρτανε ταύτην τὴν αἰτίαν εἰπών, εἰ δ' ἑωρακώς,
ἄτοπον τὸ ἔτι νομίζειν αἰτίαν εἶναι τὴν τῆς ὑστέρας
θερμότητα ἢ ψυχρότητα· ἄμφω γὰρ ἂν ἐγίνετο
ἢ θήλεα ἢ ἄρρενα, νῦν δὲ τοῦτ' οὐχ ὁρῶμεν
συμβαῖνον.

Λέγοντί τε τὰ μόρια διεσπάσθαι τοῦ γινομένου
5 (τὰ μὲν γὰρ ἐν τῷ ἄρρενί φησιν εἶναι τὰ δ' ἐν
τῷ θήλει, διὸ καὶ τῆς ἀλλήλων ὁμιλίας ἐπιθυμεῖν)
ἀναγκαῖον καὶ τῶν τοιούτων διῃρῆσθαι τὸ μέγεθος
καὶ γίνεσθαι σύνοδον, ἀλλ' οὐ διὰ ψύξιν ἢ θερμασίαν.
ἀλλὰ περὶ μὲν τῆς τοιαύτης αἰτίας [τοῦ σπέρματος][1]
τάχ' ἂν εἴη πολλὰ λέγειν· ὅλως γὰρ ἔοικεν ὁ τρόπος
10 τῆς αἰτίας πλασματώδης εἶναι. εἰ δ' ἔστι περὶ
σπέρματος οὕτως ἔχον ὥσπερ τυγχάνομεν εἰρη-
κότες, καὶ μήτ' ἀπὸ παντὸς ἀπέρχεται μήθ' ὅλως
τὸ ἀπὸ τοῦ ἄρρενος παρέχει τοῖς γινομένοις ὕλην
μηδεμίαν, καὶ πρὸς τοῦτον καὶ πρὸς Δημόκριτον,
15 καὶ εἴ τις ἄλλος οὕτω τυγχάνει λέγων, ὁμοίως
ἀπαντητέον. οὔτε γὰρ διεσπασμένον ἐνδέχεται τὸ

---

[1] secl. Platt, qui post θερμασίαν supra transfert.

---

[a] See quotation, 722 b 12, 764 b 17 and context.
[b] For μέγεθος = σῶμα, cf. G. & C. 321 b 16; and 765 a 13

been amply observed by us from dissections in all the Vivipara, both in the land-animals and in the fishes. Now if Empedocles had not detected this, it is understandable that he should have made the mistake of assigning the cause he did ; if on the other hand he had detected it, it is extraordinary that he should still continue to think that the cause is the heat and cold of the uterus, since according to his theory the twins should both turn out male, or both female ; whereas in actual fact we do not observe this to occur.

Also, he says that the parts of the creature which gets formed are " torn asunder " [a] ; some, he says, are in the male and some in the female, and that also explains why they desire intercourse with each other. If so, necessity requires that the physical substance [b] of these parts [c] as well as of the others is " torn asunder " and that a junction takes place, not that the difference is due to cooling or heating. However, discussion of a cause of this sort might well prove lengthy, as the whole cast of this cause seems to be a product of the imagination. If on the other hand the truth about semen is as we have actually stated—*i.e.*, that it is not drawn from the whole body and that the secretion from the male provides no material at all for the creatures which get formed—then we must take up our stand against Empedocles and against Democritus and against anyone else who maintains this position, because (*a*) it is impossible

below. μέγεθος thus means something which has size, *i.e.*, a physical body or substance. Empedocles, says Aristotle, is inconsistent in saying (*a*) that the physical substance of the parts is present as such in the parents to begin with, and (*b*) that the formation of the sexual parts is due to the action of heat and cold.

[c] Viz., testes and uterus.

764 b

σῶμα τοῦ σπέρματος¹ εἶναι, τὸ μὲν ἐν τῷ θήλει τὸ
δ' ἐν τῷ ἄρρενι, καθάπερ Ἐμπεδοκλῆς φησιν εἰπὼν

ἀλλὰ διέσπασται μελέων φύσις, ἡ μὲν ἐν
ἀνδρός . . .,

οὔτ' ἐξ ἑκατέρου πᾶν ἀποκρινόμενον, τῷ κρατῆσαί
20 τι μέρος ἄλλου μέρους γίνεσθαι τὸ μὲν θῆλυ τὸ
δ' ἄρρεν. ὅλως δὲ τό γε τὴν τοῦ μέρους ὑπεροχὴν
κρατήσασαν ποιεῖν θῆλυ βέλτιον μὲν ἢ μηθὲν
φροντίσαντα τὸ θερμὸν αἰτιᾶσθαι μόνον, τὸ μέντοι
συμβαίνειν ἅμα καὶ τὴν τοῦ αἰδοίου μορφὴν ἑτέραν
δεῖται λόγου πρὸς τὸ συνακολουθεῖν ἀεὶ ταῦτ'
25 ἀλλήλοις. εἰ γὰρ ὅτι σύνεγγυς, καὶ τῶν λοιπῶν
ἕκαστον ἔδει μορίων ἀκολουθεῖν· ἑτέρῳ γὰρ ἕτερον
ἐγγὺς τῶν νικώντων, ὥστε ἅμα θῆλύ τ' ἂν ἦν καὶ
τῇ μητρὶ ἐοικός, ἢ ἄρρεν καὶ τῷ πατρί. ἔτι ἄτοπον
καὶ τὸ μόνον ταῦτ' οἴεσθαι δεῖν γίγνεσθαι τὰ
μόρια, καὶ μὴ τὸ σύνολον μεταβεβληκέναι σῶμα,

---

¹ τοῦ σπέρματος velit secludere Platt.

ᵃ See above, note on μέγεθος, l. 7.
ᵇ Perhaps " of the semen " should be deleted.
ᶜ Cf. 722 b 12.
ᵈ This is Democritus's view. Empedocles had said that
each parent supplied only *half* the tale of the parts ;
Democritus said that each parent supplied a *full* tale of
parts. See also note on pangenesis, 721 b 9.
ᵉ *i.e.*, one sexual part over the other ; see 764 a 10, 11.
ᶠ *i.e.*, the conformation of the part employed in intercourse
as well as the conformation of the uterus : in all cases they
both exhibit a difference from the corresponding parts in
males, the penis and the testes respectively.
ᵍ *e.g.*, why no individual is found having uterus and penis.
378

that the physical substance *a* of the semen *b* exists
" torn asunder," one part in the male and the other
in the female, as Empedocles alleges—

> But torn asunder waits
> The substance of the limbs ; part is in man's . . . *e*

and (*b*) it is impossible that a complete tale *d* of parts
is secreted off from each of the parents and that a
male or female embryo is formed according as one
part prevails over another part.*e* Considering the
matter generally : To hold that the superiority of one
part prevails and that this is what makes the embryo
female is certainly better than saying that heat alone
is the cause without having stopped to think about
it ; but the fact that at the same time the conformation of the pudendum as well *f* is different requires an
explanation to show why these parts are always of a
piece with each other.*g* If the answer is " Because
they are in close proximity," then every one of the
remaining parts ought to be all of a piece as well,*h*
since while the parts are gaining the mastery *i* any
one of them is close to any other, so that on that
showing all the characteristics should go together,
i.e., the offspring, if female, should also take after its
mother, and if male after its father.*j* Besides, it is
fantastic to imagine that these parts alone can be
formed, without the whole body also having under-

h *i.e.*, as well as the sexual parts; *e.g.*, if the offspring has
sexual parts resembling those of its father—*i.e.*, male ones—
then it ought to resemble its father in all its other parts too.
 i This refers to the " prevailing " mentioned above, l. 21,
etc.
 j *i.e.*, the offspring should take after the parent whose sex
has determined its own, and take after it not only in respect
of sexual parts but in all other respects as well. But of
course this is not borne out by the facts.

30 καὶ μάλιστα καὶ πρῶτον τὰς φλέβας, περὶ ἃς ὡς
περὶ ὑπογραφὴν τὸ σῶμα περίκειται τὸ τῶν σαρκῶν. ἃς οὐ διὰ τὴν ὑστέραν εὔλογον γενέσθαι
ποιάς τινας, ἀλλὰ μᾶλλον δι' ἐκείνας τὴν ὑστέραν·
ὑποδοχὴ γὰρ αἵματός τινος ἑκάτερον, προτέρα δ'
ἡ τῶν φλεβῶν. τὴν δὲ κινοῦσαν ἀρχὴν ἀναγκαῖον
35 ἀεὶ προτέραν εἶναι καὶ τῆς γενέσεως αἰτίαν τῷ
ποιὰν εἶναί τινα. συμβαίνει μὲν οὖν ἡ διαφορὰ
τῶν μερῶν τούτων πρὸς ἄλληλα τοῖς θήλεσι καὶ
τοῖς ἄρρεσιν, ἀλλ' οὐκ ἀρχὴν οἰητέον οὐδ' αἰτίαν

εἶναι ταύτην, ἀλλ' ἑτέραν, κἂν εἰ μηθὲν ἀποκρίνεται σπέρμα μήτε ἀπὸ τοῦ θήλεος μήτ' ἀπὸ τοῦ
ἄρρενος, ἀλλ' ὅπως δή ποτε συνίσταται [τὸ σπέρμα]¹
τὸ γιγνόμενον.

  Ὁ δ' αὐτὸς λόγος καὶ πρὸς τοὺς λέγοντας τὸ
5 μὲν ἄρρεν ἀπὸ τῶν δεξιῶν εἶναι τὸ δὲ θῆλυ ἀπὸ
τῶν ἀριστερῶν ὥσπερ καὶ πρὸς Ἐμπεδοκλέα καὶ
πρὸς Δημόκριτον. εἴτε γὰρ μηδεμίαν ὕλην συμβάλλεται τὸ ἄρρεν, οὐθὲν ἂν λέγοιεν οἱ λέγοντες οὕτως·
εἴτε καὶ συμβάλλεται, καθάπερ φασίν, ὁμοίως ἀναγκαῖον ἀπαντᾶν καὶ πρὸς τὸν Ἐμπεδοκλέους λόγον,
10 ὃς διορίζει τὸ θῆλυ πρὸς τὸ ἄρρεν θερμότητι καὶ ψυχρότητι τῆς ὑστέρας. οἱ δὲ τὸ αὐτὸ τοῦτο² ποιοῦσι,
τοῖς δεξιοῖς καὶ τοῖς ἀριστεροῖς ὁρίζοντες, ὁρῶντες

---

¹ secl. Platt : τὸ κύημα coni. A.-W.
² τοῦτο PSY*Z, om. Bekker per errorem.

ᵃ See 716 b 2 ff. and 766 a 24 ff.   ᵇ Cf. 743 a 2, n.
  ᶜ This is the statement of the general rule of which the
foregoing is an example ; Aristotle makes a similar criticism
(of Empedocles) for putting the cart before the horse at P.A.
640 a 20 ff., e.g., ἀγνοῶν . . . ὅτι τὸ ποιῆσαν πρότερον ὑπῆρχεν:
the whole context is apposite.
  ᵈ συμβαίνει: it happens κατὰ συμβεβηκός, not καθ' αὑτό: it

gone a change,[a] and first and foremost the blood-vessels, on to which the fleshy structure of the body has been applied all round, as on to a framework.[b] And it is reasonable to suppose not that the blood-vessels have been formed to be of a particular character on account of the uterus, but rather that the uterus has been so formed on account of them, since although each is a receptacle of blood in some form, the blood-vessels are prior to the uterus; and the motive principle must of necessity be prior always and be the cause of the process of formation in virtue of possessing a particular character.[c] So then, this difference of the sexual parts as between males and females is a contingent phenomenon [d] : we must not look upon it as being a " principle " or a cause : this function is fulfilled by something else, even though no semen at all is discharged either by the female or by the male and whatever the manner may really be by which the forming creature takes shape.

The same argument which we used against Empedocles and Democritus holds good against those who allege that the male comes from the right side and the female from the left [e] : thus if the male contributes no material at all, then those who take this view are of course talking nonsense ; if on the other hand it does contribute something, as they assert, we have to counter them in the same way that we countered Empedocles' argument which draws the line as between male and female by reference to the heat and coldness of the uterus. They make the same mistake as he does, in drawing the line by

is an " accidental," not an " essential," characteristic. For the sentiment, see 766 b 2 ff.

  [e] e.g., Anaxagoras ; see 763 b 33.

διαφέροντα τὸ θῆλυ καὶ τὸ ἄρρεν καὶ μορίοις ὅλοις,
ὧν διὰ τίν' αἰτίαν ὑπάρξει τοῖς ἐκ τῶν ἀριστερῶν,
τοῖς δ' ἐκ τῶν δεξιῶν οὐχ ὑπάρξει τὰ σῶμα τὸ
15 τῆς ὑστέρας; ἂν γὰρ ἔλθῃ μὲν μὴ σχῇ δὲ τοῦτο
τὸ μόριον, ἔσται θῆλυ οὐκ ἔχον ὑστέραν καὶ ἄρρεν
ἔχον, ἂν τύχῃ. [ἔτι δ' ὅπερ εἴρηται καὶ πρότερον,
ὦπται καὶ θῆλυ ἐν τῷ δεξιῷ μέρει τῆς ὑστέρας καὶ
ἄρρεν ἐν τῷ ἀριστερῷ καὶ ἄμφω ἐν τῷ αὐτῷ μέρει,
20 καὶ τοῦτ' οὐχ ὅτι ἅπαξ ἀλλὰ πλεονάκις, ἢ τὸ ἄρρεν
μὲν ἐν τοῖς δεξιοῖς, τὸ θῆλυ δ' ἐν τοῖς ἀριστεροῖς·
οὐχ ἧττον δὲ ἀμφότερα γίνεται ἐν τοῖς δεξιοῖς].[1]
παραπλησίως δέ τινες πεπεισμένοι τούτοις εἰσὶ καὶ
λέγουσιν ὡς τὸν δεξιὸν ὄρχιν ἀποδουμένοις ἢ τὸν
ἀριστερὸν συμβαίνει τοῖς ὀχεύουσιν ἀρρενοτοκεῖν
25 ἢ θηλυτοκεῖν· οὕτω γὰρ καὶ Λεωφάνης ἔλεγεν.
ἐπί τε τῶν ἐκτεμνομένων τὸν ἕτερον ὄρχιν τὸ αὐτὸ
τοῦτο συμβαίνειν τινές φασιν, οὐκ ἀληθῆ λέγοντες,
ἀλλὰ μαντευόμενοι τὸ συμβησόμενον ἐκ τῶν εἰκό-
των, καὶ προλαμβάνοντες ὡς οὕτως ἔχον πρὶν
γινόμενον οὕτως ἰδεῖν, ἔτι δ' ἀγνοοῦντες ὡς οὐθὲν
30 συμβάλλεται πρὸς τὴν γένεσιν τῆς ἀρρενογονίας
καὶ θηλυγονίας τὰ μόρια ταῦτα τοῖς ζῴοις. τούτου
δὲ σημεῖον ὅτι πολλὰ τῶν ζῴων αὐτά τε θήλεα καὶ
ἄρρενά ἐστι, καὶ γεννᾷ τὰ μὲν θήλεα τὰ δ' ἄρρενα,

---

[1] ἢ τὸ ἄρρεν . . . γίνεται ἐν τοῖς δεξιοῖς secl. Platt ; om.
Σ ; credo equidem etiam ἔτι δ' ὅπερ huc usque secl., nam
argumento aliena : cf. 764 a 1.

---

[a] Lit., "body of the uterus," drawing special attention to
the fact of its physical existence : cf. μέγεθος above, 764 b 7.
[b] This sentence, which has nothing to do with the argument, must be deleted.
[c] Leophanes is quoted by Theophrastus, De caus. plant. II.

reference to right and left, although they can see
for themselves that male and female differ in fact by
the entirety of the parts concerned. By what cause,
then, will the uterus[a] be present in those which
come from the left side but not in those which come
from the right? Supposing one comes ⟨from the
left⟩ without having got this part, there will be a
female without a uterus—or if it so chance, a male
with one! [Again, as has in fact been said before,
a female embryo has actually been observed in the
right part of the uterus, and a male one in the left
part, and both male and female in the self-same part,
and that not once but several times over; or the
male one on the right side, and the female on the
left, and no less both are formed on the right side].[b]
There are some who are firmly convinced of a similar
view to this, and maintain that males who copulate
with the right or left testicle tied up produce male
or female offspring respectively: this used in fact
to be maintained by Leophanes.[c] Some allege that
the same occurs in the case of those who have one
testis excised. This statement is untrue, and is a
mere piece of guesswork on their part. They start
from probabilities and guess what will occur; they
prejudge that it is so before they see it happen.
Added to which they do not know that these parts
of animals contribute nothing at all to generation so
far as producing male and female offspring is concerned; and a proof that this is so is that many
animals, although they are themselves male and
female and generate male and female offspring,

4. 11 ; and the fact that in Aëtius' *Placita* V. 7. 5 (*Doxogr.*
420 a 7) he comes between Anaxagoras and Leucippus may
give a rough indication of his date.

ὄρχεις οὐκ ἔχοντα, καθάπερ τὰ μὴ ἔχοντα πόδας,
οἷον τό τε τῶν ἰχθύων γένος καὶ τὸ τῶν ὄφεων.

35 Τὸ μὲν οὖν θερμότητα καὶ ψυχρότητα αἰτίαν
οἴεσθαι τοῦ ἄρρενος καὶ τοῦ θήλεος, καὶ τὸ τὴν

ἀπόκρισιν ἀπὸ τῶν δεξιῶν γίνεσθαι ἢ τῶν ἀριστερῶν, ἔχει τινὰ λόγον· θερμότερα γὰρ τὰ δεξιὰ
τοῦ σώματος τῶν ἀριστερῶν, καὶ τὸ σπέρμα τὸ
πεπεμμένον θερμότερον, τοιοῦτον δὲ τὸ συνεστός,
γονιμώτερον δὲ τὸ συνεστὸς μᾶλλον. ἀλλὰ λίαν
5 τὸ λέγειν οὕτω πόρρωθέν ἐστιν ἅπτεσθαι τῆς
αἰτίας, δεῖ δ' ὅτι μάλιστα προσάγειν ἐκ τῶν ἐνδεχομένων ἐγγὺς τῶν πρώτων αἰτίων.

Περὶ μὲν οὖν ὅλου τε[1] τοῦ σώματος καὶ τῶν
μορίων, τί τε ἕκαστόν ἐστι καὶ διὰ τίν' αἰτίαν, εἴρηται πρότερον ἐν ἑτέροις. ἀλλ' ἐπεὶ τὸ ἄρρεν καὶ
10 τὸ θῆλυ διώρισται δυνάμει τινὶ καὶ ἀδυναμίᾳ (τὸ
μὲν γὰρ δυνάμενον πέττειν καὶ συνιστάναι τε καὶ
ἐκκρίνειν σπέρμα ἔχον τὴν ἀρχὴν τοῦ εἴδους ἄρρεν·
λέγω δ' ἀρχὴν οὐ τὴν τοιαύτην ἐξ ἧς ὥσπερ ὕλης
γίνεται τοιοῦτον οἷον τὸ γεννῶν, ἀλλὰ τὴν κινοῦσαν
πρώτην, ἐάν τ' ἐν αὐτῷ ἐάν τ' ἐν ἄλλῳ τοῦτο
15 δύνηται ποιεῖν· τὸ δὲ δεχόμενον μὲν ἀδυνατοῦν δὲ

---

[1] τε PY : om. vulg.

---

[a] See 716 b 14 f.
[b] Thus the semen which comes from the right side will be hotter.   [c] Cf. above, 747 a 5 ff.
[d] And therefore, of course, capable of producing males.
[e] Compare the method described in Physics, 184 a 10 ff.
[f] In the Parts of Animals and in the first book of the Generation of Animals.
[g] Dynamis : see Introd. § 30.

possess no testes—as is the case with the animals that have no feet, *e.g.*, the tribes of fishes and serpents.[a]

Now the opinion that the cause of male and female is heat and cold, and that the difference depends upon whether the secretion comes from the right side or from the left, has a modicum of reason in it, because the right side of the body is hotter than the left[b]; hotter semen is semen which has been concocted; the fact that it has been concocted means that it has been set and compacted,[c] and the more compacted semen is, the more fertile it is.[d] All the same, to state the matter in this way is attempting to lay hold of the cause from too great a distance, and we ought to come as closely to grips as we possibly can with the primary causes.[e]

We have dealt already elsewhere[f] with the body as a whole and with its several parts, and have stated what each one is, and on account of what cause it is so. But that is not all, for (1) the male and the female are distinguished by a certain ability[g] and inability.[h] Male is that which is able to concoct, to cause to take shape, and to discharge, semen[i] possessing the " principle " of the " form "; and by " principle " I do not mean that sort of principle out of which, as out of matter, an offspring is formed belonging to the same kind as its parent, but I mean the *proximate motive principle*, whether it is able to act thus[j] in itself or in something else. Female is that which receives the semen, but is unable to cause

The fundamental distinction between male and female.

---

[h] Thus much has already been stated at 716 a 18 ff., but Aristotle now develops it more fully.

[i] With this passage *cf.* the discussion at 724 a 29 ff.

[j] *i.e.*, act as the cause of movement.

συνιστάναι καὶ ἐκκρίνειν θῆλυ), ἔτι εἰ¹ πᾶσα πέψις
ἐργάζεται θερμῷ, ἀνάγκη [καὶ]² τῶν ζῴων τὰ ἄρρενα
τῶν θηλέων θερμότερα εἶναι· διὰ γὰρ ψυχρότητα
καὶ ἀδυναμίαν πολυαιμεῖ κατὰ τόπους τινὰς τὸ
θῆλυ μᾶλλον. καὶ ἔστιν αὐτὸ τοὐναντίον σημεῖον ἢ
20 δι' ἥνπερ αἰτίαν οἴονταί τινες τὸ θῆλυ θερμότερον
εἶναι τοῦ ἄρρενος, διὰ τὴν τῶν καταμηνίων πρόεσιν·
τὸ μὲν γὰρ αἷμα θερμόν, τὸ δὲ πλεῖον ἔχον μᾶλλον.
ὑπολαμβάνουσι δὲ τοῦτο γίνεσθαι τὸ πάθος δι'
ὑπερβολὴν αἵματος καὶ θερμότητος, ὥσπερ ἐνδεχόμενον αἷμα εἶναι πᾶν ὁμοίως, ἅπερ μόνον ὑγρὸν
25 ᾗ καὶ τὴν χρόαν αἱματῶδες, καὶ οὐκ ἔλαττον
γινόμενον καὶ καθαρώτερον τοῖς εὐτροφοῦσιν. οἱ
δ' ὥσπερ τὸ κατὰ τὴν κοιλίαν περίττωμα, τὸ
πλεῖον τοῦ ἐλάττονος οἴονται σημεῖον εἶναι θερμῆς
φύσεως μᾶλλον. καίτοι τοὐναντίον ἐστίν. ὥσπερ
γὰρ καὶ ἐκ τῆς πρώτης τροφῆς ἐκ πολλῆς ὀλίγον
30 ἀποκρίνεται τὸ χρήσιμον ἐν ταῖς περὶ τοὺς καρποὺς
ἐργασίαις, καὶ τέλος οὐθὲν μέρος τὸ ἔσχατον πρὸς
τὸ πρῶτον πλῆθός ἐστιν, οὕτω πάλιν καὶ ἐν τῷ
σώματι διαδεχόμενα τὰ μέρη ταῖς ἐργασίαις, τὸ
τελευταῖον πάμπαν μικρὸν ἐξ ἁπάσης γίνεται³ τῆς
τροφῆς. τοῦτο δὲ ἐν μέν τισιν αἷμά ἐστιν, ἐν δέ
35 τισι τὸ ἀνάλογον.

Ἐπεὶ δὲ τὸ μὲν δύναται τὸ δ' ἀδυνατεῖ ἐκκρί-

---

¹ ἐπεὶ δὲ coni. Platt ; fort. ἔτι ἐπεὶ scribendum.
² secl. Platt, om. Z.   ³ ἐκκρίνει Btf.

ᵃ Cf. 725 a 17 f.

semen to take shape or to discharge it. And (2) all concoction works by means of heat. Assuming the truth of these two statements, it follows of necessity that (3) male animals are hotter than female ones, since it is on account of coldness and inability that the female is more abundant in blood in certain regions of the body. And this abundance of blood is a piece of evidence which goes to prove the opposite of the view held by some people, who suppose that the female must be hotter than the male, on account of the discharge of menstrual fluid : blood, they argue, is hot, so that which has more blood in it is hotter. They suppose, however, that this condition occurs owing to excess of blood and heat, as though it were possible for anything and everything to be equally blood if only it is fluid and bloodlike in colour, without allowing for the possibility of its becoming less in quantity and purer in animals that are well-nourished. They apply the same standard here as they do to the residue in the intestine : if there is more of it they imagine that is a sign of a hotter nature. Yet in fact the opposite is the truth. Take a parallel case, that of fruit. Here the nourishment in its first stage is large in quantity,[a] but the useful product resulting from it through the various stages of its treatment is small, and in the end the final result is nothing in proportion compared with the original bulk. So too in the body, the various parts receive the nourishment in turn at the different stages of its treatment, and the final product resulting from all that amount of nourishment is quite small. In some, this is blood ; in others, its counterpart.

Now as the one sex is able and the other is unable Determination of sex

765 b

766 a

ναι¹ τὸ περίττωμα καθαρόν, ἁπάσῃ δὲ δυνάμει ὄργα-
νόν τί ἐστι, καὶ τῇ χεῖρον ἀποτελούσῃ ταὐτὸ καὶ
τῇ βέλτιον, τὸ δὲ θῆλυ καὶ τὸ ἄρρεν, πλεοναχῶς
λεγομένου τοῦ δυνατοῦ καὶ τοῦ ἀδυνάτου, τοῦτον ἀντίκειται τὸν τρόπον, ἀνάγκη ἄρα² καὶ τῷ
θήλει καὶ τῷ ἄρρενι εἶναι ὄργανον³· τῷ μὲν οὖν
5 ἡ ὑστέρα τῷ δ' ὁ περίνεός ἐστιν. ἅμα δ' ἡ
φύσις τήν τε δύναμιν ἀποδίδωσιν ἑκάστῳ καὶ
τὸ ὄργανον· βέλτιον γὰρ οὕτως. διὸ ἕκαστοι οἱ
τόποι ἅμα ταῖς ἐκκρίσεσι γίνονται καὶ ταῖς δυνά-
μεσιν, ὥσπερ οὔτ'⁴ ὄψις ἄνευ ὀφθαλμῶν οὔτ'
ὀφθαλμὸς τελειοῦται ἄνευ ὄψεως, καὶ κοιλία καὶ
10 κύστις ἅμα τῷ δύνασθαι τὰ περιττώματα γίνεσθαι.
ὄντος δὲ τοῦ αὐτοῦ ἐξ οὗ τε γίνεται καὶ αὔξεται,
τοῦτο δ' ἐστὶν ἡ τροφή, ἕκαστον ἂν γίνοιτο τῶν
μορίων ἐκ τοιαύτης ὕλης ἧς δεκτικόν ἐστι, καὶ
τοιούτου περιττώματος. ἔτι δὲ γίνεται πάλιν, ὡς
φαμέν, ἐκ τοῦ ἐναντίου πως. τρίτον δὲ πρὸς τού-
15 τοις ληπτέον ὅτι εἴπερ ἡ φθορὰ εἰς τοὐναντίον, καὶ
τὸ μὴ κρατούμενον ὑπὸ τοῦ δημιουργοῦντος ἀνάγκη
μεταβάλλειν εἰς τοὐναντίον. τούτων δ' ὑποκει-

---

¹ ἐκκρῖναι vulg. (ἐκκρίναι Oᵇ*): ἐκκρίνεται PSYZ (exit Σ):
γίνεσθαι coni. Btf.: τὸ secl. A.-W.
² ἄρα quod conieceram Oᵇ exhibet* (ergo Σ): γὰρ E*, vulg.,
seclusit Platt: οὖν Aldus, A.-W. (sed ἀναγκαῖον coni. A.-W.).
³ ὄργανον S*PYZ; ὄργανα E*, vulg.
⁴ οὔτ' P: οὔθ' ἡ vulg.

ᵃ i.e., here " able " means " can do it better," "unable "
means " can do it less well."
ᵇ Cf. 716 a 23 ff.
ᶜ Cf. 716 a 32, and H.A. 493 b 9 "the part between the
thigh and the buttock is the perineos."

to secrete the residue in a pure condition ; and as in the
there is an instrument for every ability or faculty, for embryo.
the one which yields its product in a more finished
condition and for the one which yields the same product in a less finished condition ; and as male and
female stand opposed in this way (" able " and " unable " being used in more senses than one [a]) ; therefore of necessity there must be an instrument [b] both
for the male and for the female ; hence the male has
the *perineos* [c] and the female has the uterus. Nature
gives each one its instrument simultaneously with its
ability, since it is *better* done thus. Hence each of
these regions of the body gets formed simultaneously
with the corresponding secretions and abilities, just
as the ability to see does not get perfected without
eyes, nor the eye without the ability to see, and
just as the gut and the bladder are perfected simultaneously with the ability to form the residues. Now
as the stuff out of which the parts are formed is the
same as that from which they derive their growth,[d]
namely the nourishment, we should expect each of
the parts to be formed out of that sort of material
and that sort of residue which it is fitted to receive.
Secondly, and on the contrary, it is, as we hold,
formed in a way out of its opposite. Thirdly, in
addition, it must be laid down that, assuming the extinction of a thing means its passing into its opposite
condition, then also that which does not get mastered
by the agent which is fashioning it must of necessity
change over into its opposite condition.[e] With these

[d] For this distinction between the grades of nourishment,
see 744 b 32 ff.
[e] This is explained at length at 768 a 1 ff. The whole of the
present passage should be read in conjunction with the later
and fuller discussion. See also 766 b 15 ff.

766 a

μένων ἴσως ἂν ἤδη μᾶλλον εἴη φανερὸν δι' ἣν
αἰτίαν γίνεται τὸ μὲν θῆλυ τὸ δ' ἄρρεν. ὅταν γὰρ
μὴ κρατῇ ἡ ἀρχὴ μηδὲ δύνηται πέψαι δι' ἔνδειαν
20 θερμότητος μηδ' ἀγάγῃ εἰς τὸ ἴδιον εἶδος τὸ αὑτοῦ,[1]
ἀλλὰ ταύτῃ ἡττηθῇ, ἀνάγκη εἰς τοὐναντίον μεταβάλλειν. ἐναντίον δὲ τῷ ἄρρενι τὸ θῆλυ, καὶ ταύτῃ
ᾗ τὸ μὲν ἄρρεν τὸ δὲ θῆλυ. ἐπεὶ δ' ἔχει διαφορὰν
ἐν τῇ δυνάμει, ἔχει καὶ τὸ ὄργανον διαφέρον· ὥστ'
εἰς τοιοῦτον μεταβάλλει. ἑνὸς δὲ μορίου ἐπικαίρου
25 μεταβάλλοντος ὅλη ἡ σύστασις τοῦ ζῴου πολὺ τῷ
εἴδει διαφέρει. ὁρᾶν δ' ἔξεστιν ἐπὶ τῶν εὐνούχων,
οἳ ἑνὸς μορίου πηρωθέντος τοσοῦτον ἐξαλλάττουσι
τῆς ἀρχαίας μορφῆς καὶ μικρὸν ἐλλείπουσι[2] τοῦ
θήλεος τὴν ἰδέαν. τούτου δ' αἴτιον ὅτι ἔνια τῶν
μορίων ἀρχαί εἰσιν· ἀρχῆς δὲ κινηθείσης πολλὰ
30 ἀνάγκη μεθίστασθαι τῶν ἀκολουθούντων.

Εἰ οὖν τὸ μὲν ἄρρεν ἀρχή τις καὶ αἴτιον, ἔστι
δ' ἄρρεν ᾗ δύναταί τι, θῆλυ δὲ ᾗ ἀδυνατεῖ, τῆς δὲ
δυνάμεως ὅρος καὶ τῆς ἀδυναμίας τὸ πεπτικὸν εἶναι

---

[1] αὑτοῦ Peck (cf. 766 b 16, 767 b 17) : αὐτοῦ vulg.
[2] ἐλλείπουσι P : λείπουσι vulg.

^a The " movement " derived from the male, the male
" principle." See 767 b 17 ff.
^b i.e., male.
^c Cf. the terminology of this and the two following chapters
with Hippocrates, π. διαίτης I. 25 ff. The following examples
may be given : I. 28 (vi. 502 Littré) ἢν ἐπικρατήσῃ τὸ ἄρσεν ;
ibid. τὸ θῆλυ μειοῦται καὶ διακρίνεται ἐς ἄλλην μοῖραν ; I. 27
(vi. 500 L.) διαλύεται ἐς τὴν μείω τάξιν.
^d See, e.g., 716 a 27 ff., 766 b 2 ff.
^e i.e., the condition of possessing the female generative
organs.
^f Cf. above, 716 b 2 ff., and 764 b 28 ff.
^g Aristotle seems to waver between asserting and denying

as our premises it may perhaps be clearer why and by what cause one offspring becomes male and another female. It is this. When the " principle " [a] is failing to gain the mastery and is unable to effect concoction owing to deficiency of heat, and does not succeed in reducing the material into its own proper form,[b] but instead is worsted in the attempt, then of necessity the material must change over into its opposite condition.[c] Now the opposite of the male is the female, and it is opposite in respect of that whereby one is male and the other female.[d] And since it differs in the ability it possesses, so also it differs in the instrument which it possesses. Hence this is the condition [e] into which the material changes over. And when one vital part changes,[f] the whole make-up of the animal differs greatly in appearance and form. This may be observed in the case of eunuchs ; the mutilation of just one part of them results in such a great alteration of their old semblance, and in close approximation to the appearance of the female. The reason for this is that some of the body's parts [g] are " principles," and once a principle has been " moved " (*i.e.*, changed), many of the parts which cohere [h] with it must of necessity change as well.

Let us assume then (1) that " the male " is a principle and is causal in its nature ; (2) that a male is male in virtue of a particular ability, and a female female in virtue of a particular inability ; (3) that the line of determination between the ability and the inability is whether a thing effects or does not effect

The ultimate source of sex is the heart.

that the sexual *parts*, as distinct from the sexes, are " principles " ; but his position is made clear by the passage 766 b 2 ff.

[h] " Are of a piece with it " : *cf.* 764 b 24, 25.

766 a

ἢ μὴ πεπτικὸν τῆς ὑστάτης τροφῆς, ὃ ἐν μὲν τοῖς
ἐναίμοις αἷμα καλεῖται ἐν δὲ τοῖς ἄλλοις τὸ ἀνά-
35 λογον, τούτου δὲ τὸ αἴτιον ἐν τῇ ἀρχῇ καὶ τῷ
μορίῳ τῷ ἔχοντι τὴν τῆς φυσικῆς θερμότητος
ἀρχήν, ἀναγκαῖον ἄρα ἐν τοῖς ἐναίμοις συνίστασθαι

766 b

καρδίαν, καὶ ἢ ἄρρεν ἔσεσθαι ἢ θῆλυ τὸ γινόμενον,
ἐν δὲ τοῖς ἄλλοις γένεσιν ⟨οἷς⟩[1] ὑπάρχει τὸ θῆλυ καὶ
τὸ ἄρρεν τὸ τῇ καρδίᾳ ἀνάλογον. ἡ μὲν οὖν ἀρχὴ
τοῦ θήλεος καὶ ἄρρενος καὶ ἡ αἰτία αὕτη καὶ ἐν
5 τούτῳ ἐστίν. θῆλυ δ' ἤδη καὶ ἄρρεν ἐστὶν ὅταν
ἔχῃ καὶ τὰ μόρια οἷς διαφέρει τὸ θῆλυ τοῦ ἄρρενος·
οὐ γὰρ καθ' ὁτιοῦν μέρος ἄρρεν οὐδὲ θῆλυ, ὥσπερ
οὐδ' ὁρῶν καὶ ἀκοῦον.

Ἀναλαβόντες δὲ πάλιν λέγομεν[2] ὅτι τὸ μὲν
σπέρμα ὑπόκειται περίττωμα τροφῆς ὂν τὸ ἔσχατον. (ἔσχατον δὲ λέγω τὸ πρὸς ἕκαστον φερόμενον.
10 διὸ καὶ ἔοικε τὸ γεννώμενον τῷ γεννήσαντι· οὐθὲν
γὰρ διαφέρει ἀφ' ἑκάστου τῶν μορίων ἀπελθεῖν ἢ
πρὸς ἕκαστον προσελθεῖν, ὀρθότερον δ' οὕτως.) διαφέρει δὲ τὸ τοῦ ἄρρενος σπέρμα, ὅτι ἔχει ἀρχὴν

---

[1] ⟨οἷς⟩ Platt, P*.　　　　　[2] λέγωμεν P.

---

[a] The bloodless animals.
[b] Cf. note on 763 b 25. This extremely important paragraph gives Aristotle's view on the seat of the distinction of
sex, and its main conclusions must be borne in mind throughout his discussion of this subject. It also serves to elucidate
the apparent contradictions in his statements elsewhere (e.g.,
716 a 28, 764 b 36, 766 a 28) as to whether or not the sexual
parts are to be considered " principles."

concoction of the ultimate nourishment (in blooded animals this is known as blood, in the bloodless ones it is the counterpart of blood) ; (4) that the reason for this lies in the " principle," *i.e.*, in the part of the body which possesses the principle of the natural heat. From this it follows of necessity that, in the blooded animals, a heart must take shape and that the creature formed is to be either male or female, and, in the other kinds *a* which have male and female sexes, the counterpart of the heart. As far, then, as the principle and the cause of male and female is concerned, this is what it is and where it is situated ; a creature, however, really is male or female only from the time when it has got the parts by which female differs from male, because it is not in virtue of some casual part that it is male or female, any more than it is in virtue of some casual part that it can see or hear.*b*

To resume then *c* : We repeat that semen has been posited to be the ultimate residue of the nourishment. (By " ultimate " I mean that which gets carried to each part of the body—and that too is why the offspring begotten takes after the parent which has begotten it, since it comes to exactly the same thing whether we speak of being drawn from every one of the parts or passing into every one of the parts, though the latter is more correct.*d*) The semen of the male, however, exhibits a difference,

Consequent difference in formation of sexual parts.

---

c The following paragraph is a short recapitulation, with additions, of the main points of the preceding argument, 765 b 8—766 b 7. (For the use of ὑπόκειται with participle, *cf.* 778 b 17 τοιόνδε ζῷον ὑπόκειται ὄν.)

d See Bk. I. 721 b 13 ff., and especially the conclusion of that discussion, 725 a 21 ff.

ἐν ἑαυτῷ τοιαύτην οἵαν κινεῖν [καὶ ἐν τῷ ζῴῳ]¹ καὶ
διαπέττειν τὴν ἐσχάτην τροφήν, τὸ δὲ τοῦ θήλεος
15 ὕλην μόνον. κρατῆσαν μὲν οὖν εἰς αὐτὸ ἄγει,
κρατηθὲν δ' εἰς τοὐναντίον μεταβάλλει ἢ εἰς φθοράν.
ἐναντίον δὲ τῷ ἄρρενι τὸ θῆλυ· θῆλυ δὲ τῇ ἀπεψίᾳ
καὶ τῇ ψυχρότητι τῆς αἱματικῆς τροφῆς. ἡ δὲ
φύσις ἑκάστῳ τῶν περιττωμάτων ἀποδίδωσι τὸ
δεκτικὸν μόριον. τὸ δὲ σπέρμα περίττωμα, τοῦτο
20 δὲ τοῖς μὲν θερμοτέροις καὶ ἄρρεσι τῶν ἐναίμων
εὔογκον τῷ πλήθει, διὸ τὰ δεκτικὰ μόρια πόροι
ταύτης τῆς περιττώσεώς εἰσι τοῖς ἄρρεσιν· τοῖς
δὲ θήλεσι δι' ἀπεψίαν πλῆθος αἱματικόν (ἀκατέρ-
γαστον γάρ), ὥστε καὶ μόριον δεκτικὸν ἀναγκαῖον
εἶναί τι, καὶ εἶναι τοῦτο ἀνόμοιον καὶ μέγεθος
25 ἔχειν. διὸ τῆς ὑστέρας τοιαύτη ἡ φύσις ἐστίν.
τούτῳ δὲ τὸ θῆλυ διαφέρει τῷ μορίῳ τοῦ ἄρρενος.

Διὰ τίνα μὲν οὖν αἰτίαν γίνεται τὸ μὲν θῆλυ τὸ
δ' ἄρρεν, εἴρηται.

II   Τεκμήρια δὲ τὰ συμβαίνοντα τοῖς εἰρημένοις.
τά τε γὰρ νέα θηλυτόκα μᾶλλον τῶν ἀκμαζόντων,

---

¹ καὶ ἐν τῷ ζῴῳ et mox τὸ δὲ τοῦ θήλεος ὕλην μόνον suspicati
sunt A.-W.; pro καὶ ἐν τῷ ζῴῳ coni. A.-W. ἐν τῷ θήλει, secl.
Btf.; pro τὸ δὲ τοῦ θήλεος ὕλην μόνον habet Σ et facere ipsum
(sc. ultimum cibum) transire ad matricem feminae. in femina
autem est creatio embrionis. cf. 765 b 10 seqq.

ᵃ The passage following has been corrupted. It should
probably read : " a principle of such a kind as to set in
movement and to concoct thoroughly the ultimate nourishment, and to cause it to pass into the uterus of the female ;
whereas the formation of the embryo takes place in the
female."   Cf. the parallel passage above, 765 b 10.

ᵇ There is no subject to this verb in the Greek ; at 766 a 18
it is " the principle " ; at 767 b 17 it is " the movement
derived from the male "—where also Aristotle explains that
394

inasmuch as the male possesses in itself a principle of such a kind [a] as to set up movement [in the animal as well] and thoroughly to concoct the ultimate nourishment, whereas the female's semen contains material only. If ⟨the male semen⟩ gains [b] the mastery, it brings ⟨the material⟩ over to itself; but if it gets mastered, it changes over either into its opposite or else into extinction. And the opposite of the male is the female, which is female in virtue of its inability to effect concoction, and of the coldness of its bloodlike nourishment. And Nature assigns to each of the residues the part which is fitted to receive it. Now the semen is a residue, and in the hotter of the blooded animals, *i.e.*, the males, this is manageable in size and amount,[c] and therefore in males the parts which receive this residual product are passages; in females, however, on account of their failure to effect concoction, this residue is a considerable volume of bloodlike substance, because it has not been matured; hence there must of necessity be here too some part fitted to receive it, different from that in the male, and of a fair size. That is why the uterus has these characteristics; and that is the part wherein the female differs from the male.[d]

We have now stated the cause why some creatures are formed as males, others as females.

And our statements are borne out by the facts.
Thus: Young parents, and those which are older too, tend to produce female offspring rather than parents

it is all one whether we say " the semen," or " the movement which causes the growth of each of the parts," or " the movement which originally sets and constitutes the fetation." *Cf.* 771 b 19 ff.

[c] Because it is more compact; see above, 765 b 3.

[d] *Cf.* 738 b 35 ff.

766 b

30 καὶ τὰ πρεσβύτερα¹· τοῖς μὲν γὰρ οὔπω τέλειον
τὸ θερμόν, τοῖς δ' ἀπολείπει. καὶ τὰ μὲν ὑγρότερα
τῶν σωμάτων καὶ γυναικικώτερα θηλυγόνα μᾶλλον,
καὶ τὰ σπέρματα τὰ ὑγρὰ τῶν συνεστηκότων.
πάντα γὰρ ταῦτα γίνεται δι' ἔνδειαν θερμότητος
φυσικῆς.

35 Καὶ τὸ βορείοις ἀρρενοτοκεῖν μᾶλλον ἢ νοτίοις
⟨διὰ ταὐτὸ συμβαίνει· ὑγρότερα γὰρ τὰ σώματα
νοτίοις,⟩² ὥστε καὶ περιττωματικώτερα. τὸ δὲ
πλεῖον περίττωμα δυσπεπτότερον· διὸ τοῖς μὲν
767 a ἄρρεσιν ὑγρότερον τὸ σπέρμα, ταῖς δὲ γυναιξὶν ἡ
τῶν καταμηνίων ἔκκρισις.

Καὶ τὸ γίνεσθαι δὲ τὰ καταμήνια κατὰ³ φύσιν
φθινόντων τῶν μηνῶν μᾶλλον διὰ τὴν αὐτὴν αἰτίαν
συμβαίνει. ψυχρότερος γὰρ ὁ χρόνος οὗτος τοῦ
5 μηνὸς καὶ ὑγρότερος διὰ τὴν φθίσιν καὶ τὴν ἀπό-
λειψιν τῆς σελήνης· ὁ μὲν γὰρ ἥλιος ἐν ὅλῳ τῷ
ἐνιαυτῷ ποιεῖ χειμῶνα καὶ θέρος, ἡ δὲ σελήνη ἐν τῷ
μηνί. [τοῦτο δ' οὐ διὰ τὰς τροπάς, ἀλλὰ τὸ μὲν αὐξανομένου συμβαίνει τοῦ φωτός, τὸ δὲ φθίνοντος.]⁴
φασὶ δὲ καὶ οἱ νομεῖς διαφέρειν πρὸς θηλυγονίαν
10 καὶ ἀρρενογονίαν οὐ μόνον ἐὰν συμβαίνῃ τὴν ὀχείαν
γίνεσθαι βορείοις ἢ νοτίοις, ἀλλὰ κἂν ὀχευόμενα

¹ τὰ πρεσβύτερα P : γηράσκοντα μᾶλλον vulg.
² supplevi ; quia corpora sunt humida quando ventus
movetur meridionalis Σ.
³ κατὰ P : τὰ κατὰ vulg.
⁴ seclusi ; om. Σ : συμβαίνει om. SY, μηνός pro φωτός S.

ᵃ Cf. H.A. 573 b 34.
ᵇ Cf. the effects of the south wind described in Hippocrates, π. ἱρῆς νούσου 13, π. ἀέρων ὑδάτων τόπων 3.
ᶜ See 777 b 24 ff.
ᵈ This explanation sounds like a gloss. Its meaning is

which are in their prime ; the reason being that in the young their heat is not yet perfected, in the older, it is failing. Also, parents which are more fluid of body and feminine tend to produce females ; this is true also of fluid semen as opposed to that which has " set " ; all these things are due to a deficiency of natural heat.

Also, the fact that when the wind is in the north[a] male offspring tend to be engendered rather than when it is in the south ⟨is due to the same cause : animals' bodies are more fluid[b] when the wind is in the south⟩ so that they are more abundant in residue as well. And the more residue there is, the more difficulty they have in concocting it ; hence the semen of the males and the menstrual discharge of the women is more fluid.

Also, the fact that the menstrual discharge in the natural course tends to take place when the moon is waning[c] is due to the same cause. That time of month is colder and more fluid on account of the waning and failure of the moon (since the moon makes a summer and winter in the course of a month just as the sun does in the course of the whole year. [This is not due to its turning at the tropics ; no, the one occurs when the moon's light is increasing, the other when it is waning.[d]]). Also, shepherds say that it makes a difference so far as the generation of males and females is concerned not only whether copulation occurs when the wind is in the north or in the south, but also whether

that whereas summer and winter result from the " turnings " of the sun, viz., the solstices, the " summer " and " winter " of the moon are not due to the moon's " turnings," but to its waxings and wanings, which are completely independent of its " turnings."

βλέπῃ πρὸς νότον ἢ βορέαν· οὕτω μικρὰν ἐνίοτε
ῥοπὴν αἰτίαν γίνεσθαι τῆς ψυχρότητος καὶ θερμότητος, ταῦτα δὲ τῆς γενέσεως.

Διέστηκε μὲν οὖν ὅλως πρὸς ἄλληλα τό τε θῆλυ
15 καὶ τὸ ἄρρεν πρὸς τὴν ἀρρενογονίαν καὶ θηλυγονίαν
διὰ τὰς εἰρημένας αἰτίας, οὐ μὴν ἀλλὰ καὶ δεῖ
συμμετρίας πρὸς ἄλληλα· πάντα γὰρ τὰ γινόμενα
κατὰ τέχνην ἢ φύσιν λόγῳ τινί ἐστιν. τὸ δὲ θερμὸν
λίαν μὲν κρατοῦν ξηραίνει τὰ ὑγρά, πολὺ δὲ ἐλ-
λεῖπον οὐ συνίστησιν, ἀλλὰ δεῖ πρὸς τὸ δημιουργού-
20 μενον ἔχειν τοῦτον τὸν[1] τοῦ μέσου λόγον· εἰ δὲ μή,
καθάπερ ἐν τοῖς ἑψομένοις προσκάει μὲν τὸ πλεῖον
πῦρ, οὐχ ἕψει δὲ τὸ ἔλαττον, ἀμφοτέρως δὲ συμβαίνει μὴ τελειοῦσθαι τὸ γινόμενον, οὕτω καὶ ἐν
τῇ τοῦ ἄρρενος μίξει καὶ τοῦ θήλεος δεῖ τῆς συμμετρίας. καὶ διὰ τοῦτο πολλοῖς καὶ πολλαῖς
25 συμβαίνει μετ᾽ ἀλλήλων μὲν μὴ γεννᾶν, διαζευχθεῖσι
δὲ γεννᾶν, καὶ ὁτὲ μὲν νέοις ὁτὲ δὲ πρεσβυτέροις
οὖσι ταύτας γίνεσθαι τὰς ὑπεναντιώσεις, ὁμοίως
περί τε γένεσιν καὶ ἀγονίαν καὶ ἀρρενογονίαν καὶ
θηλυγονίαν. διαφέρει δὲ καὶ χώρα χώρας εἰς
ταῦτα καὶ ὕδωρ ὕδατος διὰ τὰς αὐτὰς αἰτίας·
30 ποιὰ γάρ τις ἡ τροφὴ γίνεται μάλιστα καὶ τοῦ
σώματος ἡ διάθεσις διά τε τὴν κρᾶσιν τοῦ περι-

---

[1] τοῦτον τὸν P*Z[1] : τοῦτον om. vulg.

---

[a] Cf. H.A. 574 a 2.
[b] Cf. 723 a 30, 772 a 17, 777 b 25, and Introd. §§ 39 f.
[c] With the following passage, cf. Hippocrates, π. ἀέρων
ὑδάτων τόπων, chh. 1-8 (ii. 12 ff. Littré), id. π. διαίτης II. 37-39.

the animals face north or south while they are copulating [a] : such a small thing thrown in on one side or the other (so they say) acts as the cause of heat and cold, and these in turn act as the cause of generation.

Male and female, then, differ generally with regard to each other in respect of the generation of male and female offspring on account of the causes which have been stated. At the same time, they must stand in a right proportional relationship to one another,[b] since everything that is formed either by art or by nature exists in virtue of some due proportion. Now if " the hot " is too powerful it dries up fluid things ; if it is very deficient it fails to make them " set " ; what it must have in relation to the object which is being fashioned, is the mean proportional, and unless it has that, the case will be the same as what happens when you are cooking : if there is too much fire it burns up your meat, if there is too little it will not cook it—either way what you are trying to produce fails to reach completion. The same applies to the mixture of the male and the female : they require the right proportional relationship, and that is the reason why it happens that many couples fail to effect generation with one another, but if they change partners they succeed ; and also that these oppositions occur sometimes in young people, sometimes among those who are older, both with regard to failure and success in generation and also with regard to the generation of male and female offspring. [c] Also, one country differs from another in these respects, and one water from another, on account of the same causes, for the quality of the nourishment especially and of the bodily condition of a person

Importance of συμμετρία.

Effect of climate.

767 a

εστῶτος ἀέρος καὶ τῶν εἰσιόντων, μάλιστα δὲ διὰ
τὴν τοῦ ὕδατος τροφήν· τοῦτο γὰρ πλεῖστον
εἰσφέρονται, καὶ ἐν πᾶσίν ἐστι τροφὴ τοῦτο, καὶ
ἐν τοῖς ξηροῖς. διὸ καὶ τὰ ἀτέραμνα ὕδατα καὶ
35 ψυχρὰ τὰ μὲν ἀτεκνίαν ποιεῖ τὰ δὲ θηλυτοκίαν.

III Αἱ δ' αὐταὶ αἰτίαι καὶ τοῦ τὰ μὲν ἐοικότα γίνεσθαι τοῖς τεκνώσασι τὰ δὲ μὴ ἐοικότα, καὶ τὰ μὲν
767 b πατρὶ τὰ δὲ μητρί, κατά τε ὅλον τὸ σῶμα καὶ
κατὰ μόριον ἕκαστον, καὶ μᾶλλον αὐτοῖς ἢ τοῖς
προγόνοις, καὶ τούτοις ἢ τοῖς τυχοῦσι, καὶ τὰ μὲν
ἄρρενα μᾶλλον τῷ πατρὶ τὰ δὲ θήλεα τῇ μητρί, τὰ
5 δ' οὐδενὶ τῶν συγγενῶν, ὅμως δ' ἀνθρώπῳ γέ τινι,
τὰ δ' οὐδ' ἀνθρώπῳ τὴν ἰδέαν¹ ἀλλ' ἤδη τέρατι.
καὶ γὰρ ὁ μὴ ἐοικὼς τοῖς γονεῦσιν ἤδη τρόπον
τινὰ τέρας ἐστίν· παρεκβέβηκε γὰρ ἡ φύσις ἐν
τούτοις ἐκ τοῦ γένους τρόπον τινά. ἀρχὴ δὲ πρώτη
τὸ θῆλυ γίνεσθαι² καὶ μὴ ἄρρεν. ἀλλ' αὕτη μὲν
ἀναγκαία τῇ φύσει, δεῖ γὰρ σώζεσθαι τὸ γένος τῶν
10 κεχωρισμένων κατὰ τὸ θῆλυ καὶ τὸ ἄρρεν· ἐν-
δεχομένου δὲ μὴ κρατεῖν ποτε τὸ ἄρρεν³ ἢ διὰ
νεότητα ἢ γῆρας ἢ δι' ἄλλην τινὰ αἰτίαν τοιαύτην,

¹ τὴν ἰδέαν] τινὶ SY.   ² γίνεσθαι P : γενέσθαι vulg.
³ τὸ ἄρρεν Rackham : τοῦ ἄρρενος vulg.

ᵃ See Introd. §§ 39 f., and Hippocrates, π. διαίτης I. passim.
For another reference to κρᾶσις in connexion with the " surrounding air," see 777 b 7.
ᵇ Cf. Hippocrates, π. ἀέρων ὑδάτων τόπων, ch. 4 (ii. 22, 2 ff.
Littré).
ᶜ Cf. 775 a 15 : the female is a " deformity," though one

depends upon the blend[a] of the surrounding air and of the foods which the body takes up, and especially upon the nourishment supplied by the water, since this is what we take most of, water being present as nourishment in everything, even in solid substances as well. Hence hard, cold water in some cases causes barrenness, in others the birth of females.[b]

The following things are due to these same causes. Some offspring take after their parents and some do not ; some after their father, some after their mother, as well in respect of the body as a whole as in respect of each of the parts, and they take after their parents more than after their earlier ancestors, and after their ancestors more than after any casual persons. Males take after their father more than their mother, females after their mother. Some take after none of their kindred, although they take after some human being at any rate ; others do not take after a human being at all in their appearance, but have gone so far that they resemble a monstrosity, and, for the matter of that, anyone who does not take after his parents is really in a way a monstrosity, since in these cases Nature has in a way strayed from the generic type. The first beginning of this deviation is when a female is formed instead of a male, though (a) this indeed is a necessity required by Nature,[c] since the race of creatures which are separated into male and female has got to be kept in being[d] ; and (b) since it is possible for the male sometimes not to gain the mastery either on account of youth or age or some other such cause, female

produced in the normal course of nature (ὥσπερ ἀναπηρίαν φυσικήν). See Introd. § 13.

[d] This is an instance of a necessity required by the Final Cause ; see 731 b 25—732 a 3.

ἀνάγκη γίνεσθαι θηλυτοκίαν ἐν τοῖς ζῴοις. τὸ δὲ
τέρας οὐκ ἀναγκαῖον πρὸς τὴν ἕνεκά του καὶ τὴν
τοῦ τέλους αἰτίαν, ἀλλὰ κατὰ συμβεβηκὸς ἀναγ-
15 καῖον, ἐπεὶ τήν γ' ἀρχὴν ἐντεῦθεν δεῖ λαμβάνειν.
εὐπέπτου μὲν γὰρ οὔσης τῆς περιττώσεως ἐν τοῖς
καταμηνίοις τῆς σπερματικῆς, καθ' αὑτὴν ποιήσει
τὴν μορφὴν ἡ τοῦ ἄρρενος κίνησις. (τὸ γὰρ γονὴν
λέγειν ἢ κίνησιν τὴν αὔξουσαν ἕκαστον τῶν μορίων
20 οὐθὲν διαφέρει, οὐδὲ τὴν αὔξουσαν ἢ τὴν συνιστᾶσαν
ἐξ ἀρχῆς· ὁ γὰρ αὐτὸς λόγος τῆς κινήσεως.) ὥστε
κρατοῦσα¹ μὲν ἄρρεν τε ποιήσει καὶ οὐ θῆλυ, καὶ
ἐοικὸς τῷ γεννῶντι ἀλλ' οὐ τῇ μητρί· μὴ κρατήσασα²
δέ, καθ' ὁποίαν ἂν μὴ κρατήσῃ δύναμιν, τὴν ἔλλειψιν ποιεῖ κατ' αὐτήν. λέγω δ' ἑκάστην δύναμιν
τόνδε τὸν τρόπον· τὸ γεννῶν ἐστιν οὐ μόνον ἄρρεν
25 ἀλλὰ καὶ τοῖον ἄρρεν, οἷον Κορίσκος ἢ Σωκράτης,
καὶ οὐ μόνον Κορίσκος ἐστὶν ἀλλὰ καὶ ἄνθρωπος.
καὶ τοῦτον δὴ τὸν τρόπον τὰ μὲν ἐγγύτερον τὰ δὲ
πορρώτερον ὑπάρχει τῷ γεννῶντι, καθὸ γεννητικόν,
ἀλλ' οὐ κατὰ συμβεβηκός, οἷον εἰ γραμματικὸς ὁ
30 γεννῶν ἢ γείτων τινός. ἀεὶ δ' ἰσχύει πρὸς τὴν
γένεσιν μᾶλλον τὸ ἴδιον καὶ τὸ καθ' ἕκαστον. ὁ
γὰρ Κορίσκος καὶ ἄνθρωπός ἐστι καὶ ζῷον· ἀλλ'

---

¹ κρατοῦσα Peck : κρατούσης vulg.
² κρατήσασα Peck : κρατῆσαν vulg.

ª This is an instance of a necessity enforced by the nature
of the Matter ; see below, 768 a 2–b 33. For these two
modes of necessity (here distinguished as ἕνεκά του and κατὰ
συμβεβηκός), cf. P.A. 642 a 33, and Introd. §§ 6 ff.
ᵇ Cf. 766 a 18, 766 b 15, 771 b 22, 772 b 32.

offspring must of necessity be produced by animals.*
As for monstrosities, they are not necessary so far
as the purposive or final cause is concerned, yet *per
accidens* they are necessary, since we must take it
that their origin at any rate is located here. Thus:
If the seminal residue in the menstrual fluid is well-
concocted, the movement derived from the male
will make the shape after its own pattern.* (It
comes to the same thing whether we say " the
semen " or " the movement which makes each of the
parts grow "; or whether we say " makes them
grow " or " constitutes and ' sets ' them from the
beginning "—because the *logos* of the movement is
the same either way.) So that if this movement
gains the mastery it will make a male and not a
female, and a male which takes after its father, not
after its mother; if however it fails to gain the
mastery, whatever be the " faculty " in respect of
which it has not gained the mastery, in that " faculty "
it makes the offspring deficient. " Faculty," as
applied to each instance, I use in the following sense.
The generative parent is not merely male, but in
addition a male with certain characteristics, *e.g.*,
Coriscus or Socrates; and it is not merely Coriscus,
but in addition a human being. And it is of course
in this sense that, of the characteristics belonging
to the generating parent, some are more closely,
some more remotely his, *qua* procreator (not *qua*
anything else he may be *per accidens*, *e.g.*, supposing he were a good scholar or somebody's next-door
neighbour); and where generation is concerned, it
is always the peculiar and individual characteristic
that exerts the stronger influence. Thus: Coriscus
is both a human being and an animal; but the

767 b

ἐγγύτερον τοῦ ἰδίου τὸ[1] ἄνθρωπος ἢ τὸ ζῷον. γεννᾷ
δὲ καὶ τὸ καθ' ἕκαστον καὶ τὸ γένος, ἀλλὰ μᾶλλον
τὸ καθ' ἕκαστον· τοῦτο γὰρ ἡ οὐσία· καὶ[2] τὸ
35 γινόμενον γίνεται μὲν καὶ ποιόν τι, ἅμα δὲ[3] τόδε
τι, καὶ τοῦθ' ἡ οὐσία. διόπερ ἀπὸ τῶν δυνάμεων
ὑπάρχουσιν αἱ κινήσεις ἐν τοῖς σπέρμασι πάντων
τῶν τοιούτων, δυνάμει δὲ καὶ τῶν προγόνων,
768 a μᾶλλον δὲ τοῦ ἐγγύτερον ἀεὶ τῶν καθ' ἕκαστόν
τινος· λέγω δὲ καθ' ἕκαστον τὸν Κορίσκον καὶ
τὸν Σωκράτην. ἐπεὶ δ' ἐξίσταται πᾶν οὐκ εἰς τὸ
τυχὸν ἀλλ' εἰς τὸ ἀντικείμενον, καὶ τὸ ἐν τῇ γενέσει
μὴ κρατούμενον ἀναγκαῖον ἐξίστασθαι καὶ γίνεσθαι
5 τὸ ἀντικείμενον καθ' ἣν δύναμιν οὐκ ἐκράτησε τὸ
γεννῶν καὶ κινοῦν. ἐὰν μὲν οὖν ᾖ ἄρρεν, θῆλυ
γίνεται, ἐὰν δὲ ᾖ Κορίσκος ἢ Σωκράτης, οὐ τῷ
πατρὶ ἐοικὸς ἀλλὰ τῇ μητρὶ γίνεται· ἀντίκειται γὰρ
ὥσπερ τῷ ὅλως[4] πατρὶ μήτηρ, καὶ τῷ καθ' ἕκαστον
γεννῶντι ἡ καθ' ἕκαστον γεννῶσα. ὁμοίως δὲ καὶ
10 κατὰ τὰς ἐχομένας δυνάμεις· ἀεὶ γὰρ εἰς τὸν ἐχόμενον μεταβαίνει μᾶλλον τῶν προγόνων, καὶ ἐπὶ

---

[1] τὸ P : ὁ vulg., om. S. [2] καὶ P : καὶ γὰρ vulg.
[3] ἅμα δὲ Rackham : ἀλλὰ vulg.
[4] ὅλως PZ², totaliter Σ : ὅλῳ vulg.

---

[a] Cf. 731 b 34, and below 768 a 1 ; and see the definition
of οὐσία given in *Cat.* 2 a 11, and the examples cited, ὁ τὶς
ἄνθρωπος, ὁ τὶς ἵππος. There are of course other usages and
meanings of οὐσία. Cf. Introd. § 16, App. A § 18.

[b] Viz., individual, human being, animal, etc.

[c] Loses and alters its character ; degenerates. The force
of ἐξίστασθαι can be seen from the phrase ἐξίστησι καὶ φθείρει
τὴν φύσιν (*Eth. Nic.* 1119 a 23) ; cf. *G. & C.* 323 b 28, *Phys.*
261 a 20 (τῆς φύσεως, τῆς οὐσίας, ἐξίστασθαι), and 725 a 28
above. [d] Cf. above, 766 a 15.

former characteristic stands closer to what is peculiar
to him than the latter does. Now both the individual and the *genus* to which it belongs are at
work in the act of generation ; but of the two the
individual takes the leading part, because this is the
really existent thing[a] ; the offspring also which
is formed, though of course it is formed so as to
possess the generic characteristics at the same time
comes to be a particular individual—and this, again,
is the really existent thing. Therefore, it is from
the "faculties" of all such things as these[b] that
the movements which are present in the semens
are derived, potentially even from ⟨the faculties⟩ of
earlier ancestors, but more specially of that which on
each occasion stands closer to some individual ; and
by individual I mean Coriscus, or Socrates. Now
everything, when it departs from type,[c] passes not
into any casual thing but into its own opposite ;
thus, applying this to the process of generation, the
⟨substance⟩ which does not get mastered must of
necessity depart from type and become the opposite[d]
in respect of that "faculty" wherein the generative
and motive agent has failed to gain the mastery.
Hence, if this is the "faculty" in virtue of which the
agent is male, then the offspring formed is female ;
if it is that in virtue of which the agent is Coriscus
or Socrates, then the offspring formed does not take
after its father but after its mother, since, just as
"mother" is the opposite of "father" as a general
term, so also the individual mother is the opposite
of the individual father. The same applies to the
"faculties" that stand next in order, since the offspring always tends to shift over to that one of its
ancestors which stands next, both on the father's side

768 a

πατέρων καὶ ἐπὶ[1] μητέρων. ἔνεισι δ' αἱ μὲν ἐνεργείᾳ τῶν κινήσεων, αἱ δὲ δυνάμει, ἐνεργείᾳ μὲν αἱ
τοῦ γεννῶντος καὶ τῶν καθόλου, οἷον ἀνθρώπου καὶ
ζῴου, δυνάμει δὲ αἱ τοῦ θήλεος καὶ τῶν προγόνων.
15 μεταβάλλει μὲν οὖν ἐξιστάμενον πρὸς τὰ ἀντικεί-
μενα, λύονται δὲ αἱ κινήσεις αἱ δημιουργοῦσαι εἰς
τὰς ἐγγύς, οἷον ἡ τοῦ γεννῶντος ἂν λυθῇ κίνησις,
ἐλαχίστη διαφορᾷ μεταβαίνει εἰς τὴν τοῦ πατρός,
δεύτερον δ' εἰς τὴν τοῦ πάππου· καὶ τοῦτον δὴ τὸν
τρόπον [καὶ ἐπὶ τῶν ἀρρένων καὶ ἐπὶ τῶν θηλειῶν][2]
20 ἡ τῆς γεννώσης εἰς τὴν τῆς μητρός, ἐὰν δὲ μὴ
εἰς ταύτην, εἰς τὴν τῆς τήθης· ὁμοίως δὲ καὶ
ἐπὶ τῶν ἄνωθεν.

Μάλιστα μὲν οὖν πέφυκεν ᾗ ἄρρεν καὶ ᾗ πατὴρ
ἅμα κρατεῖν καὶ κρατεῖσθαι· μικρὰ γὰρ ἡ διαφορά,
ὥστ' οὐκ ἔργον ἅμα συμβῆναι ἀμφότερα· ὁ γὰρ
Σωκράτης ἀνὴρ τοιόσδε τις.[3] διὸ ὡς ἐπὶ τὸ πολὺ
25 τὰ μὲν ἄρρενα τῷ πατρὶ ἔοικεν, τὰ δὲ θήλεα τῇ
μητρί, ἅμα γὰρ εἰς ἄμφω ἔκστασις ἐγένετο, ἀντί-

---

[1] ἐπὶ P : om. vulg.   [2] secl. A.-W.
[3] ὁ γὰρ . . . τις secl. A.-W. : ἀνὴρ om. S.

---

[a] Aristotle now introduces the distinction between ἐξίστασθαι καὶ μεταβάλλειν (" departing from type and changing
over ") and λύεσθαι (" relapsing "): as will be seen, the
result of the former process is that the embryo acquires a
characteristic opposite to that of the original movement (this
process has been clearly described already) ; the result of
the latter process (not so far described) is that the embryo
acquires a characteristic which belonged to one of its ancestors.
(The explanation of these two processes is given below at
768 b 15 ff.)
[b] The semen, the movement derived from the male parent.
Cf. 766 a 17.
[c] See 768 a 2 above.

and the mother's. Some of the movements (those of the male parent and those of general kinds, *e.g.*, of human being and animal) are present in ⟨the semen⟩ *in actuality*, others (those of the female and those of ancestors) are present *potentially*. [a] Now when (a) it [b] departs from type,[c] it *changes over* into its opposites; but when (b) the movements which are fashioning the embryo *relapse*, they relapse into those which stand quite near them ; for example, if the movement of the male parent relapses, it shifts over to that of his father—a very small difference—and in the second instance to that of his grandfather. And in this way too [not only on the male side but also on the female] the movement of the female parent shifts over to that of her mother, and if not to that, then to that of her grandmother ; and so on with the more remote ancestors.

(1) Usually the natural course of events is that when ⟨the movement of the male parent⟩ [d] gains the mastery—and when it is mastered—it will do so both *qua* male and *qua* individual father,[e] since the difference between the two ⟨faculties⟩ is a small one, and so there is no difficulty in their both coinciding (for Socrates is a man who, while (a) he has the characteristics of a class,[f] (b) is also an individual). Hence for the most part males take after their father—and females after their mother, since a departure from type takes place in both directions [g]

[d] See above, 766 b 15.

[e] Care must be taken to distinguish the use of " father " applied (a) to the male parent *qua* a particular individual, and (b) to the father of the male parent.

[f] *i.e.*, is " male." For τοιόσδε, τοιοσδί, *cf. Met.* 1077 b 20 ff.

[g] *i.e.*, from " male " into " female," and from " father " into " mother."

768 a

κεῖται δὲ τῷ μὲν ἄρρενι τὸ θῆλυ τῷ δὲ πατρὶ
ἡ μήτηρ, ἡ δ' ἔκστασις εἰς τἀντικείμενα. ἐὰν δ'
ἡ μὲν ἀπὸ τοῦ ἄρρενος κρατήσῃ κίνησις, ἡ δ' ἀπὸ
τοῦ Σωκράτους μὴ κρατήσῃ, ἢ αὕτη μὲν ἐκείνη δὲ
30 μή, τότε συμβαίνει γίνεσθαι ἄρρενά τε μητρὶ
ἐοικότα καὶ θήλεα πατρί. ἐὰν δὲ λυθῶσιν αἱ κινήσεις, καὶ ᾗ μὲν ἄρρεν μείνῃ, ἡ δὲ τοῦ Σωκράτους
λυθῇ εἰς τὴν τοῦ πατρός, ἔσται ἄρρεν τῷ πάππῳ
ἐοικὸς ἢ τῶν ἄλλων τινὶ τῶν ἄνωθεν προγόνων
[κατὰ τοῦτον τὸν λόγον].[1] κρατηθέντος[2] δὲ ᾗ ἄρρεν,[3]
35 θῆλυ ἔσται, καὶ ἐοικὸς μάλιστα μὲν τῇ μητρί, ἐὰν
δὲ καὶ αὕτη λυθῇ ἡ κίνησις, μητρὶ μητρὸς ἢ ἄλλῃ
768 b τινὶ τῶν ἄνωθεν ἔσται ἡ ὁμοιότης κατὰ τὸν αὐτὸν
λόγον. ὁ δ' αὐτὸς τρόπος καὶ ἐπὶ τῶν μορίων·
καὶ γὰρ τῶν μορίων τὰ μὲν τῷ πατρὶ ἔοικε πολλάκις, τὰ δὲ τῇ μητρί, τὰ δὲ τῶν προγόνων τισίν·
5 ἔνεισι γὰρ καὶ τῶν μορίων αἱ μὲν ἐνεργείᾳ κινήσεις
αἱ δὲ δυνάμει, καθάπερ εἴρηται πολλάκις. καθόλου
δὲ δεῖ λαβεῖν ὑποθέσεις, μίαν μὲν τὴν εἰρημένην,
ὅτι ἔνεισι τῶν κινήσεων αἱ μὲν δυνάμει αἱ δ'
ἐνεργείᾳ, ἄλλας δὲ δύο, ὅτι κρατούμενον μὲν ἐξίσταται εἰς τὸ ἀντικείμενον, λυόμενον δὲ εἰς τὴν
ἐχομένην κίνησιν, καὶ ἧττον μὲν λυόμενον εἰς τὴν

---

[1] om. PS ; seclusi. [2] κρατηθέντα Υ.
[3] post ἄρρεν addunt codd. ἢ (ἢ om. P) θῆλυ, τῶν προγόνων
τινὶ ἐοικός PYZ ; amplius κρατηθείσης δὲ καὶ (καὶ om. Z) τῆς
τοῦ προγόνου κινήσεως PSYZ.

---

[a] See 768 a 3.
[b] i.e., the movement derived from that particular individual male.   [c] Cf. 772 b 36.

simultaneously, and the opposite of "male" is "female" and the opposite of "father" is "mother," departure from type always being into opposites.[a] But (2) if the movement that comes from "the male" gains the mastery and the movement that comes from Socrates does not, or the other way round, then the result is that male offspring taking after their mother are formed and female ones taking after their father. Supposing (3) the movements *relapse* : if (i) the male "faculty" stands fast but the movement from Socrates[b] relapses into that of his father, then the offspring will be male and take after its grandfather or some other more remote ancestor [according to this principle] ; if (ii) the male-faculty gets mastered, the offspring will be female, and usually will take after the mother ; but supposing this movement also relapses, it will take after the mother's mother or some other more remote ancestor on the same principle. Precisely the same scheme holds good with the various parts of the body ; very often, of course, some parts take after the father and some after the mother, and others after some of the ancestors, since the movements belonging to the parts[c] as well are present in ⟨the seminal substance⟩, some of them *in actuality*, some *potentially*, as has often been stated. We must lay down as general principles that which we stated just now, for one (viz., that some of the movements are present in ⟨the seminal substance⟩ *potentially*, others *in actuality*), and also two others : (*a*) that which *gets mastered* departs from type and passes into its opposite ; (*b*) that, however, which *relapses* passes into the movement next to it in order : if it relapses a little, into the movement

Absence of family resemblance.

409

10 ἐγγύς, μᾶλλον δὲ εἰς τὴν πορρώτερον. τέλος δ'
οὕτως συγχέονται ὥστε μηθενὶ ἐοικέναι τῶν οἰκείων καὶ συγγενῶν, ἀλλὰ λείπεσθαι τὸ κοινὸν
μόνον καὶ εἶναι ἄνθρωπον. τούτου δ' αἴτιον ὅτι
πᾶσιν ἀκολουθεῖ τοῦτο τοῖς καθ' ἕκαστον· καθόλου
γὰρ ὁ ἄνθρωπος, ὁ δὲ Σωκράτης πατήρ, καὶ ἡ
15 μήτηρ ἥτις ποτ' ἦν, τῶν καθ' ἕκαστον.

Αἴτιον δὲ τοῦ μὲν λύεσθαι τὰς κινήσεις ὅτι τὸ
ποιοῦν καὶ πάσχει ὑπὸ τοῦ πάσχοντος (οἷον τὸ τέμ-
νον ἀμβλύνεται ὑπὸ τοῦ τεμνομένου καὶ τὸ θερμαῖνον ψύχεται ὑπὸ τοῦ θερμαινομένου, καὶ ὅλως
τὸ κινοῦν ἔξω τοῦ πρώτου ἀντικινεῖταί τινα
20 κίνησιν, οἷον τὸ ὠθοῦν ἀντωθεῖταί πως καὶ ἀντι-
θλίβεται τὸ θλῖβον· ἐνίοτε δὲ καὶ ὅλως ἔπαθε
μᾶλλον ἢ ἐποίησεν, καὶ ἐψύχθη μὲν τὸ θερμαῖνον,
ἐθερμάνθη δὲ τὸ ψῦχον, ὁτὲ μὲν οὐθὲν ποιῆσαν,
ὁτὲ δὲ ἧττον ἢ παθόν· εἴρηται δὲ περὶ αὐτῶν ἐν
τοῖς περὶ τοῦ ποιεῖν καὶ πάσχειν διωρισμένοις, ἐν
25 ποίοις ὑπάρχει τῶν ὄντων τὸ ποιεῖν καὶ πάσχειν).
ἐξίσταται δὲ τὸ πάσχον καὶ οὐ κρατεῖται ἢ δι'
ἔλλειψιν δυνάμεως τοῦ πέττοντος καὶ κινοῦντος, ἢ
διὰ πλῆθος καὶ ψυχρότητα τοῦ πεττομένου καὶ
διοριζομένου· τῇ μὲν γὰρ κρατοῦν τῇ δὲ οὐ κρα-

---

[a] The species is " consequent " to every individual ; cf.
Topics 128 b 4 ὡς γένους ὄντος τοῦ ἀεὶ ἀκολουθοῦντος.
[b] See *G. & C.* 324 a 31 ff.
[c] Not extant. But see *G. & C.* 324 a 33 ff.
[d] *Cf.* 766 b 15.

which is close by, if more, into that which is further removed. In the end, they become so confused that the product does not take after any of its family or kindred, and all that remains is what is common to the race—*i.e.*, it is just a human being. The reason for which is that *all* particular individuals are accompanied[a] by this characteristic: since " human being " is general, whereas Socrates who is the father, and the mother whoever she may be, are to be classed as particular individuals.

(1) The reason why the movements *relapse* is that the agent in its turn gets acted upon by that upon which it acts (*e.g.*, a thing which cuts gets blunted by the thing which is cut, and a thing which heats gets cooled by the thing which is heated, and, generally, any motive agent, except the " prime mover," gets moved somehow itself in return,[b] *e.g.*, that which pushes gets pushed somehow in return, and that which squeezes gets squeezed in return; sometimes the extent to which it gets acted upon is greater than that to which it is acting—a thing which heats may get cooled, or one which cools may get heated, sometimes (*a*) without having acted at all, sometimes (*b*) having acted less than it has been acted upon. These matters have been discussed in the treatise on *Acting and being acted upon*,[c] where it is stated in what sorts of things acting and being acted upon occur). (2) The reason, however, why that which is acted upon departs from type and does not *get mastered* is either (*a*) deficient potency in the concocting and motive agent, or (*b*) the bulk and coldness of that which is being concocted and articulated; since ⟨the motive agent⟩,[d] gaining the mastery at one place but not at another, causes the embryo that is

The mechanics of λύεσθαι and μεταβάλλειν.

Uneven development.

411

τοῦν ποιεῖ πολύμορφον τὸ συνιστάμενον, οἷον ἐπὶ
30 τῶν ἀθλητῶν συμβαίνει διὰ τὴν πολυφαγίαν· διὰ
πλῆθος γὰρ τροφῆς οὐ δυναμένης τῆς φύσεως κρατεῖν, ὥστ' ἀνάλογον αὔξειν καὶ διανέμειν[1] ὁμοίως[2]
τὴν τροφήν,[3] ἀλλοῖα γίνεται τὰ μέρη, καὶ σχεδὸν
ἐνίοθ' οὕτως ὥστε μηθὲν ἐοικέναι τῷ πρότερον.
παραπλήσιον δὲ τούτῳ καὶ τὸ νόσημα τὸ καλού-
35 μενον σατυριᾶν· [καὶ γὰρ ἐν τούτῳ διὰ ῥεύματος ἢ
πνεύματος ἀπέπτου πλῆθος εἰς[4] μόρια τοῦ προσώπου παρεμπεσόντος τοῦ[5] ζῴου, καὶ σατύρου
φαίνεται τὸ πρόσωπον.][6]

Διὰ τίνα μὲν οὖν αἰτίαν θήλεα καὶ ἄρρενα γίνεται,
καὶ τὰ μὲν ἐοικότα τοῖς γονεῦσι, θήλεά τε θήλεσι
καὶ ἄρρενα ἄρρεσι, τὰ δ' ἀνάπαλιν, θήλεά τε τῷ
πατρὶ καὶ ἄρρενα τῇ μητρί, καὶ ὅλως τὰ μὲν τοῖς
5 προγόνοις ἐοικότα[7] τὰ δ' οὐθενί, καὶ ταῦτα καὶ καθ'
ὅλον τὸ σῶμα καὶ τῶν μορίων ἕκαστον, διώρισται
περὶ πάντων.

Εἰρήκασι δέ τινες τῶν φυσιολόγων καὶ ἕτερα[8]
περὶ τούτων, διὰ τίν' αἰτίαν ὅμοια καὶ ἀνόμοια
γίγνεται τοῖς γονεῦσιν. δύο δὴ τρόπους λέγουσι
τῆς αἰτίας. ἔνιοι μὲν γάρ φασιν, ἀφ' ὁποτέρου

---

[1] διανέμειν S, Aldus, Platt : διαμένειν vulg.
[2] ὁμοίως E : ὁμοίαν vulg.
[3] τροφήν E, Aldus : μορφήν vulg.
[4] εἰς τὰ Aldus.
[5] ἄλλου Oᵇ*m Z², A.-W. in textu : του Z¹ : pro τοῦ ζῴου coni. ζῴου του A.-W. cf. *Pol.* 1302 b 39.
[6] corrupta et fort. secludenda ; pro καὶ γὰρ . . . πρόσωπον *quoniam accidit ex* [con]*descensu ad membrum maris cum vento generato ex cibo indigesto* Σ.
[7] ἐοικότα P : ἔοικε vulg.
[8] ἕτερα Platt, *quod causae . . . sunt aliae* Σ : ἕτερόν τι P : ἕτεροι vulg.

taking shape to turn out diversiform. This is just
what happens to athletes through eating an excessive amount; in their case, owing to the great bulk of
nourishment there is, Nature cannot gain the mastery
over it so as to bring about well-proportioned growth
and distribute the nourishment evenly throughout ;
the result is that the parts turn out ill-assorted, and
sometimes even bear hardly any resemblance at all
to what they were like before. Similar to this is
the disease which is known as satyriasis ; [in this
too, a large bulk of unconcocted flux or *pneuma* finds
its way into parts of the face of the animal, and in
consequence the face actually appears like that of a
satyr.] [a]

We have now expounded the cause of all the following : why male and female offspring are formed ; why
some take after their parents, female after female
and male after male, and others the other way
round, females taking after their father and males
after their mother ; and generally why some take
after their ancestors and some after none of them,
in respect both of the body as a whole and of each
of its parts.

Certain of the physiologers, however, have treated Earlier
of these matters on different lines, explaining other- theories
wise the cause why offspring are formed similar and of resem-
dissimilar to their parents. The cause is presented blance
by them in two ways. (1) Some say that the off- examined.
spring which is formed takes more closely after that

---

[a] This sentence is probably a marginal note which has
crept into the text ; in any case it is corrupt, and "unconcocted *pneuma*" is meaningless. Scot has no mention of
animal or face ; see critical note. The disease seems to be
elephantiasis.—With b 30-37 however *cf. Pol.* 1302 b 35 ff.

10 ἂν ἔλθῃ σπέρμα πλέον, τούτῳ γίγνεσθαι μᾶλλον
ἐοικός, ὁμοίως παντί τε πᾶν καὶ μέρει μέρος, ὡς
ἀπιόντος ἀφ᾽ ἑκάστου τῶν μορίων σπέρματος· ἂν
δ᾽ ἴσον ἔλθῃ ἀφ᾽ ἑκατέρου, τοῦτο δ᾽ οὐδετέρῳ
γίγνεσθαι ὅμοιον. εἰ δὲ τοῦτ᾽ ἐστὶ ψεῦδος καὶ μὴ
ἀπὸ παντὸς ἀπέρχεται, δῆλον ὡς οὐδὲ τῆς ὁμοιό-
15 τητος καὶ ἀνομοιότητος αἴτιον ἂν εἴη τὸ λεχθέν.
ἔτι δὲ πῶς ἅμα θῆλυ μὲν πατρὶ ἐοικὸς ἄρρεν δὲ
μητρὶ ἐοικός, οὐκ εὐπόρως δύνανται διορίζειν· οἱ
μὲν γὰρ ὥσπερ Ἐμπεδοκλῆς λέγοντες ἢ Δημόκριτος περὶ τοῦ θήλεος καὶ ἄρρενος τὴν αἰτίαν
ἄλλον τρόπον ἀδύνατα λέγουσιν· οἱ δὲ τῷ πλείον
20 ἢ ἔλαττον ἀπιέναι ἀπὸ τοῦ ἄρρενος ἢ θήλεος, καὶ
διὰ τοῦτο γίγνεσθαι τὸ μὲν θῆλυ τὸ δ᾽ ἄρρεν, οὐκ
ἂν ἔχοιεν ἀποδεῖξαι τίνα τρόπον τό τε θῆλυ τῷ
πατρὶ ἐοικὸς ἔσται καὶ τὸ ἄρρεν τῇ μητρί· ἅμα
γὰρ ἐλθεῖν πλέον ἀπ᾽ ἀμφοτέρων ἀδύνατον. ἔτι
δὲ διὰ τίν᾽ αἰτίαν ἐοικὸς γίνεται τοῖς προγόνοις ὡς
25 ἐπὶ τὸ πολὺ[1] καὶ τοῖς ἄποθεν; οὐ γὰρ ἀπ᾽ ἐκείνων
γ᾽ ἀπελήλυθεν οὐθὲν τοῦ σπέρματος. ἀλλὰ μᾶλλον
οἱ τὸν λειπόμενον τρόπον λέγοντες περὶ τῆς ὁμοιότητος καὶ τἆλλα βέλτιον καὶ τοῦτο λέγουσιν. εἰσὶ
γάρ τινες οἵ φασι τὴν γονὴν μίαν οὖσαν οἷον
πανσπερμίαν εἶναί τινα πολλῶν· ὥσπερ οὖν[2] εἴ τις

---

[1] ὡς ἐπὶ τὸ πολὺ fort. secludendum.
[2] οὖν] ἂν S.

---

[a] See 764 a—765 a.
[b] e.g., Alcmeon; see Diels 24 A 14.

parent from which the larger portion of the semen comes, and that the whole of the offspring takes after the whole of the parent, and part after part (this assumes that semen is drawn from each of the parts) ; if the same amount comes from each of the two, then, they say, the offspring formed resembles neither. But if this is untrue (as it is), *i.e.*, if the semen is *not* drawn from the whole of the body, then, clearly, the reason they give for the similarity and dissimilarity of the offspring cannot be true either. Further, they cannot explain with any ease how it is that at the same time a female offspring takes after the father and a male offspring after the mother ; for those who state the cause of male and female as Empedocles or Democritus state it,[a] make statements which on another score are impossible ; while those[b] who maintain that it all depends upon whether more or less semen comes from either the male or the female, and that this is why one offspring is formed as a male, and another as a female, these people, I am sure, are not in a position to show how the female is going to take after the father and the male after the mother, since it is impossible for *more* semen to come from *both* parents at one and the same time. And further, for what cause is it that the offspring for the most part takes after its ancestors, even distant ones ? Surely no portion at all of the semen has come from them, anyway. (2) One more type of explanation of the resemblance remains to be mentioned, and those who adopt it give a better account of the matter, including this particular question. There are some who hold that the semen, though a unity, is as it were a " seed-aggregate " consisting of a large number of ingredients ; it is as though someone were to mix and

769 a

30 κεράσειε πολλοὺς χυμοὺς εἰς ἓν ὑγρόν, κἄπειτ'
ἐντεῦθεν λαμβάνοι, [καὶ]¹ δύναιτ' ἂν λαμβάνειν μὴ
ἴσον ἀεὶ ἀφ' ἑκάστου, ἀλλ' ὁτὲ μὲν τοῦ τοιοῦδε
πλέον ὁτὲ δὲ τοῦ τοιοῦδε, ὁτὲ δὲ τοῦ μὲν λαβεῖν
τοῦ δὲ μηθὲν λαβεῖν—τοῦτο συμβαίνειν² καὶ ἐπὶ τῆς
γονῆς πολυμιγοῦς οὔσης· ἀφ' οὗ γὰρ ἂν τῶν
35 γεννώντων πλεῖστον ἐγγένηται, τούτῳ γίνεσθαι τὴν
μορφὴν ἐοικός. οὗτος δὲ ὁ λόγος οὐ σαφὴς μὲν

769 b κ αὶ πλασματίας ἐστὶ πολλαχῇ, βούλεται δὲ καὶ
βέλτιον λέγειν μὴ ἐνεργείᾳ ὑπάρχειν, ἀλλὰ κατὰ
δύναμιν, ἣν λέγει πανσπερμίαν· ἐκείνως μὲν γὰρ
ἀδύνατον, οὕτως δὲ δυνατόν·

Οὐ ῥᾴδιον δὲ οὐδὲ τρόπον ἕνα τῆς αἰτίας ἀπο-
διδόντας τὰς αἰτίας εἰπεῖν περὶ πάντων, τοῦ τε
5 γίνεσθαι θῆλυ καὶ ἄρρεν, καὶ διὰ τί τὸ μὲν θῆλυ
τῷ πατρὶ πολλάκις ὅμοιον τὸ δ' ἄρρεν τῇ μητρί,
καὶ πάλιν τῆς πρὸς τοὺς προγόνους ὁμοιότητος,
ἔτι δὲ διὰ τίν' αἰτίαν ὁτὲ μὲν ἄνθρωπος μὲν τούτων
δ' οὐθενὶ προσόμοιος, ὁτὲ δὲ προϊὸν οὕτως τέλος
οὐδὲ ἄνθρωπος ἀλλὰ ζῷόν τι μόνον φαίνεται τὸ
10 γιγνόμενον, ἃ δὴ καὶ λέγεται τέρατα.

Καὶ γὰρ ἐχόμενον τῶν εἰρημένων ἐστὶν εἰπεῖν
περὶ τῶν τοιούτων τὰς αἰτίας. τέλος γὰρ τῶν μὲν
κινήσεων λυομένων, τῆς δ' ὕλης οὐ κρατουμένης,
μένει τὸ καθόλου μάλιστα· τοῦτο δ' ἐστὶ τὸ ζῷον.

---

¹ secl. A.-W.　　　　² συμβαίνει PSYZ.

ᵃ Because it can be restated in Aristotelian terminology,
as he goes on to show.

416

blend together a large number of juices into one fluid, and then take off some of this mixture ; in doing so he could take off not always an equal amount of each juice, but sometimes more of this one, sometimes more of that, and sometimes he might take some of one and nothing at all of another : So, they say, it is with the semen, which is a mixture of a large number of ingredients ; and in appearance the offspring takes after that parent from whom the largest amount is derived. This theory is obscure, and at many points a sheer fabrication. At the same time, it aims at a more satisfactory [a] statement, viz., that this " seed-aggregate " is something that exists not *in actuality*, but only *potentially*, since it cannot exist *in actuality*, whereas it can exist *potentially*.

Still it is not easy, by stating a single mode of cause, to explain the causes of everything,—(1) why male and female are formed, (2) why female offspring often resembles the father and male offspring the mother, and again (3) the resemblance borne to ancestors, and further (4) what is the cause why sometimes the offspring is a human being yet bears no resemblance to any ancestor, sometimes it has reached such a point that in the end it no longer has the appearance of a human being at all, but that of an animal only—it belongs to the class of monstrosities, as they are called.

And indeed this is what comes next to be treated after what we have already dealt with—the causes of monstrosities, for in the end, when the movements ⟨that came from the male⟩ relapse and the material ⟨that came from the female⟩ does not get mastered, what remains is that which is most " general," and this is the ⟨merely⟩ " animal." People say that the

Monstrosities.

417

τὸ δὲ γιγνόμενον κριοῦ κεφαλήν φασιν ἢ βοὸς ἔχειν,
15 καὶ ἐν τοῖς ἄλλοις ὁμοίως ἑτέρου ζώου, μόσχον
παιδὸς κεφαλὴν ἢ πρόβατον βοός. ταῦτα δὲ πάντα
συμβαίνει μὲν διὰ τὰς προειρημένας αἰτίας, ἔστι
δ' οὐθὲν ὧν λέγουσιν, ἀλλ' ἐοικότα μόνον· ὅπερ
γίγνεται καὶ μὴ πεπηρωμένων. διὸ πολλάκις οἱ
20 σκώπτοντες εἰκάζουσι τῶν μὴ καλῶν ἐνίους τοὺς
μὲν αἰγὶ φυσῶντι πῦρ, τοὺς δ' οἷ[1] κυρίττοντι.
φυσιογνώμων δέ τις ἀνῆγε πάσας[1] εἰς δύο ἢ τριῶν
ζώων[2] ὄψεις, καὶ συνέπειθε πολλάκις λέγων. ὅτι
δ' ἐστὶν ἀδύνατον γίγνεσθαι τέρας τοιοῦτον, ἕτερον
ἐν ἑτέρῳ ζῶον, δηλοῦσιν οἱ χρόνοι τῆς κυήσεως
πολὺ διαφέροντες ἀνθρώπου καὶ προβάτου καὶ
25 κυνὸς καὶ βοός· ἀδύνατον δ' ἕκαστον γενέσθαι μὴ
κατὰ τοὺς οἰκείους χρόνους.

Τὰ μὲν οὖν τοῦτον τὸν τρόπον λέγεται τῶν
τεράτων, τὰ δὲ τῷ πολυμερῆ τὴν μορφὴν ἔχειν,
πολύποδα καὶ πολυκέφαλα γινόμενα.

Πάρεγγυς δ' οἱ λόγοι τῆς αἰτίας καὶ παραπλήσιοι
τρόπον τινά εἰσιν οἵ τε περὶ τῶν τεράτων καὶ οἱ
30 περὶ τῶν ἀναπήρων ζῴων· καὶ γὰρ τὸ τέρας
ἀναπηρία τίς ἐστιν.

IV   Δημόκριτος μὲν οὖν ἔφησε γίγνεσθαι τὰ τέρατα
διὰ τὸ δύο γονὰς πίπτειν,[3] τὴν μὲν πρότερον ὁρμή-
σασαν[4] τὴν δ' ὕστερον, †καὶ[5] ταύτην ἐξελθοῦσαν[6]

---

[1] πάντας P.
[2] ἢ τριῶν ζώων P : ζώων ἢ τριῶν vulg.
[3] ⟨συμ⟩πίπτειν Diels.   [4] et non egredientem add. Gul.
[5] ὑφ' ἧς καὶ P (a quo et hanc egredientem Gul., teste Bussemaker).   [6] ἐπελθοῦσαν Diels.

offspring which is formed has the head of a ram or an ox; and similarly with other creatures, that one has the head of another, *e.g.*, a calf has a child's head or a sheep an ox's head. The occurrence of all these things is due to the causes I have named; at the same time, in no case are they what they are alleged to be, but resemblances only, and this of course comes about even when there is no deformation involved. Thus, humorists often compare those whose strong point is not good looks in some cases with a fire-spouting-goat, in others with a butting ram; and there was a physiognomist who in his lectures used to show how all people's faces could be reduced to those of two or three animals, and very often he carried conviction with his audience. It is however impossible for a monstrosity of this type to be formed (*i.e.*, one animal within another), as is shown by the gestation-periods of man, sheep, dog, and ox, which are widely different, and none of these animals can possibly be formed except in its own proper period.

This, then, is one sort of " monstrosity " we hear spoken of. There are others which qualify for the name in virtue of having additional parts to their body, being formed with extra feet or extra heads.

The account of the cause of monstrosities is very close and in a way similar to that of the cause of deformed animals, since a monstrosity is really a sort of deformity.

Now Democritus [a] explained the formation of monstrosities thus. Two semens fall into the uterus, one of them having started forth earlier and the other later, †and the second when it has gone out goes

IV
Redun-
dancy of
parts.

---

[a] See Diels, *Vorsokr.*[5] 68 A 146.

769 b

ἐλθεῖν[1] εἰς τὴν ὑστέραν†[2] ὥστε συμφύεσθαι καὶ ἐπαλλάττειν τὰ μόρια. [ταῖς δ᾿ ὄρνισιν ἐπεὶ συμβαίνει
35 ταχεῖαν γίνεσθαι τὴν ὀχείαν ἀεί, τά τ᾿ ὠὰ καὶ
τὴν χρόαν αὐτῶν ἐπαλλάττειν φησίν.][3] εἰ δὲ συμ-

770 a βαίνει ἐξ ἑνὸς σπέρματος πλείω γίνεσθαι καὶ μιᾶς
συνουσίας, ὅπερ φαίνεται, βέλτιον μὴ κύκλῳ περι-
ιέναι παρέντας τὴν σύντομον· τοῖς γὰρ τοιούτοις
μάλιστ᾿ ἀναγκαῖον τοῦτο συμβαίνειν ὅταν μὴ
διακριθῶσιν ἀλλ᾿ ἅμα τὰ σπέρματα ἔλθωσιν. εἰ
5 μὲν οὖν αἰτιάσασθαι δεῖ τὴν ἀπὸ τοῦ ἄρρενος γονήν,
τοῦτον ἂν τὸν τρόπον εἴη λεκτέον· ὅλως δὲ μᾶλλον
τὴν αἰτίαν οἰητέον ἐν τῇ ὕλῃ καὶ τοῖς συνισταμένοις κυήμασιν εἶναι   διὸ καὶ γίνονται τὰ τοιαῦτα
τῶν τεράτων ἐν μὲν τοῖς μονοτόκοις σπάνια
πάμπαν, ἐν δὲ τοῖς πολυτόκοις μᾶλλον, καὶ μάλιστ᾿
10 ἐν ὄρνισι, τῶν δ᾿ ὀρνίθων ἐν ταῖς ἀλεκτορίσιν·
αὗται γὰρ πολυτοκοῦσιν, οὐ μόνον τῷ πολλάκις
τίκτειν ὥσπερ τὸ τῶν περιστερῶν γένος, ἀλλὰ καὶ
τῷ πολλὰ ἅμα ἔχειν κυήματα καὶ πᾶσαν ὥραν
ὀχεύεσθαι. διόπερ καὶ πολλὰ δίδυμα τίκτουσιν·

---

[1] εὐθὺς pro ἐλθεῖν E.
[2] loc. corrupt. monet Platt.   *quia duo spermata cadunt in
matricem, et prius cadit unum sperma et permansit et non
exivit (et non egredientem* habet Gul. vers.), *deinde con-
tinuatur cum secundo spermate remanente etiam in matrice,
et sic,* etc. Σ.
[3] seclusi. locum sensu carere monet Platt.   pro ἀεὶ Aldus
habet ἀφεῖλε.   inter ἀεὶ et ὠὰ spatium longum Ζ[1], in quo τ à
Ζ[2].   credo haec de avibus dicta ex adnot. quae ad 770 a 9
seqq., al. locc., spectaverit irrepsisse ; conferas 717 b 29.

---

[a] This sentence, as Platt points out, is corrupt.   The
general sense is clear.   I have given Scot's translation in
the *apparatus criticus.*

into the uterus,† *a* with the result that the parts grow on to one another and get thrown into disorder. [In the case of birds, since copulation is a quick business with them always, the eggs and their colour as well, he says, get thrown into disorder.]*b* But if it is a fact that several offspring are formed from one semen and from one act of copulation, as is evidently the case, we should do better not to neglect the shortest route and go a long way round, since in cases of this sort it is absolutely necessary that this should happen when the semens have not been separated but proceed together.*c* Now if we are really obliged to refer the cause to the semen that comes from the male, then, I suppose these are the lines on which we should make our explanation ; but from every point of view we ought preferably to hold that the seat of the cause is the material*d* and in the fetations as they take shape. And that too explains why monstrosities of this sort, while they occur very seldom in animals that produce one offspring only, occur oftener in those that are prolific, and most of all in birds, and specially in the common fowl.*e* This species is prolific, not only in laying eggs frequently, as the pigeon tribe does, but also in carrying many fetations at once and in copulating at every season of the year. Hence also fowls lay many twin-eggs,

b This sentence (which may be a note on 770 a 15 ff.) seems to be from the same author as the interpolation at 717 b 29 : the speed of birds' copulation obviously was a favourite point with him, but it has nothing to do either with this passage or with that in Bk. I. In the present passage, birds are introduced later by Aristotle (a 10).

c And this is a contingency for which Democritus's explanation does not allow.   *d* Supplied by the female.

e For monstrosities, see references, p. xi.

15 συμφύεται γὰρ διὰ τὸ πλησίον ἀλλήλων εἶναι τὰ
κυήματα, καθάπερ ἐνίοτε πολλὰ τῶν περικαρπίων.
τούτων δὲ ὅσων μὲν ἂν αἱ λέκιθοι διορίζωνται
κατὰ τὸν ὑμένα, δύο γίνονται νεοττοὶ κεχωρισμένοι,
περιττὸν οὐδὲν ἔχοντες· ὅσων δὲ συνεχεῖς καὶ μὴ
διείργει μηθέν, ἐκ τούτων οἱ νεοττοὶ γίνονται
20 τερατώδεις, σῶμα μὲν καὶ κεφαλὴν μίαν ἔχοντες,
σκέλη δὲ τέτταρα καὶ πτέρυγας, διὰ τὸ τὰ μὲν
ἄνωθεν ἐκ τοῦ λευκοῦ γίνεσθαι καὶ πρότερον,
ταμιευομένης ἐκ τῆς λεκίθου τῆς τροφῆς αὐτοῖς,
τὸ δὲ κάτω μόριον ὑστερίζειν μέν, τὴν δὲ τροφὴν
εἶναι μίαν καὶ ἀδιόριστον.

Ἤδη δὲ καὶ ὄφις ὦπται δικέφαλος διὰ τὴν
25 αὐτὴν αἰτίαν· ᾠοτοκεῖ γὰρ καὶ πολυτοκεῖ καὶ τοῦτο
τὸ γένος. σπανιώτερον δὲ τὸ τερατῶδες ἐπ'
αὐτῶν διὰ τὸ σχῆμα τῆς ὑστέρας· στοιχηδὸν γὰρ
κεῖται τὸ πλῆθος τῶν ᾠῶν διὰ τὸ μῆκος αὐτῆς.
καὶ περὶ τὰς μελίττας καὶ τοὺς σφῆκας οὐδὲν
γίνεται τοιοῦτον· ἐν κεχωρισμένοις γὰρ κυτταρίοις
30 ὁ τόκος ἐστὶν αὐτῶν. περὶ δὲ τὰς ἀλεκτορίδας
τοὐναντίον συμβέβηκεν, ᾗ καὶ δῆλον ὡς ἐν τῇ ὕλῃ
τὴν αἰτίαν δεῖ νομίζειν τῶν τοιούτων· καὶ γὰρ
τῶν ἄλλων ἐν τοῖς πολυτόκοις μᾶλλον. διὸ ἐν
ἀνθρώπῳ ἧττον· ὡς γὰρ ἐπὶ τὸ πολὺ μονότοκον
ἐστὶ καὶ τελειογόνον, ἐπεὶ καὶ τούτων ἐν οἷς τόποις
35 πολύγονοι αἱ γυναῖκές εἰσι, τοῦτο συμβαίνει μᾶλ-

---

[a] i.e., yolk only, not white as well; and as there are two
yolks these parts are formed double. For the distinction
between "nutritive" (i.e., formative) and "growth-promoting" nourishment, see 744 b 32 ff. Cf. also 751 b 2 ff.

[b] Not huddled up together.

since the fetations, on account of being situated close to each other, grow on to each other, just as many fruits sometimes do. Of these twin-eggs, those in which the yolks are kept apart by the membrane develop into two separate chicks, and there is nothing extraordinary about them ; those in which the yolks are continuous, with nothing to hold them apart, give rise to chicks that are monstrosities : they have one body and one head, but four legs and wings, the reason for which is that the upper parts of the body are formed out of the white and before the rest, the nourishment being dispensed to them from the store in the yolk, whereas the lower part (a) is formed afterwards, (b) its nourishment is uniform and homogeneous.[a]

A snake, too, has been seen with two heads, and the cause is the same—this also is a class of animal which lays eggs and is prolific. Monstrosities occur less frequently, however, with snakes owing to the shape of their uterus, in which, on account of its length, the numerous eggs lie one after another in a row.[b] Nothing of this kind occurs with bees and wasps, because their offspring are laid in separate cells. With the common fowl, however, the opposite is the case—a fact which clearly goes to show that we are bound to hold that the cause of such things is in the material,[c] since with other animals too they occur more frequently in those that are prolific. Hence they occur less frequently in human beings, for the offspring which these produce is as a rule one in number, and it is perfected by the time of birth, since even in this species the occurrence of monstrosities is more common in those regions where the women are

[c] Not in the semen.

λον, οἷον περὶ Αἴγυπτον. ἐν δὲ ταῖς αἰξὶ καὶ τοῖς
προβάτοις γίνεται μᾶλλον· πολυτοκώτερα γάρ ἐστιν.
ἔτι δὲ μᾶλλον ἐν τοῖς πολυσχιδέσιν· πολυτόκα γάρ

ἐστι τὰ τοιαῦτα¹ τῶν ζῴων καὶ οὐ τελειογόνα,
καθάπερ ἡ κύων· τὰ γὰρ πολλὰ τίκτει τυφλὰ τούτων. δι᾽ ἣν δ᾽ αἰτίαν τοῦτο συμβαίνει καὶ δι᾽ ἣν
αἰτίαν πολυτοκοῦσιν, ὕστερον λεκτέον. ἀλλὰ προ-
ωδοποίηται τῇ φύσει [πρὸς]² τὸ τερατοτοκεῖν τῷ³
5 μὴ γεννᾶν ὅμοια διὰ τὴν ἀτέλειαν· ἔστι δὲ καὶ τὸ
τέρας τῶν ἀνομοίων. διόπερ ἐπαλλάττει τοῦτο τὸ
σύμπτωμα τοῖς τοιούτοις τὴν φύσιν. ἐν γὰρ τούτοις μάλιστα γίνεται καὶ τὰ μετάχοιρα καλούμενα.
ταῦτα δ᾽ ἐστὶ κατά τι πεπονθότα τερατῶδες· τὸ
γὰρ ἐκλείπειν ἢ προσεῖναί τι τερατῶδες. ἔστι γὰρ
10 τὸ τέρας τῶν παρὰ φύσιν [τι],⁴ παρὰ φύσιν δ᾽ οὐ
πᾶσαν ἀλλὰ τὴν ὡς ἐπὶ τὸ πολύ· περὶ γὰρ τὴν
ἀεὶ καὶ τὴν ἐξ ἀνάγκης οὐθὲν γίνεται παρὰ φύσιν,
ἀλλ᾽ ἐν τοῖς ὡς ἐπὶ τὸ πολὺ μὲν οὕτω γινομένοις,
ἐνδεχομένοις δὲ καὶ ἄλλως, ἐπεὶ καὶ τούτων ἐν
ὅσοις συμβαίνει παρὰ τὴν τάξιν μὲν ταύτην, ἀεὶ
15 μέντοι μὴ τυχόντως, ἧττον εἶναι δοκεῖ τέρας διὰ
τὸ καὶ τὸ παρὰ φύσιν εἶναι τρόπον τινὰ κατὰ

---

¹ sic PSY*Z (πολυτοκώτερα Z) : ἔστι γὰρ τὰ τ. π. vulg.
² secl. Btf.   ³ τῷ A.-W. : τω Y : τὸ vulg.   ⁴ τι om. P.

ᵃ Cf. *H.A.* 584 b 7, 31 ; the passage in Hippocrates, π.
ἀέρων ὑδάτων τόπων 12 (ii. 54 Littré) τά τε κτήνεα τίκτειν τε
πυκνότατα καὶ ἐκτρέφειν κάλλιστα may refer to Egypt and
Libya.   ᵇ Ch. 6 below.   ᶜ 771 a 18 ff.
ᵈ Viz., which produce imperfect offspring.
ᵉ See 749 a 2.   ᶠ Cf. 772 a 35, etc.   ᵍ See Introd. § 9.

prolific—in Egypt, for instance.[a] Monstrosities occur
more frequently in goats and sheep, because they
are more prolific ; and still more frequently in the
fissipede animals, because animals of this sort are
prolific and the offspring is not perfected when born
(e.g., the dog)—most of these creatures' young, of
course, are born blind. The cause why this occurs [b]
and the cause why they are prolific [c] must be stated
later. But the way to the production of monstrosities
has been already prepared for Nature by the fact that
they generate offspring which, owing to its imperfect
state, is unlike its parents :—for monstrosities come
under the class of offspring which is unlike its parents.
And that is why this particular accident extends its
range to affect animals of that nature,[d] and, to bear
this out, it is among these animals especially that
metachoira [e] as they are called occur. These *metachoira* are creatures which have in some respect
undergone some "monstrous" affection, since the
lack of any part or the presence of an extra part is
such an affection. A monstrosity, of course, belongs
to the class of "things contrary to Nature," although
it is contrary not to Nature in her entirety but only
to Nature *in the generality of cases.[f]* So far as concerns the Nature which is *always [g]* and is *by necessity*,
nothing occurs contrary to that ; no ; unnatural
occurrences are found only among those things which
occur as they do *in the generality of cases*, but which
may occur otherwise. Why, even in those instances
of the phenomena we are considering, what occurs
is contrary to this particular order, certainly, but
it never happens in a merely random fashion ; and
therefore it seems less of a monstrosity because
even that which is contrary to Nature is, in a

φύσιν, ὅταν μὴ κρατήσῃ τὴν κατὰ τὴν ὕλην ἡ
κατὰ τὸ εἶδος φύσις. διόπερ οὔτε τὰ τοιαῦτα
τέρατα λέγουσιν, οὔτ' ἐν τοῖς ἄλλοις ἐν ὅσοις εἴωθέ
τι γίνεσθαι, καθάπερ ἐν τοῖς περικαρπίοις. ἔστι
20 γάρ τις ἄμπελος ἣν καλοῦσί τινες κάπνεον, ἥν,[1] ἂν
ἐνέγκῃ μέλανας βότρυας,[2] οὐ κρίνουσι τέρας διὰ τὸ
πλειστάκις εἰωθέναι ταύτην τοῦτο ποιεῖν. αἴτιον
δ' ὅτι μεταξὺ λευκῆς ἐστι τὴν φύσιν καὶ μελαίνης,
ὥστ' οὐ πόρρωθεν ἡ μετάβασις οὐδ' ὡσπερανεὶ
παρὰ φύσιν· οὐ γὰρ εἰς ἄλλην φύσιν.

25   Ἐν δὲ τοῖς πολυτόκοις ταῦτα[3] συμβαίνει διὰ[4]
τὸ[5] τὴν πολυτοκίαν ἐμποδίζειν[6] τὰς τελειώσεις
ἀλλήλων καὶ τὰς κινήσεις τὰς γεννητικάς.

   Περὶ δὲ τῆς πολυτοκίας καὶ τοῦ πλεονασμοῦ τοῦ
τῶν μερῶν, καὶ τῆς ὀλιγοτοκίας καὶ μονοτοκίας
30 καὶ τῆς ἐνδείας τῶν μερῶν, ἀπορήσειεν ἄν τις.
γίνεται γὰρ ἐνίοτε τὰ μὲν πλείους ἔχοντα δακτύλους, τὰ δ' ἕνα μόνον, καὶ περὶ τὰ ἄλλα μέρη τὸν
αὐτὸν τρόπον· καὶ γὰρ πλεονάζει καὶ κολοβὰ
γίνεται, τὰ δὲ καὶ δύο ἔχοντα αἰδοῖα, τὸ μὲν ἄρρενος
τὸ δὲ θήλεος, καὶ ἐν ἀνθρώποις καὶ μάλιστα περὶ
35 τὰς αἶγας. γίνονται γὰρ ἃς καλοῦσι τραγαίνας
διὰ τὸ θήλεος καὶ ἄρρενος ἔχειν αἰδοῖον· ἤδη δὲ
καὶ κέρας αἲξ ἔχουσα ἐγένετο πρὸς τῷ σκέλει.

---

[1] ἥ Sus.   [2] βότρυας PZ : βότρυς vulg.
[3] ταῦτα A.-W. (ταῦτά τε Aldus) : ταῦτά τε vulg. :   τε
om. Z[1] : τέρατα coni. A.-W.   [4] διὰ Z[1] : καὶ διὰ vulg.
[5] τὸ suprascr. Z[2] : om. vulg.   [6] ἐμποδίζει PZ[2].

---

[a] As it can be represented as a case of one " nature "
failing to control another " nature," it can be termed " in
accordance with nature."   See Introd. § 14.
[b] Cf. Theophrastus, Hist. Plant. II. 3. 2, where it is stated
that the μάντεις do not consider the vagaries of this plant

way, in accordance with Nature (*i.e.*, whenever the
" formal " nature has not gained control over the
" material " nature).[a] Hence, people do not call
things of this sort monstrosities any more than they
do in the other cases where something occurs habitually—as happens with fruit. Thus, there is a certain
sort of vine—" smoky "[b] is the name some people
give it ;—and if it bears black grapes they do not
reckon it as a monstrosity, because it often and
habitually does this. The reason is that it is intermediate in its nature between white and black, and
so the alteration is quite small and not really contrary
to nature, because it is not an alteration to a different
nature.

These things, then, occur in the case of the animals
which produce numerous young, because the numerous offspring which are produced hamper each other's
being brought to perfection and also the movements
which effect generation.

A puzzle may be raised about this production of
numerous offspring and the redundance of parts, and
the production of few or one offspring and the
deficiency of parts : sometimes animals are born
having too many toes, some having one only ; and
the same with the other parts : some have too many ;
some are mutilated ; some actually have two organs
of generation, one male and the other female. This
happens with human beings, and with goats especially. Goats are born which are called *tragainai*[c] on
account of their possessing both male and female
organs of generation. We have also had an instance
of a goat being born that had a horn on its leg. Altera-

to be sufficiently unusual or unnatural to be of any terato-
logical significance.   [c] Hermaphrodites.

ARISTOTLE

γίνονται δὲ μεταβολαὶ καὶ πηρώσεις καὶ περὶ τὰ
ἐντὸς μόρια, τῷ ἢ μὴ ἔχειν ἔνια ἢ κεκολοβωμένα
ἔχειν καὶ πλείω καὶ μεθεστῶτα τοὺς τόπους.
καρδίαν μὲν οὖν οὐθὲν πώποτε ἐγένετο ζῷον οὐκ
5 ἔχον, σπλῆνα δ' οὐκ ἔχον, καὶ δύο ἔχον, καὶ νεφρὸν
ἕνα· ἧπαρ δ' οὐκ ἔχον μὲν οὐθέν, οὐχ ὅλον δὲ
ἔχον. ταῦτα δὲ πάντα ἐν τοῖς τελειωθεῖσι καὶ
ζῶσιν. εὑρίσκεται καὶ χολὴν οὐκ ἔχοντα, πεφυ-
κότα ἔχειν· τὰ δὲ πλείους ἔχοντα μιᾶς. ἤδη δ'
ἐγένετο καὶ μεθεστηκότα κατὰ τόπον, τὸ μὲν ἧπαρ
ἐν τοῖς ἀριστεροῖς, ὁ δὲ σπλὴν ἐν τοῖς δεξιοῖς.
10 καὶ ταῦτα μὲν ἔν γε τετελεσμένοις ὦπται τοῖς
ζῴοις, ὥσπερ εἴρηται· ἐν δὲ τοῖς τικτομένοις[1]
ἔχοντα πολλὴν καὶ παντοδαπὴν ταραχήν. τὰ μὲν
οὖν μικρὸν παρεκβαίνοντα τὴν φύσιν ζῆν εἴωθεν,
τὰ δὲ πλεῖον οὐ ζῆν, ὅταν ἐν τοῖς κυρίοις τοῦ
ζῆν γένηται τὸ παρὰ φύσιν.

Ἡ δὲ σκέψις ἐστὶν ἡ περὶ τούτων πότερον τὴν
15 αὐτὴν αἰτίαν δεῖ νομίζειν τῆς μονοτοκίας καὶ τῆς ἐνδείας τῶν μερῶν καὶ τοῦ πλεονασμοῦ καὶ τῆς πολυτοκίας ἢ μὴ τὴν αὐτήν.

Πρῶτον μὲν οὖν διὰ τί τὰ μέν ἐστι πολυτόκα
τὰ δὲ μονοτόκα, τοῦτ' ἄν τις δόξειεν εὐλόγως
θαυμάζειν. τὰ γὰρ μέγιστα μονοτόκα τῶν ζῴων
20 ἐστίν, οἷον ἐλέφας κάμηλος ἵππος καὶ τὰ μώνυχα·
τούτων δὲ τὰ μὲν μείζω τῶν ἄλλων, τὰ δὲ πολὺ

---

[1] sic Bekker: γεννωμένοις O[b] marg.*: in filiis Σ: εἰρημένοις
PSYZ.

---

[a] i.e., have passed beyond the embryonic stage, have
reached the end of their period of development.
[b] For a discussion of this see P.A. Bk. IV, ch. 2.

tions and deformations occur in respect of the inward parts too ; animals either lack certain parts, or have them in a mutilated form, or have too many of them, or in the wrong places. No animal, it is true, has ever been born without a heart, but there have been animals without a spleen, and with two spleens, and with one kidney ; none without any liver at all, but certainly with an incomplete one. These phenomena are found in animals that are perfect [a] and living. We find, also, animals with no gall-bladder which naturally should have one [b] ; others with more than one. Instances have occurred of organs in the wrong places : the liver on the left side and the spleen on the right. These things, as I said, have been observed among animals which have reached perfect growth ; among newly born animals instances have been seen exhibiting great and varied confusion. Those which depart only slightly from the natural usually live ; those which depart more than that do not—*i.e.*, when their unnatural conformation lies in the parts that control the creature's life.

The point about these which we have to consider is the following. Ought we to hold that one and the same cause is responsible for the production of a single offspring and the deficiency in the parts, and also for the production of many offspring and the redundancy in the parts, or not ?

Consider, then, first of all, the fact that some animals produce many offspring, others a single one only. Surely surprise at this is very reasonable, as it is the largest of the animals which produce one only, *e.g.*, the elephant, the camel, the horse and those with uncloven hoofs ; of these, some are larger than

(b) Number of offspring.

429

διαφέρει κατὰ τὸ μέγεθος. κύων δὲ καὶ λύκος
καὶ τὰ πολυσχιδῆ πάντα σχεδὸν πολυτόκα,[1] καὶ τὰ
μικρὰ τῶν τοιούτων, οἷον τὸ τῶν μυῶν γένος. τὰ
δὲ διχηλὰ ὀλιγοτόκα πλὴν ὑός· αὕτη δὲ τῶν
25 πολυτόκων ἐστίν. εὔλογον γὰρ τὰ μὲν μεγάλα
πλείω δύνασθαι γεννᾶν καὶ σπέρμα φέρειν πλεῖον.
αἴτιον δ' αὐτὸ τὸ θαυμαζόμενον τοῦ μὴ θαυμάζειν·
διὰ γὰρ τὸ μέγεθος οὐ πολυτοκοῦσιν· ἡ γὰρ τροφὴ
καταναλίσκεται τοῖς τοιούτοις εἰς τὴν αὔξησιν τοῦ
σώματος· τοῖς δ' ἐλάττοσιν ἀπὸ τοῦ μεγέθους ἡ
30 φύσις ἀφελοῦσα[2] πρὸς τὸ περίττωμα προστίθησι
τὸ σπερματικὸν τὴν ὑπεροχήν. ἔτι δὲ τὸ γεννῆσαν
σπέρμα πλεῖον μὲν τὸ τοῦ μείζονος ἀναγκαῖον
εἶναι, μικρὸν δὲ τὸ τῶν ἐλαττόνων. πολλὰ μὲν
οὖν[3] μικρὰ γένοιτ' ἂν ἐν ταὐτῷ, μεγάλα δὲ πολλὰ
χαλεπόν. [τοῖς δὲ μέσοις μεγέθεσι τὸ μέσον
35 ἀπέδωκεν ἡ φύσις. τοῦ μὲν οὖν τὰ μὲν εἶναι
μεγάλα τῶν ζῴων τὰ δ' ἐλάττω τὰ δὲ μέσα πρό-
771 b τερον εἰρήκαμεν τὴν αἰτίαν· μονοτόκα δέ, τὰ δ'
ὀλιγοτόκα, τὰ δὲ πολυτόκα τῶν ζῴων ἐστίν.][4] ὡς
μὲν ἐπὶ τὸ πολὺ τὰ μὲν μώνυχα μονοτόκα, τὰ δὲ
διχηλὰ ὀλιγοτόκα, τὰ δὲ πολυσχιδῆ πολυτόκα.
τούτου δ' αἴτιον ὅτι ὡς ἐπὶ τὸ πολὺ τὰ μεγέθη

---

[1] σ. π. P : π. σ. vulg.
[2] ἀφελοῦσα PS : ἀφαιροῦσα vulg.
[3] οὖν PSY : οὖν καὶ vulg.   [4] seclusi : om. Σ.

the other animals, some are really outstanding in respect of size. The dog, on the other hand, and the wolf, and practically all the fissipede animals produce many offspring; even small animals of this class do so, such as the mouse family. The cloven-hoofed animals produce few offspring, except the pig, which is among those that produce many. As I said, this is surprising, because we might have expected the large animals to be able to generate more offspring and to produce more semen. But the very thing that surprises us is the reason why we should not be surprised. Their size is the very reason why they do not produce many offspring, because in animals of this sort the nourishment gets used up to supply the growth of the body, whereas in the case of the smaller animals, Nature takes away from their size and adds the surplus on to the seminal residue. Further, the generative semen of a larger animal must of necessity be greater in bulk,[a] and that of the lesser ones small. Also, though many small ones may very well be formed in one place, it is difficult for many large ones to be. [To the intermediate sizes Nature has allotted the intermediate number. As for the fact that some animals are large, some smaller, and some intermediate, we have stated the cause of this earlier.][b] For the most part it is the solid-hoofed animals which produce a single offspring, the cloven-hoofed animals which produce few, and the fissipede animals which produce many. The reason for this is that for the most part the distinction of

[a] But this *pro rata* merely; so that a large animal has no net advantage over a small one in this respect.

[b] The preceding words seem to be irrelevant; those which follow immediately in the Greek cannot be construed, and I have omitted them from the translation.

5 διώρισται κατὰ τὰς διαφορὰς ταύτας. οὐ μὴν ἔχει
γ' οὕτως ἐπὶ πάντων· αἴτιον γὰρ μέγεθος καὶ
μικρότης τῶν σωμάτων τῆς ὀλιγοτοκίας καὶ πολυτοκίας, ἀλλ' οὐ τὸ μώνυχον ἢ πολυσχιδὲς ἢ διχηλὸν
εἶναι τὸ γένος. τούτου δὲ μαρτύριον· ὁ γὰρ ἐλέφας
μέγιστον τῶν ζῴων, ἔστι δὲ πολυσχιδές, ἥ τε
10 κάμηλος διχηλὸν τῶν λοιπῶν μέγιστον ὄν. οὐ
μόνον δ' ἐν τοῖς πεζοῖς ἀλλὰ καὶ ἐν τοῖς πτηνοῖς
καὶ ἐν τοῖς πλωτοῖς τὰ μὲν μεγάλα ὀλιγοτόκα
ἐστὶ τὰ δὲ μικρὰ πολυτόκα, διὰ τὴν αὐτὴν αἰτίαν.
ὁμοίως δὲ καὶ τῶν φυτῶν οὐ τὰ μέγιστα φέρει
πλεῖστον καρπόν.

15 Διὰ τί μὲν οὖν τῶν ζῴων τὰ μὲν πολυτόκα τὰ
δ' ὀλιγοτόκα τὰ δὲ μονοτόκα[1] τὴν φύσιν ἐστίν,
εἴρηται· τῆς δὲ νῦν ῥηθείσης ἀπορίας μᾶλλον ἄν
τις εὐλόγως[2] θαυμάσειεν ἐπὶ τῶν πολυτοκούντων,
ἐπειδὴ φαίνεται πολλάκις ἀπὸ μιᾶς ὀχείας κυϊσκό-
μενα τὰ τοιαῦτα τῶν ζῴων. τὸ δὲ σπέρμα τὸ
20 τοῦ ἄρρενος, εἴτε συμβάλλεται πρὸς τὴν ὕλην
μόριον γινόμενον τοῦ κυήματος καὶ τῷ τοῦ θήλεος
σπέρματι μιγνύμενον, εἴτε καὶ μὴ τοῦτον τὸν
τρόπον, ἀλλ' ὥσπερ φαμὲν συνάγον καὶ δημιουργοῦν τὴν ὕλην τὴν ἐν τῷ θήλει καὶ τὸ
περίττωμα τὸ σπερματικόν, καθάπερ ὁ ὀπὸς τὴν
ὑγρότητα τοῦ γάλακτος, διὰ τίνα ποτ' αἰτίαν οὐχ
25 ἓν ἀποτελεῖ ζῷον μέγεθος ἔχον, ὥσπερ ἐνταῦθα
ὁ ὀπός,[3] ⟨ἀλλ' ἐν τούτῳ τῷ περιττώματι πλείω

---

[1] τὰ δὲ μονοτόκα P : om. vulg.   [2] εὐλόγως P : om. vulg.
[3] ὥσπερ . . . ὀπός fortasse secludenda.

physical condition when the fetations are becoming sizable is that the growth of the fetation needs more nourishment than that afforded by the residue. There are some few women who are in better physical condition during pregnancy. This occurs with those whose bodies contain but small amounts of residue, and as a result this is completely used up together with the nourishment that goes to the embryo.

We now have to treat of the *mola uteri*,[a] as it is called. This occurs in women occasionally only, but it does occur in some during pregnancy. They bring forth a " *mola.*" It has been known to happen, in the case of a woman who has had intercourse and thinks she has conceived, that her figure has increased to begin with, and all the rest has proceeded as expected, but when the time for her delivery was at hand, she has neither brought anything to birth nor yet has the size of her girth decreased ; instead, she has continued in that condition for three or four years, till she was seized with dysentery which brought her to a dangerous pass, and then she has produced a fleshy mass, known as a " *mola.*" Sometimes, also, this condition lasts on into old age and persists until death. In such instances the objects which make their way out of the body are so hard that it is difficult to cut them in two even by means of an iron edge. Well, I have spoken in the *Problems*[b] of the cause of this occurrence ; the case of the fetation in the womb is exactly the same as that of meat, when it is undercooked ; and it is due not to heat, as some people allege, but rather to weakness of heat (because it looks as though Nature in these cases suffers from

VII
Mola uteri

[b] This reference cannot be found.

νατεῖν καὶ οὐ δύνασθαι τελειῶσαι οὐδ' ἐπιθεῖναι
5 τῇ γενέσει πέρας· διὸ καὶ συγκαταγηράσκει ἢ
πολὺν ἐμμένει χρόνον· οὔτε γὰρ ὡς τετελεσμένον[1]
οὔθ' ὡς πάμπαν ἀλλότριον ἔχει τὴν φύσιν)· τῆς
γὰρ σκληρότητος ἡ ἀπεψία αἰτία· ἀπεψία γάρ τις
καὶ ἡ μώλυνσίς[2] ἐστιν.

Ἀπορίαν δ' ἔχει, διὰ τί ποτ' ἐν τοῖς ἄλλοις οὐχὶ
10 γίνεται ζῴοις, εἰ μή τι πάμπαν λέληθεν. αἴτιον δὲ
δεῖ νομίζειν ὅτι μόνον ὑστερικόν ἐστι γυνὴ τῶν
ἄλλων ζῴων, καὶ περὶ τὰς καθάρσεις πλεονάζει
καὶ οὐ δύναται πέττειν αὐτάς· ὅταν οὖν ἐκ δυσ-
πέπτου ἰκμάδος συστῇ τὸ κύημα, τότε γίνεται ἡ
καλουμένη μύλη ἐν ταῖς γυναιξὶν εὐλόγως ἢ μάλιστα
ἢ μόναις.

VIII 15 Τὸ δὲ γάλα γίνεται τοῖς θήλεσιν ὅσα ζῳοτοκεῖ
ἐν αὑτοῖς χρήσιμον μὲν εἰς τὸν χρόνον τὸν τοῦ
τόκου, τῆς γὰρ τροφῆς χάριν αὐτὸ τῆς θύραζε
ἐποίησεν ἡ φύσις τοῖς ζῴοις, ὥστ' οὔτ' ἐλλείπειν
αὐτὸ ἐν τῷ χρόνῳ τούτῳ οὐθὲν οὔθ' ὑπερβάλλειν
οὐθέν· ὅπερ καὶ φαίνεται συμπῖπτον, ἂν μή τι
20 γένηται παρὰ φύσιν. τοῖς μὲν οὖν ἄλλοις ζῴοις,
διὰ τὸ τὸν χρόνον ἕνα τῆς κυήσεως εἶναι, πρὸς
τοῦτον ἀπαντᾷ τὸν καιρὸν ἡ πέψις αὐτοῦ· τοῖς δ'
ἀνθρώποις ἐπεὶ πλείους οἱ χρόνοι, κατὰ τὸν πρῶτον
ἀναγκαῖον ὑπάρχειν· διὸ πρὸ τῶν ἑπτὰ μηνῶν
ἄχρηστον τὸ γάλα ταῖς γυναιξί, τότε δ' ἤδη γίνεται

---

[1] τετελεσμένον P : τετελειωμένον vulg.   [2] μολ. codd.

---

[a] χρήσιμον μέν, because although it *serves a purpose*, it
is also (ll. 25 ff.) due to *necessity* in the sense that its formation follows inevitably from the circumstances, as Aristotle
explains.
[b] See 772 b 5 ff. and *H.A.* 584 a 33.

some inability, and is unable to complete her work
and to bring the process of formation to its consum-
mation ; that is why the *mola* lasts on into old age or
at any rate for a considerable time, for in its nature
it is neither a finished product nor yet something
wholly alien) ; since the cause of its hardness is the
lack of concoction, just as underdone meat is another
instance of lack of concoction.

But there is a puzzle here. Why is it that this
phenomenon does not occur in the other animals ?
(unless of course it does, but has entirely escaped
observation). We must take the reason to be that
alone of all animals women are liable to uterine affections ; they produce an excess of menstrual evacuations and cannot concoct them ; and so, when the
fetation has been " set," formed out of a liquid which
is difficult to concoct, then what is called the *mola* is
produced ; and thus it is not surprising that this takes
place chiefly in women if not exclusively in them.

Milk is produced towards the time of parturition VIII
in those female animals which are internally vivipar- Milk.
ous, and it is (1) of a useful and serviceable quality,[a]
for Nature has provided animals with it so that they
may nourish their young externally, and she has so
arranged that it is neither deficient nor excessive in
any way at that time ; this we actually observe to
obtain unless some accident contrary to nature
occurs. In the case of the other animals, as there is
but a single period of gestation, the concoction of the
milk coincides with that ; in man, however, as there
are more periods than one,[b] the milk must of necessity be available at the earliest of the possible dates ;
hence in women the milk, which is useless until seven
months are up, at that point becomes useful and

25 χρήσιμον. εὐλόγως δὲ συμβαίνει καὶ διὰ τὴν ἐξ
ἀνάγκης αἰτίαν πεπεμμένον εἰς τοὺς τελευταίους
χρόνους· τὸ μὲν γὰρ πρῶτον ἡ τοῦ τοιούτου
περιττώματος ἀπόκρισις εἰς τὴν τῶν ἐμβρύων.
ἀναλίσκεται γένεσιν· πάντων δ' ἡ τροφὴ τὸ
γλυκύτατον καὶ πεπεμμένον, ὥστ' ἀφαιρουμένης
30 τῆς τοιαύτης δυνάμεως ἀνάγκη τὸ λοιπὸν ἁλμυρὸν
γίνεσθαι καὶ δύσχυμον. τελεουμένων δὲ τῶν κυημάτων πλέον τὸ περίττωμα τὸ περιγινόμενον
(ἔλαττον γὰρ τὸ ἀναλισκόμενον)¹ καὶ γλυκύτερον,
οὐκ ἀφαιρουμένου ὁμοίως τοῦ εὐπέπτου. ²οὐ γὰρ
ἔτι εἰς πλάσιν τοῦ ἐμβρύου γίγνεται ἡ δαπάνη, ἀλλ'
35 εἰς μικρὰν αὔξησιν, ὥσπερ ἑστηκὸς ἤδη διὰ τὸ

776 b τέλος ἔχειν τὸ ἔμβρυον· ἔστι γάρ τις καὶ κυήματος τελείωσις. διόπερ ἐξέρχεται καὶ μεταβάλλει
τὴν γένεσιν, ὡς ἔχον τὰ αὑτοῦ καὶ οὐκέτι λαμβάνει
τὸ μὴ αὑτοῦ, ἐν ᾧ καιρῷ γίνεται τὸ γάλα χρήσιμον.

Εἰς δὲ τὸν ἄνω τόπον καὶ τοὺς μαστοὺς συλ-
5 λέγεται διὰ τὴν ἐξ ἀρχῆς τάξιν τῆς συστάσεως.
τὸ μὲν γὰρ ἄνω τοῦ ὑποζώματος τὸ κύριον τοῦ
ζῴου³ ἐστί, τὸ δὲ κάτω τόπος⁴ τῆς τροφῆς καὶ τοῦ
περιττώματος, ὅπως ὅσα πορευτικὰ τῶν ζῴων ἐν

---

¹ sic interpunxit Bussemaker.
² pro οὐ γὰρ ἔτι . . . χρήσιμον 776 b 3 habet Σ quoniam
non indigetur ea. non ergo accipitur in illo tempore quod
accipiebatur ante ex lacte. vide 777 a 22-27.
³ τῆς ζωῆς coni. Btf.   ⁴ τόπος P : om. vulg.

ᵃ Cf. P.A. 676 a 35.
ᵇ Aristotle here notes correctly that growth proceeds long
after differentiation has ceased.
ᶜ i.e., as well as a creature which has reached an indepen-

serviceable. But the fact that it is fully concocted at the final stages is due also (2) to another cause—the *necessary* cause, which is what we should expect, for, to begin with, the secretion of this particular residue is used up for the formation of the embryos ; and in every animal the nourishment is the sweetest ingredient they possess and the most concocted, so that when this sweet substance is drawn off, what remains is bound to be briny and ill-savoured.[a] When, however, the fetations are approaching their completion, then there is more surplus residue, because less of it is being used up, and it is sweeter, since the well-concocted residue is no longer being drawn off to the same extent : it is no longer being expended upon the moulding of the embryo, but upon the small growth which it is making,[b] as though the embryo had by now, being completed, reached a stationary point (since a fetation, too, has its point of completion.)[c] That is why it makes its way out, and changes over to another process of formation as now possessing all that belongs to it, and it no longer takes what does not belong to it[d] ; and that is the time when the milk becomes serviceable.

The milk collects in the upper part of the body, in the breasts, and this is accounted for by the original order of the body's construction. The part of the body above the diaphragm is the controlling part of the animal. (The part below is the place for the nourishment and the residue, in order that those animals which move about may have within them a

___

dent state of existence ; and even the wind has its γένεσις and φθίσις (778 a 2), where see note; and also *cf.* 737 b 9.

[d] This remark is obscure, and the sentence may be an interpolation. See the parallel passage, 777 a 22 ff.

αὐτοῖς ἔχοντα τὴν τῆς τροφῆς αὐτάρκειαν μετα-
βάλλῃ τοὺς τόπους. ἐντεῦθεν δὲ καὶ ἡ σπερματικὴ
10 περίττωσις ἀποκρίνεται διὰ τὴν εἰρημένην αἰτίαν
ἐν τοῖς κατ᾽ ἀρχὰς λόγοις. ἔστι δὲ τό τε τῶν
ἀρρένων περίττωμα καὶ τὰ καταμήνια τοῖς θήλεσιν
αἱματικῆς φύσεως. τούτου δ᾽ ἀρχὴ καὶ τῶν
φλεβῶν ἡ καρδία· αὕτη δ᾽ ἐν τοῖς μορίοις τούτοις.
15 διὸ πρῶτον ἐνταῦθα ἀναγκαῖον γίγνεσθαι τὴν
μεταβολὴν ἐπίδηλον τῆς τοιαύτης περιττώσεως.
διόπερ αἵ τε φωναὶ μεταβάλλουσι καὶ τῶν ἀρρένων
καὶ τῶν θηλειῶν, ὅταν ἄρχωνται σπέρμα φέρειν
(ἡ γὰρ ἀρχὴ τῆς φωνῆς ἐντεῦθεν· ἀλλοία δὲ γίνεται
ἀλλοίου γινομένου τοῦ κινοῦντος), καὶ τὰ περὶ τοὺς
μαστοὺς αἴρεται καὶ τοῖς ἄρρεσιν ἐπιδήλως, μᾶλλον
20 δὲ τοῖς θήλεσιν· διὰ γὰρ τὸ κάτω τὴν ἔκκρισιν
γίνεσθαι πολλὴν κενὸς ὁ τόπος γίνεται ὁ τῶν
μαστῶν αὐταῖς καὶ σομφός. ὁμοίως δὲ καὶ τοῖς
κάτω τοὺς μαστοὺς ἔχουσιν. γίνεται μὲν οὖν
ἐπίδηλος καὶ ἡ φωνὴ καὶ τὰ περὶ τοὺς μαστοὺς καὶ
ἐν τοῖς ἄλλοις ζῴοις τοῖς ἐμπείροις περὶ ἕκαστον
25 γένος, ἐπὶ δὲ τῶν ἀνθρώπων διαφέρει πλεῖστον.
αἴτιον δὲ τὸ πλείστην εἶναι τὴν περίττωσιν τοῖς
θήλεσι τούτοις τῶν θηλέων καὶ τοῖς ἄρρεσι τῶν
ἀρρένων ὡς κατὰ μέγεθος [ταῖς μὲν τὴν τῶν καταμηνίων, τοῖς δὲ τὴν τοῦ σπέρματος πρόεσιν].[1] ὅταν
οὖν μὴ λαμβάνῃ μὲν τὸ ἔμβρυον τὴν τοιαύτην

---

[1] glossema : om. Σ.

---

[a] See 738 b 12 ff., 747 a 20.　　　　[b] i.e., upper.

sufficient independent supply of nourishment and be able to go about from place to place.) It is from here, too, that the seminal residue is drawn : the reason is given in the earlier chapters of our discussion.[a] Both the residue in males and the menstrual fluid in females are of a bloodlike nature ; now the source of the blood and of the blood-vessels is the heart, which is situated in these [b] parts ; therefore of necessity it is here that the change which this sort of residue undergoes must be first of all apparent. For this reason the voice of both male and female undergoes a change when they begin to produce semen, because the source of the voice is there,[c] and the voice changes its quality when that which provides its movement does so ; and further, the parts around the breasts rise up plainly in males as well as in females, though more so in the latter, since, as there is a plentiful excretion of matter downwards in females, the region of the breasts becomes empty and spongy ; and similarly in the case of those animals whose breasts are down below. Of course, this change in the voice and in the region of the breasts makes itself evident in the other animals as well—to those who have experience of each particular kind ; but the change is greatest in human beings. The reason is that women produce more residue than any other female animal, and so do men than other male animals, in proportion to their size [this refers to the excretion of menstrual fluid and of semen respectively]. Thus, when the embryo no longer absorbs

---

[c] The heart, which is the ἀρχή of the organism, is also in particular the source of all physical sexual characteristics ; see 766 a 30 ff., and note on 763 b 27. *Cf.* 787 b 15 *et preced.* See also App. B § 31.

776 b

30 ἀπόκρισιν, κωλύῃ δὲ θύραζε βαδίζειν, ἀναγκαῖον
εἰς τοὺς κενοὺς τόπους ἀθροίζεσθαι τὸ[1] περίττωμα
πᾶν, ὅσοιπερ ἂν ὦσιν ἐπὶ τῶν αὐτῶν πόρων. ἔστι
δ' ἑκάστοις τοιοῦτος ὁ τῶν μαστῶν τόπος δι'
ἀμφοτέρας τὰς αἰτίας ἕνεκά τε τοῦ βελτίστου
γεγονὼς τοιοῦτος καὶ ἐξ ἀνάγκης· ἐνταῦθα δὲ
ἤδη συνίσταται καὶ γίνεται πεπεμμένη τροφὴ τοῖς
35 ζῴοις. τῆς δὲ πέψεως ἔστι μὲν λαβεῖν τὴν εἰρη-

777 a
μένην αἰτίαν, ἔστι δὲ τὴν ἐναντίαν· εὔλογον γὰρ
καὶ μεῖζον ὂν τὸ ἔμβρυον πλείω λαμβάνειν τροφήν,
ὥστε ἔλαττον περιγίνεσθαι περὶ τὸν χρόνον τοῦτον·
πέττεται δὲ θᾶττον τὸ ἔλαττον.

Ὅτι μὲν οὖν ἐστι τὸ γάλα τὴν αὐτὴν ἔχον φύσιν
5 τῇ ἀποκρίσει ἐξ ἧς γίνεται ἕκαστον, δῆλον, εἴρηται
δὲ καὶ πρότερον. ἡ γὰρ αὐτὴ ὕλη ἡ τρέφουσα καὶ
ἐξ ἧς συνιστᾷ τὴν γένεσιν ἡ φύσις. ἔστι δὲ τοῦτο
ἡ αἱματικὴ ὑγρότης τοῖς ἐναίμοις· τὸ γὰρ γάλα
πεπεμμένον αἷμά ἐστιν, ἀλλ' οὐ διεφθαρμένον.
Ἐμπεδοκλῆς δ' ἢ οὐκ ὀρθῶς ὑπελάμβανεν ἢ οὐκ
10 εὖ μετήνεγκε ποιήσας ὡς τὸ γάλα[2]

μηνὸς ἐν ὀγδοάτου δεκάτῃ πύον ἔπλετο λευκόν.

σαπρότης γὰρ καὶ πέψις ἐναντίον, τὸ δὲ πύον
σαπρότης τις ἐστίν, τὸ δὲ γάλα τῶν πεπεμμένων.
οὐ γίνονται δὲ οὔτε θηλαζομέναις αἱ καθάρσεις

---

[1] τούτοις τὸ Z : τοῦτο τὸ A.-W.
[2] [τὸ γάλα] Diels : τὸ αἷμα Kranz.

---

[a] *Cf.* Hippocrates, π. φύσιος παιδίου 21 (vii. 512 Littré) καὶ
ἐς τὰς μήτρας δὲ ὀλίγον ἔρχεται διὰ τῶν αὐτέων φλεβῶν·
τείνουσι γὰρ ἐς τοὺς μαζοὺς καὶ ἐς τὰς μήτρας φλέβια ταὐτά τε
καὶ παραπλήσια ἄλλα.

this residual secretion but at the same time prevents it from making its way out, the whole of the residue is bound to collect in the empty spaces which are situated on the same passages.[a] In each kind of animal the place around the breasts is just such an empty space, and it is so for both of the two possible reasons : it was formed such as it is (a) for the sake of the *best*, and (b) by *necessity*. And it is precisely here that the concocted nourishment for the young animals takes shape and is formed. As for its concoction : to explain that, either the reason stated [b] may be taken, or the opposite one, since it is just as reasonable to adopt the view that as the embryo is bigger it takes more nourishment, so that there is less nourishment left over at this particular time ; and a smaller amount takes less time to concoct.

It is clear that milk is possessed of the same nature as the secretion out of which each animal is formed (this has in fact been stated already) [c] : the material which supplies nourishment and the material out of which Nature forms and fashions the animal are one and the same.[d] And this material, in the case of blooded animals, is the bloodlike liquid, since milk is *concocted*, not *decomposed*, blood. As for Empedocles, either he was mistaken, or else his metaphor was a bad one, when he wrote [e] how the milk is formed

On the eighth moon's tenth day, a whitish pus.

No ; putrefaction and concoction are opposites, and pus is a putrefaction, whereas milk is to be classed as something concocted. In the natural course of

---

[b] *i.e.*, that the embryo requires less nourishment.
[c] At 739 b 26.   [d] *Cf.* 744 b 35.
[e] Diels, *Vorsokr.*[5] 31 B 68.

κατὰ φύσιν, οὔτε συλλαμβάνουσι θηλαζόμεναι· κἂν
15 συλλάβωσιν, ἀποσβέννυται τὸ γάλα διὰ τὸ τὴν
αὐτὴν εἶναι φύσιν τοῦ γάλακτος καὶ τῶν κατα-
μηνίων· ἡ δὲ φύσις οὐ δύναται πολυχοεῖν οὕτως
ὥστ᾽ ἐπαμφοτερίζειν, ἀλλ᾽ ἂν ἐπὶ θάτερα γένηται
ἡ ἀπόκρισις, ἀναγκαῖον ἐπὶ θάτερα ἐκλείπειν, ἐὰν
μὴ γίνηταί ⟨τι⟩[1] βίαιον καὶ παρὰ τὸ ὡς ἐπὶ τὸ πολύ.
20 τοῦτο δ᾽ ἤδη παρὰ φύσιν· ἐν γὰρ τοῖς μὴ ἀδυνάτοις
ἄλλως ἔχειν ἀλλ᾽ ἐνδεχομένοις τὸ κατὰ φύσιν ἐστὶ
τὸ ὡς ἐπὶ τὸ πολύ.

Καλῶς δὲ διώρισται τοῖς χρόνοις καὶ ἡ γένεσις
ἡ τῶν ζῴων· ὅταν γὰρ διὰ τὸ μέγεθος μηκέτι
ἱκανὴ ᾖ τῷ κυουμένῳ ἡ διὰ τοῦ ὀμφαλοῦ τροφή,
ἅμα[2] τὸ γάλα γίνεται χρήσιμον [πρὸς τὴν γινομένην
25 τροφήν],[3] καὶ οὐκ εἰσιούσης διὰ τοῦ ὀμφαλοῦ τροφῆς,[4] συμπίπτουσιν αἱ φλέβες περὶ ἃς ὁ καλούμενος
ὀμφαλός ἐστι χιτών, καὶ διὰ ταῦτα καὶ τότε
συμβαίνει θύραζε ἡ ἔξοδος.

IX Ἐπὶ κεφαλὴν δ᾽ ἡ γένεσίς ἐστι τοῖς ζῴοις πᾶσιν
ἡ κατὰ φύσιν διὰ τὸ τὰ ἄνω τοῦ ὀμφαλοῦ μείζω
30 ἔχειν ἢ τὰ κάτω. καθάπερ οὖν ἐν ζυγοῖς ἠρτημένα[5]
ἐξ αὐτοῦ ῥέπει ἐπὶ τὸ βάρος. ἔχει δὲ τὰ μείζω
πλεῖον βάρος.

X Οἱ δὲ χρόνοι τῆς κυήσεως ἑκάστῳ τῶν ζῴων
ὡρισμένοι τυγχάνουσιν ὡς μὲν ἐπὶ τὸ πολὺ κατὰ
τοὺς βίους· τῶν γὰρ χρονιωτέρων[6] καὶ τὰς γενέσεις
35 εὔλογον εἶναι χρονιωτέρας. οὐ μὴν τοῦτό γ᾽ ἐστὶν

───────────

[1] τι Peck.   [2] ἅμα Platt : ἀλλὰ vulg., secl. A.-W.
[3] seclusi ; om. Σ : πρὸς τὴν τοῦ γενομένου τροφήν coni. A.-W.
ἀλλὰ . . . γίγνηται (Ζ²) . . . γενησομένην . . . συμπίπτωσιν
coni. Btf.   (cum vv. 22-27 conferas 776 a 33 seqq.)
[4] εἴσεισι διὰ τοῦ ὀμφαλοῦ ἡ τροφή P.

events, no menstrual evacuations take place during the suckling period, nor do women conceive then; and if they do conceive, the milk dries up, because the nature of the milk is the same as that of the menstrual fluid, and Nature cannot produce a plentiful enough supply to provide both; so that if the secretion takes place in one direction it must fail in the other, unless some violence is done contrary to what is normal. And that *ipso facto* means something contrary to Nature, because in the case of things which admit and do not exclude the possibility of being other than they are, " normal " and " natural " are identical.

In the actual birth of the young animals we have another instance of good timing. When the nourishment that passes through the umbilical cord is no longer sufficient for the fetus, owing to its size, at that same time the milk is becoming serviceable, and when no nourishment is entering by way of the umbilical cord, then the blood-vessels to which the cord acts as a sheath collapse; and for these reasons and at that time the exit of the fetus takes place.

The natural manner of birth for all animals is head IX first, because they have a larger bulk above the umbilical cord than below it, so that they are suspended from it, as it might be in a balance, and the heavier side (*i.e.*, the larger parts) goes down. *Animals born head foremost.*

The period of gestation is of a definite length for X each of the animals, and normally the periods are proportionate to the animals' span of life; after all, we should expect those which have a longer life-span to take longer over their formation than others. *Length of gestation-period.*

---

⁵ hic in Z spatium xi vel xii litterarum.
⁶ χρονιωτέρων P : χρονίων vulg.

777 b

αἴτιον, ἀλλ' ὡς ἐπὶ τὸ πολὺ τοῦτο συμβέβηκεν· τὰ
γὰρ μείζω καὶ τελειότερα τῶν ἐναίμων ζῴων καὶ
ζῶσι πολὺν χρόνον, οὐ μέντοι τὰ μείζω πάντα
μακροβιώτερα. πάντων γὰρ ἄνθρωπος πλεῖστον[1] ζῇ
χρόνον, πλὴν ἐλέφαντος, ὅσων ἀξιόπιστον ἔχομεν
5 τὴν πεῖραν· ἔλαττον δ' ἐστὶ τὸ γένος τὸ τῶν
ἀνθρώπων ἢ τὸ τῶν λοφούρων καὶ πολλῶν ἄλλων.
αἴτιον δὲ τοῦ μὲν εἶναι μακρόβιον ὁτιοῦν ζῷον τὸ
κεκρᾶσθαι παραπλησίως πρὸς τὸν περιέχοντα ἀέρα,
καὶ δι' ἄλλα συμπτώματ' ἄττα φυσικά, περὶ ὧν
ὕστερον ἐροῦμεν, τῶν δὲ χρόνων τῶν περὶ τὴν
10 κύησιν τὸ μέγεθος τῶν γεννωμένων· οὐ γὰρ ῥᾴδιον ἐν ὀλίγῳ χρόνῳ λαμβάνειν τὴν τελείωσιν τὰς
μεγάλας συστάσεις οὔτε ζῴων οὔτε τῶν ἄλλων ὡς
εἰπεῖν οὐθενός. διόπερ ἵπποι καὶ τὰ συγγενῆ ζῷα
τούτοις ἐλάττω ζῶντα χρόνον κύει πλείω χρόνον·
τῶν μὲν γὰρ ἐνιαύσιος ὁ τόκος, τῶν δὲ δεκάμηνος
15 ὁ πλεῖστος. διὰ τὴν αὐτὴν δ' αἰτίαν πολυχρόνιος
καὶ ὁ τῶν ἐλεφάντων ἐστὶ τόκος· διετὴς γὰρ ἡ
κύησις διὰ τὴν ὑπερβολὴν τοῦ μεγέθους.

Εὐλόγως δὲ πάντων οἱ χρόνοι καὶ τῶν κυήσεων
καὶ[2] γενέσεων καὶ τῶν βίων μετρεῖσθαι βού-
λονται κατὰ φύσιν περιόδοις.[3] λέγω δὲ περίοδον

---

[1] πλεῖστον P : πλείω vulg.   [2] καὶ P*Z : καὶ τῶν vulg.
[3] ὅλαις add. P.

---

[a] This was apparently a popular term meaning " bushy-
tailed " ; see *H.A.* 491 a 1 where " the *lophouroi* as they are
called " are the horse, the ass, the mule, etc.   *Cf.* 755 b 19.

[b] *Cf.* 767 a 30 ff., and Hippocrates, π. ἀέρων ὑδάτων τόπων,
chh. 1-6 ; and for " blend," *idem*, π. διαίτης I. 32, and Introd.
§ 40.   *Cf.* 777 b 28, n.

[c] See *De long. et brev. vit.* 466 a 15 ff., *P.A.* 677 a 35 ff.

Still, this is not the *reason* for it ; only, this is what in fact normally occurs. The larger and more perfect of the blooded animals do certainly live a long time, but not all the larger ones are also longer-lived. Man is the longest-lived of them all except the elephant, so far as we have any reliable experience ; but human beings are smaller than the *lophouroi* [a] and many others. The reason why any animal is long-lived really is that its " blend " is about the same in comparison with the air which is around it,[b] and there are other contributory factors inherent in its nature, which will be mentioned later on.[c] The reason for the various times of gestation is the size of the creatures which are generated. It is not easy for any large structure, be it an animal or anything else, almost, to reach its perfection in a short time. Hence horses and kindred animals, though they live a shorter time than men, have a longer time of gestation : in horses birth occurs at the end of a year, in the others, generally, after ten months. And for the same reason it takes a long time in elephants, whose gestation lasts two years owing to their excessive size.

[d] In all cases, as we should expect, the times of gestation and formation [e] and of lifespan aim, according to nature,[f] at being measured by " periods." By a " period " I mean day and night and month and

[d] The following important paragraph is not fully intelligible without reference to Aristotle's theory of the universe and of movement. A collection of passages from other treatises relevant to this will be found in App. A and App. B § 11, which will provide the best commentary on the present passage.　　　　　　　　　　　　　[e] Or " birth."
[f] But Nature cannot always succeed in her aim ; see 778 a 5 below.

Periods of animals governed by cosmic periods.

ἡμέραν καὶ νύκτα καὶ μῆνα καὶ ἐνιαυτὸν καὶ τοὺς
20 χρόνους τοὺς μετρουμένους τούτοις, ἔτι δὲ τὰς τῆς
σελήνης περιόδους. εἰσὶ δὲ περίοδοι σελήνης
πανσέληνός τε καὶ φθίσις[1] καὶ τῶν μεταξὺ χρόνων
αἱ διχοτομίαι· κατὰ γὰρ ταύτας συμβάλλει πρὸς
τὸν ἥλιον· ὁ γὰρ μεὶς κοινὴ περίοδός ἐστιν ἀμφοτέρων. ἔστι δὲ ἡ σελήνη ἀρχὴ διὰ τὴν πρὸς τὸν
25 ἥλιον κοινωνίαν καὶ τὴν μετάληψιν τὴν τοῦ φωτός·
γίνεται γὰρ ὥσπερ ἄλλος ἥλιος ἐλάττων· διὸ
συμβάλλεται εἰς πάσας τὰς γενέσεις καὶ τελειώσεις.

---

[1] πανσέληνός τε καὶ φθίσις P : πανσέληνοί τε καὶ φθίσεις vulg.

---

[a] *i.e.*, full moon, new moon, first quarter and last quarter.
The meaning of συμβάλλει is obscure. The word occurs
twice in *Meteor.*, once (345 b 6) in an astronomical context,
and once (376 b 24) in connexion with the rainbow, but
neither passage helps to elucidate the present statement. It
must, however, have some reference to the fact that the
month is a " joint period " of moon and sun (see note below),
so the rendering I have given may be offered as at any
rate not inappropriate. The importance here attached to
the " bisections " of the times is found again in Theophr.
De signis 6, where it is said that times and seasons (*e.g.*, the
year, the month, the day) are delimited by their bisections
(αἱ διχοτομίαι διορίζουσι τὰς ὥρας), the bisections of the month
being the full moons, the eighth days and the fourth days
(τὸν μῆνα ἕκαστον . . . διχοτομοῦσι . . . αἵ τε πανσέληνοι καὶ αἱ
ὀγδόαι καὶ αἱ τετράδες, § 8) ; and changes of weather tend to
coincide with these divisions (§ 9).

[b] *Periodos* is really a circuit or cycle.

[c] This phrase, which he translates " the month being a
period common to both," is excised by Platt on the ground
that it gives no sense, and that " a period common to both
sun and moon would be one which contained both the solar
and lunar periods exactly." The phrase is, however, in
Scot ; and, as it can be satisfactorily explained in view of
the context, it must be retained. The explanation is this :
the month, taken in the sense of a lunation, *i.e.*, the period
from one new moon to another, or the time required by the

478

year and the times which are measured by these ; also the moon's " periods " which are : full moon and waning moon, and the bisections of the intervening times,[a] since these are the points at which it stands in a definite " aspect " with the sun, the month being a joint period [b] of both moon and sun.[c] The moon is a " principle " on account of its association with the sun and its participation in the sun's light, being as it were a second and lesser sun,[d] and therefore is a contributory factor in all processes of

moon to go through all its phases once, is, literally and properly speaking, not a private period of the *moon's*, but, as Aristotle says, a joint period of the *moon and sun*, since it is the moon's position relative to the sun which determines how much of the moon's disk is illuminated. If the moon were self-luminous, there would be no phases, and therefore there could be no " phase-period." This is made even more clear if we consider that the moon does in fact possess a " period " proper to itself, pertaining to the moon's own actual motion, and not to the mere illumination of its surface by another body, and it is a period which differs in length from the lunation or " phase-period "—a fact which was probably better known to Aristotle than to some moderns. This is the period known in astronomy as the " sidereal period," *i.e.*, the time taken by the moon to return again to its same apparent position among the stars—*not* to return into conjunction with the sun. The duration of this period is roughly 27 days 8 hours, as against an average of 29 days 13 hours for the " phase-period." Aristotle is therefore quite correct in stating that the " month," by which, as the context clearly shows, he means the " phase-period," is a joint period of the sun and the moon. (I should, perhaps, apologize to astronomers for the un-astronomical term " phase-period," which I have used instead of " synodic period " in order to emphasize the point that phases are an *incidental* phenomenon, and not an *essential* concomitant of a synodic period.)

[d] This statement reappears in Theophr. *De vent.* 17 ἡ σε λήνη . . . οἷον ἀσθενὴς ἥλιός ἐστι, and *cf. id. De signis temp.* 5, where the moon is described as " the sun of the night."

777 b

αἱ¹ γὰρ θερμότητες καὶ ψύξεις μέχρι συμμετρίας
τινὸς ποιοῦσι τὰς γενέσεις, μετὰ δὲ ταῦτα² τὰς
30 φθοράς· τούτων δ' ἔχουσι τὸ πέρας καὶ τῆς ἀρχῆς
καὶ τῆς τελευτῆς αἱ τούτων κινήσεις τῶν ἄστρων.
ὥσπερ γὰρ καὶ θάλατταν καὶ πᾶσαν ὁρῶμεν τὴν
τῶν ὑγρῶν φύσιν ἱσταμένην καὶ μεταβάλλουσαν
κατὰ τὴν τῶν πνευμάτων κίνησιν καὶ στάσιν, τὸν
δ' ἀέρα καὶ τὰ πνεύματα κατὰ τὴν τοῦ ἡλίου καὶ
35 τῆς σελήνης περίοδον, οὕτω καὶ τὰ ἐκ τούτων
778 a φυόμενα καὶ τὰ ἐν τούτοις ἀκολουθεῖν ἀναγκαῖον·
κατὰ λόγον γὰρ ἀκολουθεῖν καὶ τὰς τῶν ἀκυρο-
τέρων περιόδους ταῖς τῶν κυριωτέρων. βίος γάρ
τις καὶ πνεύματός ἐστι καὶ γένεσις καὶ φθίσις.
τῆς δὲ τῶν ἄστρων τούτων περιφορᾶς τάχ' ἂν
5 ἕτεραί τινες εἶεν ἀρχαί. βούλεται μὲν οὖν ἡ φύσις
τοῖς τούτων ἀριθμοῖς ἀριθμεῖν τὰς γενέσεις καὶ τὰς
τελευτάς, οὐκ ἀκριβοῖ δὲ διά τε τὴν τῆς ὕλης

---

¹ αἱ P : καὶ vulg.　　　　² ταύτας S.

ᵃ Cf. Phys. 246 b 4 τὰς μὲν γὰρ τοῦ σώματος, οἷον ὑγίειαν
καὶ εὐεξίαν, ἐν κράσει καὶ συμμετρίᾳ θερμῶν καὶ ψυχρῶν
τίθεμεν ἢ αὐτῶν πρὸς αὐτὰ τῶν ἐντὸς ἢ πρὸς τὸ περιέχον (cf.
777 b 7, and 767 a 30 ff.)· ὁμοίως δὲ . . . καὶ τὰς ἄλλας ἀρετὰς
καὶ κακίας. The language used in the context of this passage
is very similar to that of Eth. Nic. Bk. II (dealing with the
doctrine of " the mean "), where it is stated that the moral
ἀρεταί also are produced and preserved by τὰ σύμμετρα
(1104 a 18), whereas they are destroyed by excess and defect,
just as the corresponding physical ἀρεταί are.
ᵇ Cf. Meteor. 339 a 21 ἔστι δ' ἐξ ἀνάγκης συνεχής πως οὗτος
[i.e., ὁ περὶ τὴν γῆν κόσμος, the sublunary world] ταῖς ἄνω
φοραῖς, ὥστε πᾶσαν αὐτοῦ τὴν δύναμιν κυβερνᾶσθαι ἐκεῖθεν . . .
ὥστε τῶν συμβαινόντων περὶ αὐτὸν πῦρ μὲν καὶ γῆν καὶ τὰ συγγενῆ
τούτοις ὡς ἐν ὕλης εἴδει τῶν γιγνομένων αἴτια χρὴ νομίζειν, . . .
τὸ δ' οὕτως αἴτιον ὡς ὅθεν ἡ τῆς κινήσεως ἀρχὴ τὴν τῶν ἀεὶ
κινουμένων αἰτιατέον δύναμιν.

generation and perfecting. As we know, it is heat and cooling in their various manifestations which up to a certain due proportion[a] bring about the generation of things, and beyond that point their dissolution ; and the limits of these processes, both as regards their beginning and their end, are controlled by the movements of these heavenly bodies.[b] Just as we observe that the sea and whatever is of a fluid nature remains settled or is on the move according as the winds are at rest or in motion, while the behaviour of the air and the winds in turn depends upon the period of the sun and moon,[c] so too the things which grow out of them and are in them are bound to follow suit (as it is only reasonable that the periods of things of inferior standing should follow those which belong to things of higher standing) since even the wind has a sort of lifespan[d]—a generation and a decline. And as for the revolution of these heavenly bodies, there may very well be other principles which lie behind them.[e] Nature's aim, then, is to measure the generations and endings of things by the measures of these bodies, but she

[c] *Cf.* 738 a 20 : the times about new moon (αἱ τῶν μηνῶν σύνοδοι) are cold because of the failing of the moon, and for the same reason they are stormier than the middle points of the month ; a precisely similar statement, using exactly the same terminology that Aristotle uses, is found twice in Theophr. *De ventis* 17 and *De signis* 5: in the latter passage the cause given is that the moon's light " fails " (ἀπολείπει) from the fourth day of the waning moon until the fourth day of the new moon, and this apparently is the time covered by αἱ σύνοδοι τῶν μηνῶν. The way in which the sun determines the weather is discussed at *Meteor.* 359 b 26 ff.

[d] *Cf.* above, 776 b 1, and Plato, *Timaeus* 91 B, C, where the course of a disease is compared with the lifespan of a living organism.   [e] See, *e.g.*, *De caelo* I, II.

ἀοριστίαν καὶ διὰ τὸ γίνεσθαι πολλὰς ἀρχάς, αἳ
τὰς γενέσεις τὰς κατὰ φύσιν καὶ τὰς φθορὰς ἐμπο-
δίζουσαι πολλάκις αἴτιαι τῶν παρὰ φύσιν συμ-
πιπτόντων εἰσίν.

10   Περὶ μὲν οὖν τῆς ἔσωθεν τροφῆς τῶν ζῴων καὶ
τῆς θύραζε γενέσεως εἴρηται, καὶ χωρὶς περὶ
ἑκάστου καὶ κοινῇ περὶ πάντων.¹

¹ περὶ δὲ (τε Υ) τῶν διαφορῶν αἷς (ἃς Ζ, αἱ Υ) διαφέρουσι τὰ
μόρια τῶν ζῴων, καὶ μάλιστα τὸ τοιοῦτο (τοιοῦτον P) συμβαίνει
περὶ τοὺς ἀνθρώπους addunt PΥΖ : amplius ΥΖ ὅσα μὲν (μὲν
οὖν Ζ) ἔχουσι μόρια τὰ ζῷα πάντα καὶ τῶν ἐντὸς καὶ τῶν ἐκτός.
totum vertit Σ, et 778 a 10 initium facit libri insequentis.

cannot bring this about exactly on account of the indeterminateness of matter and the existence of a plurality of principles which impede the natural processes of generation and dissolution and so are often the causes of things occurring contrary to Nature.

Very well : we have now spoken of the nourishment of animals within the parent, and of their birth and exit into the outer world ; and we have dealt with each kind separately as well as generally with them all.[a]

[a] Some MSS. have an addition here, for which see opposite.

E

I Περὶ δὲ τῶν παθημάτων οἷς διαφέρουσι τὰ
μόρια τῶν ζῴων θεωρητέον νῦν. λέγω δὲ τὰ
τοιαῦτα παθήματα τῶν μορίων, οἷον γλαυκότητα
ὀμμάτων καὶ μελανίαν, καὶ φωνῆς ὀξύτητα καὶ
20 βαρύτητα, καὶ χρώματος [ἢ σώματος] καὶ τριχῶν
ἢ πτερῶν διαφοράς.[1] τυγχάνει δὲ τῶν τοιούτων
ἔνια μὲν ὅλοις[2] ὑπάρχοντα τοῖς γένεσιν, ἔνια[3] δ'
ὅπως ἔτυχεν, οἷον μάλιστ' ἐπὶ τῶν ἀνθρώπων
τοῦτο συμβέβηκεν. ἔτι δὲ κατὰ τὰς τῶν ἡλικιῶν[4]
μεταβολὰς τὰ μὲν πᾶσιν ὁμοίως ὑπάρχει τοῖς
25 ζῴοις, τὰ δ' ὑπεναντίως, ὥσπερ περί τε φωνὰς
καὶ περὶ τριχῶν χρόαν· τὰ μὲν γὰρ οὐ πολιοῦται
πρὸς τὸ γῆρας ἐπιδήλως, ὁ δ' ἄνθρωπος μάλιστα
τοῦτο πάσχει τῶν ἄλλων ζῴων. καὶ τὰ μὲν εὐθὺς
ἀκολουθεῖ γενομένοις, τὰ δὲ προϊούσης τῆς ἡλικίας
γίνεται δῆλα καὶ γηρασκόντων. περὶ δὲ[5] τούτων
30 καὶ τῶν τοιούτων πάντων οὐκέτι τὸν αὐτὸν τρόπον
δεῖ νομίζειν εἶναι τῆς αἰτίας. ὅσα γὰρ μὴ τῆς φύσεως [ἔργα][6] κοινῇ[7] μηδ' ἴδια τοῦ γένους ἑκάστου,

---

[1] ἢ σώματος secl. Bekker, χρώματος ἢ δέρματος coni. Platt;
fortasse scribendum χρώματος μεταβολὰς (alterationem coloris
Σ). mox ἢ καὶ πτερῶν SY ; διαφοράς P, διαφοράν vulg.
[2] ὀλίγοις P.   [3] ἔνια Peck (idem Richards) : ἐνίοις vulg.
[4] τῶν ἡλικιῶν PZ : τῆς ἡλικίας vulg. : ἡλικίας SY.
[5] δὴ P.   [6] om. Z ; secl. A.-W.   [7] κοινὰ Btf.

BOOK V

We must now study the "conditions" in respect
of which the parts of animals differ. I mean such
conditions of the parts as the following : blue and
dark colour of the eyes, high and deep *a* pitch of the
voice, and differences of colour and of hair or feathers.
Some of these conditions are found throughout certain classes of animals ; some occur irregularly, and
a striking instance of this is afforded by the human
species. Further, there are some conditions, accompanying the changes in the times of life, which occur
in all animals alike, but there are others which are
divergent in different animals, as, *e.g.*, those which
have to do with the voice and the colour of the hair :
thus, some animals do not go noticeably grey towards
old age, whereas man is affected by this condition
more than any other animal. Again, some of these
conditions come on immediately after birth, others
make themselves noticed as age advances, or in old
age. When we come to consider these conditions
and all others like them, we must not suppose that
the same sort of cause is operative as before, for
there are certain conditions which are not characteristics belonging to Nature in general, nor peculiarities
proper to this or that particular class of animal ; and
whatever the quality of such conditions may be, in

I Παθήματα
(secondary
sex-char-
acteristics).

---

a See 787 b 1, n.

485

τούτων οὐθὲν ἔνεκά του τοιοῦτον οὔτ' ἔστιν οὔτε
γίνεται. ὀφθαλμὸς μὲν γὰρ ἔνεκά του, γλαυκὸς
δ' οὐχ ἔνεκά του, πλὴν ἂν ἴδιον ᾖ τοῦ γένους τοῦτο
τὸ πάθος. οὔτε δ' ἐπ' ἐνίων πρὸς τὸν λόγον
35 συντείνει τὸν τῆς οὐσίας, ἀλλ' ὡς ἐξ ἀνάγκης

γιγνομένων εἰς τὴν ὕλην καὶ τὴν κινήσασαν ἀρχὴν
ἀνακτέον τὰς αἰτίας. ὥσπερ γὰρ ἐλέχθη κατ'
ἀρχὰς ἐν τοῖς πρώτοις λόγοις, οὐ διὰ τὸ γίγνεσθαι
ἕκαστον ποιόν τι, διὰ τοῦτο ποιόν τι ἐστίν, ὅσα
τεταγμένα καὶ ὡρισμένα ἔργα τῆς φύσεώς ἐστιν,
5 ἀλλὰ μᾶλλον διὰ τὸ εἶναι τοιαδὶ γίγνεται τοιαῦτα·
τῇ γὰρ οὐσίᾳ ἡ γένεσις ἀκολουθεῖ καὶ τῆς οὐσίας
ἕνεκά ἐστιν, ἀλλ' οὐχ αὕτη τῇ γενέσει. οἱ δ'
ἀρχαῖοι φυσιολόγοι τοὐναντίον ᾠήθησαν. τούτου
δ' αἴτιον ὅτι οὐχ ἑώρων πλείους οὔσας τὰς αἰτίας,
ἀλλὰ μόνον τὴν τῆς ὕλης καὶ τὴν τῆς κινήσεως,
10 καὶ ταύτας ἀδιορίστως, τῆς δὲ τοῦ λόγου καὶ τῆς
τοῦ τέλους ἀνεπισκέπτως εἶχον.

Ἔστι μὲν οὖν ἕκαστον ἕνεκά του, γίνεται δ' ἤδη

---

[a] *i.e.*, serves no *purpose*, is not on account of any Final
Cause.—In view of the discoveries of modern genetics,
Aristotle's clear-cut distinction may be somewhat misleading;
but it will always remain true that some characteristics are
more " trivial " than others. Whether the genes control
individual characters such as the possession of blue eyes
instead of brown, as well as *specific* characters such as the
possession of red feathers instead of black, and *phyletic*
characters such as the possession of a liver instead of a hepato-
pancreas—is still uncertain ; but it is likely that they do.

[b] The *logos* defines the thing's essence, see Introd. § 10;
and *cf.* below, 778 b 17 τοιόνδε ζῷον ὑπόκειται ὄν, and the
context.

[c] *i.e.*, the Material Cause and the Motive Cause. *Cf.*
Bk. II, *init.* and Introd. § 6.

[d] See *P.A.* I. 640 a 10 ff.

no instance is either its existence or its formation " for the sake of something." [a]   Thus, the existence and the formation of an eye is " for the sake of something," but its being blue is not—unless this condition is a peculiarity proper to the particular class of animal.   And further, in some cases this condition has nothing to do with the *logos* [b] of the animal's being ; instead of that, we are to assume that these things come to be *by necessity*, and so their causes must be referred back to the matter and to the source which initiated their movement. [c]   Remember what was said at the beginning, at the outset of our discussion. [d]   So far as the regular, definite products of Nature's hand are concerned, whatever a thing may be as regards its quality, the reason why each thing *is* of such or such a quality is not because it gets formed such while it develops ; the truth is that things get formed such because they *are* such, [e] for of course the process of formation takes its lead from the being, and is for the sake of that ; the being does not take its lead from the process. [f]   The old physiologers, however, thought the opposite, because they did not see that the causes were numerous ; they recognized only the Material Cause and the Motive Cause (and even these they did not clearly distinguish), whereas they paid no attention to the Formal Cause and the Final Cause.

Each thing, then, *is* " for the sake of something," [g]

---

[e] *Cf.* Dante, *Paradiso* xx. 78, quoted on p. 1.

[f] οὐσία here is no doubt, in the first place, the individual existing thing which the process is destined to produce (see 736 b 27, n., and 767 b 34 ff.) ; but we may also remember the use of οὐσία with reference to the essential nature of a thing, as in the phase λόγος τῆς οὐσίας, l. 35 above.

[g] *i.e.*, on account of some Final Cause.

διά τε ταύτην τὴν αἰτίαν καὶ διὰ τὰς λοιπὰς ὅσαπερ
ἐν τῷ λόγῳ ἐνυπάρχει τῷ ἑκάστου ἢ ἐστιν ἕνεκά
του ἢ οὗ ἕνεκα. τῶν δὲ μὴ τοιούτων, ὅσων ἐστὶ
γένεσις, ἤδη τούτων τὸ αἴτιον ἐν τῇ κινήσει δεῖ
15 καὶ τῇ γενέσει ζητεῖν, ὡς ἐν αὐτῇ τῇ συστάσει τὴν
διαφορὰν λαμβανόντων. ὀφθαλμὸν μὲν γὰρ ἐξ
ἀνάγκης ἕξει (τοιόνδε γὰρ ζῷον ὑπόκειται ὄν),
τοιόνδε δὲ ὀφθαλμὸν ἐξ ἀνάγκης μέν, οὐ τοιαύτης
δ᾽ ἀνάγκης, ἀλλ᾽ ἄλλον τρόπον, ὅτι τοιονδὶ ἢ
τοιονδὶ ποιεῖν πέφυκε καὶ πάσχειν.
20 Διωρισμένων δὲ τούτων λέγωμεν περὶ τῶν
ἐφεξῆς συμβαινόντων. πρῶτον μὲν οὖν ὅταν
γένωνται τὰ παιδία πάντων, μάλιστα τῶν ἀτελ⟨ῆ
τικτόντ⟩ων,[1] καθεύδειν εἴωθε, διὰ τὸ καὶ ἐν τῇ
μητρί, ὅταν λάβῃ πρῶτον αἴσθησιν, καθεύδοντα
διατελεῖν. ἔχει δ᾽ ἀπορίαν περὶ τῆς ἐξ ἀρχῆς
γενέσεως, πότερον ἐγρήγορσις ὑπάρχει τοῖς ζῴοις
25 πρότερον ἢ ὕπνος. διὰ γὰρ τὸ φαίνεσθαι προϊούσης
τῆς ἡλικίας ἐγειρόμενα μᾶλλον, εὔλογον τοὐναντίον
ἐν τῇ ἀρχῇ τῆς γενέσεως ὑπάρχειν, τὸν ὕπνον,
ἔτι δὲ διὰ τὸ τὴν μετάβασιν ἐκ τοῦ μὴ εἶναι εἰς

---

[1] corrupt. agnovit Platt : correxi (cf. 779 a 24) : ἀτελῶν
vulg. : et maxime filii qui pariuntur incompleti Σ.

---

[a] i.e., the Final Cause.

[b] τοιόνδε here = ὁρατικόν or ὀφθαλμὸν ἔχον; to use the terminology of a few lines above, τὸ ὁρατικὸν εἶναι ἐν τῷ λόγῳ
ἐνυπάρχει τῷ τοῦ ζῴου.

[c] And since the animal ὑπόκειται to be e.g. ὁρατικόν,
the sort of necessity which requires it to be ὁρατικόν is necessity ἐξ ὑποθέσεως (see Introd. § 7), the necessity which is
implied by the Final Cause. For ὑπόκειται see also 766 b 8.

while as regards their *process of formation*, all those characteristics which are contained in its *logos*, or are subservient to some end, or are an end in themselves —these come to be formed on account of this Cause [a] as well as the remaining Causes. Other characteristics, however, are formed during the process which do not fall under the headings just given, and the cause of them is to be looked for in the movement, *i.e.*, the process of formation—we must assume that they acquire their differences within the actual process of construction. Thus (to take an example) X will *of necessity* possess an eye (because that characteristic [b] is included in the essence of the animal as posited),[c] and it will—also *of necessity*— possess a particular sort of eye, but the latter is a different mode of necessity from the former,[d] and is derived from the fact that it is naturally constituted to act and to be acted upon in this or that way.[e]

Having settled these points we may proceed to those which immediately follow. First then : the habit of the young of all animals, especially those of animals which bring forth their young imperfect, once they have been born, is to sleep, because they are in fact continually asleep within the parent from the time that they first acquire sensation. There is, however, a puzzle concerning their original formation, which is this : which state exists first in animals, sleep or waking ? From the fact that, as we see, they become more awake the older they get, it seems reasonable to suppose that the opposite state, sleep, is the one that exists at the beginning of their formation—and also from the fact that the transition from

Sleep.

[d] *i.e.*, the necessity implied by the Motive and Material Causes. See Introd. § 7.   [e] *Cf.* App. B §§ 8 ff.

778 b

τὸ εἶναι διὰ τοῦ μεταξὺ γίνεσθαι· ὁ δ' ὕπνος εἶναι
30 δοκεῖ τὴν φύσιν τῶν τοιούτων, οἷον τοῦ ζῆν καὶ
τοῦ μὴ ζῆν μεθόριον, καὶ οὔτε μὴ εἶναι παντελῶς
ὁ καθεύδων οὔτ' εἶναι. τῷ γὰρ ἐγρηγορέναι τὸ
ζῆν μάλισθ' ὑπάρχει διὰ τὴν αἴσθησιν. εἰ δ' ἐστὶν
ἀναγκαῖον ἔχειν αἴσθησιν τὸ ζῷον, καὶ τότε πρῶτόν
ἐστι ζῷον ὅταν αἴσθησις γένηται πρῶτον, τὴν μὲν
35 ἐξ ἀρχῆς διάθεσιν οὐχ ὕπνον ἀλλ' ὅμοιον ὕπνῳ δεῖ
νομίζειν, οἷανπερ ἔχει καὶ τὸ τῶν φυτῶν γένος·

779 a κ αὶ γὰρ συμβέβηκε κατὰ τοῦτον τὸν χρόνον τὰ
ζῷα φυτοῦ βίον ζῆν. τοῖς δὲ φυτοῖς ὑπάρχειν ὕπνον
ἀδύνατον· οὐθεὶς γὰρ ὕπνος ἀνέγερτος, τὸ δὲ τῶν
φυτῶν πάθος τὸ ἀνάλογον τῷ ὕπνῳ ἀνέγερτον.
5 καθεύδειν μὲν οὖν τὰ ζῷα τὸν πλείω χρόνον ἀναγκαῖον διὰ τὸ τὴν αὔξησιν καὶ τὸ βάρος ἐπικεῖσθαι
τοῖς ἄνω τόποις (εἰρήκαμεν δὲ τὴν αἰτίαν τοῦ
καθεύδειν τοιαύτην οὖσαν ἐν ἑτέροις)· ἀλλ' ὅμως
ἐγειρόμενα φαίνεται καὶ ἐν τῇ μήτρᾳ (δῆλον δὲ
γίνεται τοῦτο ἐν ταῖς ἀνατομαῖς καὶ ἐν τοῖς ᾠο-
10 τοκοῦσιν),[1] εἶτ' εὐθὺς καθεύδουσι καὶ καταφέρονται
πάλιν. διόπερ καὶ ἐξελθόντα τὸν πολὺν διάγει
χρόνον καθεύδοντα.

Καὶ ἐγρηγορότα μὲν οὐ γελᾷ τὰ παιδία, καθεύδοντα δὲ καὶ δακρύει καὶ γελᾷ. συμβαίνουσι γὰρ
καὶ καθεύδουσιν αἰσθήσεις τοῖς ζῴοις, οὐ μόνον

[1] ᾠοτοκουμένοις Z.

---

a Cf. De somno et vig. 457 a 3 ff.   See also P.A. 686 b 2 ff.,
G.A. 741 b 28 ff.

b See P.A. 653 a 10 ff., De somno et vig. 455 b 28 ff.,
especially 456 b 17 ff.   Sleep is caused by the upper

not-being to being is effected through the inter-mediate state, and sleep would appear to be by its nature a state of this sort, being as it were a border-land between living and not living : a person who is asleep would appear to be neither completely non-existent nor completely existent : for of course it is to the waking state *par excellence* that life pertains, and that in virtue of sensation. On the other hand, assuming it is necessary that an animal should possess sensation, and that it is first an animal at the moment it has first acquired sensation, we ought to regard its original state not as being sleep but something re-sembling sleep—the sort of state that plants also are in ; indeed the fact is that at this stage animals are living the life of a plant. Sleep, however, cannot possibly pertain to plants, because there is no sleep from which there is not an awaking, and there is no awaking from the condition in plants which is ana-logous to sleep. Anyway, young animals must of necessity sleep for the greater part of the time be-cause the burden of their growth and the consequent weight is laid upon the upper regions of the body.[a] (We have explained elsewhere [b] that such is the cause of sleep.) All the same, animals are clearly found to wake even within the uterus, as is shown by dissections and by the case of the Ovipara ; after-wards they immediately drop off and fall asleep again. That is why after birth as well they spend most of their time asleep.

Infants do not laugh while they are awake, but they both laugh and weep while they are asleep, for of course sensations occur in animals during sleep as

regions of the body becoming weighed down by various hot substances which are carried up to them.

τὰ καλούμενα ἐνύπνια, ἀλλὰ καὶ παρὰ τὸ ἐνύπνιον,
15 καθάπερ τοῖς ἀνισταμένοις καθεύδουσι καὶ πολλὰ
πράττειν ἄνευ τοῦ ἐνυπνιάζειν. εἰσὶ γάρ τινες οἳ
καθεύδοντες ἀνίστανται καὶ πορεύονται βλέποντες
ὥσπερ οἱ¹ ἐγρηγορότες. τούτοις γὰρ γίνεται τῶν
συμβαινόντων αἴσθησις, οὐκ ἐγρηγορόσι μέν, οὐ
μέντοι ὡς ἐνύπνιον. τὰ δὲ παιδία ἐοίκασιν, ὥσπερ
20 ἀνεπιστήμονα τοῦ ἐγρηγορέναι, διὰ συνήθειαν ἐν
τῷ καθεύδειν αἰσθάνεσθαι καὶ ζῆν. προϊόντος δὲ
τοῦ χρόνου, καὶ τῆς αὐξήσεως εἰς τὰ² κάτω μετα-
βαινούσης, ἐγείρονταί τε μᾶλλον ἤδη, καὶ τὸν
πλείω χρόνον οὕτω διάγουσιν. μᾶλλον δὲ τῶν
ἄλλων ζῴων ἐν ὕπνῳ τὸ πρῶτον διατελοῦσιν·
25 ἀτελέστατα γὰρ γεννᾶται τῶν τετελεσμένων, καὶ
τὴν αὔξησιν ἔχοντα μάλιστα ἐπὶ τὸ ἄνω μέρος
τοῦ σώματος.

Γλαυκότερα δὲ τὰ ὄμματα τῶν παιδίων εὐθὺς
γενομένων³ ἐστὶ πάντων, ὕστερον δὲ μεταβάλλει
πρὸς τὴν ὑπάρχειν μέλλουσαν φύσιν αὐτοῖς· ἐπὶ
δὲ τῶν ἄλλων ζῴων οὐ συμβαίνει τοῦτ' ἐπιδήλως.
30 τούτου μὲν οὖν αἴτιον τὸ μονόχροα τὰ ὄμματα τῶν
ἄλλων εἶναι μᾶλλον, οἷον οἱ βόες μελανόφθαλμοι,
τὸ δὲ τῶν προβάτων ὑδαρὲς πάντων, τῶν δὲ
χαροπὸν ὅλον τὸ γένος ἢ γλαυκόν, ἔνια δ' αἰγωπά,
καθάπερ καὶ τὸ τῶν αἰγῶν αὐτὸ πλῆθος. τὰ δὲ
τῶν ἀνθρώπων ὄμματα πολύχροα συμβέβηκεν

---

¹ οἱ om. PZ¹.   ² τὰ PSY*Z ; τὸ Bekker per errorem.
³ γενομένων P : γεννωμένων vulg.

ᵃ Man produces his young " perfect " (see 770 a 33) ; the
492

well as in waking hours, and this includes not only what we call dreams but something more besides; thus persons who get up while they are asleep do quite a number of things without dreaming at all. There are those who get up while asleep and walk about and can see as well as anyone awake. The reason is that they are aware through their senses of what is going on, and though they are not awake, still this awareness is different from that of a dream. Infants, it would seem, have not yet acquired the art of being awake, if we may put it so, and thus both their sensations and their life go on during their sleep by force of habit. As time wears on, and the scene of their growth shifts its ground to the lower parts of the body, at this stage they wake up more and spend the greater part of their time awake. To begin with, however, infants spend more time asleep than any other animal, because they are born in a more imperfect condition than any other perfected [a] animal and have made their advance in growth chiefly in the upper part of the body.

The eyes of all infants are bluish immediately after birth; later on they change over to the colour which is going to be their natural colour for life. In the other animals this does not occur noticeably, and the reason is that their eyes exhibit more singleness of colour: thus, cattle have dark eyes; all sheep have pallid [b] eyes; another class of animal will all have greyish-blue, or blue, eyes; some have "goat's-eyes," [c] as indeed the majority of goats themselves have. The eyes of human beings, however, show

Colour of Eyes.

fissipede animals, such as the dog, produce them "imperfect," e.g., they are born blind.
[b] Lit., "watery." [c] i.e., yellow.

35 εἶναι· καὶ γὰρ γλαυκοὶ καὶ χαροποὶ καὶ μελαν-
όφθαλμοί τινές εἰσιν, οἱ δ' αἰγωποί. ὥστε τὰ μὲν
ἄλλα ὥσπερ¹ οὐδ' ἀλλήλων διαφέρουσιν, οὕτως οὐδ'
αὐτὰ αὑτῶν· οὐ γὰρ πέφυκε πλείους μιᾶς ἴσχειν
χρόας.² μάλιστα δὲ τῶν ἄλλων ζῴων ἵππος πολύχρων ἐστίν· καὶ γὰρ ἑτερόγλαυκοί τινες αὐτῶν
5 γίνονται. τοῦτο δὲ τῶν μὲν ἄλλων οὐθὲν πάσχει
ζῴων ἐπιδήλως, ἄνθρωποι δὲ γίνονταί τινες ἑτερόγλαυκοι.

Τοῦ μὲν οὖν τἆλλα ζῷα νέα ὄντα καὶ πρεσβύτερα
μηθὲν ἐπίδηλον μεταβάλλειν, ἐπὶ δὲ τῶν παιδίων
τοῦτο συμβαίνειν, ἱκανὴν οἰητέον αἰτίαν εἶναι καὶ
10 ταύτην, ὅτι τῶν μὲν μονόχρων τῶν δὲ πολύχρων
τὸ μόριόν ἐστιν· τοῦ δὲ γλαυκότερα καὶ μὴ χρόαν
ἄλλην ἴσχειν αἴτιον ὅτι ἀσθενέστερα τὰ μόρια τῶν
νέων, ἀσθένεια δέ τις ἡ γλαυκότης.

Δεῖ δὲ λαβεῖν καθόλου περὶ τῆς διαφορᾶς τῶν
ὀμμάτων, διὰ τίν' αἰτίαν τὰ μὲν γλαυκὰ τὰ δὲ
15 χαροπὰ τὰ δ' αἰγωπὰ τὰ δὲ μελανόμματ' ἐστίν.
τὸ μὲν οὖν ὑπολαμβάνειν τὰ μὲν γλαυκὰ πυρώδη,
καθάπερ Ἐμπεδοκλῆς φησί, τὰ δὲ μέλανα πλεῖον
ὕδατος ἔχειν ἢ πυρός, καὶ διὰ τοῦτο τὰ μὲν ἡμέρας
οὐκ ὀξὺ βλέπειν, τὰ γλαυκά, δι' ἔνδειαν ὕδατος,
θάτερα δὲ νύκτωρ δι' ἔνδειαν πυρός, οὐ λέγεται
20 καλῶς, εἴπερ μὴ πυρὸς τὴν ὄψιν θετέον ἀλλ' ὕδατος

---

¹ ὥστε τὰ μὲν ἄλλα διόπερ (ὣς supra π Ζ²) Ζ : διὸ ὥσπερ Υ :
διὸ τὰ μὲν ἄλλα ὥσπερ A.-W. : διὸ καὶ ὥσπερ vulg.
² sic Platt, Btf. : πλείω μιᾶς ἴσχειν χρόας (vel χρόας) ΡΖ :
πλείω μιᾶς ἴσχειν vulg.

ᵃ i.e., they do not vary at different times. Or it may mean,
" are not odd-coloured."
ᵇ Lit., " blue in one eye."

in practice a multiplicity of colour ; some are blue, some greyish-blue, some dark, some yellow. Hence in the case of the other animals, just as the individuals of any class do not differ from each other, so they do not differ from themselves,[a] the reason in both cases being that they are not naturally constituted to have more than one colour. The greatest multiplicity of colour, however, among the other animals is found in the horse ; indeed in some horses the two eyes are of odd colours.[b] No other animal is noticeably affected in this way, though some human beings are.

Well, then, for the fact that in the other animals, young or old, no noticeable change occurs, whereas in infants a change does occur, we must consider simply this to be a sufficient cause, viz., that in animals this part is single-coloured, in human beings multicoloured ; while for the fact that the young have bluish eyes and not some other colour, the reason is that their parts are weaker than those of adults, and blueness is a form of weakness.

We must now determine the general question of why eyes differ, and what is the cause why some are blue, some greyish-blue, some yellow, some dark. There is a theory, stated by Empedocles, that blue eyes are fiery in composition, while dark ones contain more water than fire, and that therefore blue eyes are not keen-sighted in the daytime owing to their deficiency of water, and the other ones suffer in the same way at night owing to their deficiency of fire. But if we ought in point of fact [c] to posit that the sight,[d] in all cases, consists of water, not of fire, then

---

[c] This is Aristotle's own theory ; see *De anima* 425 a 4 ; *De sensu* 438 a 5, 13 ff., b 5. For details, see App. B § 28.

[d] *i.e.*, the organ of sight, as often in this discussion.

779 b

πᾶσιν. ἔτι δ' ἐνδέχεται τῶν χρωμάτων τὴν αἰτίαν
ἀποδοῦναι καὶ κατ' ἄλλον τρόπον· ἀλλ' εἴπερ ἐστὶν
ὥσπερ ἐλέχθη πρότερον ἐν τοῖς περὶ τὰς αἰσθήσεις
καὶ τούτων ἔτι πρότερον ἐν τοῖς περὶ ψυχῆς διωρισμένοις, καὶ ὅτι ὕδατος, καὶ δι' ἣν αἰτίαν ὕδατος
25 ἀλλ' οὐκ ἀέρος ἢ πυρὸς τὸ αἰσθητήριον τοῦτ' ἐστί,
ταύτην αἰτίαν ὑποληπτέον εἶναι τῶν εἰρημένων. οἱ
μὲν γὰρ ἔχουσι τῶν ὀφθαλμῶν πλέον ὑγρόν, οἱ δ'
ἔλαττον τῆς συμμέτρου κινήσεως, οἱ δὲ σύμμετρον.
τὰ μὲν οὖν ἔχοντα τῶν ὀμμάτων πολὺ τὸ ὑγρὸν
μελανόμματά ἐστι διὰ τὸ μὴ εὐδίοπτ' εἶναι τὰ
30 πολλά, γλαυκὰ δὲ τὰ ὀλίγον, καθάπερ φαίνεται καὶ
ἐπὶ τῆς θαλάττης· τὸ μὲν γὰρ εὐδίοπτον αὐτῆς
γλαυκὸν φαίνεται, τὸ δ' ἧττον ὑδατῶδες, τὸ δὲ μὴ
διωρισμένον διὰ βάθος μέλαν καὶ κυανοειδές. τὰ
δὲ μεταξὺ τῶν ὀμμάτων τούτων τῷ μᾶλλον ἤδη
διαφέρει καὶ ἧττον.

35 Τὴν δ' αὐτὴν αἰτίαν οἰητέον καὶ τοῦ τὰ μὲν
γλαυκὰ μὴ εἶναι ὀξυωπὰ τῆς ἡμέρας, τὰ δὲ

780 a μελανόμματα τῆς νυκτός. τὰ μὲν γὰρ γλαυκὰ δι'
ὀλιγότητα τοῦ ὑγροῦ κινεῖται μᾶλλον ὑπὸ τοῦ
φωτὸς καὶ τῶν ὁρατῶν, ᾗ ὑγρὸν καὶ ᾗ διαφανές.
ἔστι δ' ἡ τούτου τοῦ μορίου κίνησις ὅρασις ᾗ
5 διαφανές, ἀλλ' οὐχ ᾗ ὑγρόν. τὰ δὲ μελανόμματα
διὰ πλῆθος τοῦ ὑγροῦ ἧττον κινεῖται. ἀσθενὲς

---

[a] See references already given in a previous note, a few
lines above.

[b] The meaning of this will be seen later, e.g., 780 a 1 ff.,
b 24. See also App. B §§ 26 ff.

sizes corresponds to these differences. At the same time, this does not hold good of all of them, because the reason for their producing few or many offspring is the size, great or small, of their bodies, not the fact that that particular kind of animal is cloven- or solid-hoofed or is fissipede. Here is a proof of this. The elephant is the biggest of the animals, but it is fissipede; the camel, which is the next biggest, is cloven-hoofed. And it is not only among the animals that walk but also among those that fly and swim that the big ones produce few offspring and the small ones produce many; and the cause is the same. Similarly, too, it is not the biggest plants that bear the most fruit.

We have stated why the nature of some animals is to produce many offspring, that of others to produce few, that of others to produce one only. So far as the puzzle which has now been mentioned is concerned, one might rather be justifiably surprised in the case of those animals which produce many offspring, in view of the fact that animals of this sort, as we see, often conceive as the result of one act of copulation. Now it may be that the semen of the male contributes to the material ⟨in the female⟩ by becoming part of the fetation and by mixing with the semen of the female; or it may be that it does not act in this way, but, as we hold, acts by concentrating and fashioning [a] the material in the female, i.e., the seminal residue, just as fig-juice [b] acts upon the fluid portion of the milk; but whichever of these views is right, what on earth is the cause why the semen does not turn out one single animal of a fair size, just as the fig-juice acts in our example, ⟨but that instead several off-

[a] Cf. 767 b 17, 772 b 32.   [b] See 737 a 15.

771 b

γίνεται;)[1] [οὐ κεχώρισται τῷ συνιστάναι[2] ποσόν
τι,[3] ἀλλ᾽ ὅσῳπερ ἂν εἰς πλεῖον ἔλθῃ καὶ πλείων,
τοσούτῳ τὸ πηγνύμενόν ἐστι μεῖζον.][4] τὸ μὲν οὖν
ἕλκειν φάναι τοὺς τόπους τῆς ὑστέρας τὸ σπέρμα,
καὶ διὰ τοῦτο πλείω γίνεσθαι, διὰ τὸ τῶν τόπων
πλῆθος καὶ τὰς κοτυληδόνας[5] οὐχ ἓν οὔσας,[6] οὐθέν
30 ἐστιν· ἐν ταὐτῷ γὰρ γίνονται τόπῳ τῆς ὑστέρας
δύο πολλάκις, ἐν δὲ τοῖς πολυτόκοις, ὅταν πληρωθῇ
τῶν ἐμβρύων, ἐφεξῆς κείμενα φαίνεται. τοῦτο δὲ
δῆλον ἐκ τῶν ἀνατομῶν ἐστιν. ἀλλ᾽ ὥσπερ καὶ
τελεουμένων τῶν ζῴων ἔστιν ἑκάστου τι μέγεθος
καὶ ἐπὶ τὸ μεῖζον καὶ ἐπὶ τὸ ἔλαττον, ὧν οὔτ᾽ ἂν
35 μεῖζον γένοιτο οὔτ᾽ ἔλαττον, ἀλλ᾽ ἐν τῷ μεταξὺ
διαστήματι τοῦ μεγέθους λαμβάνουσι πρὸς ἄλληλα

772 a

τὴν ὑπεροχὴν καὶ τὴν ἔλλειψιν, καὶ γίνεται μείζων
ὁ δ᾽ ἐλάττων ἄνθρωπος καὶ τῶν ἄλλων ζῴων
ὁτιοῦν, οὕτω καὶ ἐξ ἧς γίνεται ὕλης σπερματικῆς,
οὐκ ἔστιν ἀόριστος οὔτ᾽ ἐπὶ τὸ πλεῖον οὔτ᾽ ἐπὶ τὸ
ἔλαττον, ὥστ᾽ ἐξ ὁποσησοῦν γίνεσθαι τῷ πλήθει.
5 ὅσα οὖν τῶν ζῴων διὰ τὴν εἰρημένην αἰτίαν πλεῖον
προΐεται περίττωμα ἢ εἰς ἑνὸς ζῴου ἀρχήν, οὐκ

---

[1] talia desideraverat Platt, ego supplevi (sed generantur
in illa materia et superfluitate multi filii Σ).
[2] τῷ συνεστάναι PZ, om. Y.   [3] τι om. SZ[1].
[4] procul dubio secludenda (cf. 772 a 22) : om. Σ.
[5] λέγουσιν addunt YSZ[1].
[6] οὐκ ἐνούσας Z[1].   credo etiam διὰ τὸ . . . οὔσας secludenda.

[a] The words supplied are necessary to complete the argument, as Platt points out ; and they are in fact preserved in
Scot's version (see app. crit.). They were no doubt ousted
from the Greek text by the additional remarks about fig-

spring are formed out of that residue)?[a] [It is not divided up owing to its causing a certain quantity of milk to set, but the more the amount of milk into which it is put and the more fig-juice there is, so much the greater is the amount that gets curdled.] It is sometimes said that the regions of the uterus draw the semen, and on that account several offspring are formed, because these regions are several in number and because the cotyledons[b] are not a unity. This theory, however, has nothing in it, because often two embryos are formed in the same region of the uterus, and in the case of animals which produce many offspring, when the uterus is full of embryos, they can be seen lying in a row. This is clear from dissections. No; what happens is this. When animals are being perfected, there is a certain size for each, a limit of bigger and smaller; none will be formed either bigger or smaller than these sizes, but the excess or deficiency of size which they acquire as compared with one another lies within this interval between the two limits, and thus it is that one human being (or any other animal) is formed bigger and another smaller. In precisely the same way, the seminal material out of which ⟨the embryo⟩ is formed is not unlimited in either direction—the amount of it can be neither bigger nor smaller than certain limits; the embryo cannot be formed out of any casual amount of it. Thus, in the case of those animals which (on account of the cause stated) discharge more residue than is requisite for the principle

juice, which appear to have formed part of a marginal note (*cf.* below 772 a 22 ff., with which passage they are obviously connected).

[b] For the cotyledons, see above, Bk. II. 745 b end.

772 a

ἐνδέχεται ἐκ ταύτης ἓν γίνεσθαι πάσης, ἀλλὰ
τοσαῦτα ὅσα τοῖς μεγέθεσιν ὥρισται τοῖς ἱκνου-
μένοις. οὐδὲ τὸ τοῦ ἄρρενος σπέρμα ἢ ἡ δύναμις
ἡ ἐν τῷ σπέρματι οὐθὲν συστήσει[1] πλέον ἢ ἔλαττον
10 τοῦ πεφυκότος. ὁμοίως τ' εἰ πλέον σπέρμα ἀφίησι
τὸ ἄρρεν ἢ δυνάμεις·πλείους ἐν διαιρουμένῳ τῷ
σπέρματι, οὐθὲν ποιήσει μεῖζον τὸ πλεῖστον, ἀλλὰ
καὶ τοὐναντίον διαφθερεῖ καταξηραῖνον. οὐδὲ γὰρ
τὸ πῦρ θερμαίνει τὸ ὕδωρ μᾶλλον, ὅσῳπερ ἂν ᾖ
πλέον, ἀλλ' ἔστιν ὅρος τις[2] τῆς θερμότητος, ἧς ὑπ-
15 αρχούσης ἐὰν αὔξῃ τις τὸ πῦρ, θερμὸν μὲν οὐκέτι
γίνεται μᾶλλον, ἐξατμίζει δὲ μᾶλλον, καὶ τέλος
ἀφανίζεται καὶ γίνεται ξηρόν. ἐπεὶ δὲ φαίνεται
συμμετρίας δεῖσθαί τινος πρὸς ἄλληλα τό τε περίττωμα τὸ τοῦ θήλεος καὶ τὸ παρὰ τοῦ ἄρρενος,
ὅσα προΐεται σπέρμα τῶν ἀρρένων, τὰ πολυτόκα
20 τῶν ζῴων εὐθὺς ἀφίησι τὸ μὲν ἄρρεν δυνάμενον
πλείω συνιστάναι μεριζόμενον, τὸ δὲ θῆλυ τοσοῦτον
ὥστε πλείους γίνεσθαι συστάσεις. (τὸ δ' ἐπὶ τοῦ
γάλακτος παράδειγμα λεχθὲν οὐχ ὅμοιόν ἐστιν· ἡ
μὲν γὰρ τοῦ σπέρματος θερμότης οὐ μόνον συνίστησι ποσὸν ἀλλὰ καὶ ποιόν[3] τι, ἡ δ' ἐν τῷ ὀπῷ
25 καὶ τῇ πυετίᾳ τὸ ποσὸν μόνον.) τοῦ μὲν οὖν πολλὰ

---

[1] συστήσει PY : συνίστησι vulg.
[2] τις P : om. vulg.   [3] ποιὸν ἀλλὰ καὶ ποσόν P.

---

[a] See Bk. I, ch. 21 and Introd. §§ 26 ff.
[b] Cf. 729 a 18.   [c] Cf. 723 a 30, 767 a 16.
[d] See 737 a 15, 771 b 24.
[e] I suspect that this parenthesis may have come from a
marginal annotation ; cf. 771 b 24 above.

of a single animal, it is not possible that the entirety of this should be used to form one embryo ; on the contrary, as many are formed as is determined by the sizes proper to those animals. Nor again will the semen of the male or the *dynamis* [a] residing in the semen put into shape anything that is greater or less than the natural size. Similarly, if the male emits more semen, or more *dynameis* in the semen (in cases where the semen gets divided up), the greatest possible amount will not make anything bigger ⟨than the natural size⟩, but on the contrary will dry the material up [b] and destroy it. The parallel case of fire and water shows this. An increase in the amount of fire does not mean that the fire increases the heat of the water in the same ratio ; on the contrary, there is a limit to the heat, and when that has been reached, you may increase the amount of fire, but the water does not continue to get hotter ; instead it evaporates more, and finally disappears and dries up. Now since, as it seems, there must be some proportional relationship [c] between the residue of the female and that which comes from the male (this applies where the males emit semen), in the case of those animals which produce many offspring the male at the outset emits semen which is able, when divided up into portions, to give shape to a number of fetations, while the female contributes enough material so that a number of fetations can take shape out of it. (The parallel instance of milk, which was cited,[d] is not comparable, since, in the case of that which the semen's heat causes to take shape, not only quantity is involved but also quality, whereas in the case of the heat in the fig-juice and the rennet, quantity alone is involved.) [e] This, then, is the reason why in those

772 a

γίνεσθαι τὰ κυήματα καὶ μὴ συνεχὲς ἓν ἐκ πάντων ἐν τοῖς πολυτόκοις τοῦτ'[1] αἴτιον, ὅτι οὐκ ἐξ
ὁποσουοῦν γίνεται κύημα, ἀλλ' ἐάν τε ὀλίγον ᾖ,
οὐκ ἔσται, ἐάν τε πολὺ λίαν· ὥρισται γὰρ ἡ δύναμις
καὶ τοῦ πάσχοντος καὶ τῆς θερμότητος τῆς ποιού-
30 σης. ὁμοίως δὲ καὶ ἐν τοῖς μονοτόκοις καὶ μεγάλοις τῶν ζῴων οὐ πολλὰ γίγνεται ἐκ πολλοῦ
περιττώματος· καὶ γὰρ ἐν ἐκείνοις ἐκ ποσοῦ τινος
ποσόν τι τὸ ἐργαζόμενόν ἐστιν. οὐ προΐεται μὲν
οὖν πλείω τοιαύτην ὕλην διὰ τὴν προειρημένην
αἰτίαν· ἣν δὲ προΐεται, τοσαύτη κατὰ φύσιν ἐστὶν
35 ἐξ ἧς ἓν γίνεται κύημα μόνον. ἐὰν δέ ποτε πλεῖον
ἔλθῃ, διτοκεῖ τότε. διὸ καὶ δοκεῖ τερατώδη τὰ
τοιαῦτ' εἶναι μᾶλλον, ὅτι γίνεται παρὰ τὸ ὡς ἐπὶ
772 b τὸ πολὺ καὶ τὸ εἰωθός. ὁ δὲ ἄνθρωπος ἐπαμφοτε-
ρίζει πᾶσι τοῖς γένεσιν· καὶ γὰρ μονοτοκεῖ καὶ
πολυτοκεῖ ποτε[2] καὶ ὀλιγοτοκεῖ, μάλιστα δὲ μονοτόκον τὴν φύσιν ἐστί, διὰ μὲν τὴν ὑγρότητα τοῦ
σώματος καὶ θερμότητα πολύτοκον, [τοῦ γὰρ σπέρ-
5 ματος ἡ φύσις ὑγρὰ καὶ θερμή,][3] διὰ δὲ τὸ μέγεθος
ὀλιγότοκον καὶ μονότοκον. διὰ δὲ τοῦτο καὶ τοὺς
τῆς κυήσεως χρόνους μόνῳ τῶν ζῴων ἀνωμάλους
εἶναι συμβέβηκεν. τοῖς μὲν γὰρ ἄλλοις εἷς ἐστιν
ὁ χρόνος, τοῖς δ' ἀνθρώποις πλείους· καὶ γὰρ
ἑπτάμηνα καὶ δεκάμηνα γεννῶνται καὶ κατὰ τοὺς
10 μεταξὺ χρόνους· καὶ γὰρ τὰ ὀκτάμηνα ζῇ μέν,
ἧττον δέ. τὸ δ' αἴτιον ἐκ τῶν νῦν λεχθέντων

---

[1] τοῦτ' P : τοῦτ' αὐτὸ vulg.
[2] ποτε hic P, post ὀλιγοτοκεῖ vulg.
[3] τοῦ . . . θερμή secl. Platt.

---

[a] Cf. 776 a 22.

animals which produce many offspring the fetations
are many in number and a single continuous one does
not result instead of many—viz., a fetation is not
formed out of any casual quantity : if there is too
little or too much, none will be formed, because there
is a definite limit set both to the *dynamis* of the
material which is acted upon and to that of the heat
which acts upon it. Similarly also in the case of those
animals which are large and produce one offspring
only, a large amount of residue does not give rise to
a large number of offspring, for the same holds good :
here too, the amount of the material and of that
which works upon it are definite. So then they do
not emit a larger amount of such material, owing to
the cause already mentioned ; and the material which
they do emit is, in the natural course, just sufficient
in amount to provide for a single fetation only. If
ever more of it is supplied, then twins are produced.
And hence, also, such creatures seem rather to be
monstrosities, because their formation is contrary to
the general rule and to what is usual. Man, however, has a footing in all the classes, producing one
offspring, or on occasion, many, or few, though most
naturally and normally one is the number : the
production of many offspring is due to fluidity of the
body and to heat, [since the nature of semen is fluid
and hot ;] of few or of one, to the size of the body.
And to this it is due also that in man alone among
the animals is the period of gestation of variable
length [a] : other animals have a single period, but
with man there are several : children are born at
seven months and ten months and at intermediate
times, and indeed eight months' babies live, though
less often than the others. The reason may be

συνίδοι τις ἄν, εἴρηται δὲ περὶ αὐτῶν ἐν τοῖς προβλήμασιν.

Καὶ περὶ μὲν τούτων διωρίσθω τὸν τρόπον τοῦτον.

Τῶν δὲ πλεοναζόντων μορίων παρὰ φύσιν τὸ αὐτὸ αἴτιον καὶ τῆς διδυμοτοκίας. ἤδη γὰρ ἐν
15 τοῖς κυήμασι συμβαίνει τὸ αἴτιον, ἐὰν πλείων ὕλη συστῇ¹ ἢ κατὰ τὴν τοῦ μορίου φύσιν· τότε γὰρ συμβαίνει μὲν μόριον μεῖζον τῶν ἄλλων ἔχειν, οἷον δάκτυλον ἢ χεῖρα ἢ πόδα ἤ τι τῶν ἄλλων ἀκρω-τηρίων ἢ μελῶν, ἢ σχισθέντος τοῦ κυήματος πλείω γίνεσθαι, καθάπερ ἐν τοῖς ποταμοῖς αἱ δῖναι· καὶ
20 γὰρ ἐν τούτοις τὸ φερόμενον ὑγρὸν καὶ κίνησιν ἔχον ἂν ⟨τινι⟩² ἀντικρούσῃ, δύο ἐξ ἑνὸς γίνονται συστάσεις, ἔχουσαι τὴν αὐτὴν κίνησιν· τὸν αὐτὸν δὲ τρόπον καὶ ἐπὶ τῶν κυημάτων συμβαίνει. προσ-φύεται δὲ μάλιστα μὲν πλησίον ἀλλήλων, ἐνίοτε δὲ καὶ πόρρω διὰ τὴν γιγνομένην ἐν τῷ κυήματι κίνησιν, μάλιστα δὲ διὰ τὸ τὴν τῆς ὕλης ὑπεροχὴν
25 ὅθεν ἀφῃρέθη ἐκεῖ ἀποδιδόναι, τὸ δ᾽ εἶδος ἔχειν ὅθεν ἐπλεόνασεν.

Ὅσα δὲ συμβαίνει τοιαῦτα ὥστε δύο ἔχειν αἰδοῖα, [τὸ μὲν ἄρρενος τὸ δὲ θήλεος,]³ ἀεὶ μὲν τῶν πλεοναζόντων γίνεται τὸ μὲν κύριον τὸ δ᾽ ἄκυρον

---

¹ πλείων ὕλη συστῇ coni. Platt, cui consentit Σ *sustentatur multa materia* : πλείω ὕλην συστήσῃ vulg.
² ἄν τινι Peck : ἄν vulg.   ³ seclusit Platt.

---

[a] This cannot be traced.
[b] *Cf.* Bk. I, chh. 21, 22 ; 767 b 18, etc.
[c] *e.g.*, the excessive material is drawn from *X* ; it settles at *Y*, and therefore begins to take the form of *Y* during the process of development ; but as there are enough *Y* already,

perceived from what has just been said ; a discussion of these matters is also to be found in the *Problems*.[a]

This, then, may be taken as the way in which we deal with this subject.

With regard to the redundance of parts which (c) Reason occurs contrary to Nature, the cause of this is the stated. same as that of the production of twins, since the cause occurs right back in the fetations, whenever more material gets "set" than the nature of the part requires : the result then is that the embryo has some part larger than the others, *e.g.*, a finger or a hand or a foot, or some other extremity or limb ; or, if the fetation has been split up, several come to be formed—just as eddies are formed in rivers ; here too, if the fluid which is being carried along and is in movement meets with any resistance, two self-contained eddies are formed out of the original one, both of which have the same movement.[b] What happens in the case of the fetations is on the same lines. The normal part and the redundant one are usually attached quite close to one another, although sometimes they are farther away because of the movement which arises in the fetation, and above all because (*a*) the excess of material recurs again at the place from which it was originally drawn off, and (*b*) the form which it has is derived from the part where it developed as a redundancy.[c]

Some creatures develop in such a way that they have two generative organs [one male, the other female]. Always, when this redundancy happens, one of the two is operative and the other inoperative,

it goes back to where it came from, viz., X ; thus a Y is formed at X.

30 τῷ κατὰ τὴν τροφὴν ἀεὶ ἀμαυροῦσθαι ἅτε παρὰ
φύσιν ὄν, προσπέφυκε δ' ὥσπερ τὰ φύματα· καὶ
γὰρ ταῦτα λαμβάνει τροφήν, καίπερ ὄντα ὑστερο-
γενῆ καὶ παρὰ φύσιν. γίνεται δὲ κρατήσαντος μὲν
τοῦ δημιουργοῦντος ὅμοια δύο καὶ κρατηθέντος
ὅλως· ἂν δὲ τῇ μὲν κρατήσῃ τῇ δὲ κρατηθῇ, τὸ
μὲν θῆλυ τὸ δὲ ἄρρεν· οὐθὲν γὰρ διαφέρει τοῦτο
λέγειν ἐπὶ τῶν μορίων ἢ ἐπὶ τοῦ ὅλου, δι' ἣν
35 αἰτίαν γίνεται τὸ μὲν θῆλυ τὸ δ' ἄρρεν. ὅσα δ'
ἐλλείποντα γίνεται τῶν τοιούτων μορίων, οἷον
ἀκρωτηρίου τινὸς ἢ τῶν ἄλλων μελῶν, τὴν αὐτὴν

δεῖ νομίζειν αἰτίαν ἥνπερ καὶ ἐὰν ὅλον[1] τὸ γινόμενον ἀμβλωθῇ, ἀμβλώσεις δὲ γίνονται πολλαὶ τῶν
κυημάτων.

[Διαφέρουσι δ' αἱ μὲν παραφύσεις τῆς πολυτοκίας
τὸν εἰρημένον τρόπον, τὰ δὲ τέρατα τούτων τῷ
πολλὰ εἶναι αὐτῶν[2] σύμφυσιν.][3] ⟨γίνονται δὲ
καὶ μεταβολαί, ἐνίοις μὲν ἐπ' ἐλαττόνων καὶ
ἀτιμοτέρων μορίων,⟩ ἐνίοις[4] δὲ καὶ τοῦτον τὸν
5 τρόπον, ἐὰν ἐπὶ μειζόνων γένωνται καὶ κυριωτέρων μορίων, οἷον ἔνια ἔχει δύο σπλῆνας καὶ

---

[1] ἥνπερ καὶ ἐὰν ὅλον P, A.-W., Platt: ὅμοιον γάρ, κἂν ὅλως
vulg.   [2] τῷ τὰ πολλὰ αὐτῶν εἶναι P.

[3] διαφέρουσι . . . σύμφυσιν secl., nam argumento haud
consona. cetera ex Σ versione supplevi: *et forte erit alteratio*
(=μεταβολή, cf. 771 a 1) *in membris parvis vilibus et in
magnis principalibus* Σ.

[4] ἐνίοις Peck : ἔνια vulg.

---

[a] *Cf.* 767 b 17.   The semen of the male, the " movement "
of the male.   [b] *Cf.* 768 b 3.

[c] The words marked for excision are probably an annotation which has ousted the text (here tentatively restored from
Scot's Latin version) ; and it may be remarked that the

since the latter, being contrary to Nature, always gets stunted so far as nourishment is concerned; however, it is attached, just as growths (or tumours) are : these, like it, secure nourishment, although the date of their origin is later than that of the creature itself and they are contrary to Nature. The result of the fashioning agent [a] having gained the mastery, or having been completely mastered, is that two similar generative organs are formed ; if it to some extent gains the mastery and to some extent gets mastered, one is formed female and the other male,— for it comes to the same thing whether we apply this explanation of why one is formed female and another male to the case of the parts or to the animal as a whole.[b] And wherever a deficiency occurs in such parts as *e.g.* an extremity or some other limb, we must take it that the cause is the same as it is if the whole of the forming creature suffers abortion— and abortions of fetations frequently occur.

[c] [Redundant growths differ from the production of numerous offspring at a birth in the way which has been stated ; monstrosities differ from redundant growths in that most monstrosities are instances of embryos growing together.] ⟨Alterations, too, occur ; in some cases they affect the smaller and less important parts,⟩ whereas others are affected in a different way, *i.e.*, if the alteration occurs in the larger parts, which have more to do with the control of the organism—*e.g.*, some have two spleens, or several

meaning borne by τέρατα is at variance from that which it bears elsewhere in the discussion. The words may be an annotation intended for 773 a 13. The lines following (down to μεθισταμένης) seem to be a similar kind of summary, though more correct, and they too may be out of place or redundant.

443

πλείους νεφρούς. ἔτι δὲ μεταστάσεις τῶν μορίων
παρατρεπομένων[1] τῶν κινήσεών εἰσι καὶ τῆς ὕλης
μεθισταμένης. ἐν δ' εἶναι τὸ ζῷον τὸ τερατῶδες
ἢ πλείω συμπεφυκότα δεῖ νομίζειν κατὰ τὴν ἀρχήν,
10 οἷον εἰ τοιοῦτόν ἐστιν ἡ καρδία μόριον, τὸ μὲν
μίαν ἔχον καρδίαν ἓν ζῷον, τὰ δὲ πλεονάζοντα
μόρια παραφύσεις, τὰ δὲ πλείω ἔχοντα δύο μὲν
εἶναι, συμπεφυκέναι δὲ διὰ τὴν τῶν κυημάτων
σύναψιν.

Συμβαίνει δὲ πολλάκις καὶ τῶν οὐ δοκούντων
ἀναπήρων εἶναι ζῴων πολλοῖς ἤδη τετελειωμένοις
15 τοὺς μὲν συμπεφυκέναι τῶν πόρων τοὺς δὲ παρ-
εκτετράφθαι. καὶ γὰρ θήλεσί τισιν ἤδη τὸ στόμα
τῶν ὑστερῶν συμπεφυκὸς διετέλεσεν, ἤδη δ' ὥρας
οὔσης τῶν καταμηνίων καὶ πόνων ἐπιγιγνομένων[2]
ταῖς μὲν αὐτόματον ἐρράγη, ταῖς δ' ὑπὸ ἰατρῶν
διῃρέθη· τὰς δὲ διαφθαρῆναι συνέπεσεν ἢ βιαίας[3]
20 γενομένης τῆς ῥήξεως ἢ γενέσθαι μὴ δυναμένης.
καὶ τῶν παίδων ἐνίοις οὐ κατὰ τὸ αὐτὸ συνέπεσε
τὸ πέρας τοῦ αἰδοίου καὶ ὁ πόρος ᾗ διέρχεται τὸ
περίττωμα τὸ ἐκ τῆς κύστεως, ἀλλ' ὑποκάτωθεν·
διὸ καὶ καθήμενοι οὐροῦσι, τῶν δὲ ὄρχεων ἀνε-
σπασμένων ἄνω δοκοῦσι τοῖς ἄποθεν ἅμα θήλεος
25 ἔχειν αἰδοῖον καὶ ἄρρενος. ἤδη δὲ καὶ ὁ τῆς ξηρᾶς
τροφῆς[4] πόρος συμπεφυκὼς ἐπί τινων ζῴων γέγονε,

---

[1] παρεκτρεπομένων P.
[2] ἐπιγιγνομένων P : γιγνομένων vulg.
[3] βιαίας P : βίᾳ vulg.
[4] fort. ⟨περιττώματος⟩ supplendum: *exitum superfluitatis sicce* Σ.

kidneys. Also, there are instances of the parts changing their position, due to diversion of the "movements" and change of position of the material. Whether an animal which is a monstrosity is to be reckoned as one or as several grown together depends upon its "principle"; thus, assuming that the heart is a part answering to this description,[a] a creature which possesses one heart will be one animal, and any supernumerary parts will be merely redundant growths; those, however, which have more than one heart we shall reckon as being two, which have grown together owing to the conjoining of the fetations.

It often happens, even with many animals that do not appear to be deformed and have actually reached complete development, that some of their passages have grown together, and that others have been diverted. We know of instances of women in whom the *os uteri* was grown together and continued so until the time arrived for the menstrual discharge to begin and pain came on; in some, the passage burst open of its own accord, in others, it was separated by physicians; and in some cases, where the opening either was forcibly made or could not be made at all, the patients succumbed. There have been instances of boys in whom the termination of the penis has not coincided with the passage through which the residue from the bladder passes out, so that the passage came too low; and on this account they sit in order to pass water, and when the testes are drawn up they seem from a distance to have both male and female generative organs. There have also been instances in certain animals, sheep and others too, where the passage ⟨for the

(d) Other irregular formations.

----

[a] Viz., the "principle."

773 a

καὶ προβάτων καὶ ἄλλων, ἐπεὶ καὶ βοῦς ἐν Περίνθῳ
ἐγένετο ᾗ διὰ τῆς κύστεως λεπτὴ διηθουμένη τροφὴ
διεχώρει, καὶ ἀνατμηθέντος τοῦ ἀρχοῦ ταχὺ πάλιν
συνεφύετο, καὶ οὐκ ἐπεκράτουν διαιροῦντες.

30 Περὶ μὲν οὖν ὀλιγοτοκίας καὶ πολυτοκίας καὶ
περὶ φύσεως[1] τῶν πλεοναζόντων ἢ ἐλλειπόντων[2]
μορίων, ἔτι δὲ περὶ τῶν τερατωδῶν, εἴρηται.

V Τῶν δὲ ζῴων τὰ μὲν ὅλως οὐκ ἐπικυΐσκεται τὰ
δ' ἐπικυΐσκεται, καὶ τῶν ἐπικυϊσκομένων τὰ μὲν
35 δύναται τὰ κυήματα ἐκτρέφειν, τὰ δὲ ποτὲ μὲν
ποτὲ δ' οὔ. τοῦ δὲ μὴ ἐπικυΐσκεσθαι αἴτιον ὅτι

773 b
μονοτόκα ἐστίν. τά τε γὰρ μώνυχα οὐκ ἐπικυΐσκεται καὶ τὰ τούτων μείζονα· διὰ γὰρ τὸ
μέγεθος τὸ περίττωμα ἀναλίσκεται εἰς τὸ κύημα.
πᾶσι γὰρ ὑπάρχει μέγεθος τούτοις σώματος, τῶν
5 δὲ μεγάλων καὶ τὰ ἔμβρυα μεγάλα κατὰ λόγον
ἐστίν· διὸ καὶ τὸ τῶν ἐλεφάντων ἔμβρυον ἡλίκον
μόσχος ἐστίν. τὰ δὲ πολυτόκα ἐπικυΐσκεται διὰ
τὸ καὶ τῶν πλειόνων[3] τοῦ ἑνὸς εἶναι θατέρῳ θάτερον
ἐπικύημα. τούτων δ' ὅσα μὲν μέγεθος ἔχει, καθάπερ ἄνθρωπος, ἐὰν μὲν ἡ ἑτέρα ὀχεία τῆς ἑτέρας
10 γένηται πάρεγγυς, ἐκτρέφει τὸ ἐπικυηθέν· ἤδη γὰρ
ὦπται τὸ τοιοῦτον συμβεβηκός. αἴτιον δὲ τὸ
εἰρημένον· καὶ γὰρ ἐν τῇ μιᾷ συνουσίᾳ πλεῖον τὸ

---

[1] περὶ φύσεως scripsi : dispositionem Σ : παρὰ φύσιν Btf. :
περὶ παραφύσεως P : παραφύσεως vulg.

[2] ἢ ἐλλειπόντων om. Σ.

[3] καὶ τῶν πλειόνων P, A.-W. : τὰ πλείονα vulg. ; sed propter
parvitatem corporis filii Σ pro διὰ . . . ἐπικύημα.

---

[a] Superfetation is a very abnormal occurrence. It happens
when a later ovum is fertilized as a result of coitus during

residue) of the solid nourishment was grown together ; in fact, in Perinthus a cow was born which used to pass finely-sifted nourishment through the bladder. They cut its anus open, but it quickly grew together again, and they did not succeed in keeping it apart.

We have now discussed the production of few offspring and many, the nature of supernumerary or deficient parts, and also monstrosities.

In some animals superfetation [a] does not occur at all, in others it does ; and among the latter some are able to complete the nourishing of the fetations, others can sometimes do it and sometimes not. The reason why in some animals superfetation does not occur is that they produce one offspring only. Thus, it does not occur in solid-hoofed animals and in larger animals than these, because on account of their size the residue goes to the fetation and gets used up. All of these have large bodies, and large animals have large embryos, proportionate to their size ; that is why the embryo of an elephant is as big as a calf. Superfetation, however, does occur in animals which produce numerous offspring at a birth, because where there are more than a single offspring one is really a superfetation upon another. Of these animals, those that are large, such as man, complete the nourishing of the second fetation, if the second copulation has taken place not long after the first; such an occurrence has in fact been observed. The reason is as already stated : Even in a single act of intercourse the semen

V
Super-
fetation.

pregnancy. The young resulting from the second coitus are usually born at the same time as those resulting from the first coitus, but are smaller. See F. H. A. Marshall, *Physiology of Reproduction*[2] (1922), 154.

ἀπιόν ἐστι σπέρμα, ὃ μερισθὲν ποιεῖ πολυτοκεῖν,
ὧν ὑστερίζει θάτερον.[1] ὅταν δ' ἤδη τοῦ κυήματος
ηὐξημένου συμβῇ γίνεσθαι τὴν ὀχείαν, ἐπικυΐσκεται
15 μέν ποτε, ὀλιγάκις μέντοι διὰ τὸ τὴν ὑστέραν
συμμύειν ὡς τὰ πολλὰ μέχρι τῶν κυουμένων ταῖς
γυναιξίν. ἂν δὲ συμβῇ ποτέ (καὶ γὰρ τοῦτ' ἤδη
γέγονεν), οὐ δύναται τελειοῦν, ἀλλὰ κυήματ' ἐκπέμπει[2] παραπλήσια τοῖς καλουμένοις ἐκτρώμασιν.
ὥσπερ γὰρ ἐπὶ τῶν μονοτόκων διὰ τὸ μέγεθος εἰς
20 τὸ προϋπάρχον τὸ περίττωμα τρέπεται πᾶν, οὕτω
καὶ τούτοις, πλὴν ἐκείνοις μὲν εὐθύς, τούτοις δ'
ὅταν αὐξηθῇ τὸ ἔμβρυον· τότε γὰρ ἔχουσι παραπλησίως τοῖς μονοτόκοις. ὁμοίως δὲ διὰ τὸ τὸν
ἄνθρωπον φύσει πολυτόκον εἶναι, καὶ περιεῖναί τι
τῷ μεγέθει τῆς ὑστέρας καὶ τοῦ περιττώματος, μὴ
25 μέντοι τοσοῦτον ὥστε ἕτερον ἐκτρέφειν, μόνα τῶν
ζῴων ὀχείαν ἐπιδέχονται κυοῦντα γυνὴ καὶ ἵππος,
ἡ μὲν διὰ τὴν εἰρημένην αἰτίαν, ἡ δ' ἵππος διά
τε τὴν τῆς φύσεως στερρότητα[3] καὶ τὸ περιεῖναί
τι τῆς ὑστέρας μέγεθος, πλέον μὲν ἢ τῷ ἑνί, ἔλαττον δὲ ἢ ὥστε ἄλλο ἐπικυΐσκεσθαι τέλειον. ἔστι
30 δὲ φύσει ἀφροδισιαστικὸν διὰ τὸ ταὐτὸ πεπονθέναι
τοῖς στερροῖς· ἐκεῖνά τε γὰρ τοιαῦτ' ἐστὶ διὰ τὸ

---

[1] ὧν . . . θάτερον haud sanum videtur.
[2] ἐκπέμπει P : ἐκπίπτει vulg.
[3] στερεότητα PSY.

---

[a] Viz., those which produce more than one offspring.
[b] See 748 a 15 ff.

discharged is more than sufficient, and this when
divided up into portions causes the production of
numerous offspring, one of which is later than another.
When, however, the fetation is already advanced in
its growth before the copulation takes place, superfetation sometimes occurs, but infrequently, because
in women the uterus generally closes up during the
time of pregnancy. But if ever it does happen (as in
fact it has been known to do), the mother cannot
bring the second one to completion, but ejects fetations that are very similar to what are known as
abortions. The situation is comparable with that in
the one-offspring animals, in which, on account of
their size, all the residue is directed to the already
existing embryo. So too it happens in these animals,[a]
except that in the former it happens straight away,
whereas in these it happens when the embryo is
already advanced in growth, because then their condition is similar to that of the one-offspring animals.
Similarly, because man is by nature an animal which
produces numerous offspring, and because there is
something over and to spare as regards the size both
of the uterus and of the residue (though not enough
to bring the nourishing of a second embryo to completion), women and mares are the only animals which
admit copulation while they are with young. In
women it is due to the reason already stated ; in
mares it is due to the barrenness of their nature,[b]
and because the size of their uterus has something
over and to spare—there is more than enough room
for one, but not sufficient for a second fetation to be
brought to completion. Also, mares are by nature
prone to sexual intercourse because they are in the
same predicament as females which are barren—

μὴ γίνεσθαι κάθαρσιν (τοῦτο δ' ἐστὶν ὥσπερ τοῖς
ἄρρεσι τὸ ἀφροδισιάσαι) καὶ ἵπποι αἱ θήλειαι
ἥκιστα προΐενται κάθαρσιν. ἐν πᾶσι δὲ τοῖς ζῳοτοκοῦσι τὰ στερρὰ τῶν θηλέων ἀφροδισιαστικὰ
διὰ τὸ παραπλησίως ἔχειν τοῖς ἄρρεσιν, ὅταν
35 συνειλεγμένον μὲν ᾖ τὸ σπέρμα, μὴ ἀποκρινόμενον

δέ. τοῖς γὰρ θήλεσιν ἡ τῶν καταμηνίων κάθαρσις
σπέρματος ἔξοδός ἐστιν· ἔστι γὰρ τὰ καταμήνια
σπέρμα ἄπεπτον, ὥσπερ εἴρηται πρότερον. διὸ
καὶ τῶν γυναικῶν ὅσαι πρὸς τὴν ὁμιλίαν ἀκρατεῖς
τὴν τοιαύτην, ὅταν πολυτοκήσωσι, παύονται τῆς
5 πτοήσεως· ἐκκεκριμένη γὰρ ἡ σπερματικὴ περίττωσις οὐκέτι ποιεῖ τῆς ὁμιλίας ταύτης ἐπιθυμίαν.
ἐν δὲ τοῖς ὄρνισιν αἱ θήλειαι τῶν ἀρρένων ἧττόν
εἰσιν ἀφροδισιαστικαὶ διὰ τὸ πρὸς τῷ ὑποζώματι
τὰς ὑστέρας ἔχειν, τὰ δ' ἄρρενα τοὐναντίον· ἀν-
εσπασμένους γὰρ ἔχει τοὺς ὄρχεις ἐντός, ὥστ' ἂν
10 ᾖ τι[1] γένος τῶν τοιούτων [ὀρνίθων][2] φύσει σπερματικόν, ἀεὶ δεῖσθαι τῆς ὁμιλίας ταύτης. τοῖς μὲν
οὖν θήλεσι τὸ κάτω καταβαίνειν τὰς ὑστέρας, τοῖς
δ' ἄρρεσι τὸ ἀνασπᾶσθαι τοὺς ὄρχεις συμβαίνει
πρὸ ὁδοῦ πρὸς τὴν ὀχείαν.

Δι' ἣν μὲν οὖν αἰτίαν τὰ μὲν οὐκ ἐπικυΐσκεται
15 παντελῶς, τὰ δ' ἐπικυΐσκεται μέν, τὰ δὲ κυήματα
ἐκτρέφει ὁτὲ μὲν ὁτὲ δ' οὔ, καὶ διὰ τίν' αἰτίαν
τὰ μὲν ἀφροδισιαστικὰ τὰ δ' οὐκ ἀφροδισιαστικὰ
τῶν τοιούτων ἐστίν, εἴρηται.

[1] τι Platt : τὸ vulg.
[2] seclusi ; ὀρνίθων τούτων P. fortasse scribendum ὥστε διὰ
τὸ τοῦτο τὸ γένος εἶναι φύσει σπερματικὸν κτλ. (et indigent
multo coitu propter multitudinem spermatis naturaliter Σ.)

since this also is a condition due to there being no evacuation (which corresponds to the emission of semen in the male), and mares discharge extremely little evacuation. Further, in all the Vivipara those females which are barren are prone to sexual intercourse, because they are in a similar condition to males when their semen is ready, collected together,[a] but is not being emitted, the evacuation of the menstrual fluid in females being the emission of semen, since, as has been stated earlier, the menstrual fluid is semen that is unconcocted. Hence, too, those women who are incontinent in the matter of sexual intercourse, cease from their passionate excitement when they have borne several children, because once the seminal residue has been expelled from the body it no longer produces the desire for this intercourse. Among birds the females are less sexually excitable than the males because their uterus is close up by the diaphragm, whereas the males, on the contrary, have their testes drawn up internally,[b] so that if any class of such creatures tends naturally to abound in semen, they are always wanting to have sexual intercourse. Thus in females it is the descent of the uterus which encourages copulation, whereas in males it is the drawing up of the testicles.

We have now stated the cause on account of which superfetation does not occur at all in some animals, why it does occur in others, and why these can sometimes bring the nourishing of the fetation to completion, sometimes not ; and what is the cause why of such animals some are prone to sexual intercourse and others not.

[a] *Cf.* 717 b 25, 718 a 6 ff.   [b] See 717 b 10 ff.

Ἔνια δὲ τῶν ἐπικυϊσκομένων καὶ πολὺν χρόνον
διαλειπούσης τῆς ὀχείας δύναται τὰ κυήματα
ἐκτρέφειν, ὅσων σπερματικόν τε τὸ γένος ἐστὶ καὶ
20 μὴ τὸ σῶμα μέγεθος ἔχει καὶ τῶν πολυτόκων
ἐστίν· διὰ μὲν γὰρ τὸ πολυτοκεῖν εὐρυχωρίαν ἔχει
τῆς ὑστέρας, διὰ δὲ τὸ σπερματικὸν εἶναι πολὺ
προΐεται περίττωμα τῆς καθάρσεως· διὰ δὲ τὸ μὴ
τὸ σῶμα μέγεθος ἔχειν, ἀλλὰ πλείονι λόγῳ τὴν
κάθαρσιν ὑπερβάλλει τῆς εἰς τὸ κύημα τροφῆς,
25 δύναταί τε συνιστάναι[1] ζῷα καὶ ὕστερον καὶ ταῦτ'
ἐκτρέφειν. ἔτι δ' αἱ ὑστέραι τῶν τοιούτων οὐ
συμμεμύκασι διὰ τὸ περιεῖναι περίττωμα τῆς
καθάρσεως. τοῦτο δὲ καὶ ἐπὶ γυναικῶν ἤδη
συμβέβηκεν· γίνεται γάρ τισι κυούσαις κάθαρσις
καὶ διὰ τέλους. ἀλλὰ ταύταις μὲν παρὰ φύσιν
30 (διὸ βλάπτει τὸ κύημα), τοῖς δὲ τοιούτοις τῶν
ζῴων κατὰ φύσιν· οὕτω γὰρ τὸ σῶμα συνέστηκεν
ἐξ ἀρχῆς, οἷον τὸ τῶν δασυπόδων· τοῦτο γὰρ ἐπικυΐσκεται τὸ ζῷον· οὔτε γὰρ τῶν μεγάλων ἐστὶ
πολυτόκον τε (πολυσχιδὲς γάρ, τὰ δὲ πολυσχιδῆ
πολυτόκα) καὶ σπερματικόν. δηλοῖ δ' ἡ δασύτης·
35 ὑπερβάλλει γὰρ τοῦ τριχώματος τὸ πλῆθος· καὶ
γὰρ ὑπὸ τοὺς πόδας καὶ ἐντὸς τῶν γνάθων τοῦτ'
ἔχει τρίχας μόνον τῶν ζῴων. ἡ δὲ δασύτης ση-
774 b   μεῖον πλήθους περιττώματός ἐστι, διὸ καὶ τῶν

---

[1] συνιστάναι A.-W. : συνίστασθαι vulg.

---

* I use (a), (b), and (c) to mark respectively the same
characteristic all through this passage for clarity of reference.

[a] Lit., "is seminal"; i.e., the males abound in semen and
the females in menstrual fluid (which is unconcocted semen).

[b] i.e., the embryos produced by way of superfetation.

Some of those animals in which superfetation occurs are able to bring to completion the nourishing of their fetations even when there is a long interval between the copulations ; these are animals which (a)* belong to some kind which is abundant in semen,[a] (b) are not large in bodily size, and (c) are among those which produce numerous offspring ; the reason being that (c)* because they produce numerous offspring their uterus is roomy, (a) because they are abundant in semen they discharge a great deal of residue by way of evacuation, (b) because they are not large in bodily size, but the evacuation exceeds by a larger measure the nourishment which goes to the fetation, they are able to cause young animals to take shape at the later stage too [b] and to bring their nourishing to completion. Also, in such animals the uterus does not close up, because there is a surplus amount of residue by way of evacuation. This has occurred to our knowledge in the case of women : in some women evacuation continues throughout the time of pregnancy. In them, however, it is contrary to nature (that is why it injures the fetation) ; but in the animals we are discussing it is natural, because that is the way in which their body took shape from the beginning. The hare is an example of this. This is an animal in which superfetation occurs, for (b)* it is not one of the large animals, (c) it produces numerous offspring (since it is fissipede, and fissipede animals produce numerous offspring), and (a) it is abundant in semen. This is shown by its hairiness. It has an excessive amount of hair ; indeed, it has hair under the feet and inside the jaws, and is the only animal which does so. This hairiness is a sign that it has a large amount of residue ; and for this

ἀνθρώπων οἱ δασεῖς ἀφροδισιαστικοὶ καὶ πολύ-
σπερμοι μᾶλλόν εἰσι τῶν λείων. ὁ μὲν οὖν δασύπους
τὰ μὲν τῶν κυημάτων ἀτελῆ πολλάκις ἔχει, τὰ
δὲ προΐεται τετελειωμένα τῶν τέκνων.

VI 5 Τῶν δὲ ζῳοτόκων τὰ μὲν ἀτελῆ προΐεται ζῷα
τὰ δὲ τετελειωμένα, τὰ μὲν μώνυχα τετελειωμένα
καὶ τὰ διχηλά, τῶν δὲ πολυσχιδῶν ἀτελῆ τὰ[1] πολλά.
τούτου δ' αἴτιον ὅτι τὰ μὲν μώνυχα μονοτόκα ἐστί,
τὰ δὲ διχηλὰ ἢ μονοτόκα ἢ διτόκα ὡς ἐπὶ τὸ πολύ,
10 ῥᾴδιον δὲ τὰ ὀλίγα ἐκτρέφειν. τῶν δὲ πολυσχιδῶν
ὅσα ἀτελῆ τίκτει, πάντα πολυτόκα· διὸ νέα μὲν
ὄντα δύναται τὰ κυήματα τρέφειν,[2] ὅταν δ' αὐξηθῇ
καὶ λάβῃ μέγεθος οὐ δυναμένου τοῦ σώματος
ἐκτρέφειν, προΐεται καθάπερ τὰ σκωληκοτόκα τῶν
ζῴων. καὶ γὰρ τούτων τὰ μὲν ἀδιάρθρωτα σχεδὸν
15 γεννᾷ, καθάπερ ἀλώπηξ ἄρκτος λέων, παραπλησίως δ' ἔνια καὶ τῶν ἄλλων· τυφλὰ δὲ πάντα
σχεδόν, οἷον ταῦτά τε καὶ ἔτι κύων λύκος θώς.
μόνον δὲ πολυτόκον ὂν ἡ ὗς τελειοτοκεῖ, καὶ
ἐπαλλάττει τοῦτο μόνον· πολυτοκεῖ μὲν γὰρ ὡς
τὰ πολυσχιδῆ,[3] διχηλὸν δ' ἐστὶ καὶ μώνυχον· εἰσὶ
20 γάρ που μώνυχες ὗες. πολυτοκεῖ μὲν οὖν διὰ τὸ

---

[1] τὰ P : om. vulg.
[2] τρέφειν PS : ἐκτρέφειν vulg.
[3] ὡς πολυσχιδῆ Z : ὡς πολυσχιδές PY.

---

[a] But see the proviso at 771 b 5 ff.
[b] i.e., in an imperfect condition.
[c] See H.A. 499 b 12. The solid-hoofed is the more unusual variety.

same reason, too, men that are hairy are more prone to sexual intercourse and have more semen than men that are smooth. As for the hare, often some of its fetations are imperfect; others of its offspring, however, it brings to birth in a perfected state.

Among the Vivipara, some bring their young to birth in a perfect, some in an imperfect, state. To the former class belong the solid-hoofed and the cloven-hoofed animals, to the latter most of the fissipede animals. The reason for this is that the solid-hoofed animals produce one at a birth, the cloven-hoofed animals produce either one or two, in general,[a] and it is an easy matter to bring the nourishing of a few to completion. Those fissipede animals which produce their offspring in an imperfect state, all produce numerous offspring, and on that account while the fetations are quite young they are able to nourish them, but once they have advanced in growth and have attained some size their bodies are unable to bring the nourishing of them to completion, and so discharge them just as the larva-producing animals do,[b] for indeed their young, like the larvae, are practically unarticulated when born, e.g., those of the fox, the bear, the lion, and similarly with some of the others ; moreover, practically all of them are blind, e.g., the ones just mentioned, and in addition those of the dog, the wolf, and the jackal. The only animal which produces numerous offspring that are perfectly formed is the pig ; thus it is the only one which has a footing in both classes : (a) it produces numerous offspring, as the fissipede animals do, but (b) it is a species which is cloven-hoofed and solid-hoofed—for solid-hoofed pigs exist, as we know.[c] It produces numerous offspring because the nourishment available for

455

τὴν εἰς τὸ μέγεθος τροφὴν εἰς τὴν σπερματικὴν
ἀποκρίνεσθαι περίττωσιν· τοῦτο γὰρ ὡς μώνυχον
ὄν¹ οὐκ ἔχει μέγεθος. ἅμα δὲ καὶ μᾶλλον, ὥσπερ
ἀμφισβητοῦν τῇ φύσει τῇ τῶν μωνύχων, διχηλόν
ἐστιν. διὰ μὲν οὖν τοῦτο καὶ μονοτοκεῖ ποτε² καὶ
25 διτοκεῖ καὶ πολυτοκεῖ τὰ πλεῖστα, ἐκτρέφει δ' εἰς
τέλος διὰ τὴν τοῦ σώματος εὐβοσίαν· ἔχει γὰρ ὡς
πίειρα γῇ φυτοῖς ἱκανὴν καὶ δαψιλῆ τροφήν.

Τίκτουσι δ' ἀτελῆ καὶ τυφλὰ καὶ τῶν ὀρνίθων
τινές, ὅσοι πολυτοκοῦσιν αὐτῶν μὴ σωμάτων
ἔχοντες μέγεθος, οἷον κορώνη κίττα στρουθοὶ
χελιδόνες, καὶ τῶν ὀλιγοτοκούντων ὅσα μὴ δαψιλῆ
30 τροφὴν συνεκτίκτει τοῖς τέκνοις, οἷον φάττα καὶ
τρυγὼν καὶ περιστερά. καὶ διὰ τοῦτο τῶν χελι-
δόνων ἐάν τις ἔτι νέων ὄντων ἐκκεντήσῃ τὰ ὄμματα,
πάλιν ὑγιάζονται· γινομένων γὰρ ἀλλ' οὐ γεγενημένων φθείρονται,³ διόπερ φύονται καὶ βλαστά-
35 νουσιν ἐξ ἀρχῆς. ὅλως δὲ προτερεῖ μὲν τῆς τε-
λειογονίας διὰ τὴν ἀδυναμίαν τοῦ ἐκτρέφειν, ἀτελῆ

¹ ὄν P : om. vulg.   ² ποτε P : om. vulg.
³ φθείρονται Y : φθείρουσι P : φθείρεται vulg.

---

^a The distinction which Aristotle makes here corresponds
to the distinction now made between nidicolous birds (those
here described) and nidifugous birds. The former are born
blind, the latter can see at birth.

^b Or, magpie.   ^c See table of birds, p. 368.

^d i.e., not enough yolk.

^e The origin of this story is not clear. It cannot be true
if " put out " means " removed," but lesser degrees of injury
might be followed by repair and recovery of function. A
somewhat similar phenomenon is the well-known " Wolffian
regeneration " in amphibia, where after removal of the lens
of the eye a new lens regenerates from the margin of the
iris, i.e., from a place other than that of its normal origin,

increase of size is secreted to yield seminal residue—since, for a solid-hoofed animal, the pig is not large in size ; at the same time and more commonly, it is cloven-hoofed, as though it were at odds with the nature of the solid-hoofed animals. On account of this, then, it not only produces sometimes one off-spring, and two, but also and for the most part it produces numerous offspring, and it brings their nourishing to completion because of its fine physical condition : it is like a rich soil which can provide plants with sufficient and indeed abundant nourishment.

The offspring of some of the birds also are hatched in an imperfect state, and blind[a] ; viz., of those which lay numerous eggs although they themselves are small in physique—e.g., the crow, the jay,[b] sparrows, and swallows[c] ; and of those birds which lay few eggs and yet do not provide in the egg abundant nourishment[d] for the chick—e.g., the ring-dove, the turtle-dove, and the pigeon. And on this account, if the eyes of a swallow are deliberately put out while the bird is still young, they recover, because the injury is inflicted during the process of their formation and not after its completion ; that is why they grow and spring up afresh.[e] In general, then, the reason why offspring are born early before their formation is perfected, is because of inability to bring their nourishing to completion ; and the reason why they are born in an imperfect state is because they

viz., the young skin. This may happen many times in succession if the experiment is repeated. The connexion between regeneration and embryonic growth is well grasped by Aristotle, but there are of course some animals, such as the newts, where the power of regeneration is retained throughout adult life (cf. H.A. 508 b 4 ff.).

774 b

775 a

δὲ γίνεται διὰ τὸ προτερεῖν. δῆλον δὲ τοῦτο καὶ
ἐπὶ τῶν ἑπταμήνων· διὰ γὰρ τὸ ἀτελῆ εἶναι πολλάκις ἔνια αὐτῶν γίνεται οὐδὲ τοὺς πόρους ἔχοντά
πω διηρθρωμένους, οἷον ὤτων καὶ μυκτήρων,
ἀλλ' ἐπαυξανομένοις διαρθροῦται, καὶ βιοῦσι πολλὰ
τῶν τοιούτων.

Γίνεται δὲ ἀνάπηρα μᾶλλον ἐν τοῖς ἀνθρώποις
5 τὰ ἄρρενα τῶν θηλέων, ἐν δὲ τοῖς ἄλλοις οὐθὲν
μᾶλλον. αἴτιον δ' ὅτι ἐν τοῖς ἀνθρώποις πολὺ
διαφέρει τὸ ἄρρεν τοῦ θήλεος τῇ θερμότητι τῆς
φύσεως, διὸ κινητικώτερά ἐστι κυούμενα τὰ ἄρρενα
τῶν θηλέων· διὰ δὲ τὸ κινεῖσθαι θραύεται μᾶλλον·
εὔφθαρτον[1] γὰρ τὸ νέον διὰ τὴν ἀσθένειαν. διὰ
10 τὴν αὐτὴν δὲ ταύτην αἰτίαν καὶ τελειοῦται τὰ
θήλεα τοῖς ἄρρεσιν οὐχ ὁμοίως· ⟨αἱ γὰρ ὑστέραι
αὐτῶν οὐχ ὁμοίως ἔχουσιν· ἐν δὲ τοῖς ἄλλοις
ζῴοις ὁμοίως τελειοῦται· οὐδὲν γὰρ ὑστερεῖ τὰ
θήλεα τῶν ἀρρένων ὥσπερ⟩[2] ἐν ταῖς γυναιξίν·
ἐν μὲν γὰρ τῇ μητρὶ ἐν πλείονι χρόνῳ διακρίνεται
τὸ θῆλυ τοῦ ἄρρενος, ἐξελθοῦσι[3] δὲ πάντα πρότερον ἐπιτελεῖται, οἷον ἥβη καὶ ἀκμὴ καὶ γῆρας,
τοῖς θήλεσιν ἢ τοῖς ἄρρεσιν· ἀσθενέστερα γάρ

---

[1] εὔφθαρτον PZ : εὔθραυστον vulg.

[2] supplevi ; *quoniam matrices earum sunt secundum
modum divisum* (leg. *diversum* ; v.l. *sunt diversae sec. modum
sorum*). *in aliis autem animalibus non apparet diversitas in
complemento creationis feminarum et masculorum quoniam
non est in feminis diminutio a maribus* Σ : *in aliis autem
animalibus similiter : nichil enim tardat femela plus mas-
culo, sicut in mulieribus* Gul. Moerb. teste Bussemaker ;
similia ex Gul. vers. suppleverat Schneider, ed. *H.A.* vol. iv.
443.

[3] ἐξελθοῦσι Peck : ἐξελθόντα PSYZ[1] : ἐξελθόντων Z[2], vulg.

are born early. This is plain, indeed, in the case of seven months' children : in some of them, when they are born, because they are imperfect, even the passages (*e.g.*, those of the ears and nostrils) are often not yet fully articulated ; as the child grows, however, they become articulated. Many such individuals survive.

In human beings, more males are born deformed than females ; in other animals, there is no preponderance either way. The reason is that in human beings the male is much hotter in its nature than the female. On that account male embryos tend to move about more than female ones,[a] and owing to their moving about they get broken more, since a young creature can easily be destroyed owing to its weakness. And it is due to this self-same cause that the perfecting of female embryos is inferior to that of male ones, ⟨since their uterus is inferior in condition.[b] In other animals, however, the perfecting of female embryos is not inferior to that of male ones : they are not any later in developing than the males, as they *are*⟩[c] in women, for while still within the mother, the female takes longer to develop than the male does[d] ; though once birth has taken place everything reaches its perfection sooner in females than in males—*e.g.*, puberty, maturity, old age—because females are weaker and colder in

---

[a] *Cf. H.A.* 584 a 26 ff.

[b] *i.e.*, it is colder, because the nature of women is colder than that of other female animals, as is stated immediately above, and below ; *cf.* also 776 a 10, where women are said to be alone in suffering from uterine affections, again owing to lack of heat, resulting in inability to concoct ; and 775 a 30 ff.

[c] See *app. crit.*   [d] *Cf. H.A.* 583 b 22 ff.

15 ἐστι καὶ ψυχρότερα τὰ θήλεα τὴν φύσιν, καὶ δεῖ
ὑπολαμβάνειν ὥσπερ ἀναπηρίαν εἶναι τὴν θηλύτητα
φυσικήν. ἔσω μὲν οὖν διακρίνεται διὰ τὴν ψυχρότητα βραδέως (ἡ γὰρ διάκρισις πέψις ἐστί,
πέττει δ' ἡ θερμότης, εὔπεπτον δὲ τὸ θερμότερον),
ἐκτὸς δὲ διὰ τὴν ἀσθένειαν ταχὺ συνάπτει πρὸς
20 τὴν ἀκμὴν καὶ τὸ γῆρας· πάντα γὰρ τὰ ἐλάττω
πρὸς τὸ τέλος ἔρχεται θᾶττον, ὥσπερ καὶ ἐν τοῖς
κατὰ τέχνην ἔργοις, καὶ ἐν τοῖς ὑπὸ φύσεως συνισταμένοις. διὰ τὸ εἰρημένον δ' αἴτιον καὶ ἐν
μὲν τοῖς ἀνθρώποις τὰ διδυμοτοκούμενα θῆλυ καὶ
ἄρρεν ἧττον σώζεται, ἐν δὲ τοῖς ἄλλοις οὐθὲν
25 ἧττον· τοῖς μὲν γὰρ παρὰ φύσιν τὸ ἰσοδρομεῖν,
οὐκ ἐν ἴσοις χρόνοις γινομένης τῆς διακρίσεως,
ἀλλ' ἀνάγκη τὸ ἄρρεν ὑστερεῖν ἢ τὸ θῆλυ προτερεῖν, ἐν δὲ τοῖς ἄλλοις οὐ παρὰ φύσιν. συμβαίνει
δὲ καὶ διαφορὰ περὶ τὰς κυήσεις ἐπί τε τῶν ἀνθρώπων καὶ ἐπὶ τῶν ἄλλων ζῴων· τὰ μὲν γὰρ
30 εὐθηνεῖ μᾶλλον τοῖς σώμασι τὸν πλεῖστον χρόνον,
τῶν δὲ γυναικῶν αἱ πολλαὶ δυσφοροῦσι περὶ τὴν
κύησιν. ἔστι μὲν οὖν αἴτιόν τι τούτου[1] καὶ διὰ
τὸν βίον· ἑδραῖαι γὰρ οὖσαι πλείονος γέμουσι
περιττώματος, ἐπεὶ ἐν οἷς ἔθνεσι πονητικὸς ὁ τῶν
γυναικῶν βίος, οὔθ' ἡ κύησις ὁμοίως ἐπίδηλός
35 ἐστι, τίκτουσί τε ῥαδίως κἀκεῖ καὶ πανταχοῦ αἱ

---

[1] τούτου Platt : τούτων vulg.

---

[a] Cf. 767 b 9, and see Introd. § 13.

their nature ; and we should look upon the female
state as being as it were a deformity, though one
which occurs in the ordinary course of nature.[a] While
it is within the mother, then, it develops slowly on
account of its coldness, since development is a sort of
concoction, concoction is effected by heat, and if a
thing is hotter its concoction is easy ; when, however,
it is free from the mother, on account of its weakness
it quickly approaches its maturity and old age, since
inferior things all reach their end more quickly, and
this applies to those which take their shape under the
hand of Nature just as much as to the products of the
arts and crafts. The reason which I have just stated
accounts also for the fact that (a) in human beings
twins survive less well if one is male and the other
female, but (b) in other animals they survive just as
well : in human beings it is contrary to nature for the
two sexes to keep pace with each other, male and
female requiring unequal periods for their development to take place ; the male is bound to be late or
the female early ; whereas in the other animals equal
speed is not contrary to nature. There is also a
difference between human beings and the other
animals with regard to gestation. Other animals are
most of the time in better physical condition, whereas
the majority of women suffer discomfort in connexion
with gestation. Now the cause of this is to some
extent attributable to their manner of life, which is
sedentary, and this means that they are full of residue ; they have more of it than the other animals.
This is borne out by the case of those tribes where
the women live a life of hard work. With such women
gestation is not so obvious, and they find delivery an
easy business. And so do women everywhere who

ARISTOTLE

εἰωθυῖαι πονεῖν· ἀναλίσκει γὰρ ὁ πόνος τὰ περιττώματα, ταῖς δ' ἑδραίαις ἐνυπάρχει πολλὰ τοιαῦτα
διὰ τὴν ἀπονίαν καὶ τὸ μὴ γίνεσθαι καθάρσεις

κυούσαις, ἥ τε ὠδὶς ἐπίπονός ἐστιν· ὁ δὲ πόνος
γυμνάζει τὸ πνεῦμα ὥστε δύνασθαι κατέχειν, ἐν ᾧ
τὸ τίκτειν ἐστὶ ῥᾳδίως ἢ χαλεπῶς. ἔστι μὲν οὖν,
ὥσπερ εἴρηται, καὶ ταῦτα συμβαλλόμενα πρὸς τὴν
διαφορὰν τοῦ πάθους τοῖς ἄλλοις ζῴοις καὶ ταῖς
5 γυναιξί, μάλιστα δ' ὅτι τοῖς μὲν αὐτῶν ὀλίγη
γίνεται κάθαρσις, τοῖς δ' οὐκ ἐπίδηλος ὅλως, ταῖς
δὲ γυναιξὶ πλείστη τῶν ζῴων, ὥστε μὴ γινομένης
τῆς ἐκκρίσεως διὰ τὴν κύησιν ταῖς μὲν ταραχὴν
παρέχει· καὶ γὰρ μὴ κυούσαις, ὅταν αἱ καθάρσεις
μὴ γίγνωνται, νόσοι συμβαίνουσιν· καὶ τὸ πρῶτον
10 δὲ ταράττονται συλλαβοῦσαι[1] μᾶλλον αἱ πλεῖσται
τῶν γυναικῶν· τὸ γὰρ κύημα κωλύειν μὲν δύναται
τὰς καθάρσεις, διὰ μικρότητα δὲ οὐδὲν ἀναλίσκει
πλῆθος τοῦ περιττώματος τὸ πρῶτον, ὕστερον δὲ
κουφίζει μεταλαμβάνον· ἐν δὲ τοῖς ἄλλοις ζῴοις διὰ
15 τὸ ὀλίγον εἶναι σύμμετρον γίνεται πρὸς τὴν αὔξησιν
τῶν ἐμβρύων, καὶ ἀναλισκομένων τῶν περιττωμάτων τῶν ἐμποδιζόντων τὴν τροφὴν εὐημερεῖ τοῖς
σώμασι μᾶλλον. καὶ ἐν τοῖς ἐνύδροις τὸν αὐτὸν
τρόπον καὶ ἐν τοῖς ὄρνισιν. ἤδη δὲ μεγάλων
γινομένων τῶν κυημάτων, ὅσοις μηκέτι συμβαίνει

---

[1] συλλαβοῦσαι P : συλλαμβάνουσαι vulg.

ᵃ Cf. H.A. 587 a 1 ff., and see De somno et vig. 456 a 16
"strength is required for causing 'movement,' and strength
462

are used to hard work. The reason is that the effort of working uses up the residues, whereas sedentary women have a great deal of such matter in their bodies owing to the absence of effort, as well as to the cessation of the menstrual discharges during gestation, and they find the pains of delivery severe. Hard work, on the other hand, gives the breath (*pneuma*) exercise, so that they can hold it [a]; and it is this which determines whether delivery is easy or difficult. All these things, then, as we have said, are in their way factors producing the difference in gestation as between women and the other animals; but the chief one is that whereas in some animals there is but little menstrual evacuation, and in others no visible evacuation at all, in women it is greater in volume than in any other animal; and the result of this is that when it is not being discharged owing to pregnancy it causes them trouble (and indeed even apart from pregnancy, when the menstrual discharge fails to take place diseases are the result); and most women are troubled in this way rather more at the beginning, just after they have conceived, because although the fetation is able to prevent the evacuation, yet as it is so small it does not at first use up any amount of the residue; afterwards, when it does take up some of it, it relieves the trouble. In the other animals, however, as there is but little of it, its amount is just right for the growth of the embryos; and as the residues which obstruct the nourishment get used up, the animals are in better physical condition. The same applies to water-animals and to birds. The reason why some animals are no longer in good

is supplied by the holding of the breath." *Cf.* also *M.A.* 703 a 18, 9; *P.A.* 659 b 18, 667 a 29; and App. B §§ 22 ff.

775 b

20 ἡ εὐτροφία τῶν σωμάτων, αἴτιον τὸ τὴν αὔξησιν
τοῦ κυήματος δεῖσθαι πλείονος ἢ[1] τῆς περιττω-
ματικῆς τροφῆς. ὀλίγαις δέ τισι τῶν γυναικῶν
βέλτιον ἔχειν τὰ σώματα συμβαίνει κυούσαις·
αὗται δ' εἰσὶν ὅσαις μικρὰ τὰ περιττώματα ἐν τῷ
σώματι, ὥστε καταναλίσκεσθαι μετὰ τῆς εἰς τὸ
ἔμβρυον τροφῆς.

VII 25 Περὶ δὲ τῆς καλουμένης μύλης ῥητέον, ἣ γίνεται
μὲν ὀλιγάκις ταῖς γυναιξί, γίνεται δέ τισι τοῦτο τὸ
πάθος κυούσαις. τίκτουσι γὰρ ὃ καλοῦσι μύλην.
ἤδη γὰρ συνέβη τινὶ γυναικὶ συγγενομένῃ τῷ
ἀνδρὶ καὶ δοξάσῃ συλλαβεῖν, τὸ μὲν πρῶτον ὅ τε
30 ὄγκος ηὐξάνετο τῆς γαστρὸς καὶ τἆλλα ἐγίγνετο
κατὰ λόγον, ἐπεὶ δὲ ὁ χρόνος ἦν τοῦ τόκου, οὔτ'
ἔτικτεν οὔτε ὁ ὄγκος ἐλάττων ἐγίνετο, ἀλλ' ἔτη
τρία ἢ τέτταρα οὕτω διετέλει, ἕως δυσεντερίας
γενομένης καὶ κινδυνεύσασα ὑπ' αὐτῆς ἔτεκε σάρκα
ἣν καλοῦσι μύλην. ἔτι δὲ καὶ συγκαταγηράσκει
καὶ συναποθνήσκει τοῦτο τὸ πάθος. τὰ δὲ θύραζε
35 ἐξιόντα τῶν τοιούτων γίνεται σκληρὰ οὕτως ὥστε
μόλις διακόπτεσθαι καὶ σιδήρῳ. περὶ μὲν οὖν τῆς
τοῦ πάθους αἰτίας εἴρηται ἐν τοῖς προβλήμασιν·

776 a πάσχει γὰρ ταὐτὸν τὸ κύημα ἐν τῇ μήτρᾳ ὅπερ ἐν
τοῖς ἑψομένοις τὰ μωλυνόμενα,[2] καὶ οὐ διὰ θερμότητα, ὥσπερ τινές φασιν, ἀλλὰ μᾶλλον δι'
ἀσθένειαν θερμότητος (ἔοικε γὰρ ἡ φύσις ἀδυ-

---

[1] ἢ PZ : om. vulg. [2] μολ. codd.

---

[a] The uterine hydatiform mole, deciduoma, etc., are
tumours of the uterine wall; they occur spontaneously and
can be produced experimentally by mechanical stimulus,
given the right glandular conditions.

464

Empedocles' statement is incorrect. And besides, another method is open for explaining the cause of the colours. But assuming the correctness of what was said earlier in the treatise *Of the Senses,* and before that in the treatise *Of the Soul,*[a] *i.e.*, that the sense-organ of sight is composed of water, and also the correctness of the cause there assigned for its being composed of water and not of air or of fire, then we should take it that the following is the cause responsible for the phenomena just described. Some eyes contain too much fluid, some too little, to suit the right movement,[b] others contain just the right amount; and so those eyes which contain a large amount of fluid are dark, because large volumes of fluid are not transparent; those which contain a small amount are blue. (Sea-water is a parallel instance. Transparent sea-water appears blue, the less transparent appears pallid, and water so deep that its depth is undetermined is dark or dark blue.) Eyes intermediate between these two extremes differ merely by " the more and less." [c]

We ought to suppose that to the same cause is due **Keenness** the fact that blue eyes are not keen-sighted during **of Sight.** the daytime nor dark eyes at night. Blue eyes, on account of the small amount of fluid in them, are unduly set in movement by the light and by visible objects, in respect both of fluidity and of transparency. It i , however, the setting in movement of this part in respect of its transparency that constitutes sight, not in respect of its fluidity.[d] Dark eyes are set in movement less owing to the amount of

[c] See Introd. § 70.
[d] For the details of Aristotle's theory of vision, see App. B §§ 26 ff.

γὰρ τὸ νυκτερινὸν φῶς· ἅμα γὰρ καὶ δυσκίνητον ἐν
τῇ νυκτὶ ὅλως γίγνεται τὸ ὑγρόν. δεῖ δὲ οὔτε
μὴ κινεῖσθαι αὐτὸ οὔτε μᾶλλον ᾖ[1] διαφανές· ἐκ-
κρούει γὰρ ἡ ἰσχυροτέρα κίνησις τὴν ἀσθενεστέραν.
10 διὸ καὶ ἀπὸ τῶν ἰσχυρῶν χρωμάτων μεταβάλλοντες οὐχ ὁρῶσι, καὶ ἐκ τοῦ ἡλίου εἰς τὸ σκότος
ἰόντες· ἰσχυρὰ γὰρ οὖσα ἡ ἐνυπάρχουσα κίνησις
κωλύει τὴν θύραθεν, καὶ ὅλως οὔτε σθένουσα οὔτε
ἀσθενὴς ὄψις τὰ λαμπρὰ δύναται ὁρᾶν διὰ τὸ
πάσχειν τι μᾶλλον καὶ κινεῖσθαι τὸ ὑγρόν. δηλοῖ
15 δὲ καὶ τὰ ἀρρωστήματα τῆς ὄψεως ἑκατέρας. τὸ
μὲν γὰρ γλαύκωμα γίνεται μᾶλλον τοῖς γλαυκοῖς,
οἱ δὲ νυκτάλωπες καλούμενοι τοῖς μελανοφθάλμοις.
ἔστι δὲ τὸ μὲν γλαύκωμα ξηρότης τις [μᾶλλον][2]
τῶν ὀμμάτων, διὸ καὶ συμβαίνει μᾶλλον[3] γηρά-
σκουσιν· ξηραίνεται γάρ, ὥσπερ καὶ τὸ ἄλλο σῶμα,
20 καὶ ταῦτα τὰ μόρια πρὸς τὸ γῆρας· ὁ δὲ νυκτάλωψ
ὑγρότητος πλεονασμός, διὸ τοῖς νεωτέροις γίνεται
μᾶλλον· ὑγρότερος γὰρ ὁ ἐγκέφαλος ὁ τούτων.
ἡ δὲ μέση τοῦ πολλοῦ καὶ τοῦ ὀλίγου ὑγροῦ βελ-

[1] ᾖ Z[1] : ᾖ ᾖ vulg.   [2] secl. A.-W.
[3] μᾶλλον om. Z.

---

[a] The movement already in progress in the eye is so strong
that it precludes any fresh movement that comes from outside from making itself felt in the eye.

[b] Dark eyes have so much fluid in them that the weakness
of the light at night cannot set them in movement (780 a 5).
—Night-blindness is also the sense of the word as defined by
Galen ; but the term seems to have been used in opposite
senses in ancient times ; e.g., in Hippocrates, Prorrh. II. 33
(ix. 64 Littré) νυκτάλωπες = οἱ τῆς νυκτὸς ὁρῶντες (though one
ms. apparently reads οὐχ ὁρῶντες) ; and see L. & S.

[c] But he has said above (779 a 28 and 779 b 11 ; repeated
below 780 b 1) that the eyes of new-born infants and young

498

fluid which they contain, for the light is weak during the night, and, in addition to that, fluid generally is not easily set in movement at night. To obtain the best results, it must avoid both (a) not being set in movement at all and also (b) being set in movement too much in respect of its transparency, because the stronger movement ousts the weaker.[a] That is why people who have been looking at strong, brilliant colours, or who go out of the sunlight into the dark, cannot see : the movement which is already present in their eyes is so strong that it precludes the movement which comes from without. And in general, neither strong sight nor weak sight can see bright things because the action undergone by the fluid in the eye is unduly intense—*i.e.*, the fluid is set in movement unduly. This is borne out by the ailments besetting either kind of sight. Cataract tends to attack the blue-eyed more than the dark-eyed, night-blindness [b] as it is called attacks the latter. Cataract is a sort of dryness of the eyes, and that is why it occurs oftener in the ageing, as these parts (the eyes), like the rest of the body, become dry towards old age. Night-blindness is superabundance of fluid, and that is why it tends to attack younger people : their brain is more fluid.[c] The best sight of all is that which is midway between a large amount and a

children are bluish ; and the reason given for blueness at 780 b 1 (and 779 b 29) is the *small* amount of fluid. At 779 b 11, however, the reason given for blueness is weakness (weakness is explained at 780 b 7 as being due to lack of concoction of the fluid) ; and at 780 b 8 undue thinness of fluid is said to " be equivalent " (τὴν αὐτὴν ἔχει δύναμιν) to a small amount of fluid. We may deduce, therefore, that a large amount of thin fluid is equivalent to a small amount of fluid ; at any rate, this seems to be the only way of reconciling Aristotle's apparently contradictory statements.

τίστη ὄψις· οὔτε γὰρ ὡς ὀλίγη οὖσα διὰ τὸ ταράτ-
τεσθαι ἐμποδίζει τὴν τῶν χρωμάτων κίνησιν, οὔτε
25 διὰ τὸ¹ πλῆθος παρέχει δυσκινησίαν.

Οὐ μόνον δὲ τὰ εἰρημένα αἴτια τοῦ ἀμβλὺ ἢ ὀξὺ
ὁρᾶν, ἀλλὰ καὶ ἡ τοῦ δέρματος φύσις τοῦ ἐπὶ τῇ
κόρῃ καλουμένη. δεῖ γὰρ αὐτὸ διαφανὲς εἶναι,
τοιοῦτον δ' ἀναγκαῖον εἶναι τὸ λεπτὸν καὶ λευκὸν
καὶ ὁμαλόν, λεπτὸν μὲν ὅπως ἡ θύραθεν εὐθυπορῇ
30 κίνησις, ὁμαλὸν δ' ὅπως μὴ ἐπισκιάζῃ ῥυτιδού-
μενον (καὶ γὰρ διὰ τοῦθ' οἱ γέροντες οὐκ ὀξὺ
ὁρῶσιν· ὥσπερ γὰρ καὶ τὸ ἄλλο δέρμα, καὶ τὸ τοῦ
ὄμματος ῥυτιδοῦταί τε καὶ παχύτερον γίνεται
γηράσκουσιν), λευκὸν δὲ διὰ τὸ τὸ μέλαν μὴ εἶναι
διαφανές· αὐτὸ γὰρ τοῦτ' ἐστὶ τὸ μέλαν, τὸ μὴ
35 διαφαινόμενον. διόπερ οὐδ' οἱ λαμπτῆρες δύνανται
φαίνειν ἐὰν ὦσιν ἐκ τοιούτου δέρματος.

780 b

Ἐν μὲν οὖν τῷ γήρᾳ καὶ ταῖς νόσοις διὰ ταύτας
τὰς αἰτίας οὐκ ὀξὺ βλέπουσι, τὰ δὲ παιδία δι'
ὀλιγότητα τοῦ ὑγροῦ γλαυκὰ φαίνεται τὸ πρῶτον.
ἑτερόγλαυκοι δὲ γίνονται μάλιστα οἱ ἄνθρωποι καὶ
οἱ ἵπποι διὰ τὴν αὐτὴν αἰτίαν δι' ἥνπερ ὁ μὲν
5 ἄνθρωπος πολιοῦται μόνον,² τῶν δ' ἄλλων ἵππος
μόνον ἐπιδήλως γηράσκων λευκαίνεται τὰς τρίχας.
ἥ τε γὰρ πολιότης ἀσθένειά τίς ἐστι τοῦ ὑγροῦ τοῦ
ἐν τῷ ἐγκεφάλῳ καὶ ἀπεψία καὶ ἡ γλαυκότης· τὸ
γὰρ λίαν λεπτὸν ἢ λίαν παχὺ τὴν αὐτὴν ἔχει δύναμιν τὸ μὲν τῷ ὀλίγῳ τὸ δὲ τῷ πολλῷ ὑγρῷ.

---

¹ τὸ Z, Aldus; om. vulg.
² μόνον Aldus, codd.*: μόνος Bekker.

ᵃ And therefore weak-sighted.
ᵇ i.e., unconcocted.

small amount of fluid, because on the one hand it is not so small in volume that it gets disturbed and so hampers the movement produced by the colours, nor on the other hand is it so large in volume that its movement is rendered difficult.

These are not the only causes of dullness and keenness of sight. In addition to them we must mention the nature of the skin upon what is known as the pupil. This skin should be transparent, a condition which must of necessity be satisfied by skin that is thin, and white, and even—thin, in order that the movement that comes from without may take a straight course ; even, so that its wrinkles shall not produce a shadow (the reason why old people do not have keen vision is that the skin in the eyes, like that elsewhere, gets wrinkled and thicker in old age) ; white, because that which is black is not transparent, non-transparency being precisely what blackness is ; and that too is why lanterns cannot give any light if they are made of black skin.

In old age and disease, then, these are the causes owing to which the sight is not keen ; in children, however, it is the small volume of fluid which makes the eyes appear blue to begin with.[a] And odd-coloured eyes occur more often in human beings and horses than other animals for the same cause that human beings are the only animals that go grey and the horse is the only one of the remainder whose hairs noticeably whiten in old age :—Greyness is a weakness, viz., a lack of concoction, of the fluid in the brain ; so is blueness of the eyes ; since unduly thin [b] fluid and unduly thick fluid are the equivalent [c] respectively of a small amount and a large amount of

---

[a] For ἔχει δύναμιν, cf. 733 b 15 784 b 14, and Introd. § 26.

10 ὅταν οὖν μὴ δύνηται ἀπαρτίσαι ἡ φύσις ὁμοίως
ἢ πέψασα τὸ ἐν ἀμφοτέροις ὑγρὸν ἢ μὴ πέψασα,
ἀλλὰ τὸ μὲν τὸ δὲ μή, τότε συμβαίνει γίνεσθαι
ἑτερογλαύκους.

Περὶ δὲ τοῦ τὰ μὲν ὀξυωπὰ εἶναι τῶν ζῴων τὰ
δὲ μή, δύο τρόποι τῆς αἰτίας εἰσίν. διχῶς γὰρ
15 λέγεται τὸ ὀξὺ σχεδόν, καὶ περὶ τὸ ἀκούειν καὶ
τὸ ὀσφραίνεσθαι ὁμοίως τοῦτ' ἔχει. λέγεται γὰρ
ὀξὺ ὁρᾶν ἓν μὲν τὸ πόρρωθεν δύνασθαι ὁρᾶν, ἓν δὲ
τὸ τὰς διαφορὰς ὅτι μάλιστα διαισθάνεσθαι τῶν
ὁρωμένων. ταῦτα δ' οὐχ ἅμα συμβαίνει τοῖς
αὐτοῖς. ὁ γὰρ αὐτὸς ἐπηλυγασάμενος[1] τὴν χεῖρα
20 ἢ δι' αὐλοῦ βλέπων τὰς μὲν διαφορὰς οὐθὲν μᾶλλον
οὐδ' ἧττον κρινεῖ[2] τῶν χρωμάτων, ὄψεται δὲ πορρώτερον. οἱ γοῦν ἐκ τῶν ὀρυγμάτων καὶ φρεάτων
ἐνίοτε ἀστέρας ὁρῶσιν. ὥστ' εἴ τι τῶν ζῴων ἔχει
μὲν προβολὴν τοῦ ὄμματος πολλήν, τὸ δ' ἐν τῇ
κόρῃ ὑγρὸν μὴ καθαρὸν μηδὲ σύμμετρον τῇ κινήσει
25 τῇ θύραθεν, μηδὲ τὸ ἐπιπολῆς δέρμα λεπτόν, τοῦτο
περὶ μὲν τὰς διαφορὰς οὐκ ἀκριβώσει τῶν χρωμάτων, πόρρωθεν δ' ἔσται ὁρατικόν (ὥσπερ εἰ καὶ
ἐγγύθεν)[3] μᾶλλον τῶν τὸ μὲν ὑγρὸν καθαρὸν ἐχόντων καὶ τὸ σκέπασμα αὐτοῦ, μὴ ἐχόντων δ' ἐπι-
σκύνιον πρὸ τῶν ὀμμάτων μηθέν. τοῦ μὲν γὰρ
30 οὕτως ὀξὺ ὁρᾶν ὥστε διαισθάνεσθαι τὰς διαφοράς,
ἐν αὐτῷ τῷ ὄμματί ἐστιν ἡ αἰτία· ὥσπερ γὰρ ἐν
ἱματίῳ καθαρῷ καὶ αἱ μικραὶ κηλῖδες ἔνδηλοι

[1] ἐπηλυγασάμενος P : -γισ- vulg.
[2] κρινεῖ Peck (idem Sus., Richards) : κρίνει vulg.
[3] ὥσπερ . . . ἐγγύθεν secl. A.-W., om. Σ : ὅσωπερ YZ[1] pro
ὥσπερ εἰ, et καὶ om. Y.

---

[a] Chiefly, as will shortly appear, the differences of colour.

fluid ; therefore, whenever Nature cannot make the
fluid in both eyes tally, either by concocting it or
by not concocting it in both, but instead of that
concocts it in one and not in the other, the result is
odd-coloured eyes.

The fact that some animals are keen-sighted and Two senses
others not is due to two sets of causes, for " keen " of "keen."
here has practically two meanings (so it has when
applied to hearing and smelling). Thus, keen sight
means (a) ability to see from a distance, (b) distinguishing as accurately as possible the differences [a] of
the objects which are seen ; and these faculties do
not occur together in the same persons. The man
who shades his eye with his hand or looks through a
tube will not distinguish any more or any less the
differences of colours, but he will see further ; at any
rate, people in pits and wells sometimes see the stars.
So that if any animal has a considerable projection
over his eyes, while the fluid in his pupils is not pure
nor suitably proportionate to the movement coming
from without, and if the skin on the surface of them
is not thin, then that animal will not have accuracy of
vision in so far as differences of colours are concerned,
but he will be able to see from a distance (just as he
would from close quarters) better than animals which
though they have pure fluid in their eyes and a pure
covering round it, yet have no projecting brow at all
in front of their eyes. The reason is that (a) the
cause of being keen-sighted enough to distinguish the
differences ⟨of colour⟩ lies in the eye itself, since just
as quite small stains are plain and distinct on a pure,
clean shirt, so quite small movements are plain and

γίνονται, οὕτως καὶ ἐν τῇ καθαρᾷ ὄψει καὶ αἱ
μικραὶ κινήσεις δῆλαι καὶ ποιοῦσιν αἴσθησιν. τοῦ
δὲ τὰ πόρρωθεν ὁρᾶν καὶ τὴν ἀπὸ τῶν πόρρωθεν
35 ὁρατῶν ἀφικνεῖσθαι κίνησιν ἡ θέσις αἰτία τῶν
ὀφθαλμῶν· τὰ μὲν γὰρ ἐξόφθαλμα οὐκ εὐωπὰ

731 a πόρρωθεν, τὰ δ' ἐντὸς ἔχοντα τὰ ὄμματα ἐν κοίλῳ
κείμενα ὁρατικὰ τῶν πόρρωθεν διὰ τὸ τὴν κίνησιν
μὴ σκεδάννυσθαι εἰς ἀχανὲς ἀλλ' εὐθυπορεῖν. οὐθὲν
γὰρ διαφέρει τὸ λέγειν ὁρᾶν, ὥσπερ τινές φασι,
τῷ τὴν ὄψιν ἐξιέναι (ἂν γὰρ μὴ ᾖ τι πρὸ τῶν
5 ὀμμάτων, διασκεδαννυμένην ἀνάγκη ἐλάττω προσ-
πίπτειν τοῖς ὁρωμένοις καὶ ἧττον τὰ πόρρωθεν
ὁρᾶν), ἢ τὸ τῇ ἀπὸ τῶν ὁρωμένων κινήσει ὁρᾶν.
ὁμοίως γὰρ ἀνάγκη καὶ τὴν ὄψιν τῇ κινήσει ὁρᾶν.
μάλιστα μὲν οὖν ἑωρᾶτο ἂν τὰ πόρρωθεν, εἰ ἀπὸ
τῆς ὄψεως εὐθὺς συνεχὴς ἦν πρὸς τὸ ὁρώμενον
10 οἷον αὐλός· οὐ γὰρ ἂν διελύετο ἡ κίνησις ἡ ἀπὸ
τῶν ὁρατῶν· εἰ δὲ μή, ὅσῳπερ ἂν ἐπὶ πλέον ἐπέχῃ,[1]
τοσούτῳ ἀκριβέστερον τὰ πόρρωθεν ὁρᾶν ἀνάγκη.

Καὶ τῆς μὲν τῶν ὀμμάτων διαφορᾶς ἔστωσαν
αὗται αἱ αἰτίαι.

II Τὸν αὐτὸν δὲ τρόπον ἔχει καὶ περὶ τὴν ἀκοὴν
15 καὶ τὴν ὄσφρησιν· ἐν μὲν γάρ ἐστι τοῦ ἀκριβῶς
ἀκούειν καὶ ὀσφραίνεσθαι τὸ τὰς διαφορὰς τῶν
ὑποκειμένων αἰσθητῶν ὅτι μάλιστα αἰσθάνεσθαι

---

[1] ἐπέχῃ Platt: ἀπέχῃ vulg.: πλέονἀπέχηι Z, sed ἐονἀπ Z²
in ras.

---

[a] *i.e.*, the substance of the eye.
[b] This theory is put forward by Timaeus in Plato, *Timaeus*
45 B ff. A similar theory seems to have been held by Empedocles.

distinct in a pure, clean sight [a] and they give rise to sense-perception. As for (b) the ability to see things at a distance, and the fact that the movement coming from objects at a distance succeeds in reaching into the eyes, the cause of this is the position of the eyes. Animals with prominent eyes do not see well from a distance, but those with sunken eyes placed in a hollowed recess are able to see things at a distance, because the movement does not get scattered into space but follows a straight course. It makes no difference to this which of the two theories of sight we adopt. Thus, if we say, as some people do, that seeing is effected " by the sight issuing forth," [b] then on this theory, unless there is something projecting in front of the eyes, the "sight" of necessity gets scattered and so less of it strikes the object, with the result that distant objects are less well seen. If we say that seeing is effected " by a movement derived from the visible object," then on this theory, the clarity with which the sight sees will of necessity vary directly as the clarity of the movement : distant objects would be seen best of all if there were a sort of continuous tube extending straight from the sight to that which is seen, for then the movement which proceeds from the visible objects would not get dissipated ; failing that, the further the tube extends, the greater is bound to be the accuracy with which distant objects are seen.

These, then, shall be the causes which we assign to explain the different sorts of eyes.

The same situation is found in connexion with II two other senses—hearing and smell—as with sight. Keenness of Smell and To hear and to smell " accurately " means (a) to Hearing. perceive as well as possible all the differences in the

505

781 a

πάσας, ἐν δὲ τὸ πόρρωθεν καὶ ἀκούειν καὶ ὀσφραίνεσθαι. τοῦ μὲν οὖν τὰς διαφορὰς κρίνειν
καλῶς τὸ αἰσθητήριον αἴτιον, ὥσπερ ἐπὶ τῆς
20 ὄψεως, ἂν ᾖ καθαρὸν αὐτό τε καὶ ἡ περὶ αὐτὸ
μῆνιγξ. [1][οἱ γὰρ πόροι τῶν αἰσθητηρίων πάντων,
ὥσπερ εἴρηται ἐν τοῖς περὶ αἰσθήσεως, τείνουσι
πρὸς τὴν καρδίαν, τοῖς δὲ μὴ ἔχουσι καρδίαν πρὸς
τὸ ἀνάλογον. ὁ[2] μὲν οὖν τῆς ἀκοῆς, ἐπεί ἐστι τὸ
αἰσθητήριον ἀέρος, ᾗ τὸ πνεῦμα τὸ σύμφυτον
25 ποιεῖται ἐνίοις μὲν τὴν σφύξιν τοῖς δὲ τὴν ἀνα-
πνοὴν [καὶ εἰσπνοήν],[3] ταύτῃ περαίνει[4]· διὸ καὶ ἡ
μάθησις γίνεται τῶν λεγομένων ὥστ᾽ ἀντιφθέγ-
γεσθαι τὸ ἀκουσθέν· οἷα γὰρ ἡ κίνησις εἰσῆλθε διὰ
τοῦ αἰσθητηρίου, τοιαύτη πάλιν, οἷον ἀπὸ χαρα-
κτῆρος τοῦ αὐτοῦ καὶ ἑνός, διὰ τῆς φωνῆς γίνεται
30 ἡ κίνησις, ὥσθ᾽ ὃ ἤκουσε, τοῦτ᾽ εἰπεῖν. καὶ
χασμώμενοι καὶ ἐκπνέοντες[5] ἧττον ἀκούουσιν ἢ
εἰσπνέοντες[6] διὰ τὸ ἐπὶ τῷ πνευματικῷ μορίῳ τὴν
ἀρχὴν[7] τοῦ αἰσθητηρίου εἶναι τοῦ τῆς ἀκοῆς, καὶ
σείεσθαι καὶ κινεῖσθαι ἅμα κινοῦντος τοῦ ὀργάνου

---

[1] sequitur (781 a 21–b 6) locus corruptus et sine dubio ex-
traneus.   vide pp. 563 sq.
[2] ὁ Aldus, vulg. : ἡ Y*PSZ.
[3] καὶ εἰσπνοήν om. Z[1], Platt : καὶ εἰσπνοήν τε S : καὶ εἰσπνοὴν
καὶ Y.
[4] haec sensu carere monet Platt : fortasse scribere malles
ᾗ τὸ πνεῦμα τὸ σύμφυτον ποιεῖ ἐν τοῖς φλεβίοις τὴν σφύξιν, ταύτῃ
περαίνει· τὸ γὰρ πνεῦμα τὸ σύμφυτον ποιεῖ ἐν μὲν τῷ αἰσθητηρίῳ
τὴν ἀναπνοήν, ὁμοίως δ᾽ ἐν τοῖς ὠσὶν τὴν ἀκοήν.   vertit Σ ὁ μὲν
οὖν τῆς ἀκοῆς κτλ. et instrumentum sensus auditus est plenum
spiritu naturali, quoniam spiritus naturalis facit in venis
motum pulsatilem, et facit in instrumento hanelitus, et
similiter facit in aure virtutem auditus.
[5] καὶ ἐκπνέοντες om. Σ.   [6] ἢ εἰσπνέοντες om. Σ.
[7] τελευτὴν SY, Z[2] in ras., Aldus.

objects perceived, (b) to hear and smell from a distance. As for (a) the ability to distinguish the differences well, the cause of this is the sense-organ, just as it is in the case of sight, i.e., it must be pure and clean itself, and so must the membrane round it.[a] [b For the passages of all the sense-organs, as is stated in the treatise Of Sensation, run to the heart, or to the counterpart of it in animals which have no heart. Now the passage of the hearing, since the sense-organ of hearing consists of air, terminates at the point where the connate pneuma causes in some the pulsation, in others, the respiration [and inspiration]. This, too, is why we are able to understand what is said and to repeat what we have heard, for whatever the character of the movement was which entered through the sense-organ, the character of the movement caused by means of the voice is the same in its turn—they might be two impressions from one and the same die. So, if you have heard a thing, you can utter it. Again, people hear less well while yawning and breathing out than they do while breathing in. The reason is that the principle of the sense-organ of hearing is situated upon the part[c] that is concerned with the pneuma, and it is shaken and set in movement when the organ sets the pneuma in movement [since the organ gets set in

[a] Cf. De anima II. 420 a 13: we can no longer hear if the membrane is damaged which encloses the air in the ear, any more than we can see if the skin on the pupil of the eye is damaged.

[b] For the difficulties involved in the following lines, see note, pp. 563 f. For the theories here assumed, see the account of Σύμφυτον Πνεῦμα, App. B, especially §§ 26 ff.

[c] Viz., the heart; see App. B §§ 31 f., and 776 b 17, 787 b 28.

781 a

τὸ πνεῦμα· [κινεῖται γὰρ κινοῦν τὸ ὄργανον.]¹ καὶ
ἐν ταῖς ὑγραῖς ὥραις καὶ κράσεσι συμβαίνει²
35 τὸ αὐτὸ πάθος,³ καὶ τὰ ὦτα πληροῦσθαι δοκεῖ

781 b πνεύματος διὰ τὸ γειτνιᾶν τὴν ἀρχὴν τῷ πνευματικῷ τόπῳ.⁴ ἡ μὲν οὖν περὶ τὰς διαφορὰς
ἀκρίβεια τῆς κρίσεως καὶ τῶν ψόφων καὶ τῶν
ὀσμῶν ἐν τῷ τὸ αἰσθητήριον καθαρὸν εἶναι καὶ τὸν
ὑμένα τὸν ἐπιπολῆς ἐστιν· πᾶσαι γὰρ αἱ κινήσεις
5 διάδηλοι, καθάπερ ἐπὶ τῆς ὄψεως, καὶ ἐπὶ τῶν
τοιούτων συμβαίνουσιν.] καὶ τὸ πόρρωθεν δὲ αἰσθάνεσθαι [τὰ δὲ μὴ αἰσθάνεσθαι]⁵ ὁμοίως συμβαίνει
ὥσπερ ἐπὶ τῆς ὄψεως. τὰ γὰρ ἔχοντα πρὸ τῶν
αἰσθητηρίων ἐπὶ πολὺ οἷον ὀχετοὺς διὰ τῶν μορίων,
ταῦτα πόρρωθεν αἰσθητικά ἐστιν. διὸ ὅσων οἱ
10 μυκτῆρες μακροί, οἷον τῶν Λακωνικῶν κυνιδίων,
ὀσφραντικά· ἄνω γὰρ ὄντος τοῦ αἰσθητηρίου αἱ
πόρρωθεν κινήσεις⁶ οὐ διασπῶνται ἀλλ' εὐθυπο-
ροῦσιν, ὥσπερ τοῖς ἐπηλυγαζομένοις⁷ πρὸ τῶν
ὀμμάτων. ὁμοίως δὲ καὶ ὅσοις τὰ ὦτα μακρὰ καὶ
ἀπογεγεισσωμένα πόρρωθεν, οἷα ἔχουσιν ἔνια τῶν
15 τετραπόδων, καὶ ἔσω τὴν ἕλικην μακράν· καὶ γὰρ
ταῦτα ἐκ πολλοῦ λαμβάνοντα τὴν κίνησιν ἀποδίδωσι πρὸς τὸ αἰσθητήριον.

Τὴν μὲν οὖν πόρρωθεν ἀκρίβειαν τῶν αἰσθήσεων

¹ seclusi: om. Σ.
² τοῦ σώματος addit Z (corporis post ὀσμῶν b 3 addit Σ).
³ lacunam statuit Platt.
⁴ sic Platt : τῇ ἀρχῇ τοῦ πνευματικοῦ τόπου vulg.: Σ vertit
et implentur aures secundum quod opilatur spiritus propter
principium instrumenti in quo est [spiritus].
⁵ aut haec secludenda (om. Z), aut (docente Platt) πόρρωθεν
δὲ ⟨τὰ μὲν⟩ αἰσθάνεσθαι scribendum.
⁶ αἱ π. κ. Platt : π. αἱ κ. vulg.
⁷ ἐπηλυγαζομένοις P : -γιζ- vulg.

movement while it is causing movement]. The same condition occurs during damp seasons and in damp climates,[a] and the ears appear to get filled with *pneuma*, because the principle is situated close by the region that is concerned with the *pneuma*. Thus, accuracy in distinguishing the differences both of sounds and smells depends upon the purity of the sense-organ and of the membrane upon its surface, for all the movements turn out plain and distinct in such cases also, just as in the case of sight.] (*b*) Perception from a distance, too, [and failure to perceive from a distance] occurs in the same way as in the case of sight. Thus, animals which have as it were channels passing through the parts concerned and projecting well out in front of the sense-organs can perceive from a distance ; and that is why animals which have long nostrils, like the Laconian hounds,[b] are keen-scented : the sense-organ is set well back in the interior, and therefore the movements which come from a distance do not get scattered but take a straight course, which is just what happens when we shade our eyes with the hand. Another similar case is that of those animals which have ears that are long and jut well out like the cornice of a house [c]—some quadrupeds have ears of this sort—and a long internal spiral passage ; these long ears, like the long noses, catch the movement a long way off and transmit it to the sense-organ.

Accuracy of perception by the senses when exer-

---

[a] Lit., " blends " ; *cf.* 767 a 31, 777 b 7.
[b] There is a long passage about Laconian hounds in *H.A.* 574 a 16 ff.   [c] *Cf. P.A.* 658 b 16.

781 b

ἥκιστα ὡς εἰπεῖν ἄνθρωπος ἔχει ὡς κατὰ μέγεθος
τῶν ζῴων, τὴν¹ δὲ περὶ τὰς διαφορὰς μάλιστα
20 πάντων εὐαίσθητον. αἴτιον δ' ὅτι τὸ αἰσθητήριον
καθαρὸν καὶ ἥκιστα γεῶδες καὶ σωματῶδες, καὶ
φύσει λεπτοδερμότατον τῶν ζῴων ὡς κατὰ μέγεθος
ἄνθρωπός ἐστιν.

Εὐλόγως δ' ἀπείργασται ἡ φύσις καὶ τὰ περὶ
τὴν φώκην· τετράπουν γὰρ ὂν καὶ ζῳοτόκον οὐχ
ἔχει ὦτα ἀλλὰ πόρους μόνον. αἴτιον δ' ὅτι ἐν
25 ὑγρῷ αὐτῇ ὁ βίος· τὸ δὲ τῶν ὤτων μόριον
πρόσκειται τοῖς πόροις πρὸς τὸ σῴζειν τὴν τοῦ
πόρρωθεν ἀέρος κίνησιν· οὐθὲν οὖν χρήσιμόν ἐστιν
αὐτῇ, ἀλλὰ καὶ τοὐναντίον ἀπεργάζοιτ' ἄν, δεχό-
μενα εἰς αὐτὰ ὑγροῦ πλῆθος.

Καὶ περὶ μὲν ὄψεως καὶ ἀκοῆς καὶ ὀσφρήσεως
εἴρηται.

III 30 Τὰ δὲ τριχώματα διαφέρουσι καὶ πρὸς αὐτὰ τοῖς
ἀνθρώποις κατὰ τὰς ἡλικίας καὶ πρὸς τὰ γένη τῶν
ἄλλων ζῴων, ὅσαπερ ἔχει τρίχας αὐτῶν. ἔχει δ'
ὅσαπερ ἐντὸς αὐτῶν ζῳοτοκεῖ πάντα σχεδόν· καὶ
γὰρ τὰ ἀκανθώδεις ἔχοντα τῶν τοιούτων τριχῶν
35 εἶδός τι ὑποληπτέον, οἷον τάς τε τῶν χερσαίων
782 a ἐχίνων καὶ εἴ τι ἄλλο τοιοῦτόν ἐστι τῶν ζῳοτόκων.
εἰσὶ δὲ διαφοραὶ τῶν τριχῶν κατά τε σκληρότητα
καὶ μαλακότητα, καὶ κατὰ μῆκος καὶ βραχύτητα,
καὶ εὐθύτητα καὶ οὐλότητα, καὶ πλῆθος καὶ ὀλι-

---

¹ τὴν Z¹ : τὰ PSYZ².

ᵃ See App. B §§ 27 ff.

cised at a distance is possessed by man to a lesser degree, in proportion to his size, than almost any other animal ; on the other hand, he is better than any of them at accurately perceiving the differences in the objects perceived. The reason is that in man the sense-organ is pure and least earthy and corporeal, and besides that, nature has given him, for his size, the thinnest skin that any animal has.

Nature has brought off a clever piece of work in the seal, too, which, although it is a viviparous quadruped, possesses no ears but passages merely. The reason is that it spends its life in a fluid medium. The ear is a part of the body which is an addition made to the passages in order to safeguard the movement of the air *a* which comes from a distance, and therefore it is no use to the seal ; indeed it would actually be a hindrance rather than a help, because it would act as a receptacle for a large volume of water.

This concludes our remarks about sight, hearing and smell.

The various kinds of growths of hair.—In human III beings these differ in the same individuals at different Varieties periods of life, and they differ also in comparison of hair. with the other animals that have hair. Practically all the animals which are internally viviparous have hair ; I say " all," because the spines which some of them have on the body must be considered as being a kind of hair, *e.g.*, the spines of the hedgehog *b* and any other such viviparous creature. Hair exhibits the following differences : it may be hard or soft, long or short, straight or curly, plentiful or

b Gk. " land-echinus," to distinguish it from the " sea-echinus " or sea-urchin.

γότητα, πρὸς δὲ τούτοις καὶ κατὰ[1] τὰς χρόας,
5 κατά τε[2] λευκότητα καὶ μελανίαν καὶ τὰς μεταξὺ
τούτων. ἐνίαις[3] δὲ τούτων τῶν διαφορῶν καὶ
κατὰ τὰς ἡλικίας διαφέρουσι νέα τε καὶ παλαιού-
μενα. μάλιστα δὲ τοῦτ᾽ ἐπίδηλον ἐπὶ τῶν ἀνθρώ-
πων· καὶ γὰρ δασύνεται μᾶλλον πρεσβύτερα
γιγνόμενα, καὶ φαλακροῦνται τῆς κεφαλῆς ἔνιοι τὰ
10 πρόσθεν. καὶ παῖδες μὲν ὄντες οὐ γίγνονται
φαλακροί, οὐδ᾽ αἱ γυναῖκες· οἱ δ᾽ ἄνδρες προϊούσης
ἤδη τῆς ἡλικίας. καὶ πολιοῦνται δὲ τὰς κεφαλὰς
γηράσκοντες οἱ ἄνθρωποι. τῶν δ᾽ ἄλλων ζῴων
οὐθενὶ τοῦθ᾽ ὡς εἰπεῖν γίνεται ἐπίδηλον, μάλιστα
δ᾽ ἵππῳ τῶν ἄλλων. καὶ φαλακροῦνται μὲν οἱ
15 ἄνθρωποι τὰ ἔμπροσθεν τῆς κεφαλῆς, πολιοὶ δὲ
πρῶτον γίνονται τοὺς κροτάφους· φαλακροῦται δ᾽
οὐθεὶς οὔτε τούτους οὔτε τὰ ὄπισθεν τῆς κεφαλῆς.
ὅσα δὲ τῶν ζῴων μὴ ἔχει τρίχας ἀλλὰ τὸ ἀνάλογον
αὐταῖς, οἷον ὄρνιθες μὲν πτερά, τὸ δὲ τῶν ἰχθύων
γένος λεπίδας, καὶ τούτοις συμβαίνει τῶν τοιούτων
20 παθημάτων ἔνια κατὰ τὸν αὐτὸν λόγον.

Τίνος μὲν οὖν ἔνεκα τὸ τῶν τριχῶν ἡ φύσις
ἐποίησε γένος τοῖς ζῴοις, εἴρηται πρότερον ἐν ταῖς
αἰτίαις ταῖς περὶ τὰ μέρη τῶν ζῴων· τίνων δ᾽
ὑπαρχόντων καὶ διὰ τίνας ἀνάγκας συμβαίνει
τούτων ἕκαστον, δηλῶσαι τῆς μεθόδου τῆς νῦν
ἐστίν.

Παχύτητος μὲν οὖν καὶ λεπτότητος αἴτιόν ἐστι[4]
25 μάλιστα τὸ δέρμα· τοῖς μὲν γὰρ παχὺ τοῖς δὲ λεπτόν,
καὶ τοῖς μὲν μανὸν τοῖς δὲ πυκνόν ἐστιν. ἔτι δὲ

---

[1] καὶ κατὰ P : καὶ S : κατὰ vulg.   [2] τε PZ, om. vulg.
[3] ἐνίαις PZ[2] : ἐνίας vulg.
[4] ἐστι PZ : om. vulg.

scanty ; beside this, it also shows differences of colour : it may be white or black or any shade between these two. Some of these differences are also exhibited by the hair according to the various times of life, youth and more advanced age. This is noticeable chiefly in the case of human beings. Thus the hair gets shaggier as age advances, and some people go bald in front. Children do not go bald, nor do women ; men do, however, when they begin to get on in years. In human beings, the hair on the head turns white as age approaches ; in other animals, however, this does not noticeably occur : the horse is the one which shows it most. Human beings go bald on the front of the head, but they go grey first on the temples ; none however goes bald either here or at the back of the head. As for those animals which have no hair but the counterpart of hair instead (thus, birds have feathers, and the fish tribe have scales)—in them some conditions of the kind described occur in a corresponding way.

We have already stated in the treatise on the *Causes of the Parts of Animals* [a] the purpose for the sake of which Nature has made hair in general and provided animals with it. The business of our present investigation is to show what are the pre-existing circumstances, what are the factors of *necessity*, on account of which the particular sorts of hair occur.

The chief cause, then, of its thickness and thinness is the skin ; which in some animals is thick, in others thin ; looseknit in some, compact in others. A con-

[a] *P.A.* 658 a 18 ; viz., for the sake of shelter and protection.

782 a

συναίτιον καὶ τῆς ἐνούσης ὑγρότητος ἡ διαφορά·
τοῖς μὲν γὰρ ὑπάρχει λιπαρὰ τοῖς δ' ὑδατώδης.
ὅλως μὲν γὰρ ἡ τοῦ δέρματος φύσις ὑπόκειται
30 γεώδης· ἐπιπολῆς γὰρ οὖσα ἐξατμίζοντος τοῦ
ὑγροῦ στερεὰ γίνεται καὶ γεώδης, αἱ δὲ τρίχες καὶ
τὸ ἀνάλογον αὐταῖς οὐκ ἐκ τῆς σαρκὸς γίνονται
ἀλλ' ἐκ τοῦ δέρματος [ἐξατμίζοντος καὶ ἀναθυμιω-
μένου ἐν αὐτοῖς τοῦ ὑγροῦ. διὸ παχεῖαι μὲν ἐκ
τοῦ παχέος, λεπταὶ δὲ[1] ἐκ τοῦ λεπτοῦ δέρματος
35 γίνονται].[2] ἂν μὲν οὖν ᾖ τὸ δέρμα μανότερον καὶ

782 b

παχύτερον, παχεῖαι διά τε τὸ πλῆθος τοῦ γεώδους
καὶ διὰ τὸ μέγεθος τῶν πόρων εἰσίν· ἂν δὲ
πυκνότερον, λεπταὶ διὰ τὴν στενότητα τῶν πόρων.
ἔτι δ' ἂν ᾖ ἡ ἰκμὰς ὑδατώδης, ταχὺ ἀναξηραινο-
μένης οὐ λαμβάνουσι μέγεθος αἱ τρίχες, ἂν δὲ
5 λιπαρά, τοὐναντίον· οὐ γὰρ εὐξήραντον τὸ λιπαρόν.
διόπερ ὅλως μὲν τὰ παχυδερμότερα παχυτριχώ-
τερα τῶν ζῴων, οὐ μέντοι τὰ μάλιστα μᾶλλον,
διὰ τὰς εἰρημένας αἰτίας, οἷον τὸ τῶν ὑῶν γένος
πρὸς τὸ τῶν βοῶν πέπονθε καὶ πρὸς ἐλέφαντα καὶ
πρὸς[3] πολλὰ τῶν ἄλλων. διὰ τὴν αὐτὴν δ' αἰτίαν
10 καὶ αἱ ἐν τῇ κεφαλῇ τρίχες τοῖς ἀνθρώποις παχύ-
ταται· τοῦ γὰρ δέρματος τοῦτο παχύτατον καὶ ἐπὶ[4]
πλείστῃ ὑγρότητι, ἔτι δ' ἔχει μανότητα πολλήν.
αἴτιον δὲ καὶ τοῦ μακρὰς [ἢ βραχείας][5] τὰς τρίχας
εἶναι τὸ μὴ εὐξήραντον εἶναι τὸ ἐξατμίζον ὑγρόν.
τοῦ δὲ μὴ εὐξήραντον εἶναι δύ' αἰτίαι, τό τε ποσὸν

---

[1] ἐκ τοῦ παχέος, λεπταὶ δὲ om. SZ¹. [2] secl. Platt.
[3] πρὸς P*Z : om. vulg. [4] ἐπὶ Z¹ : ἐν vulg.
[5] seclusi ; om. Σ.

tributory cause is the difference of the fluid present in it : in some this is greasy, in others watery. In general, of course, the fundamental nature of the skin is earthy in substance : being on the surface of the body it becomes solid and earthy as the fluid evaporates off. Now the hair and its counterparts are formed not out of the flesh but out of the skin [as the fluid in them evaporates and exhales ; thus thick hair is formed out of thick skin and thin hair out of thin skin].[a] If, then, the skin tends to be looseknit and thick, the hair is thick both on account of the large amount of earthy matter and on account of the size of the passages ; but if the skin tends to be compact, the hair is thin on account of the narrowness of the passages. Further, if the moisture is watery, it quickly dries off and the hair does not attain to any size, though it does if the moisture is greasy, because greasy matter does not readily dry off. Thus, generally speaking, thick-skinned animals have thick hair[b] ; but it is not true that the thickest-skinned have thicker hair than ⟨the others in the same category⟩, for the causes mentioned, an example being afforded by the pig tribe when compared with that of oxen, or with the elephant and many other animals. For the same cause, too, our hair is thickest on the head : the skin there is thickest and situated over the largest amount of fluid,[c] and besides that it is very loosely knit. And the reason why the hair is long [or short] is that the fluid which evaporates is not easily dried off. There are two causes which prevent it being easily dried off : one is its quantity, the other its

---

[a] These words are deleted by Platt as partly unintelligible and as not fitting in with what follows.
[b] But see 783 a 2.　　　　　[c] Viz., the brain.

15 καὶ τὸ ποιόν· ἄν τε γὰρ πολὺ ᾖ τὸ ὑγρόν, οὐκ
εὐξήραντον, καὶ ἂν λιπαρόν. καὶ διὰ τοῦτο τοῖς
ἀνθρώποις αἱ ἐκ τῆς κεφαλῆς τρίχες μακρόταται·
ὁ γὰρ ἐγκέφαλος ὑγρὸς καὶ ψυχρὸς ὢν πολλὴν
παρέχει δαψίλειαν τοῦ ὑγροῦ.

Εὐθύτριχα δὲ καὶ οὐλότριχα γίνεται διὰ τὴν ἐν
20 ταῖς θριξὶν ἀναθυμίασιν. ἂν μὲν γὰρ ᾖ καπνώδης,
θερμὴ οὖσα καὶ ξηρὰ οὔλην τὴν τρίχα ποιεῖ.
κάμπτεται γὰρ διὰ τὸ δύο φέρεσθαι φοράς· τὸ
μὲν γὰρ γεῶδες κάτω, τὸ δὲ θερμὸν ἄνω φέρεται. εὐκάμπτου δ᾽ οὔσης[1] δι᾽ ἀσθένειαν στρέ-
φεται· τοῦτο δ᾽ ἐστὶν οὐλότης τριχός. ἐνδέχεται
μὲν οὖν οὕτω λαβεῖν τὴν αἰτίαν, ἐνδέχεται δὲ καὶ
25 διὰ τὸ ὀλίγον ἔχειν τὸ ὑγρόν, πολὺ δὲ τὸ γεῶδες,
ὑπὸ τοῦ περιέχοντος ξηραινομένας συσπᾶσθαι.
κάμπτεται γὰρ τὸ εὐθύ, ἐὰν ἐξατμίζηται, καὶ
συντρέχει ὥσπερ ἐπὶ τοῦ πυρὸς καομένη θρίξ,[2]
ὡς οὔσης τῆς οὐλότητος συσπάσεως δι᾽ ἔνδειαν
ὑγροῦ ὑπὸ τῆς τοῦ περιέχοντος θερμότητος. ση-
30 μεῖον δ᾽ ὅτι καὶ σκληρότεραι αἱ οὖλαι τρίχες τῶν
εὐθειῶν εἰσιν· τὸ γὰρ ξηρὸν σκληρόν. εὐθύτριχα
δὲ ὅσα ὑγρότητ᾽ ἔχει πολλήν· ῥέον γὰρ ἀλλ᾽ οὐ
στάζον προέρχεται ἐν ταύταις τὸ ὑγρόν. καὶ διὰ
τοῦτο οἱ μὲν ἐν τῷ Πόντῳ Σκύθαι καὶ Θρᾷκες
εὐθύτριχες· καὶ γὰρ αὐτοὶ ὑγροὶ καὶ ὁ περιέχων

---

[1] οὔσης Peck : ὄντος vulg.
[2] θρίξ PZ[1], A.-W. : ἡ θρίξ vulg.

---

[a] For this and other subjects dealt with in Book V, see
H. Diller, *Wanderarzt und Aitiologe*, pp. 115 ff., *cf.* 50 ff.
et passim.
[b] According to Aristotle, there were two sorts of " ex-

quality. Thus, if there is a great deal of the fluid, and also if it is greasy, it does not easily dry off. And that is why the hair on our heads is the longest : the brain, being fluid and cold, provides fluid in large abundance.

Straight hair and curly hair [a] is due to the exhalation in it : if this exhalation is smoky,[b] being hot and dry it makes the hair curly ; for the hair gets bent because it is subjected to the impulse of two directional motions—the earthy constituent urges its way downwards, the hot constituent upwards ; and as the hair will easily bend on account of its weakness, it gets twisted ; that is what curliness of the hair really is. Well, that is one cause that may be assigned for it : here is another. It may equally well be that, owing to its containing but little fluid as against a great deal of earthy matter, the hair gets dried by its environment and so contracts. Anything that is straight bends if its vapour is drawn off, and shrinks up like a hair burning on the fire, which would imply that the curliness of hair is a contraction due to lack of fluid caused by the heat of its environment. In favour of this is the fact that curly hair is also harder than straight hair, and of course anything dry is hard. Animals that contain a great deal of fluid have straight hair, because in their hair the fluid advances in a continuous stream and not drop by drop. That is why the Scythians by the Black Sea and the Thracians have straight hair : both their constitution and the environing air are fluid (moist).

halation " : the " smoky," a compound of Air and Earth, which is hot and dry ; and the " aqueous," which is cold and moist. For further details see *De sensu* 443 a 21 ff., *Meteor.* 360 a 22 ff., *cf. G.A.* 784 b 10.

35 αὐτοὺς ἀὴρ ὑγρός· Αἰθίοπες δὲ καὶ οἱ ἐν τοῖς
θερμοῖς οὐλότριχες· ξηροὶ γὰρ οἱ ἐγκέφαλοι καὶ
ὁ ἀὴρ ὁ περιέχων.

Ἔστι δ' ἔνια τῶν παχυδέρμων λεπτότριχα διὰ
τὴν εἰρημένην αἰτίαν πρότερον· ὅσῳ γὰρ ἂν λεπτό-
τεροι οἱ πόροι ὦσιν, τοσούτῳ λεπτοτέρας ἀναγκαῖον
5 γίνεσθαι τὰς τρίχας. διὸ τὸ τῶν προβάτων γένος
τοιαύτας ἔχει τὰς τρίχας· τὸ γὰρ ἔριον τριχῶν
πλῆθός ἐστιν. ἔστι δ' ἔνια τῶν ζῴων ἃ μαλακὴν
μὲν ἔχει τὴν τρίχα, ἧττον δὲ λεπτήν, οἷον τὸ τῶν
δασυπόδων πρὸς τὸ τῶν προβάτων πέπονθεν. τῶν
γὰρ τοιούτων ἐπιπολῆς ἡ θρὶξ τοῦ δέρματος. διὸ
10 μῆκος οὐκ ἴσχει, ἀλλὰ συμβαίνει παραπλήσιον ὥσπερ τὰ ἀπὸ τῶν¹ λίνων² ξυόμενα· καὶ γὰρ ταῦτα
μῆκος μὲν οὐθὲν ἴσχει, μαλακὰ δ' ἐστὶ καὶ οὐ
δέχεται πλοκήν. τὰ δ' ἐν τοῖς ψυχροῖς πρόβατα
τοὐναντίον πέπονθε τοῖς ἀνθρώποις· οἱ μὲν γὰρ
Σκύθαι μαλακότριχες, τὰ δὲ πρόβατα τὰ Σαυρο-
15 ματικὰ σκληρότριχα. τούτου δ' αἴτιον ταὐτὸ καὶ
ἐπὶ τῶν ἀγρίων πάντων. ἡ γὰρ ψυχρότης σκληρύνει διὰ τὸ ξηραίνειν πηγνύουσα· ἐκθλιβομένου γὰρ
τοῦ θερμοῦ συνεξατμίζει τὸ ὑγρόν, καὶ γίνονται καὶ
αἱ τρίχες καὶ τὸ δέρμα γεῶδες καὶ σκληρόν. αἴτιον
δὲ τοῖς μὲν ἀγρίοις ἡ θυραυλία, τοῖς δ' ὁ τόπος
20 τοιοῦτος ὤν. σημεῖον δὲ καὶ τὸ ἐπὶ τῶν ποντίων
ἐχίνων συμβαῖνον, οἷς χρῶνται πρὸς τὰς στραγ-
γουρίας. καὶ γὰρ οὗτοι διὰ τὸ ἐν ψυχρᾷ εἶναι
τῇ θαλάττῃ διὰ τὸ βάθος (καθ' ἑξήκοντα γὰρ καὶ

¹ τῶν PSY*Z : om. Bekker per errorem.
² λίνων fortasse scrib. monet Platt.

On the other hand, Ethiopians and people who live in hot regions have curly hair, because both their brain and the environing air are dry.

Some, however, of the thick-skinned animals have fine hair owing to the cause previously mentioned [a] : the finer the passages are, the finer of necessity must the hairs be. That is why all sheep have fine hair, wool being just a very large number of hairs. There are some animals whose hair is soft, yet not so fine ; this is true of hares, for instance, in comparison with sheep. In such animals the hair is on the surface of the skin ; and so it is not long, but turns out to be very much on a par with the scrapings that come off linen cloth, which have no length worth mentioning, but are soft and cannot be used for weaving. In cold climates sheep and human beings exhibit opposite " conditions " from each other : thus the Scythians have soft hair, but Sarmatian [b] sheep have hard hair, the reason for which is the same as it is in all wild animals. The cold congeals them and so dries them, and this makes them hard : in other words, the fluid evaporates at the same time as the heat is expelled, and both hair and skin become earthy and hard. Thus with wild animals the reason is that they live in the open air ; but in other cases it is the nature of their situation which is responsible. This is shown by what occurs in the case of the sea-urchins which are used as a remedy for cases of strangury. These creatures, although small in themselves, have long, hard spines, because the seawater they live in is cold on account of its being so deep (60 fathoms or even

_____

[a] See 782 b 1.
[b] Sarmatia is the territory between the Vistula and the Don, part of modern Poland and Russia.

ἔτι πλειόνων γίγνονται ὀργυιῶν), αὐτοὶ μὲν μικροί,
τὰς δὲ ἀκάνθας μεγάλας ἔχουσι καὶ σκληράς,
25 μεγάλας μὲν διὰ τὸ ἐνταῦθα τὴν τοῦ σώματος
τετράφθαι αὔξησιν (ὀλιγόθερμοι γὰρ ὄντες καὶ οὐ
πέττοντες τὴν τροφὴν πολὺ τὰ περίττωμα ἔχουσιν, αἱ
δ' ἄκανθαι καὶ αἱ τρίχες καὶ τὰ τοιαῦτα γίνονται
ἐκ περιττώματος), σκληρὰς δὲ καὶ λελιθωμένας
διὰ τὴν ψυχρότητα καὶ τὸν πάγον. τὸν αὐτὸν δὲ
30 τρόπον καὶ τἆλλα τὰ φυόμενα σκληρότερα συμβαίνει γίνεσθαι καὶ γεωδέστερα καὶ λιθωδέστερα
τὰ ἐν τοῖς προσβόρροις[1] τῶν πρὸς νότον καὶ τὰ
προσήνεμα τῶν ἐν κοίλοις· ψύχεται γὰρ πάντα
μᾶλλον, καὶ ἐξατμίζει τὸ ὑγρόν. σκληρύνει μὲν
οὖν καὶ τὸ θερμὸν καὶ τὸ ψυχρόν· ἐξατμίζεσθαι
35 γὰρ ὑπ' ἀμφοτέρων συμβαίνει τὸ ὑγρόν, ὑπὸ μὲν
τοῦ θερμοῦ καθ' αὑτό, ὑπὸ δὲ τοῦ ψυχροῦ κατὰ
συμβεβηκός (μετὰ τοῦ θερμοῦ γὰρ συνεξέρχεται·
οὐθὲν γὰρ ὑγρὸν ἄνευ θερμοῦ). ἀλλὰ τὸ μὲν
783 b ψυχρὸν οὐ μόνον σκληρύνει ἀλλὰ καὶ πυκνοῖ, τὸ
δὲ θερμὸν μανότερον ποιεῖ.

Διὰ τὴν αὐτὴν δ' αἰτίαν καὶ πρεσβυτέρων γιγνομένων τοῖς μὲν τρίχας ἔχουσι σκληρότεραι γίγνονται αἱ τρίχες, τοῖς δὲ πτερωτοῖς καὶ λεπιδωτοῖς
5 τὰ πτερὰ καὶ αἱ λεπίδες. τὰ γὰρ δέρματα γίνεται
σκληρότερα καὶ παχύτερα πρεσβυτέρων γιγνο-
μένων· ξηραίνεται γάρ, καὶ τὸ γῆράς ἐστι κατὰ
τοὔνομα γεηρὸν διὰ τὸ ἀπολείπειν τὸ θερμὸν καὶ
μετ' αὐτοῦ τὸ ὑγρόν.

---

[1] προσβόρροις A.-W. : προσβόροις PSZ[1] : πρὸς βορρᾶν vulg.

[a] This is an important statement, and should be noted in
connexion with Aristotle's theories of the part played by
fluid and heat both in nourishment and in spontaneous

more is the depth at which they are found). Their spines are long because the growth of the body is diverted to them, since, as the creatures possess but little heat, they cannot concoct the nourishment, and so contain a great deal of residue; and it is out of residue that spines and hair and the like are formed. Their spines are hard and petrified on account of the cold and its congealing effect. And in the same way plants, too, are harder, and earthier, and more petrified if they grow where the aspect is northerly, or in a windy situation, than if they grow where the aspect is southerly, or in a sheltered spot. It is because they all get more chilled, and their fluid evaporates. Hardening, then, is brought about by both cold and heat : the effect of both is to cause the fluid to evaporate : it is evaporated by heat *per se*, but by cold *per accidens*—in the latter case the fluid accompanies the heat when it makes its exit, as there is no fluid without its heat.[a] There is this difference, however : cold causes compression as well as hardening, whereas heat lightens a thing's consistency.[b]

For the same cause hair, feathers and scales in the various animals respectively become harder as they get on in years : it is because their skins grow harder and thicker then, and that is due to their drying up, and old age or to " get on in years " is something earthy (as the similarity of the word with yearth, the old form of " earth," shows),[c] and this is due to the fact that the heat is failing and with it the fluid.

generation. See also *P.A.* 652 b 8 ff. and App. B § 11 and § 17 and note.

[b] This hardly agrees with Aristotle's statements elsewhere (*e.g.*, 765 b 1 ff.) about the thickening effects of concoction.

[c] This is a piece of " etymology " comparable with that of the original Greek : *gēras* (old age), *gēēron* (earthy).

Φαλακροῦνται δ' ἐπιδήλως οἱ ἄνθρωποι μάλιστα
10 τῶν ζῴων. ἔστι δέ τι καθόλου τὸ τοιοῦτον πάθος·
καὶ γὰρ τῶν φυτῶν τὰ μὲν ἀείφυλλα τὰ δὲ φυλ-
λοβολεῖ, καὶ τῶν ὀρνίθων οἱ φωλεύοντες ἀπο-
βάλλουσι τὰ πτερά. τοιοῦτον δέ τι πάθος καὶ ἡ
φαλακρότης ἐστὶν ἐπὶ τῶν ἀνθρώπων, ὅσοις συμβαίνει φαλακροῦσθαι· κατὰ μέρος μὲν γὰρ ἀπορρεῖ
15 καὶ τὰ φύλλα τοῖς φυτοῖς πᾶσι καὶ τὰ πτερὰ καὶ
αἱ τρίχες τοῖς ἔχουσιν, ὅταν δ' ἀθρόον γένηται
τὸ πάθος, λαμβάνει τὰς εἰρημένας ἐπωνυμίας·
φαλακροῦσθαί τε γὰρ λέγεται καὶ φυλλορροεῖν.[1]
αἴτιον δὲ τοῦ πάθους ἔνδεια ὑγρότητος θερμῆς,
τοιοῦτον δὲ μάλιστα τῶν ὑγρῶν τὸ λιπαρόν· διὸ
20 καὶ τῶν φυτῶν τὰ λιπαρὰ ἀείφυλλα μᾶλλον.
ἀλλὰ περὶ μὲν τούτων ἐν ἄλλοις τὸ αἴτιον λεκτέον·
καὶ γὰρ ἄλλα συναίτια τούτου τοῦ[2] πάθους αὐτοῖς.
γίνεται δὲ τοῖς μὲν φυτοῖς ἐν τῷ χειμῶνι τὸ
πάθος (αὕτη γὰρ ἡ μεταβολὴ κυριωτέρα τῆς
ἡλικίας), καὶ τοῖς φωλεύουσι δὲ τῶν ζῴων (καὶ
25 γὰρ ταῦτα ἧττον τῶν ἀνθρώπων ὑγρὰ καὶ θερμὰ
τὴν φύσιν ἐστίν)· οἱ δ' ἄνθρωποι ταῖς ἡλικίαις
χειμῶνα καὶ θέρος ἄγουσιν. διὸ πρὶν ἀφροδισιάζειν οὐ γίνεται φαλακρὸς οὐδείς· τότε δὲ τοῖς
τοιούτοις τὴν φύσιν μᾶλλον. φύσει γάρ ἐστιν ὁ
ἐγκέφαλος ψυχρότατον τοῦ σώματος, ὁ δ' ἀφρο-
30 δισιασμὸς καταψύχει· καθαρᾶς γὰρ καὶ φυσικῆς

---

[1] ⟨καὶ πτερορροεῖν⟩ adduct A.-W., Bekkerum secuti;
melius πτερορρυεῖν Btf.; om. codd.; fort. φυλλο⟨βολεῖν καὶ
πτερο⟩ρρυεῖν scrib.
[2] τούτου τοῦ Z : τοῦ τοιούτου vulg.

---

[a] The Gk. has " shedding of leaves," but as there is no
one English word for this, and as all three are referred to in

Of all animals human beings are the ones which go bald most noticeably ; but still baldness is a general and widespread condition. Thus, although some plants are evergreen, others shed their leaves, and birds which hibernate shed their feathers. Baldness, in those human beings whom it affects, is a comparable condition to these. Of course, a partial and gradual shedding of leaves takes place in all plants, and of feathers and hair in those animals that have them ; but it is when the shedding affects the whole of the hair, feathers, etc., at once that the condition is described by the terms already mentioned (baldness, moulting,[a] etc.). The cause of this condition is a deficiency of hot fluid, the chief hot fluid being greasy fluid, and that is why greasy plants tend more to be evergreen than others. However, we shall have to deal with the cause of this condition so far as plants are concerned in another treatise, since in their case there are other contributory causes of it. Now in plants this condition occurs in winter : this seasonal change overrides in importance the change in the time of life. The same is true of the hibernating animals ; they too are in their nature less fluid and less hot than human beings. For human beings, however, it is the seasons of life which play the part of summer and winter ; and that is why no one goes bald before the time of sexual intercourse, and also why that is the time when those who are naturally prone to intercourse go bald. The reason is that the effect of sexual intercourse is to cool, as it is the excretion of some of the pure, natural heat, and the

the context, I have kept the point by substituting " moulting ": the Berlin edition and others actually insert the word for " moulting " into the Gk. text.

783 b

θερμότητος ἀπόκρισίς ἐστιν. εὐλόγως οὖν ὁ
ἐγκέφαλος αἰσθάνεται πρῶτον· τὰ γὰρ ἀσθενῆ καὶ
φαύλως ἔχοντα μικρᾶς αἰτίας καὶ ῥοπῆς ἐστιν.
ὥστ' ἄν τις ἀναλογίσηται ὅτι αὐτός τε ὀλιγόθερμος
ὁ ἐγκέφαλος, ἔτι δ' ἀναγκαῖον τὸ πέριξ δέρμα
35 τοιοῦτον εἶναι μᾶλλον, καὶ τούτου τὴν τῶν τριχῶν
φύσιν, ὅσῳ πλεῖστον ἀφέστηκεν, εὐλόγως ἂν δόξειε
τοῖς σπερματικοῖς περὶ ταύτην τὴν ἡλικίαν συμβαίνειν φαλακροῦσθαι. διὰ τὴν αὐτὴν δ' αἰτίαν
784 a καὶ τῆς κεφαλῆς τὸ πρόσθιον μόνον γίνονται φα-
λακροὶ καὶ τῶν ζῴων οἱ ἄνθρωποι μόνοι, τὸ μὲν
πρόσθιον, ὅτι ἐνταῦθα ὁ ἐγκέφαλος, τῶν δὲ ζῴων
μόνον, ὅτι πολὺ πλεῖστον ἔχει ἐγκέφαλον καὶ
μάλιστα ὑγρὸν ὁ¹ ἄνθρωπος. καὶ αἱ γυναῖκες οὐ
5 φαλακροῦνται· παραπλησία γὰρ ἡ φύσις τῇ τῶν
παιδίων· ἄγονα γὰρ σπερματικῆς ἐκκρίσεως ἀμφότερα. καὶ εὐνοῦχος οὐ γίνεται φαλακρὸς διὰ τὸ
εἰς τὸ θῆλυ μεταβάλλειν. καὶ τὰς ὑστερογενεῖς
τρίχας ἢ οὐ φύουσιν ἢ ἀποβάλλουσιν, ἂν τύχωσιν
ἔχοντες οἱ εὐνοῦχοι, πλὴν τῆς ἥβης· καὶ γὰρ αἱ
10 γυναῖκες τὰς μὲν οὐκ ἔχουσι, τὰς δ' ἐπὶ τῇ ἥβῃ
φύουσιν. ἡ δὲ πήρωσις αὕτη ἐκ τοῦ ἄρρενος εἰς
τὸ θῆλυ μεταβολή ἐστιν.

Τοῦ δὲ τὰ μὲν φωλεύοντα πάλιν δασύνεσθαι καὶ
τὰ φυλλοβολήσαντα πάλιν φύειν φύλλα, τοῖς δὲ
φαλακροῖς μὴ ἀναφύεσθαι πάλιν, αἴτιον ὅτι τοῖς
15 μὲν αἱ ὧραι τροπαί εἰσι τοῦ σώματος μᾶλλον,
ὥστ' ἐπεὶ μεταβάλλουσιν αὗται, μεταβάλλει καὶ
τὸ φύειν καὶ τὸ ἀποβάλλειν τοὺς μὲν τὰ πτερὰ

¹ ὁ Z: om. vulg.

brain is by its nature the coldest part of the body ;
thus, as we should expect, it is the first part to feel
the effect : anything that is weak and poorly needs
only a slight cause, a slight momentum, to make it
react. So that if you reckon up (*a*) that the brain
itself has very little heat, (*b*) that the skin surrounding
it must of necessity have even less, and (*c*) that the
hair, being the furthest off of the three, must have
even less still, you will expect persons who are plentiful in semen to go bald at about this time of life.
And it is owing to the same cause that it is on the
front part of the head only that human beings go
bald, and that they are the only animals which do so
at all ; *i.e.*, they go bald in front because the brain is
there,[a] and they alone do so, because they have by
far the largest brain of all and the most fluid. Women
do not go bald because their nature is similar to that
of children : both are incapable of producing seminal
secretion. Eunuchs, too, do not go bald, because of
their transition into the female state, and the hair
that comes at a later stage they fail to grow at all,
or if they already have it, they lose it, except for the
pubic hair : similarly, women do not have the later
hair, though they do grow the pubic hair. This
deformity constitutes a change from the male state
to the female.

The reason why the hair does not grow again in
cases of baldness, although hair and feathers grow
again on hibernating animals and leaves on deciduous
trees, is that in the case of the animals and trees the
seasons are the turning-points of their lives more
⟨than in the case of man⟩, and so when there is a
change of season, then they follow suit and grow or

[a] See *P.A.* 656 b 12.

784 a

καὶ τὰς τρίχας, τὰ δὲ φύλλα τὰ φυτά. τοῖς δ'
ἀνθρώποις κατὰ τὴν ἡλικίαν γίνεται χειμὼν καὶ
θέρος καὶ ἔαρ καὶ μετόπωρον, ὥστ' ἐπειδὴ[1] αἱ
20 ἡλικίαι οὐ μεταβάλλουσιν, οὐδὲ τὰ πάθη τὰ διὰ
ταύτας μεταβάλλει, καίπερ τῆς αἰτίας ὁμοίας
οὔσης.

Καὶ περὶ μὲν τἆλλα πάθη τὰ τῶν τριχῶν σχεδὸν
εἴρηται.

IV   Τῶν δὲ χρωμάτων αἴτιον τοῖς μὲν ἄλλοις ζῴοις,
καὶ τοῦ μονόχροα εἶναι καὶ τοῦ ποικίλα, ἡ τοῦ
25 δέρματος φύσις· τοῖς δ' ἀνθρώποις οὐδὲν πλὴν τῶν
πολιῶν οὐ τῶν διὰ γήρας ἀλλὰ τῶν διὰ νόσον·
ἐν γὰρ τῇ καλουμένῃ λεύκῃ λευκαὶ γίνονται αἱ
τρίχες· ἐὰν δ' αἱ τρίχες ὦσι λευκαί, οὐκ ἀκολουθεῖ τῷ δέρματι ἡ λευκότης. αἴτιον δ' ὅτι αἱ
τρίχες ἐκ τοῦ δέρματος φύονται· ἐκ νενοσηκότος
30 οὖν καὶ λευκοῦ τοῦ δέρματος καὶ ἡ θρὶξ συννοσεῖ,
νόσος δὲ τριχὸς πολιότης ἐστίν. ἡ δὲ δι' ἡλικίαν
τῶν τριχῶν πολιότης γίνεται δι' ἀσθένειαν καὶ
ἔνδειαν θερμότητος. καὶ γὰρ ἡλικία πᾶσα ῥέπει
ἀποκλίνοντος τοῦ σώματος, καὶ ἐν τῷ γήρᾳ, ἐπὶ
ψύξιν· τὸ γὰρ γῆρας ψυχρὸν καὶ ξηρόν ἐστιν. δεῖ
35 δὲ νοῆσαι τὴν εἰς ἕκαστον μόριον ἀφικνουμένην
τροφὴν ὅτι πέττει μὲν ἡ ἐν ἑκάστῳ[2] οἰκεία θερμό-

784 b

της, ἀδυνατούσης δὲ φθείρεται καὶ πήρωσις γίνεται
ἢ νόσος. ἀκριβέστερον δὲ περὶ τῆς τοιαύτης αἰτίας
ὕστερον λεκτέον ἐν τοῖς περὶ αὐξήσεως καὶ τροφῆς.

[1] ἐπειδὴ Z : ἐπεὶ vulg.
[2] ἐν ἑκάστῳ PZ : om. vulg.

---

[a] Cf. 783 b 7, and De long. et brev. vit. 466 a 21; but
according to Hippocrates, π. διαίτης I. 33 (vi. 512 Littré), the
aged are ψυχροὶ καὶ ὑγροί.

shed their feathers or hair or leaves. In man, however, the spring, summer, autumn and winter of his life are not seasons according to the calendar but seasons of his own age ; so that, as these do not go through the cycle of change, neither do the conditions which depend on them ; although the cause which controls the change of conditions is a similar one in his case too.

I think we have now discussed all the conditions that affect hair, except that of colour.

In the rest of the animals, the reason for the various colours of the hair, and for its being single-coloured or variegated, is the nature of the skin. In man, however, this reason operates only in the case of the greyness of the hair due to disease (as when the hair becomes white during leprosy), not that due to old age, and if the hair is white, the whiteness does not derive from the skin. The reason is that the hair grows out of the skin, and thus when the skin out of which it grows is diseased and white the hair is itself affected by disease, and disease of hair is greyness. On the other hand, the greyness which is due to age is the result of weakness and deficiency of heat. Every age of life tends to gravitate into chilliness when the body's vigour declines, and especially when this happens in old age, since old age is cold and dry.[a] We must bear in mind that the nourishment which reaches each part of the body is concocted by the heat in each part proper to it ; and if this heat is unable to do its work the part suffers damage, and deformity or disease is the result. A more detailed account of this cause will have to be given in the treatise *Of Growth and Nutrition.*[b] In those persons

IV

Colour of hair.

Greyness.

Not extant.

ὅσοις οὖν τῶν ἀνθρώπων ὀλιγόθερμός ἐστιν ἡ τῶν
5 τριχῶν φύσις καὶ πλείων ἡ εἰσιοῦσα ὑγρότης ἐστί,
τῆς οἰκείας θερμότητος ἀδυνατούσης πέττειν σήπεται ὑπὸ τῆς ἐν τῷ περιέχοντι θερμότητος.
γίνεται δὲ σῆψις ὑπὸ θερμότητος μὲν πᾶσα, οὐ τῆς
συμφύτου δέ, ὥσπερ εἴρηται ἐν ἑτέροις. ἔστι δ᾽
ἡ σῆψις καὶ ὕδατος καὶ γῆς καὶ τῶν σωματικῶν
10 πάντων τῶν τοιούτων, διὸ καὶ τῆς γεώδους ἀτμίδος, οἷον ὁ λεγόμενος εὐρώς· καὶ γὰρ ὁ εὐρώς
ἐστι σαπρότης γεώδους ἀτμίδος. ὥστε καὶ ἡ ἐν
ταῖς θριξὶ τοιαύτη οὖσα τροφὴ οὐ πεττομένη
σήπεται, καὶ γίνεται ἡ καλουμένη πολιά. λευκὴ
δέ, ὅτι καὶ ὁ εὐρὼς μόνον τῶν σαπρῶν ὡς εἰπεῖν
λευκόν ἐστιν. αἴτιον δὲ τούτου ὅτι πολὺν ἔχει
15 ἀέρα· πᾶσα γὰρ ἡ γεώδης ἀτμὶς ἀέρος ἔχει δύναμιν
παχέος. ὥσπερ γὰρ ἀντεστραμμένον τῇ πάχνῃ ὁ
εὐρώς ἐστιν· ἂν μὲν γὰρ παγῇ ἡ ἀνιοῦσα ἀτμίς,
πάχνη γίνεται, ἐὰν δὲ σαπῇ, εὐρώς. διὸ καὶ
ἐπιπολῆς ἐστιν ἄμφω· ἡ γὰρ ἀτμὶς ἐπιπολῆς. καὶ
εὖ δὴ οἱ ποιηταὶ ἐν ταῖς κωμῳδίαις μεταφέρουσι
20 σκώπτοντες, τὰς πολιὰς καλοῦντες γήρως εὐρῶτα
καὶ πάχνην. τὸ μὲν γὰρ τῷ γένει τὸ δὲ τῷ εἴδει
ταὐτόν ἐστιν, ἡ μὲν πάχνη τῷ γένει (ἀτμὶς γὰρ
ἄμφω), ὁ δὲ εὐρὼς τῷ εἴδει (σῆψις γὰρ ἄμφω).
σημεῖον δ᾽ ὅτι τοιοῦτόν ἐστιν· καὶ γὰρ ἐκ νόσων
25 πολλοῖς πολιαὶ ἀνέφυσαν, ὕστερον δ᾽ ὑγιασθεῖσι
μέλαιναι ἀντὶ τούτων. αἴτιον δ᾽ ὅτι ἐν τῇ ἀρρω-

ᵃ At *Meteor.* 379 a 16 ff. See App. B § 11, add. note.
ᵇ See 782 b 20, note.

where the nature of the hair has but little heat and the fluid which enters it is unduly plentiful, the heat proper to the hair is unable to do its work and the hair is putrefied by the heat present in the environment. All putrefaction, of course, is caused by heat, but not by the innate heat. This has been stated elsewhere.[a] Water and earth and all such corporeal bodies are liable to putrefaction, and therefore the earthy vapour [b] is liable to it as well; an example of this is what is called mould : mould is in fact the putrefaction of earthy vapour. So too the nourishment in the hair, being of this kind, putrefies if it does not get concocted, and what is called greyness results. It is white, because mould too is white. This is practically the only putrefied substance which is white, and the reason for that is that it contains a good deal of air : actually all earthy vapour is the equivalent [c] of thick air. In fact, mould is as it were the " opposite number " of hoar-frost, since if the vapour which rises up gets congealed, hoar-frost is the result ; if it gets putrefied, mould. And that is why both occur on the surface, because vapour is on the surface. So we see that the poets use a good metaphor in their comedies when they jokingly call white hairs the " mould " and " hoar-frost of age " : one of them is generically, the other specifically, the same as greyness : hoar-frost is the same generically (both being vapour), mould is the same specifically (both being putrefactions). Here is a sure sign that this is its character : there are many instances of people having grown grey hair as an aftermath of disease, but later on when they were restored to health dark hair took its place. The reason is that

For ἔχει δύναμιν cf. 780 b 9, and Introd. § 26.

784 b

στία, ὥσπερ καὶ τὸ ὅλον¹ σῶμα ἐν ἐνδείᾳ φυσικῆς
θερμότητός ἐστιν, οὕτω καὶ τῶν [ἄλλων]² μορίων
καὶ τὰ πάνυ μικρὰ μετέχει τῆς ἀρρωστίας ταύτης,
περίττωμα δὲ πολὺ ἐγγίνεται ἐν τοῖς σώμασι καὶ
30 ἐν³ τοῖς μορίοις· διόπερ ἡ ἐν ταῖς σαρξὶν ἀπεψία
ποιεῖ τὰς πολιάς. ὑγιάναντες δὲ καὶ ἰσχύσαντες
πάλιν μεταβάλλουσι, καὶ γίνονται ὥσπερ ἐκ γε-
ρόντων νέοι· διὸ καὶ τὰ πάθη συμμεταβάλλουσιν.
ὀρθῶς δ' ἔχει καὶ λέγειν τὴν μὲν νόσον γῆρας
ἐπίκτητον, τὸ δὲ γῆρας νόσον φυσικήν· ποιοῦσι
γοῦν νόσοι τινὲς ταὐτὰ ἅπερ καὶ τὸ γῆρας.
35 Τοὺς δὲ κροτάφους πολιοῦνται πρῶτον. τὰ μὲν

785 a γὰρ ὄπισθεν κενὰ ὑγρότητός ἐστι διὰ τὸ μὴ ἔχειν
ἐγκέφαλον, τὸ δὲ βρέγμα πολλὴν ἔχει ὑγρότητα·
τὸ δὲ πολὺ οὐκ εὔσηπτον. αἱ δ' ἐν τοῖς κροτάφοις
τρίχες οὔθ' οὕτως ὀλίγον ἔχουσιν ὑγρὸν ὥστε πέττειν, οὔτε πολὺ ὥστε μὴ σήπεσθαι· μέσος γὰρ ὢν
5 ὁ τόπος ἀμφοτέρων ἐκτὸς ἀμφοτέρων τῶν παθῶν
ἐστιν.

Περὶ μὲν οὖν τῆς τῶν ἀνθρώπων πολιότητος
εἴρηται τὸ αἴτιον.

V Τοῖς δ' ἄλλοις ζῴοις τοῦ μὴ γίνεσθαι διὰ τὴν
ἡλικίαν ταύτην τὴν μεταβολὴν ἐπιδήλως τὸ αὐτὸ
αἴτιον ὅπερ εἴρηται καὶ ἐπὶ τῆς φαλακρότητος·
10 ὀλίγον γὰρ ἔχουσι καὶ ⟨ἧττον⟩⁴ ὑγρὸν τὸν ἐγκέφαλον, ὥστε μὴ ἐξαδυνατεῖν τὸ θερμὸν πρὸς τὴν

---

¹ ὅλον Em*, Aldus, A.-W. : ἄλλο vulg.; cf. 780 a 19.
² ἄλλων secl. Btf.
³ ἐν PZ : om. vulg.: καὶ ἐν om. S.
⁴ ἧττον coni. Bekker, ut videtur ; om. PSYZ.

ᵃ See 784 a 35, b 6, 786 a 20, and Introd. § 62.
ᵇ See 784 a 2, n.

during a period of infirmity just as the whole body is afflicted by a deficiency of natural heat,[a] so the parts, including even the very small ones, share in this infirmity ; also, a great deal of residue is formed in the body and in its parts : hence the lack of concoction in the flesh produces grey hairs. But when health and strength is restored, people accomplish a change, as it might be old men renewing their youth, and, in consequence, the conditions also accomplish a corresponding change. In fact, we might justifiably go so far as to describe disease as " adventitious old age " and old age as " natural disease " ; at any rate, some diseases produce the same effects as old age does.

The temples are the first part to go grey, and the reason is this. The back of the head, since it contains no brain,[b] is empty of fluid. The *bregma*[c] contains a great deal ; but a large volume of fluid does not easily putrefy. On the other hand, the hair on the temples has neither a small enough amount of fluid to secure concoction for it, nor a large enough amount for it to avoid putrefaction, as this region of the head is intermediate between the two extremes, and therefore stands outside both of these two conditions.

We have now given the reason for greyness so far as man is concerned.

The reason why this change does not noticeably V occur on account of age in the other animals is the same as the one already given in the case of baldness : their brain is small and ⟨less⟩ fluid,[d] thus the heat does not become completely unable to effect concoc-

---

[c] See 744 a 25, n.
[d] The insertion of " less " is necessary to the sense : man's brain is the most fluid of all (see 784 a 4).

πέψιν. τοῖς δ᾽ ἵπποις [αὐτῶν]¹ ἐπισημαίνει μάλιστα
ὧν ἴσμεν ζῴων, ὅτι λεπτότατον τὸ ὀστοῦν ὡς κατὰ
μέγεθος ἔχουσι τὸ² περὶ τὸν ἐγκέφαλον τῶν ἄλλων.
τεκμήριον δ᾽ ὅτι καίριος ἡ πληγὴ³ εἰς τὸν τόπον
15 τοῦτον γίνεται αὐτοῖς· διὸ καὶ Ὅμηρος οὕτως
ἐποίησεν

ἵνα⁴ τε πρῶται τρίχες ἵππων
κρανίῳ ἐμπεφύασι, μάλιστα δὲ καίριόν ἐστιν.

ῥᾳδίως οὖν ἐπιρρεούσης τῆς ὑγρότητος διὰ τὴν
λεπτότητα τοῦ ὀστοῦ, τῆς δὲ θερμότητος ἐλλει-
πούσης διὰ τὴν ἡλικίαν, ἐπιπολιοῦνται αἱ τρίχες
αὗται. καὶ αἱ πυρραὶ δὲ θᾶττον πολιοῦνται τρίχες
20 τῶν μελαινῶν· ἔστι γὰρ καὶ ἡ πυρρότης ὥσπερ
ἀρρωστία τριχός, τὰ δ᾽ ἀσθενῆ γηράσκει πάντα
θᾶττον. μελαντέρας δὲ γίνεσθαι γηρασκούσας
λέγεται τὰς γεράνους. αἴτιον δ᾽ ἂν εἴη τοῦ πάθους
τὸ φύσει ὑγροτέραν⁵ αὐτῶν εἶναι τὴν τῶν πτερῶν
φύσιν, πλέον τε γηρασκόντων εἶναι τὸ ὑγρὸν ἐν
25 τοῖς πτεροῖς ἢ ὥστε εὔσηπτον⁶ εἶναι.

Ὅτι δὲ γίγνεται ἡ πολιὰ σήψει τινί, καὶ ὅτι οὐκ
ἔστιν, ὥσπερ οἴονταί τινες, αὔανσις, σημεῖον τοῦ
προτέρου ῥηθέντος⁷ τὸ τὰς σκεπαζομένας τρίχας
πίλοις ἢ καλύμμασι πολιοῦσθαι θᾶττον (τὰ γὰρ

---

¹ secl. Bekker: αὐτῶν PSYZ² (αὐτο Z¹): τοῦτο coni. A.-W.;
causa autem proprie apparet in equis Σ, unde videtur olim
αἴτιον ἐπισημαίνειν scriptum fuisse.
² τὸ Z : om. vulg.
³ ἡ πληγὴ PZ : ἡ πληγὴ ἡ vulg.
⁴ ὅθι text. Hom.
⁵ ὑγροτέραν A.-W. : λευκοτέραν vulg. : λεπτοτέραν Btf.
⁶ εὔσηπτον Platt : εὐσηπτότερον vulg.
⁷ τοῦ προτέρου ῥηθέντος secl. A.-W., om. Σ.

tion. Of all the animals known to us, it is most marked in the horse, the reason being that in the horse the bone which surrounds the brain is, in proportion to the animal's size, thinner than that of any other animal. A proof of this is that a blow delivered on this spot is fatal to a horse. Homer's lines [a] fit in with this too :

> Where on a horse's skull his hairs first grow,
> And where he suffers his most fell and fatal blow.

Therefore, since the thinness of the bone makes it easy for the stream of fluid to flow to the hair at this place, and as the heat begins to fail on account of age, the result is that this hair goes grey. Reddish hair goes grey more quickly than black, as redness too is a sort of infirmity of the hair, and everything that is weak ages more quickly.[b] Cranes, however, so it is alleged, go darker as they get older. If this allegation is true, the reason for this condition would be that the nature of their feathers is more fluid, and that as the birds grow old the fluid in their feathers is too plentiful to putrefy easily.[c]

Here are proofs (a) that greyness is produced by putrefaction of some sort, and (b) that it is not, as some people imagine, a process of withering. Proof of (a). Hair that is protected by hats or other coverings goes grey more quickly, the reason being that the effect of the wind blowing is to prevent putrefac-

---

[a] *Iliad* VIII. 83-84.
[b] See 775 a 19 ff.
[c] See above, 785 a 2.

785 a

πνεύματα κωλύει τὴν σῆψιν, ἡ δὲ σκέπη ἄπνοιαν
30 ποιεῖ), καὶ τὸ βοηθεῖν τὴν ἄλειψιν τὴν τοῦ ὕδατος
καὶ τοῦ ἐλαίου μιγνυμένων. τὸ μὲν γὰρ ὕδωρ
ψύχει, τὸ δ' ἔλαιον μιγνύμενον κωλύει ξηραίνεσθαι
ταχέως· τὸ γὰρ ὕδωρ εὐξήραντον. ὅτι δ' οὐκ
ἔστιν αὔανσις, οὐδ' ὥσπερ ἡ πόα αὐαινομένη λευκαίνεται, οὕτω καὶ ἡ θρίξ, σημεῖον ὅτι φύονται
35 εὐθέως ἔνιαι πολιαί[1]· αὖον δ' οὐθὲν φύεται. λευ-
καίνονται δὲ καὶ ἐπ' ἄκρου πολλαί· ἐν γὰρ τοῖς
ἐσχάτοις καὶ λεπτοτάτοις ἐλαχίστη θερμότης ἐγ-
785 b γίνεται.

Τοῖς δ' ἄλλοις ζῴοις ὅσοις γίνονται λευκαὶ αἱ
τρίχες, φύσει ἀλλ' οὐ πάθει συμβαίνει γίνεσθαι
τοῦτο. αἴτιον δὲ τῶν χρωμάτων τὸ δέρμα τοῖς
ἄλλοις· τῶν μὲν γὰρ λευκῶν λευκὸν τὸ δέρμα, τῶν
5 δὲ μελάνων μέλαν, τῶν δὲ ποικίλων καὶ γιγνομένων
ἐκ συμμίξεως τῇ μὲν λευκὸν τῇ δὲ μέλαν φαίνεται
ὄν. ἐπὶ δὲ τῶν ἀνθρώπων οὐθὲν αἴτιον τὸ δέρμα·
καὶ γὰρ οἱ λευκοὶ σφόδρα μελαίνας ἔχουσιν. αἴτιον
δ' ὅτι λεπτότατον πάντων δέρμα ὅ[2] ἄνθρωπος ἔχει
ὡς κατὰ μέγεθος, διόπερ οὐθὲν ἰσχύει πρὸς τὴν
10 τῶν τριχῶν μεταβολήν, ἀλλὰ διὰ τὴν ἀσθένειαν τὸ
δέρμα καὶ μεταβάλλει αὐτὸ τὴν χρόαν, καὶ γίνεται
ὑπὸ ἡλίων καὶ πνευμάτων μελάντερον· αἱ δὲ τρίχες
οὐθὲν συμμεταβάλλουσιν. ἐν δὲ τοῖς ἄλλοις τὸ
δέρμα χώρας ἔχει δύναμιν διὰ τὸ πάχος· διὸ αἱ

---

[1] ἔνιαι πολιαί conieceram, quod et ipsi codd.* habent:
ἔνιοι πολιοί Bekker (per errorem, ut vid.*).
[2] ὅ Z: om. vulg.

tion, and the protection keeps off the wind. Also, it
is an assistance if the hair is anointed with a mixture
of oil and water. This is because, although the
water cools it, the oil which is mixed with it prevents
the hair from drying off quickly, water being easily
dried off. (*b*) The following proves that greyness is
not a form of withering, and that when hair goes
white it is not due to withering, as it is in the case of
grass. Some hairs are grey from the very beginning
of their growth, and nothing begins its growth in a
withered condition. In many instances, too, hairs
go white at the tip ; this is because very little heat
gets into parts which are at the extreme end and
very thin.

In certain of the other animals white hairs make
their appearance ; but this is natural and not due to
any affection. The reason of the colours in these
other animals is the skin : thus, if they are white,
the skin is white ; if black, the skin is black ; if
piebald, made up of a mixture of colour, the skin is,
we find, white in some places and black in others.
In the case of human beings, however, the skin has
nothing whatever to do with it, for even people with
white skin have intensely black hair. The reason for
this is that, for his size, man has the thinnest skin of
all animals, and on that account it has no power at
all to effect any change in the hair ; instead of that,
the skin, by reason of its own weakness, changes its
colour itself, and also is darkened by the action of
the sun and the wind, while the hair undergoes no
simultaneous change at all. With the other animals,
the skin, on account of its thickness, possesses the
character of the region in which the animal lives ;
and that is why the hair changes in accordance with

μὲν τρίχες κατὰ τὰ δέρματα μεταβάλλουσι, τὰ δὲ
15 δέρματα οὐθὲν κατὰ τὰ πνεύματα καὶ τὸν ἥλιον.

VI Τῶν δὲ ζῴων τὰ μέν ἐστι μονόχροα (λέγω δὲ
μονόχροα ὧν τὸ γένος ὅλον ἐν χρῶμα ἔχει, οἷον
λέοντες πυρροὶ πάντες· καὶ τοῦτο καὶ ἐπ' ὀρνίθων
καὶ ἐπ' ἰχθύων ἐστὶ καὶ τῶν ἄλλων ζῴων ὁμοίως),
20 τὰ δὲ πολύχροα μέν, ὁλόχροα δέ (λέγω δὲ ὧν τὸ
σῶμα ὅλον τὴν αὐτὴν ἔχει χρόαν, οἷον βοῦς ἐστιν
ὅλος λευκὸς καὶ ὅλος μέλας), τὰ δὲ ποικίλα. τοῦτο
δὲ διχῶς, τὰ μὲν τῷ γένει, ὥσπερ πάρδαλις καὶ
ταώς, καὶ τῶν ἰχθύων ἔνιοι, οἷον αἱ καλούμεναι
θρᾶτται· τῶν δὲ τὸ μὲν γένος ἅπαν οὐ ποικίλον,
25 γίνονται δὲ ποικίλοι, οἷον βόες καὶ αἶγες, καὶ ἐν
τοῖς ὄρνισιν, οἷον αἱ περιστεραί· καὶ ἄλλα δὲ γένη
τὸ αὐτὸ πάσχει τῶν ὀρνίθων. μεταβάλλει δὲ τὰ
ὁλόχροα πολλῷ μᾶλλον τῶν μονοχρόων, καὶ εἰς
τὴν ἀλλήλων χρόαν τὴν ἁπλῆν, οἷον ἐκ λευκῶν
μέλανα καὶ ἐκ μελάνων λευκά, καὶ μεμιγμένα ἐξ
30 ἀμφοτέρων, διὰ τὸ ὅλῳ τῷ γένει ὑπάρχειν ἐν τῇ
φύσει τὸ μὴ μίαν ἔχειν χρόαν· εὐκίνητον γὰρ
ὑπάρχει ἐπ' ἀμφότερα τὸ γένος, ὥστε καὶ εἰς
ἄλληλα μεταβάλλειν καὶ ποικίλλεσθαι μᾶλλον. τὰ
δὲ μονόχροα τοὐναντίον· οὐ γὰρ μεταβάλλει, ἂν
μὴ διὰ πάθος, καὶ τοῦτο σπάνιον· ἤδη γὰρ ὦπται
35 καὶ πέρδιξ λευκὴ καὶ κόραξ καὶ στρουθὸς καὶ
ἄρκτος. συμβαίνει δὲ ταῦτα, ὅταν ἐν τῇ γενέσει

---

[a] A fish called *thritta* is mentioned at *H.A.* 621 b 16 (and
fragment 285, 1528 a 40), which is supposed to be the shad.
[b] Aristotle's diagnosis is essentially correct. Albinism is
not " natural," but an " affection " due to absence of pigment.

the skin in the various instances, whereas the skin does not change at all in accordance with the winds and the sun.

Of the animals, some are single-coloured (by which I mean that the whole class has a single colour only, *e.g.*, all lions are tawny ; and a similar thing obtains in the case of birds, fish, and the other animals) ; others are many-coloured, yet at the same time whole-coloured (by which I mean that the whole body is of the same colour, *e.g.*, an ox is white all over, or dark all over) ; others still are variegated. " Variegated " has two meanings : (*a*) as referred to a class of animals—like the leopard, and peacock, and certain fishes, for instance the *thratta*,[a] as it is called ; (*b*) sometimes the class as a whole is not variegated, but variegated individuals are found : examples are, oxen and goats, and certain birds, *e.g.*, pigeons, and there are other classes of birds where this same condition is found. Change of colour is much commoner among the whole-coloured animals than among the single-coloured, both (*a*) the reciprocal change between the individual colours ⟨found in the class⟩, *i.e.*, one simple colour changes into another, *e.g.*, white animals produce black ones and black ones white ; and also (*b*) the change which results in a mixture of the two. The reason for this is that it is a natural attribute of the whole class not to have one single colour : the class is mobile in both directions, and so provides more examples of interchange of colours and also of variegation. The single-coloured animals behave in the opposite way to this : they do not change, unless owing to some affection, and then but rarely ; thus, cases have been observed of a white partridge,[b] raven, sparrow, and bear. These results occur when the

786 a

διαστραφῇ· εὔφθαρτον γὰρ καὶ εὐκίνητον τὸ μικρόν, τὸ δὲ γιγνόμενον τοιοῦτον· ἐν μικρῷ γὰρ ἡ
ἀρχὴ τοῖς γιγνομένοις.

Μάλιστα δὲ μεταβάλλουσι καὶ τὰ φύσει ὁλόχροα[1]
μὲν ὄντα, τῷ γένει δὲ πολύχροα, διὰ τὰ ὕδατα· τὰ
5 μὲν γὰρ θερμὰ λευκὴν ποιεῖ τὴν τρίχα, τὰ δὲ
ψυχρὰ μέλαιναν, ὥσπερ καὶ ἐπὶ τῶν φυτῶν. αἴτιον
δ' ὅτι τὰ θερμὰ πνεύματος πλέον ἔχει ἢ ὕδατος,
ὁ δ' ἀὴρ διαφαινόμενος λευκότητα ποιεῖ, καθάπερ
καὶ τὸν ἀφρόν. διαφέρει μὲν οὖν, ὥσπερ καὶ τὰ
δέρματα τὰ διὰ πάθος λευκὰ τῶν διὰ τὴν φύσιν,
10 οὕτω καὶ ἐν ταῖς θριξὶν ἥ τε διὰ νόσον ἢ καὶ
ἡλικίαν καὶ ἡ διὰ φύσιν λευκότης τῶν τριχῶν τῷ
τὸ αἴτιον ἕτερον εἶναι· τὰς μὲν γὰρ ἡ φυσικὴ
θερμότης ποιεῖ λευκάς, τὰς δ' ἡ ἀλλοτρία. τὸ δὲ
λευκὸν ὁ ἀτμιδώδης ἀὴρ παρέχεται ἐγκατακλειό-
μενος ἐν πᾶσιν. διὸ καὶ ὅσα μὴ μονόχροά ἐστι,
15 τὰ ὑπὸ τὴν γαστέρα πάντα λευκότερά ἐστιν. καὶ
γὰρ θερμότερα καὶ ἡδυκρεώτερα πάντα τὰ λευκὰ
ὡς εἰπεῖν ἐστι διὰ τὴν αὐτὴν αἰτίαν· ἡ μὲν γὰρ
πέψις γλυκέα ποιεῖ, τὴν δὲ πέψιν τὸ θερμόν. ἡ
δ' αὐτὴ αἰτία καὶ τῶν μονοχρόων μέν, μελάνων δ'
ἢ λευκῶν· θερμότης γὰρ καὶ ψυχρότης αἰτία τῆς
20 φύσεως τοῦ δέρματος καὶ τῶν τριχῶν· ἔχει γὰρ
ἕκαστον τῶν μορίων θερμότητα οἰκείαν.

---

[1] ὁλόχροα Ζ²m (non E*), Aldus, A.-W.: μονόχροα Ζ¹,
vulg.: *et alteratio colorum generum animalium que sunt naturaliter multorum colorum erit multociens propter* etc. Σ.

[a] Cf. 775 a 9.
[b] Cf. 735 b 8—736 a 20.
[c] See 784 b 7, n.
[d] Cf. 784 a 34, b 6, 27, and Introd. § 62.

creature suffers some distortion during the process of its formation, for, since the beginning of things that pass through such a process is on a small scale, they are small at that time, and what is small can easily be given a different turn and spoilt.[a]

The ones that change most are those which, though whole-coloured by nature, belong to a class which is many-coloured. This is due to the varieties of water involved. Hot water makes the hair white, cold water makes it dark, which is exactly what happens in the case of plants. The reason is that the hot ones contain more *pneuma* than they do water, and it is the air shining through that causes the whiteness, just as it makes froth white.[b] Therefore, just as there is a difference between skins that are white by nature and those that are white owing to some affection, so there is a difference between the whiteness of hair which is due to nature and that which is due to disease or age—and the difference lies in the fact that the cause is different. In the former case, the whiteness is caused by the natural heat, in the latter, by extraneous heat.[c] It is the vaporous air shut up inside them which produces whiteness in all things ; and that, too, is why those animals which are not single-coloured are all whiter under the belly than elsewhere. Thus too practically all white animals are hotter and tastier for the same cause : their good flavour is produced by concoction, and concoction is produced by heat. And the same cause holds also in the case of those animals which, being single-coloured, are either dark or white ; since it is heat and cold which are the cause of the nature of the skin and of the hair, each of the parts of the body having its own proper heat.[d]

786 a

Ἔτι δ' αἱ γλῶτται διαφέρουσι τῶν ἁπλῶν τε
καὶ ποικίλων καὶ τῶν ἁπλῶν μὲν διαφερόντων δέ,
οἷον λευκῶν καὶ μελάνων. αἴτιον δὲ τὸ εἰρημένον
πρότερον, ὅτι τὰ δέρματα ποικίλα τῶν ποικίλων,
25 καὶ τῶν λευκοτρίχων καὶ τῶν μελανοτρίχων τῶν
μὲν λευκὰ τῶν δὲ μέλανα. τὴν δὲ γλῶτταν δεῖ
ὑπολαβεῖν ὥσπερ ἕν μόριον τῶν ἐξωτερικῶν εἶναι,
μὴ ὅτι ἐν τῷ στόματι σκεπάζεται, ἀλλ' οἷον χεῖρα
ἢ πόδα· ὥστ' ἐπεὶ τῶν ποικίλων τὸ δέρμα οὐ
μονόχρων, καὶ τοῦ ἐπὶ τῇ γλώττῃ δέρματος τοῦτ'
αἴτιον.

30 Μεταβάλλουσι δὲ τὰ χρώματα καὶ τῶν ὀρνίθων
τινὲς καὶ τῶν τετραπόδων τῶν ἀγρίων ἔνια κατὰ
τὰς ὥρας. αἴτιον δ' ὅτι ὥσπερ οἱ ἄνθρωποι κατὰ
τὴν ἡλικίαν μεταβάλλουσι, τοῦτ' ἐκείνοις συμβαίνει κατὰ τὰς ὥρας· μείζων γὰρ διαφορὰ αὕτη τῆς
κατὰ τὴν ἡλικίαν τροπῆς.

35 Εἰσὶ δὲ καὶ τὰ παμφαγώτερα ποικιλώτερα ὡς
ἐπὶ τὸ πλεῖστον[1] εἰπεῖν εὐλόγως, οἷον αἱ μέλιτ-

786 b

ται μονόχροα μᾶλλον ἢ αἱ ἀνθρῆναι καὶ σφῆκες·
εἰ γὰρ αἱ τροφαὶ αἴτιαι τῆς μεταβολῆς, εὐλόγως
αἱ ποικίλαι τροφαὶ παντοδαπωτέρας ποιοῦσι τὰς
κινήσεις καὶ τὰ περιττώματα τῆς τροφῆς, ἐξ ὧν
5 καὶ τρίχες καὶ πτερὰ[2] καὶ δέρματα γίνεται.

Καὶ περὶ μὲν χρωμάτων[3] καὶ τριχῶν διωρίσθω
τὸν τρόπον τοῦτον.

VII   Περὶ δὲ φωνῆς, ὅτι τὰ μὲν βαρύφωνα τῶν ζώων

---

[1] πλεῖστον Z : πλῆθος vulg.   [2] πτερὰ Z : πτίλα vulg.
[3] χρώματος YZ : δερμάτων P : δερμάτων χρώματος coni.
A.-W.

Further, the tongues of animals differ : those of
the simple-coloured [a] animals, those of the variegated
ones, and those of the ones which, though simple-
coloured yet differ among themselves (as, *e.g.*, dark
and white)—the tongues of these are all different.
The reason is that which has been stated already,
viz., that the skins of variegated animals are variegated, the skins of white-haired ones are white and
of dark ones dark. The tongue we should look upon
as being, as it were, one of the external parts of
the body, comparable, *e.g.*, with the hand or foot, dis-
regarding the fact that it is being covered in by the
mouth. So that, as the skin of the variegated animals
is not single-coloured, this will be the reason responsible for the skin on the tongue as well.

Some birds and some wild quadrupeds change their
colour according to the seasons of the year. The
reason is that, just as human beings change according
to their age, so these change according to the seasons,
because this constitutes a greater difference so far as
they are concerned than the change according to age. *Seasonal change of colour.*

Speaking generally, the more omnivorous animals
are more variegated, as we should expect (for instance, bees are more single-coloured than hornets
and wasps), for of course if the various sorts of
nourishment they take are the causes of the change,
we shall expect to find that variegated kinds of
nourishment make the movements which the nourishment undergoes and the residues which result from
it more variegated, and it is out of the residues that
hair, feathers, and skin are formed. *Effect of diet on colour.*

This concludes our account of the various colours,
and the various kinds of hair.

With regard to the voice : some animals have a *VII Voice.*

ἐστί, τὰ δ' ὀξύφωνα, τὰ δ' εὔτονα καὶ πρὸς ἀμφοτέρας ἔχοντα τὰς ὑπερβολὰς συμμέτρως, ἔτι δὲ
10 τὰ μὲν μεγαλόφωνα τὰ δὲ μικρόφωνα, καὶ λειότητι
καὶ τραχύτητι καὶ εὐκαμψίᾳ καὶ ἀκαμψίᾳ διαφέροντα ἀλλήλων, ἐπισκεπτέον διὰ τίνας αἰτίας
ὑπάρχει τούτων ἕκαστον. περὶ μὲν οὖν ὀξύτητος
καὶ βαρύτητος τὴν αὐτὴν αἰτίαν οἰητέον εἶναι
ἥνπερ ἐπὶ τῆς μεταβολῆς ἣν μεταβάλλει νέα ὄντα
15 καὶ πρεσβύτερα. τὰ μὲν γὰρ ἄλλα πάντα νεώτερα
ὄντα ὀξύτερον φθέγγεται, τῶν δὲ βοῶν οἱ μόσχοι
βαρύτερον. τὸ δ' αὐτὸ συμβαίνει καὶ ἐπὶ τῶν
ἀρρένων καὶ θηλειῶν· ἐν μὲν γὰρ τοῖς ἄλλοις
γένεσι τὸ θῆλυ ὀξύτερον φθέγγεται τοῦ ἄρρενος
(μάλιστα δ' ἐπίδηλον ἐπὶ τῶν ἀνθρώπων τοῦτο·
20 μάλιστα γὰρ τούτοις ταύτην τὴν δύναμιν ἀπο-
δέδωκεν ἡ φύσις διὰ τὸ λόγῳ χρῆσθαι μόνους τῶν
ζῴων, τοῦ δὲ λόγου ὕλην εἶναι τὴν φωνήν), ἐπὶ
δὲ τῶν βοῶν τοὐναντίον· βαρύτερον γὰρ αἱ θήλειαι
φθέγγονται τῶν ταύρων. τίνος μὲν οὖν ἕνεκα
25 ψόφος, τὰ μὲν ἐν τοῖς περὶ αἰσθήσεως εἴρηται, τὰ
φωνὴν ἔχει τὰ ζῷα, καὶ τί ἐστι φωνὴ καὶ ὅλως ὁ
δ' ἐν τοῖς περὶ ψυχῆς. ἐπεὶ δὲ βαρὺ μέν ἐστιν
ἐν τῷ βραδεῖαν εἶναι τὴν κίνησιν, ὀξὺ δ' ἐν τῷ
ταχεῖαν, τοῦ[1] βραδέως ἢ ταχέως πότερον τὸ κινοῦν
αἴτιον ἢ τὸ κινούμενον, ἔχει τινὰ ἀπορίαν. φασὶ
γάρ τινες τὸ μὲν πολὺ βραδέως κινεῖσθαι τὸ δ'
30 ὀλίγον ταχέως, καὶ ταύτην αἰτίαν εἶναι τοῦ τὰ μὲν
βαρύφωνα εἶναι τὰ δ' ὀξύφωνα, λέγοντες μέχρι
τινὸς καλῶς, ὅλως δ' οὐ καλῶς. τῷ μὲν γὰρ

---

[1] τοῦ Y, Platt, Hayduck : τοῦ δὲ vulg. : τοῦ δὴ Ob*Z², Btf.

---

deep [a] voice, others a high-pitched voice, others a well-pitched voice, suitably proportionate between the two extremes ; some, too, have big voices, others small ones ; also they differ in respect of being smooth, or rough, flexible and inflexible. So we must consider what are the causes to which each of these is due. With regard to the pitch, the same cause is to be held responsible as that which controls the change which they undergo in passing from youth to age. All animals when younger have a higher voice, except calves, which have a deeper one. The same occurs as between male and female as well : in all animals (except cattle) the female has a higher voice than the male, and this is especially noticeable in human beings, for Nature has given them this faculty in an exceptional degree because they alone among the animals use the voice for rational speech, of which the voice is the " material." In cattle the reverse obtains : cows have a deeper voice than bulls. We have explained partly in the treatise *Of Sensation*,[b] partly in that *Of the Soul*,[c] for what purpose animals have a voice, and what " voice " is, and generally what sound is. But since deepness of pitch consists in the movement being slow, and height of pitch in its being fast, the question is whether the speed is caused by that which initiates or that which experiences the movement, and this is somewhat puzzling. Some people hold that the movement of a large volume is slow and that of a small volume fast, and that this is the cause why some animals have deep voices and others high ones. Up to a point this statement is satisfactory, but not completely so. It is, of course, correct to say that,

[c] See 419 b 3—420 b 23.

γένει ὀρθῶς ἔοικε λέγεσθαι τὸ βαρὺ ἐν μεγέθει τινὶ
εἶναι τοῦ κινουμένου. εἰ γὰρ τοῦτο, καὶ μικρὸν
καὶ βαρὺ φθέγξασθαι οὐ ῥᾴδιον, ὁμοίως δὲ οὐδὲ
35 μέγα¹ καὶ ὀξύ. καὶ δοκεῖ γενναιοτέρας εἶναι

φύσεως ἡ βαρυφωνία, καὶ ἐν τοῖς μέλεσι τὸ βαρὺ
τῶν συντόνων βέλτιον· τὸ γὰρ βέλτιον ἐν ὑπεροχῇ,
ἡ δὲ βαρύτης ὑπεροχή τις. ἀλλ' ἐπειδή ἐστιν
ἕτερον τὸ βαρὺ καὶ ὀξὺ ἐν φωνῇ μεγαλοφωνίας
καὶ μικροφωνίας (ἔστι γὰρ καὶ ὀξύφωνα μεγαλό-
5 φωνα, καὶ μικρόφωνα βαρύφωνα ὡσαύτως), ὁμοίως
δὲ καὶ κατὰ τὸν μέσον τόνον τούτων· περὶ ὧν
τίνι ἄν τις ἄλλῳ διορίσειεν (λέγω δὲ μεγαλοφωνίαν
καὶ μικροφωνίαν) ἢ πλήθει καὶ ὀλιγότητι τοῦ
κινουμένου; εἰ οὖν κατὰ τὸν λεγόμενον ἔσται
διορισμὸν τὸ ὀξὺ καὶ βαρύ, συμβήσεται τὰ αὐτὰ
10 εἶναι βαρύφωνα καὶ μεγαλόφωνα καὶ ὀξύφωνα καὶ
μικρόφωνα. τοῦτο δὲ ψεῦδος. αἴτιον δ' ὅτι τὸ
μέγα καὶ τὸ μικρὸν καὶ τὸ πολὺ καὶ τὸ ὀλίγον τὰ
μὲν ἁπλῶς λέγεται, τὰ δὲ πρὸς ἄλληλα. μεγαλόφωνα μὲν οὖν ἐστιν ἐν τῷ πολὺ ἁπλῶς εἶναι τὸ
κινούμενον, μικρόφωνα δὲ τῷ ὀλίγον, βαρύφωνα
15 δὲ καὶ ὀξύφωνα ἐν τῷ πρὸς ἄλληλα ταύτην ἔχειν
τὴν διαφοράν. ἐὰν μὲν γὰρ ὑπερέχῃ τὸ κινούμενον
τῆς τοῦ κινοῦντος ἰσχύος, ἀνάγκη βραδέως φέρεσθαι τὸ φερόμενον, ἂν δ' ὑπερέχηται, ταχέως. τὸ

---

¹ μέγα coni. A.-W. : *vociferatio vocis magne acute est in-
possibilis* Σ : βαρὺ vulg., Z² in ras., prima litt. Z¹ fortasse
fuerat μ.

ᵃ This, as appears from the next sentence, means the
amount producing the movement as compared with the
amount undergoing it.

in general, deepness depends upon a certain size of that which is set in movement; but if the statement were wholly true, it would not be easy to utter a noise simultaneously small and deep, nor, similarly, large and high. Further, a deep voice seems to be the mark of a nobler nature, and in melodies, too, that which is deep-pitched is better than the high-pitched, since deepness is a form of superiority, and it is in superiority that betterness resides. In fact, however, deep and high pitch of the voice is a different matter from largeness and smallness of the voice, for some animals which have high-pitched voices are large-voiced, and in the same way some which have deep-pitched voices are small-voiced; and the same applies to the intermediate pitch between the two. And what other means is there for defining largeness and smallness of voice apart from the volume of that which is set in movement? So then, if high and deep pitch are to be distinguished according to the definition mentioned above, the result will be that any animal which has a deep voice will also have a large one, and any which has a high voice will also have a small one. And this is not true. The reason is that the terms " large," " small," and " large amount," " small amount " are sometimes used in an *absolute* sense, sometimes *relatively* to each other. If an animal has a large voice, this is because the amount of that which is set in movement is large *absolutely*, if small, the amount is small *absolutely*; whereas high pitch and low pitch are due to the amounts [a] involved being large and small *relatively* to each other. Thus, if that which is set moving exceeds the strength of that which sets it moving, then that which is propelled is bound to go slowly; if it is exceeded, it

787 a

δ' ἰσχῦον διὰ τὴν ἰσχὺν ὁτὲ μὲν πολὺ κινοῦν βραδεῖαν ποιεῖ τὴν κίνησιν, ὁτὲ δὲ διὰ τὸ κρατεῖν
20 ταχεῖαν. κατὰ τὸν αὐτὸν δὲ λόγον καὶ τῶν κινούν-
των τὰ ἀσθενῆ τὰ μὲν πλείω κινοῦντα τῆς δυνάμεως βραδεῖαν ποιεῖ τὴν κίνησιν, τὰ δὲ δι'
ἀσθένειαν ὀλίγον κινοῦντα ταχεῖαν.

Αἱ μὲν οὖν αἰτίαι τῶν ἐναντιώσεων αὗται, τοῦ
μήτε πάντα τὰ νέα ὀξύφωνα εἶναι μήτε βαρύφωνα,
25 μήτε τὰ πρεσβύτερα, μήτε τὰ ἄρρενα καὶ θήλεα,
πρὸς δὲ τούτοις καὶ τοῦ τοὺς κάμνοντας ὀξὺ
φθέγγεσθαι καὶ τοὺς εὖ τὸ σῶμα ἔχοντας, ἔτι δὲ
καὶ γέροντας γινομένους μᾶλλον ὀξυφωνοτέρους
γίνεσθαι, τῆς ἡλικίας ἐναντίας οὔσης τῇ τῶν νέων.

Τὰ μὲν οὖν πλεῖστα νεώτερα ὄντα καὶ θήλεα δι'
30 ἀδυναμίαν ὀλίγον κινοῦντα ἀέρα ὀξύφωνά ἐστιν·
ταχὺ γὰρ ὁ ὀλίγος φέρεται, τὸ δὲ ταχὺ ὀξὺ ἐν
φωνῇ. οἱ δὲ μόσχοι καὶ αἱ βόες αἱ θήλειαι, οἱ
μὲν διὰ τὴν ἡλικίαν, αἱ δὲ διὰ τὴν φύσιν τῆς
θηλύτητος, οὐκ ἰσχυρὸν ἔχουσι τὸ μόριον ᾧ κινοῦσι,

787 b

πολὺ δὲ κινοῦντα βαρύφθογγά ἐστιν· βαρὺ γὰρ τὸ
βραδέως φερόμενον, ὁ δὲ πολὺς ἀὴρ φέρεται
βραδέως. πολὺν δὲ κινοῦσι ταῦτα, τὰ δ' ἄλλ'
ὀλίγον, διὰ τὸ τὸ ἀγγεῖον δι' οὗ πρῶτον φέρεται
τὸ πνεῦμα, τούτοις μὲν διάστημ' ἔχειν μέγα καὶ

---

[a] The Greek word includes both meanings; and this
circumstance explains a good deal of what Aristotle says in
the present discussion.

will travel quickly. So then, the movement which a strong agent produces is sometimes slow (*i.e.*, when, in virtue of its strength, it is moving a large amount), and sometimes fast (*i.e.*, when the agent has the upper hand). In accordance with the same line of argument, in some cases the movement which a weak agent produces is slow (*i.e.*, when the agent is setting in movement an amount which is too large for its strength), in other cases the movement is fast (*i.e.*, when owing to the agent's weakness the amount which it sets moving is small).

Such, then, are the causes to which these contrarieties are due. We have shown (*a*) why neither young, nor old, nor male nor female animals all have high-pitched voices or all have deep voices ; (*b*) why sick and healthy alike speak in a high-pitched voice ; and (*c*) why, as men reach old age, the pitch of their voice rises, although old age is the opposite of youth.

On account of their debility, most animals when young, and most females, set but a small amount of air in movement and therefore have high-pitched voices, because a small amount is propelled at a fast speed, and where the voice is concerned fast means high. In calves, however, owing to their age, and in cows, owing to the nature of femininity, the part by means of which they set ⟨the air⟩ in movement is not strong, and as they set a large amount of it in movement, they have deep voices, for a large amount of air travels slowly, and anything that travels slowly is heavy (deep).[a] A large amount ⟨of air⟩ is set in movement by these animals, but only a small amount by the others, the reason being that in the former the vessel through which their breath first travels has a large opening and is therefore forced to set a large

5 πολὺν ἀναγκάζεσθαι ἀέρα κινεῖν, τοῖς δ' ἄλλοις
εὐταμίευτον εἶναι. προϊούσης δὲ τῆς ἡλικίας
ἰσχύει μᾶλλον τοῦτο τὸ μόριον τὸ κινοῦν ἐν ἑκάστοις, ὥστε μεταβάλλουσιν εἰς τοὐναντίον, καὶ τὰ
μὲν ὀξύφωνα βαρυφωνότερα γίνεται αὐτὰ αὑτῶν,
τὰ δὲ βαρύφωνα ὀξυφωνότερα· διόπερ οἱ ταῦροι
10 ὀξυφωνότεροι τῶν μόσχων καὶ τῶν θηλειῶν βοῶν.
ἔστι μὲν οὖν πᾶσιν ἡ ἰσχὺς ἐν τοῖς νεύροις, διὸ
καὶ τὰ ἀκμάζοντα ἰσχύει μᾶλλον. ἄναρθρα γὰρ
τὰ νέα μᾶλλον καὶ ἄνευρα. ἔτι δὲ τοῖς μὲν νέοις
οὔπω ἐπιτέταται, τοῖς δὲ γεγηρακόσιν[1] ἤδη ἀνεῖται[2]
ἡ συντονία· διὸ ἄμφω ἀσθενῆ καὶ ἀδύνατα πρὸς
15 τὴν κίνησιν. μάλιστα δ' οἱ ταῦροι νευρώδεις, καὶ
ἡ καρδία[3]· διόπερ σύντονον ἔχουσι τοῦτο τὸ μόριον
ᾧ κινοῦσι τὸ πνεῦμα, ὥσπερ χορδὴν τεταμένην
νευρίνην. δηλοῖ δὲ τοιαύτη τὴν φύσιν οὖσα ἡ
καρδία τῶν βοῶν τῷ καὶ ὀστοῦν ἐγγίνεσθαι ἐν
ἐνίαις αὐτῶν· τὰ δ' ὀστᾶ ζητεῖ τὴν τοῦ νεύρου
φύσιν.

20 Ἐκτεμνόμενα δὲ πάντα εἰς τὸ θῆλυ μεταβάλλει,
καὶ διὰ τὸ ἀνίεσθαι τὴν ἰσχὺν τὴν νευρώδη ἐν τῇ
ἀρχῇ ὁμοίαν ἀφίησι φωνὴν τοῖς θήλεσιν. ἡ δ'
ἄνεσις παραπλησία γίνεται ὥσπερ ἂν εἴ τις χορδὴν
κατατείνας σύντονον ποιήσειε τῷ ἐξάψαι τι βάρος,
25 οἷον δὴ ποιοῦσιν αἱ τοὺς ἱστοὺς ὑφαίνουσαι· καὶ
γὰρ αὗται τὸν στήμονα κατατείνουσι προσάπτουσαι
τὰς καλουμένας λαιάς. οὕτω γὰρ καὶ ἡ τῶν

---

[1] γεγηρακόσαν Z[1], A.-W. : γηράσκουσιν vulg.
[2] ἀνεῖται PZ, A.-W. : ἀνίεται vulg.
[3] καὶ ἡ καρδία seclusit Btf. Σ tamen vertit *et tauri proprie sunt fortiorum nervorum et cordis.*

amount of air in movement, whereas in the latter the breath is under better control. In every animal, as age advances, this part which sets ⟨the air⟩ in movement becomes stronger, so that a change-over[a] to the opposite is effected : high-pitched voices become deeper than they were, and deep-pitched ones higher. That is why bulls have higher-pitched voices than calves and cows. Now in all animals their strength lies in their sinews, and that actually is why animals in their prime are stronger than the others : young ones are less well articulated and less well supplied with sinews, and furthermore, their sinews have not yet become taut, whereas in ones that are aged their tautness has slackened off. Hence both young and old are weak and powerless so far as producing movement is concerned. Bulls however, being especially sinewy, have especially sinewy hearts; hence this part, by which they set the breath in movement, is taut, just like a sinewy string stretched tight. Bulls' hearts are shown to be sinewy by the fact that in some of them a bone[b] actually occurs, and bones seek the nature of sinew.[c]

All animals when castrated change over to the female state, and as their sinewy strength is slackened at its source they emit a voice similar to that of females. This slackening may be illustrated in the following way. It is as though you were to stretch a cord and make it taut by hanging some weight on to it, just as women do who weave at the loom ; they stretch the warp by hanging stone weights[d] on to it.

---

[a] For μεταβάλλειν see 766 a 17 ff., 768 a 15 ff.

[b] See also *P.A.* 666 b 19.

[c] This is a literal translation of the Greek. See 744 b 25, 36 ff., and Introd. § 64.

[d] *Cf.* 717 a 35. Lit., " what are called ' laiai ' (stones)."

ὄρχεων φύσις προσήρτηται πρὸς τοὺς σπερματικοὺς πόρους, οὗτοι δ᾽ ἐκ τῆς φλεβός ἧς ἡ ἀρχὴ
ἐκ τῆς καρδίας πρὸς αὐτῷ τῷ κινοῦντι τὴν φωνήν.
διόπερ[1] καὶ τῶν σπερματικῶν πόρων μεταβαλ-
30 λόντων πρὸς τὴν ἡλικίαν ἐν ᾗ ἤδη δύνανται τὸ
σπέρμα ἐκκρίνειν, συμμεταβάλλει καὶ τοῦτο τὸ
μόριον. τούτου δὲ μεταβάλλοντος καὶ ἡ φωνὴ
μεταβάλλει, μᾶλλον μὲν τοῖς ἄρρεσιν, συμβαίνει
δὲ ταὐτὸ καὶ ἐπὶ τῶν θηλειῶν, ἀλλ᾽ ἀδηλότερον,

καὶ γίνεται ὃ καλοῦσί τινες τραγίζειν, ὅταν ἀνώμαλος ᾖ ἡ φωνή. μετὰ δὲ ταῦτα καθίσταται εἰς
τὴν τῆς ἐπιούσης ἡλικίας βαρύτητα ἢ ὀξυφωνίαν.
ἀφαιρουμένων δὲ τῶν ὄρχεων ἀνίεται ἡ τάσις τῶν
πόρων, ὥσπερ ἀπὸ τῆς χορδῆς καὶ τοῦ στήμονος
5 ἀφαιρουμένου τοῦ βάρους. τούτου δ᾽ ἀνιεμένου
καὶ ἡ ἀρχὴ ἡ κινοῦσα τὴν φωνὴν ἐκλύεται κατὰ
τὸν αὐτὸν λόγον. διὰ μὲν οὖν ταύτην τὴν αἰτίαν
τὰ ἐκτεμνόμενα μεταβάλλει εἰς τὸ θῆλυ τήν τε
φωνὴν καὶ τὴν ἄλλην μορφήν, διὰ τὸ συμβαίνειν
ἀνίεσθαι τὴν ἀρχὴν ἐξ ἧς ὑπάρχει τῷ σώματι ἡ
10 συντονία, ἀλλ᾽ οὐχ ὥσπερ τινὲς ὑπολαμβάνουσιν
αὐτοὺς τοὺς ὄρχεις εἶναι σύναμμα πολλῶν ἀρχῶν·
ἀλλὰ μικραὶ μεταστάσεις μεγάλων αἰτίαι γίνονται,
οὐ δι᾽ αὐτάς, ἀλλ᾽ ὅταν συμβαίνῃ ἀρχὴν συμμεταβάλλειν. αἱ γὰρ ἀρχαὶ μεγέθει οὖσαι μικραὶ τῇ
δυνάμει μεγάλαι εἰσίν· τοῦτο γάρ ἐστι τὸ ἀρχὴν
15 εἶναι, τὸ αὐτὴν μὲν αἰτίαν εἶναι πολλῶν, ταύτης
δ᾽ ἄλλο ἄνωθεν μηθέν.

---

[1] διόπερ P : διὸ vulg.

This is the way in which the testes are attached to the seminal passages, which in their turn are attached to the blood-vessel which has its starting-point at the heart near the part which sets the voice in movement.[a] And so, as the seminal passages undergo a change at the approach of the age when they can secrete semen, this part undergoes a simultaneous change. And as this changes, so too does the voice—to a greater extent in males, but the same happens with females as well, though the change there is less obvious ; and one result of this is that, as we say, the voice " is breaking " [b] during the time that it is uneven. After that, it settles down into the deep or high pitch belonging to the age of life which is to succeed. If the testes are removed, the tautness of the passages is slackened, just as when the weight is removed from the cord or from the warp ; and as this slackens, the source (or principle) which sets the voice in movement is correspondingly loosened. This then is the cause on account of which castrated animals change over to the female condition both as regards the voice and the rest of their form : it is because the principle from which the tautness of the body is derived is slackened. The reason is not, as some people suppose, that the testes themselves are a ganglion of many principles. No ; small alterations are the causes of big ones, not in virtue of themselves, but when it happens that a principle changes at the same time.[c] The principles, though small in size, are great in power : that is what it means to be a principle—something which is itself a cause of many things, while there is nothing more ultimate which is the cause of it.

[b] Lit., " ' bleating like a goat ' as some people call it."
[c] Cf. 716 b 3, etc.

788 a

Τῷ¹ δὲ φύσει τὰ μὲν τοιαῦτα συνίστασθαι τῶν
ζῴων ὥστε βαρύφωνα εἶναι, τὰ δ' ὀξύφωνα,
συμβάλλεται καὶ ἡ θερμότης τοῦ τόπου καὶ ἡ
ψυχρότης. τὸ μὲν γὰρ θερμὸν πνεῦμα διὰ παχύ-
20 τητα ποιεῖ βαρυφωνίαν, τὸ δὲ ψυχρὸν διὰ λεπτότητα τοὐναντίον. δῆλον δὲ τοῦτο καὶ ἐπὶ τῶν
αὐλῶν· οἱ γὰρ θερμοτέρῳ τῷ πνεύματι χρώμενοι,
καὶ τοιοῦτον προϊέμενοι οἷον οἱ αἰάζοντες, βαρύτερον αὐλοῦσιν. τῆς δὲ τραχυφωνίας αἴτιον, καὶ
τοῦ λείαν εἶναι τὴν φωνήν, καὶ πάσης τῆς τοιαύτης
25 ἀνωμαλίας, τὸ τὸ μόριον καὶ τὸ ὄργανον δι' οὗ
φέρεται ἡ φωνὴ ἢ τραχὺ ἢ λεῖον εἶναι ἢ ὅλως
ὁμαλὸν ἢ ἀνώμαλον (δῆλον δ' ὅταν ὑγρότης τις
ὑπάρχῃ περὶ τὴν ἀρτηρίαν ἢ τραχύτης γένηται ὑπό
τινος πάθους· τότε γὰρ καὶ ἡ φωνὴ γίνεται ἀνώμαλος)· τῆς δ' εὐκαμψίας,² ἂν μαλακὸν ἢ σκληρὸν
30 ᾖ τὸ ὄργανον· τὸ μὲν γὰρ μαλακὸν δύναται
ταμιεύεσθαι καὶ παντοδαπὸν γίνεσθαι, τὸ δὲ
σκληρὸν οὐ δύναται. καὶ τὸ μὲν μαλακὸν καὶ
μικρὸν δύναται καὶ μέγα φθέγγεσθαι, διὸ καὶ ὀξὺ
καὶ βαρύ· ταμιεύεται γὰρ ῥᾳδίως τοῦ πνεύματος,
καὶ αὐτὸ γινόμενον ῥᾳδίως μέγα καὶ μικρόν· ἡ δὲ
σκληρότης ἀταμίευτον.

788 b

Περὶ μὲν οὖν φωνῆς ὅσα μὴ πρότερον ἐν τοῖς
περὶ αἰσθήσεως διώρισται καὶ ἐν τοῖς περὶ ψυχῆς,
τοσαῦτ' εἰρήσθω.

VIII Περὶ δὲ ὀδόντων, ὅτι μὲν οὐχ ἑνὸς χάριν, οὐδὲ
πάντα τοῦ αὐτοῦ ἕνεκεν τὰ ζῷα ἔχουσιν, ἀλλὰ
5 τὰ μὲν διὰ τὴν τροφήν, τὰ δὲ καὶ πρὸς ἀλκὴν καὶ

¹ τῷ Aldus : τοῦ Z, vulg.
² <καὶ τῆς ἀκαμψίας> Bonitz.

ᵃ P.A. 655 b 8 ff., 661 b 1 ff.

The heat and cold of their place of habitation is another factor contributing to the fact that the natural construction of some animals is such that they have deep voices, and of others, that they have high voices. Breath that is hot produces deepness (heaviness) of voice, owing to its thickness ; breath that is cold produces the opposite result, owing to its thinness. This is plain in the case of musical pipes as well : people who blow comparatively hot breath into the pipe—*i.e.*, if they breathe it out as though they were saying " Ah ! "—play a deeper note. The reason for roughness and smoothness of voice and all unevenness of that sort is that the part or organ through which the voice travels is rough, or smooth, or, to put it generally, is even or uneven. This is apparent when there is any fluid about in the trachea, or if there is any roughness due to an affection : in such circumstances the voice becomes uneven too. Flexibility depends upon whether the organ is soft or hard, since anything that is soft can be controlled and made to assume all sorts of shapes, whereas anything hard cannot. Thus this organ if it is soft can utter a small sound or a large one, and therefore a high one or a deep one as well, because it controls the breath easily, as it easily becomes large or small itself. Hardness on the other hand cannot ⟨so⟩ be controlled.

This will be a sufficient account of those points concerning the voice which we have not already settled in the treatises *Of Sensation* and *Of the Soul.*

We have already said,[a] on the subject of the teeth, that their existence is not for one purpose only, nor do they exist for the same purpose in all animals : some have teeth on account of nourishment, some for self-defence and ⟨some⟩ for rational

VIII
Teeth.

553

πρὸς τὸν ἐν τῇ φωνῇ λόγον, εἴρηται πρότερον· διότι
δ' οἱ μὲν πρόσθιοι γίνονται πρότερον οἱ δὲ γόμφιοι
ὕστερον, καὶ οὗτοι μὲν οὐκ ἐκπίπτουσιν, ἐκεῖνοι δ'
ἐκπίπτουσι καὶ φύονται πάλιν, τοῖς περὶ γενέσεως
λόγοις τὴν αἰτίαν συγγενῆ δεῖ νομίζειν.

10 Εἴρηκε μὲν οὖν περὶ αὐτῶν καὶ Δημόκριτος, οὐ
καλῶς δ' εἴρηκεν. οὐ γὰρ ἐπὶ πάντων σκεψάμενος
καθόλου λέγει τὴν αἰτίαν. φησὶ γὰρ ἐκπίπτειν
μὲν διὰ τὸ πρὸ ὥρας γίνεσθαι τοῖς ζῴοις· ἀκμαζόντων γὰρ ὡς εἰπεῖν φύεσθαι κατά γε φύσιν.
τοῦ δὲ πρὸ ὥρας γίνεσθαι τὸ θηλάζειν αἰτιᾶται.
15 καίτοι θηλάζει γε καὶ ὗς, οὐκ ἐκβάλλει δὲ τοὺς
ὀδόντας· ἔτι δὲ τὰ καρχαρόδοντα θηλάζει μὲν
πάντα, οὐκ ἐκβάλλει δ' ἔνια αὐτῶν πλὴν τοὺς
κυνόδοντας, οἷον οἱ λέοντες. τοῦτο μὲν οὖν ἥμαρτε
καθόλου λέγων, οὐ σκεψάμενος τὸ συμβαῖνον ἐπὶ
πάντων. δεῖ δὲ τοῦτο ποιεῖν· ἀνάγκη γὰρ τὸν
20 λέγοντα καθόλου τι λέγειν περὶ πάντων. ἐπεὶ δὲ
τὴν φύσιν ὑποτιθέμεθα, ἐξ ὧν ὁρῶμεν ὑποτιθέ-
μενοι, οὔτ' ἐλλείπουσαν οὔτε μάταιον οὐθὲν ποιοῦ-
σαν τῶν ἐνδεχομένων περὶ ἕκαστον, ἀνάγκη δὲ
τοῖς μέλλουσι λαμβάνειν τροφὴν μετὰ τὴν τοῦ
γάλακτος ἀπόλαυσιν ἔχειν ὄργανα πρὸς τὴν ἐρ-
25 γασίαν τῆς τροφῆς—εἰ οὖν συνέβαινεν, ὡς ἐκεῖνος
λέγει, πρὸς ἥβην, ἐνέλειπεν ἂν ἡ φύσις τῶν
ἐνδεχομένων αὐτῇ τι ποιεῖν, καὶ τὸ τῆς φύσεως

---

[a] This is repeated from *H.A.* 501 b 4, but it is incorrect.
[b] Lit., " which are saw-toothed." See *P.A.* 661 b 19.
[c] Also stated at *H.A.* 579 b 11. Other animals' habits in
teeth-shedding are noticed at *H.A.* 501 b 1 ff., 575 a 5.

speech. But why are the front teeth formed first and the molars afterwards ? And why are the molars not shed, whereas the front teeth are, and grow again ? We must take it to be appropriate to examine the cause of these things in a treatise on Generation.

Now Democritus has treated of these matters, but his treatment is not correct, because he assigns a cause to apply generally although he has not undertaken an exhaustive investigation of the facts. He says that the reason why animals shed their teeth is that they are formed prematurely, since it is when animals are in their prime or thereabouts that they grow their teeth according to *nature*. Suckling is the cause he names for their being formed prematurely. Still, the pig suckles, yet does not shed its teeth [a]; and so do all the animals with sharp interfitting teeth,[b] but some of them (*e.g.*, the lion [c]) do not shed any teeth except the canine ones. Democritus, then, made this mistake because he made a general statement without investigating the facts in all cases ; but this is precisely what we ought to do, because whenever anyone makes a general statement it must apply to all cases. Now the assumption we make— and it is an assumption founded upon what we observe —is that Nature neither defaults nor does anything idly in respect of the things that are possible in every case ; and further, if an animal is going to get any nourishment after the period of its suckling is over, it must of necessity possess instruments with which to deal with its nourishment. So that if this took place, as Democritus says, about the time of maturity, Nature would be defaulting in one of the things which it is possible for her to do, and we should have Nature

788 b

ἔργον ἐγίγνετ᾽ ἂν παρὰ φύσιν. τὸ γὰρ βίᾳ παρὰ
φύσιν, βίᾳ δέ φησι συμβαίνειν τὴν γένεσιν τῶν
ὀδόντων. ὅτι μὲν οὖν τοῦτ᾽ οὐκ ἀληθές, φανερὸν
ἐκ τούτων καὶ τοιούτων ἄλλων.

30 Γίνονται δὲ πρότερον οὗτοι τῶν πλατέων πρῶτον
μὲν ὅτι καὶ τὸ ἔργον τὸ τούτων πρότερον (πρότερον
γάρ ἐστι τοῦ λεᾶναι τὸ διελεῖν, εἰσὶ δ᾽ ἐκεῖνοι μὲν
ἐπὶ τῷ λεαίνειν, οὗτοι δ᾽ ἐπὶ τῷ διαιρεῖν), ἔπειθ᾽
ὅτι τὸ ἔλαττον, κἂν ἅμα ὁρμηθῇ, θᾶττον γίνεσθαι
πέφυκε τοῦ μείζονος. εἰσὶ δ᾽ ἐλάττους οὗτοι τῷ

789 a μεγέθει τῶν γομφίων, τῷ τὸ[1] ὀστοῦν τῆς σια-
γόνος ἐκεῖ μὲν πλατὺ εἶναι, πρὸς δὲ τῷ στόματι
στενόν. ἐκ μὲν οὖν τοῦ μείζονος πλείω ἀναγκαῖον
ἐπιρρεῖν τροφήν, ἐκ δὲ τοῦ στενωτέρου ἐλάττω.[2]

Τὸ δὲ θηλάζειν αὐτὸ μὲν οὐθὲν συμβάλλεται, ἡ
5 δὲ τοῦ γάλακτος θερμότης ποιεῖ θᾶττον βλαστάνειν
τοὺς ὀδόντας. σημεῖον δ᾽ ὅτι καὶ αὐτῶν τῶν
θηλαζόντων τὰ θερμοτέρῳ γάλακτι χρώμενα τῶν
παιδίων ὀδοντοφυεῖ θᾶττον· αὐξητικὸν γὰρ τὸ
θερμόν.

Ἐκπίπτουσι δὲ γενόμενοι τοῦ μὲν[3] βελτίονος
10 χάριν, ὅτι ταχὺ ἀμβλύνεται τὸ ὀξύ· δεῖ οὖν ἑτέρους
διαδέχεσθαι πρὸς τὸ ἔργον. τῶν δὲ πλατέων οὐκ
ἔστιν ἀμβλύτης, ἀλλὰ τῷ χρόνῳ τριβόμενοι λεαί-
νονται μόνον. ἐξ ἀνάγκης δ᾽ ἐκπίπτουσιν, ὅτι τῶν
μὲν ἐν πλατείᾳ τῇ σιαγόνι καὶ ἰσχυρῷ ὀστῷ αἱ

---

[1] τῷ τὸ Platt : καὶ τῷ τὸ coni. A.-W. (καὶ τῷ τὸ Paris.
Suppl. Gr. 333*) : καὶ τό vulg. (ῷ supra τ Z²).
[2] sic Platt : ἐκ δὲ τοῦ ἐλάττονος στενωτέραν vulg.
[3] γενόμενοι τοῦ μὲν] γ᾽ ἔνιοι τούτων τοῦ (οἱ Z² in ras., τοῦ et
βελτίονος Z² in lacuna) Z : γ᾽ ἔνιοι μὲν τοῦ μὴ S.

working contrary to Nature [a] (because he says that the formation of the teeth is brought about by force, and " by force " means " contrary to Nature "). So then it is apparent, both from these considerations and others like them, that this view is untrue.

The teeth of which we are speaking are formed earlier than the flat teeth (1) because the work they have to perform comes earlier : breaking up (which is the purpose of these teeth) comes before grinding (which is the business of the flat ones) ; (2) because a smaller thing naturally forms more quickly than a larger one, even if they both start off together, and these teeth are smaller in size than the molars, because the jawbone at that point is flat, whereas it is narrow by the mouth ; and, of necessity, a larger amount of nourishment will flow out from the larger part, and a smaller amount from the narrower.[b]

Suckling, in itself, contributes nothing to the formation of the teeth, though the warmth of the milk makes them come through more quickly. A proof of this is that within the actual class of those which suckle, those young ones which get hotter milk grow their teeth quicker, because that which is hot tends to promote growth.

After having been formed, these teeth are shed (a) *for the sake of the better*, the reason being that anything sharp quickly gets blunted, and so a fresh relay of teeth is needed to carry on the work. (The flat ones, on the other hand, cannot get blunted ; they only get worn down in the course of time by friction.) They are shed also (b) *as a result of necessity*, because, whereas the roots of the grinders are situated in the wide part of the jaw and upon good strong

---

[a] But see Introd. § 14.   [b] *i.e.*, to form the teeth.

ῥίζαι εἰσί, τῶν δὲ προσθίων ἐν λεπτῷ, διὸ ἀσθενεῖς
15 καὶ εὐκίνητοι. φύονται δὲ πάλιν, ὅτι ἐν φυομένῳ
ἔτι τῷ ὀστῷ ἡ ἐκβολὴ γίνεται καὶ ἔτι ὥρας οὔσης
γίνεσθαι ὀδόντας. τούτου δὲ σημεῖον ὅτι καὶ οἱ
πλατεῖς φύονται πολὺν χρόνον· οἱ γὰρ τελευταῖοι
ἀνατέλλουσι περὶ τὰ εἴκοσιν ἔτη, ἐνίοις δ' ἤδη καὶ
γηράσκουσι γεγένηνται οἱ ἔσχατοι παντελῶς διὰ
20 τὸ πολλὴν εἶναι τροφὴν ἐν τῇ εὐρυχωρίᾳ τοῦ ὀστοῦ.

τὸ δὲ πρόσθιον διὰ τὴν λεπτότητα ταχὺ λαμβάνει
τέλος, καὶ οὐ γίνεται περίττωμα ἐν αὐτῷ, ἀλλ'
εἰς τὴν αὔξησιν ἀναλίσκεται ἡ τροφὴ τὴν οἰκείαν.

Δημόκριτος δὲ τὸ οὗ ἕνεκα ἀφεὶς λέγειν, πάντα
ἀνάγει εἰς ἀνάγκην οἷς χρῆται ἡ φύσις, οὖσι μὲν
5 τοιούτοις, οὐ μὴν ἀλλ' ἕνεκά τινος οὖσι, καὶ τοῦ
περὶ ἕκαστον βελτίονος χάριν. ὥστε γίνεσθαι μὲν
οὐθὲν κωλύει οὕτω καὶ ἐκπίπτειν, ἀλλ' οὐ διὰ
ταῦτα, ἀλλὰ διὰ τὸ τέλος· ταῦτα δ' ὡς κινοῦντα
καὶ ὄργανα καὶ ὡς ὕλη αἴτια, ἐπεὶ καὶ τὸ τῷ
πνεύματι ἐργάζεσθαι τὰ πολλὰ εἰκὸς ὡς ὀργάνῳ·
10 οἷον γὰρ ἔνια πολύχρηστά ἐστι τῶν περὶ τὰς
τέχνας, ὥσπερ ἐν τῇ χαλκευτικῇ ἡ σφύρα καὶ ὁ
ἄκμων, οὕτως καὶ τὸ πνεῦμα ἐν τοῖς φύσει συν-
εστῶσιν. ὅμοιον δ' ἔοικε τὸ λέγειν τὰ αἴτια ἐξ

---

[a] " The ' for the sake of which.' "
[b] See Introd. § 6.
[c] i.e., " of necessity," a result of mere mechanical causation.
[d] Cf. above, 741 b 37, 742 a 16, and App. B §§ 7 ff.

bone, those of the front teeth are in a thin part, and in consequence the teeth are weak and can easily be removed. They grow a second time, because they are shed while the bone is still growing and while the age for growing teeth is still going on. A proof of this is that even the flat teeth take a long time growing : the last of them are cut at about twenty years of age ; in fact, some people have been quite aged before their last teeth finished growing. The reason for this is that there is a great deal of nourishment in the wide part of the bones. The front part, however, quickly reaches its completion owing to its thinness, and no residue finds a place in it ; instead of that, the nourishment is consumed to supply that part's own growth.

Democritus, however, omitted to mention the Final Cause,[a] and so all the things which Nature employs he refers to necessity. It is of course true that they are determined by necessity, but at the same time they are for the sake of some purpose, some Final Cause, and for the sake of that which is *better* in each case.[b] And so there is nothing to prevent the teeth being formed and being shed in the way he says [c] ; but it is not on that account that it happens, but on account of the Final Cause, the End ; those other factors are causes *qua* causing movement, *qua* instruments, and *qua* material, since in fact it is probable that Nature makes the majority of her productions by means of *pneuma*[d] used as an instrument. *Pneuma* serves many uses in the things constructed by Nature, just as certain objects do in the arts and crafts, *e.g.*, the hammer and anvil of the smith. But to allege that the causes are of the *necessary* type is on a par with

559

ἀνάγκης κἂν εἴ τις διὰ τὸ μαχαίριον οἴοιτο τὸ
ὕδωρ ἐξεληλυθέναι μόνον τοῖς ὑδρωπιῶσιν, ἀλλ᾽
15 οὐ διὰ τὸ ὑγιαίνειν οὗ ἕνεκα τὸ μαχαίριον ἔτεμεν.

Περὶ μὲν οὖν ὀδόντων, διότι οἱ μὲν ἐκπίπτουσι
καὶ γίνονται πάλιν, οἱ δ᾽ οὔ, καὶ ὅλως διὰ τίν᾽
αἰτίαν γίνονται, εἴρηται. εἴρηται δὲ καὶ περὶ τῶν
20 ἄλλων τῶν κατὰ τὰ μόρια παθημάτων, ὅσα γίνεσθαι συμβαίνει μὴ ἕνεκά του ἀλλ᾽ ἐξ ἀνάγκης καὶ
διὰ τὴν αἰτίαν τὴν κινητικήν.

supposing that when water has been drawn off from a dropsical patient the reason for which it has been done is the lancet, and not the patient's health, *for the sake of which* the lancet made the incision.

We have now dealt with the subject of the teeth, and we have stated why some of them are shed and grow a second time and why some of them do not, and generally, to what cause their being formed is due. We have also dealt with the other conditions which affect the parts of the body, conditions which occur not for the sake of any Final Cause but *of necessity* and on account of the Motive Cause.

ADDITIONAL NOTES ON THE TEXT

I add here four textual annotations for which there was
no room in the body of the work.

I. 719 a 2 ff.   The mss. and editions have various readings,
and several proposals have been made for emendation.

Bekker has :   τὸν αὐτὸν τρόπον τὰ πλεῖστα γίγνεται ὅνπερ ἐν
                  τοῖς ὄρνισιν (ὀρνιθίοις SYZ)· καταβαίνει γὰρ
                  κάτω, καὶ . . .
Z :   . . . γίνεται ὅνπερ . . . καὶ. καταβαίνει
                  κάτω . . .
P :   . . . γιγνόμενον ὅνπερ . . . καταβαίνει κάτω . . .
S :   . . . γίνεται ὥσπερ . . . καταβαίνει κάτω . . .
        (Hence Y must be the authority for γὰρ.)
Aldus :   γίνεται ὅνπερ . . . ὀρνιθίοις καταβαίνει κάτω . . .
A.-W. coni.:   ⟨ἡ τελείωσις⟩ γίνεται ὅνπερ ἐν τοῖς ὄρνισιν ⟨τὰ
                  ᾠὰ⟩ καταβαίνει κάτω . . .
Susemihl coni.:   . . . ὄρνισιν ⟨ἡ τελείωσις· τὰ δ' ᾠὰ⟩ καταβαίνει
                  κάτω.

If loss of this sort is likely, which I doubt, a more probable
emendation would be καταβαίνει γὰρ κάτω ⟨τὰ ᾠά⟩, καὶ . . .
But I suspect that the corruption is more serious, for Scot
reads: *et similiter multis ovis avium ; ⟨et quedam animalia
ovant interius, et exit ab eis animal parvum ; et cum pervenit
tempus partus⟩ descendunt ⟨ova⟩ ad partem inferiorem apud
iuncturas et exit ab eis animal sicut accidit animalibus ge-
nerantibus animalia ex prima creatione.*   The Greek original
of the words in brackets has disappeared from our text.

II. 738 a 8 ff.   I suspect that the original reading here was
τοῖς περιττώμασι τοῖς τ' ἀχρήστοις ⟨καὶ τοῖς χρησίμοις⟩, and
that the rest of our present text is part of a gloss, for τῇ
τε . . . ὑγρᾷ cannot be construed, and the reference to blood
seems to consider blood as a " residue," which is incorrect.
If my suggestion is right, the gloss will have ousted the
reference to useful residues from our text, and the reference to
useless ones from Scot's ultimate original, for Scot reads
*omnia ista habent membra recipiencia superfluitatem qua
indigent* (his regular equivalent for χρήσιμος) *sicut sanguis qui
habet locum in venis ; ergo ipse vadit in ea sicut in vasa.*
Clearly, too, Scot incorporates more of the latter part of the

GENERATION OF ANIMALS

gloss than the Greek text does, and the reference to *vasa* (=ἀγγεῖα) leads me to think that the gloss was founded on a misunderstanding of the passage at *P.A.* 650 a 33 (*q.v.*). The blood-vessels are often described as ἀγγεῖα in *P.A.*; *cf. G.A.* 740 a 23.

II. 746 a 34. Here Bitterauf, following the suggestion of Bussemaker, proposes to insert ⟨καὶ θώων⟩ after καὶ λύκων on the strength of William's and Scot's versions. The latter reads *in canibus et vulpibus et lupis et in genere quod dicitur grece comez* (Buss. and Btf. give *comex*). This is supported by the fact that at 774 b 17 Scot translates κύων λύκος θώς *canis et lupus et animal quod dicitur grece noz.* (Such variation in the spelling of proper and other unusual names is not infrequent in Scot.) At 742 a 9 θώς is not represented in Scot's version.

(According to A.-W., θώς, usually translated "jackal," is most probably the civet or genet: see D. W. Thompson, *H.A.* 580 a 29, n.)

V. 781 a 10 οἱ γὰρ πόροι . . . 781 b 5 συμβαίνουσιν. The main arguments against this passage being an original and genuine part of the text may be stated as follows :

(1) The introductory γάρ introduces no real explanation or expansion of the preceding statement. The passage is in fact completely extraneous to the argument.

(2) The reference to *De sensu* at 781 a 21 is incorrect, as A.-W. point out. There is no such clear statement in *De sensu*; at 439 a 1 the αἰσθητήριον of touch and taste is said to be πρὸς τῇ καρδίᾳ, but nothing is said to suggest that sight and smell have any further connexion beyond their connexion with the brain. At *P.A.* 656 a 29, on the other hand, there is a more exact reference to *De sensu*: "The correct view, that the ἀρχή of the senses is the region around the heart, has already been defined in the treatise *Of Sensation*, where also I show why it is that *two of the senses, touch and taste*, are evidently (φανερῶς) connected to the heart." Shortly before (656 a 20 ff.) Aristotle has stated that the brain is not the cause of any of the sensations ; it is ἀναίσθητος.

(3) The passage is concerned exclusively with that part of the mechanism of hearing which is internal, not with the superficial sense-organ, whereas the reason given for accuracy of hearing and smelling is concerned only with the superficial sense-organ (just as the similar argument for sight,

which is referred to, is concerned only with the eye itself and the skin on it).

(4) The passage has nothing whatever to say about smell.

(5) It concludes with a mere repetition of 781 a 18-20, to the effect that accuracy depends upon the purity of the organ and its membrane, ignoring the whole of the intervening discussion about the internal mechanism.

(6) The reference to a place where the connate *pneuma* causes " in some " pulsation and " in others " respiration and inspiration is, as Platt points out, meaningless, for no animal respires unless it has a heart.

The inference would appear to be that the passage, though probably of Aristotelian origin, has been corrupted, and that, so far as Book V is concerned, it began as a marginal annotation, intended to supply an account of the inner mechanism of sensation, etc., which would supplement the account of the mechanism of the superficial sense-organs of hearing and smell which no doubt originally stood here in the text. No such account, however, is there now ; and it seems reasonable to suppose that it has been ousted and supplanted by the passage which now stands there.

To understand the background of the passage, the reader may find it useful to refer to the account of Aristotle's theory of hearing in App. B §§ 29 ff., which I have compiled from various passages here and elsewhere in his works. I have suggested in the critical note some corrections, based on Scot's Latin version, which may help to bring the text into agreement with Aristotle's doctrine as ascertained from these other passages.

For the sake of completeness, I give the remainder of Scot's translation between the two passages already quoted in the *app. crit.* : [*et*] *propter hoc addiscuntur res per* (v.l. *propter*) *sensum auditus, quoniam sicut sermo intrat per sensum auditus, ita exit per linguam* [*et*] *per motum vocis. manifestum est ergo quod homo dicit* (v.l. *discit*) *quod audit. et cum homo gannit debilitatur auditus, quoniam principium instrumenti sensus istius est positum super membrum in quo est spiritus, et movetur cum eo quando spiritus movebitur instrumento in quo est. et hoc accidens accidit temporibus humide complexionis.*

The passage is discussed at considerable length by F. Susemihl, *Rhein. Mus.* XL (1885), 583 ff.

GENERATION OF ANIMALS

ADDITIONAL NOTE FOR II. 741 b 2, III. 762 b 24 ff.

The first modern work on the breeding migration of the The European eel (*Anguilla vulgaris*) is that of Grassi [a] and Common Calandruccio, who, following some previous work on Eel. the reproductive organs, made observations of eels in the Mediterranean, and showed that *Leptocephalus*, already known and described as a different animal, was the larval form of the eel. The whole subject has been fully worked out by Schmidt [b] in recent years. The facts are these. During the time when eels live in fresh water, their reproductive organs do not reach maturity, as Aristotle pointed out; but after a number of years, which may vary from five to twenty, the body takes on a metallic sheen (" silver eels ") and the fish set out on their migration to their breeding-places in the deep waters between the West Indies and Bermudas. The eggs float in the sea, and the larvae are carried by the ocean currents eastwards across the Atlantic : upon arrival at the Continental shelf two and a half years later they metamorphose into elvers, and these then move up into the estuaries and rivers of Europe, sometimes passing over damp grass to isolated pools. During the period of growth which follows, they are yellowish and greenish in colour (" yellow eels "). The old eels never return to fresh waters. The story (mentioned by Aristotle) of the development of eels out of horsehair worms was current until recent times.

ADDITIONAL NOTE FOR III. 757 a 2 ff.

Aristotle discusses the hyena both here and at *H.A.* VI. The Hyena. 579 b 15 ff.

An important piece of research on the spotted hyena recently carried out in Tanganyika Territory by L. Harrison Matthews [c] has established that externally the female of

---

[a] G. B. Grassi, *Proc. Roy. Soc.* LX (1897), 260-271.

[b] J. Schmidt (of Copenhagen), *The Breeding Places of the Eel, Phil. Trans. Roy. Soc.* (B) CCXI (1922), 179-208; see also *id., Nature*, CXI (1923), 51-54, CXIII (1924), 12; and W. Heape, *Emigration, Migration, and Nomadism*, 1931.

[c] *Reproduction in the Spotted Hyena* (Crocuta crocuta), in *Phil. Trans. Roy. Soc.* (B) CCXXX (1939), 1-78.

the spotted hyena closely resembles the male: it has a peniform clitoris, similar in form and position to the penis of the male, and scrotal pouches closely simulating those of the male. Indeed the male and non-parous female are indistinguishable externally. Matthews points out that Aristotle did not distinguish between spotted and striped hyenas: the legend " relates to the spotted hyena, but Aristotle's refutation of it to the striped, the genital anatomy of which he correctly describes " (Matthews refers to the description in *H.A.*). Of 103 specimens collected by Matthews, 63 were males ; this is a lower percentage than that given by the hunter with whom Aristotle discussed the subject: he found ten out of eleven were males, but these may have been striped hyenas.

APPENDIX A

MOVEMENT IN THE UPPER COSMOS AND IN THE LOWER COSMOS ; THE HEAVENLY BODIES ; γένεσις AND φθορά ; TIME, PERIODS, CYCLES

(SUPPLEMENT TO BOOK II, *init.* AND BOOK IV, *fin.*)

It will be seen that the terminology of the two passages above mentioned reappears in the following account, much of which is taken verbatim from the several passages to which reference is given. I have not thought it necessary to draw attention to all the parallels, as these will be obvious to the reader who has the passages of G.A. before him.

(1) *Met.* Λ 1069 a 30 ff. There are three kinds of οὐσία : **Three kinds of being.**

    (1) sensible (αἰσθητή) $\begin{cases} (a) \text{ eternal } (\mathring{a}ï\delta\iota o\varsigma) ; \\ (b) \text{ perishable } (\phi\theta a\rho\tau\acute{o}\varsigma), \ e.g., \\ \quad \text{animals and plants} ; \end{cases}$

    (2) immutable (ἀκίνητος).

Immutable οὐσία is the οὐσία of the unmoved mover (see below, § 3) ;

sensible and eternal οὐσία belongs to the " heaven " and the heavenly bodies (the stars and planets, including the Sun and Moon) ;

sensible and perishable οὐσία belongs to the things of the sublunary world (Earth, Air, etc., and the organisms made out of them, animals, plants, etc.).[a]

(2) *De caelo*, *e.g.*, 268-269, 289 a, 300 a 20 ff., etc. There are five natural substances which compose the physical universe : **Five elements.**

    Aither, whose nature it is to move eternally in a circle ; this is the substance out of which the whole of the Upper Cosmos is made, viz., the " first heaven " (the outermost shell or sphere) in which the stars are

[a] See also App. A § 18.

567

APPENDIX A

fixed, and also the planetary "heavens" together with the planets themselves which they carry ;

Fire, Air, Water and Earth, whose natural movement is rectilinear (*e.g.*, Air moves naturally outwards from the centre, Earth moves naturally towards the centre ; hence they would if left to themselves[a] arrange themselves in concentric strata, with Fire outermost, next to the innermost " heaven " ; after that Air, then Water, and Earth at the centre). These are the substances out of which all the Lower Cosmos, the sublunary world, is composed.

The Unmoved Mover and the φορά of the First Heaven. (3) *Met.* Λ 1072 a, b. The ultimate source of all movement is the Unmoved Mover, which is pure, self-thinking thought, or God ; and since the " actuality " of thought is life, we can say that ζωὴ καὶ αἰὼν συνεχὴς καὶ ἀίδιος ὑπάρχει τῷ θεῷ. This " first principle " causes movement without itself being in movement ; it is therein analogous to objects of desire or of thought, which κινεῖ οὐ κινούμενα[b] ; in fact, it κινεῖ ὡς ἐρώμενον (it causes movement by being an object of love).[c] Upon this first principle the Heaven and Nature depend. What it first sets in movement is the πρῶτον κινούμενον, the *primum mobile*, viz., the " first heaven " or outermost sphere ; and since this movement is an unceasing movement, so the first heaven will be ἀίδιος. This movement, then, is one and eternal ; it is simple φορά, simple uniform circular movement.

Movement in the Upper Cosmos. (4) All other things beside the Unmoved Mover which produce movement do so in virtue of being themselves in movement (κινούμενα τἆλλα κινεῖ). Thus the " first heaven " communicates movement to the inner " heavens," the whole system of concentric spheres, which are in contact with each other ; and the movements of these, although still continuous and eternal, are no longer uniform, because they are the resultants of more revolutions than one.[d]

[a] As in fact they are not (see § 12; *cf.* § 9). Nor, according to Aristotle, are the elements occupying their "proper" places when acting as the components of living bodies (*De caelo* II. 288 b 17 ff.).

[b] *Cf.* App. B § 1.

[c] *Cf.* Dante, *Paradiso*, vers. ult., *l' amor che move il sole e l' altre stelle.*

[d] It is not necessary here to give details of the system of spheres as worked out by Aristotle, based on the mathematical theories of Eudoxus and Callippus.

APPENDIX A

(5) In the " region about the centre," *i.e.*, the Lower Cosmos **Movement** or sublunary world, there is no circular movement at all **in the** as such. The form in which movement is found here is **Lower Cosmos.** in the " movements," *i.e.*, transformations of the four sublunary " simple " bodies, Fire, Air, Water, Earth, and in the " movements " of living creatures, animals and plants, viz., γένεσις and φθορά, " alteration," growth and diminution.*a* " Movement " is mediated to the things in the Lower Cosmos through the heavenly bodies, chiefly the Sun, as is stated at the end of *G.A.* IV.

(6) *Meteor.* I. 339 a 28. We should regard Fire, Earth, etc., **The** as the " material " causes of phenomena in the sublunary **"Causes"** world ; but the cause in the sense of the origin of move- **of things in** ment (the " motive " cause) is to be found in the *dynamis* **the Lower** of the eternally moving bodies.*b* **Cosmos.**

(7) *Ibid.* 340 b, 341 a. The " first element " (*alias* the **The** " fifth element," viz., *aither* ; see 737 a 1, n.) and the **heavenly** bodies in it revolve in a circle, and as they do so, that **bodies;** portion of the Lower Cosmos which is next to the *aither* **heat.** gets inflamed and produces heat. Thus, although not made of Fire, and although not themselves hot, the heavenly bodies produce heat by their mere movement. Aristotle explains this more fully at *De caelo* II. 289 a 29, when he says that the heat and light which proceed from them are produced by the friction set up in the Air by their φορά (*cf.* § 9 *fin.* below). The Sun, which is con- sidered to be the hottest of them all, is really white (λευκός), not fiery in colour. The Sun's φορά is sufficient to produce warmth and heat : it is fast enough and near enough, whereas the φορά of the stars though fast is distant, and the Moon's though near is slow (*cf. De caelo* II. 289 a 20-34).

(8) *Ibid.* 346 b, 359 b. Rain and winds are explained as **Rain and** being caused by the Sun's approaching and receding in **winds.** its φορά. When it approaches it draws up the moist exhalation ; when it retires this vapour cools and congeals again into water ; hence there is more rain during winter and during the night. It also draws up the dry exhala- tion, and this is the substance which makes the winds.

(9) It is pointed out in *De caelo* II. 286 b 2 that in order to **Function** **of the other**

a See Introd. §§ 47 ff., κίνησις.
b Quoted in Greek at 777 b 31, n.

APPENDIX A

heavenly φοραί in causing (a) γένεσις;

account for the transformations of the four " elements " Fire Air Water Earth, *i.e.*, for the γένεσις of them out of one another, some additional φορά or φοραί beside that of the " Whole " (or the πρῶτον κινούμενον) is required : if this were the only φορά, no transformation would take place and the four elements would be static.

And with regard to the γένεσις of living things, Aristotle describes in other treatises more strikingly and in fuller detail than he does in *G.A.* the important part played by these other φοραί (*i.e.*, those of the heavenly bodies). Thus in *Phys.* II. 194 b 13 we read ἄνθρωπος ἄνθρωπον γεννᾷ καὶ ἥλιος[a]; and at *Met.* Λ 1071 a 13 ff. the " causes " of a man are listed as (a) the " elements," viz., (i) his *matter* (Fire and Earth),[b] and (ii) his own *form* (ἴδιον εἴδος) ; also (b) something external, viz., his father ; and besides these (c) the Sun and the circle of the ecliptic (ὁ λοξὸς κύκλος)—and these last stand to him neither as matter, nor as form, nor as privation, nor as being identical with him in form, but as κινοῦντα, *i.e.*, " efficient " or " motive " causes (*cf.* §§ 5 and 6 above). *Cf.* also *G.A.* II. 737 a 3 : the heat of the Sun and the heat of animals as contained in semen is able to cause generation, whereas Fire cannot.

(b) γένεσις (10) and φθορά.

The whole question of γένεσις and φθορά is more fully discussed at the end of the treatise *G. & C.* (II. chh. 10 and 11), where the meaning of the statements about the Sun and the ecliptic is explained. Here Aristotle states that γένεσις is continuous because the circular revolution of the " first heaven " is eternal (ἡ κατὰ τὴν φορὰν κίνησις is ἀΐδιος) ; and this φορά produces γένεσις by bringing τὸ γεννητικόν (the generative agent, viz., the Sun) nearer and by taking it further away. This φορά however is a *single* movement (as we saw above, § 3), and therefore will only explain γένεσις ; it will not also ex-

---

[a] This would not, however, have sounded so strange to a Greek ; *cf. G.A.* 716 a 17 οὐρανὸν δὲ καὶ ἥλιον . . . ὡς γεννῶντας καὶ πατέρας προσαγορεύουσιν.—It is a statement which caught the fancy of the Middle Ages, and is quoted by Dante (from the Latin translation of *Physics* II) in his *De monarchia* I. 9 *init.* ; *cf. Paradiso* XXII. 116 *quegli ch' è padre d' ogni mortal vita.*

[b] Aristotle regularly takes these two as the elements *par excellence*, standing for all four (see *De caelo* III. 298 a 29, 298 b 8)—because Fire " has not heaviness " and Earth " has not lightness " (IV. 311 b 27). *Cf.* App. B §§ 20, 22, 23.

plain φθορά. Thus γένεσις-and-φθορά is to be explained *not* as being due to the primary φορά (*i.e.*, the φορά of the " first heaven "), but as being due to the φορὰ κατὰ τὸν λοξὸν κύκλον—the movement along the circle of the ecliptic, which is tilted. This, like the other, possesses continuity ; but also it is *double*, not *single*. Thus we may say that the *continuity* is caused by the φορά of " the Whole " (*i.e.*, the " first heaven " ; the primary φορά), while the *alternation* is produced by the inclination of the ecliptic, which makes the Sun alternately approach and retreat. When the Sun approaches it will cause γένεσις, when it retreats it will cause φθορά.

(11) Now in consequence of this, *natural* (κατὰ φύσιν) γένεσις and φθορά occupy equal times for their accomplishment. Hence *both the times and the lives of all several things have a " number " and by that number they are delimited . . . and every life and time is measured by a period . . .* : *for some, this period is the year ; for others, the period, which is the measure, is greater, for others, smaller* (διὸ καὶ οἱ χρόνοι καὶ οἱ βίοι ἑκάστων ἀριθμὸν ἔχουσι καὶ τούτῳ διορίζονται . . . καὶ πᾶς βίος καὶ χρόνος μετρεῖται περιόδῳ . . . · τοῖς μὲν γὰρ ὁ ἐνιαυτός, τοῖς δὲ μείζων, τοῖς δὲ ἐλάττων ἡ περίοδός ἐστι τὸ μέτρον). He then repeats that *natural* γένεσις and φθορά occupy an equal time ; but, he adds, in point of fact things often φθείρεται in a shorter time than this ; for since matter is *uneven* (ἀνώμαλος ; *cf.* his statement in *G.A.* IV *fin.* about its " indeterminateness "), the γενέσεις of things are uneven too, some being quicker and some slower than they should be ; and as a result of this the φθορά of other things is affected, because the γένεσις of one set of things is the φθορά of another. (See also App. B §§ 7-11.)

Γένεσις and φθορά governed by " periods."

(12) Γένεσις-and-φθορά is continuous, and shall never fail. The reason is that Nature always strives after τὸ βέλτιον, and being is better than not-being ; but since being cannot be possessed by all things because they are too far away from the ἀρχή (*i.e.*, from God, the Unmoved Mover), God " filled in " the Whole in the manner that remained open, viz., by making γένεσις continuous ; that was the way to ensure that as *far as possible* there should be an unbroken chain of " being " throughout the universe; for the next best thing to " being " is that

Continuous γένεσις a second-best to eternal being.

γένεσις should be continually going on (τὸ γίνεσθαι ἀεὶ τὴν γένεσιν); and the cause of this is the circular φορά; for this is the only continuous form of movement. Hence also the things which get transformed into each other (viz., the " simple bodies," such as Water, Air, Fire) imitate the circular φορά : Water is transformed into Air, Air into Fire, Fire into Water, and we say that their γένεσις has come round a full " circle." (So, too, rectilinear φορά is continuous in virtue of its imitating circular φορά.) And this also provides a solution of the problem, Why is it that the " simple bodies," in spite of their natural tendency to make each for its own proper place in the universe, have not during the enormous stretches of time which have passed become separated out each into its own proper place, into concentric layers (see § 2) ? The reason is that they are continually being transformed to and fro one into the other, and the cause of their transformations is the φορά —i.e., the *double* φορά.

Measurement of φορά; measurement of time. (13) *Phys.* IV. 219 b 3 ff. We cognize movement by means of some body which is in movement; so too we cognize φορά by means of some body which is φερόμενον : that is how we cognize the " before-and-after " factor in movement, for it is the " now " (*i.e.*, the moment at which the body is observed to be at some particular point in its course) which is " most cognizable." And just as φορά and the φερόμενον are thus closely allied, so too are the ἀριθμός [a] of the φορά and the ἀριθμός of the φερόμενον. Now time is the ἀριθμός of the φορά. We see then that time is not movement, but it is " the aspect of movement whereby movement has an ἀριθμός," *i.e.*, the aspect of movement whereby movement can be numerated or counted (ᾗ ἀριθμὸν ἔχει ἡ κίνησις) : time is *that which is counted*, not that by which we count (τὸ ἀριθμούμενον, not ᾧ ἀριθμοῦμεν); time is an ἀριθμός which is counted, not an ἀριθμός which we use as a means for counting (220 b 8). Time is the ἀριθμός of continuous movement generally (223 b 1 ; *cf. G. & C.* II. 337 a 23), not of any movement in particular ; nevertheless, what we usually mean by time, and what really

[a] This meaning of ἀριθμός is of course quite distinct from that in §§ 15-17 below.

APPENDIX A

has the best claim to the name, is the ἀριθμός of the circular movement (ἡ κύκλῳ φορά), because the ἀριθμός of this even, uniform, circular revolution is " most cognizable " (223 b). And as everything is measured by some standard which is cognate to it (*e.g.*, horses are measured or counted by the unit " a horse," see 220 b and 223 b), so time is measured by " a time," viz., by a determinate length of time ; and the time taken by the sphere of the universe to revolve is the " measure " *par excellence* : all other movements are measured by that movement, and time too is measured by that movement (*cf. De caelo* II. 287 a 23 ff., *Phys.* VIII. 265 b 8 ff.). Hence human affairs and all other things which have a *natural* movement and γένεσις and φθορά are spoken of as being a " cycle " : they are all discriminated by time, and their beginning and their end occur as it were according to some " period " (223 b). And further, since a movement may be the same over and over again, so too may time, *e.g.*, year, spring, autumn (220 b 12).

(14) *G. & C.* II. 338 a 1 ff. If a thing's " being " is " necessary " (*i.e.*, *absolutely* necessary ; see Introd. §§ 7-9), then it is eternal (ἀΐδιος) ; and if it is eternal, then its " being " is " necessary." [a] And also, if a thing's γένεσις is " necessary," then its γένεσις is eternal ; and if its γένεσις is eternal, " necessary." Thus, if a thing's γένεσις is *absolutely*, not *conditionally*, " necessary," its γένεσις must of necessity be cyclical and return upon itself (ἀνακυκλεῖν καὶ ἀνακάμπτειν). [Proof of this.— Γένεσις must be either limited or not limited. We agree it not limited. If it is not limited, it must be either rectilinear or cyclical. If it is to be eternal, it cannot be rectilinear ; hence it must be cyclical.] Thus it is in circular movement and in circular γένεσις that we find *absolute* necessity. This fits in with the doctrine (proved on other and independent grounds) that circular movement (*i.e.*, the movement of the Heavens) is eternal ; for it is the movements which belong to this eternal movement, and the movements which are caused by it, which γίνονται and εἰσίν " of necessity." That which is moving round in a circle is always setting other things in movement, so that their movement too must be circular.

Γένεσις cyclical.

---

[a] Eternal being and eternal γένεσις are mentioned at *G.A.* 742 b 27, 31.

573

Thus the upper φορά is a circular movement, hence the Sun's is too, hence the seasons γίνονται cyclically, hence τὰ ὑπὸ τούτων (cf. *G.A.* IV. 777 b 35—778 a 2) γίνονται cyclically. Thus Water → Air → Water ; cloud entails rain, rain entails cloud.

Γένεσις (15) *Ibid.* So far, so good. Why then do not men and
cyclical in animals apparently show this cyclical movement ? Why
either of do they not return upon themselves, so that the same
two modes. individual γίνεται a second time ? In other words, why is it not " necessary " that you should γίνεσθαι if your father does, although it *is* necessary that if you do, he should ? This looks like rectilinear, not cyclic, γένεσις. Well, we must make a distinction and say that there are two ways in which things " return upon themselves " : some (*a*) do it *numerically* (ἀριθμῷ, *i.e.* the individual is numerically identical) ; others (*b*) do it *specifically* only (εἴδει μόνον, *i.e.*, the specific form, not the individual, is identical). The difference depends upon the character of the οὐσία (see § 1) which is experiencing the " movement " : if (*a*) the οὐσία is " imperishable," then obviously they will be the same ἀριθμῷ as well as εἴδει ; if (*b*) their οὐσία is " perishable," then they recur εἴδει only, not ἀριθμῷ. That is why when Water γίνεται from Air, and Air from Water, it is the same εἴδει only, not ἀριθμῷ. Nothing, in fact, whose οὐσία γίνεται, *i.e.*, nothing whose οὐσία is subject to γένεσις and φθορά, whose οὐσία is such that it admits of not-being, can remain same and identical ἀριθμῷ.

(16) The meaning of the last preceding paragraph will be clearer when we recall which are the things whose οὐσία is " imperishable," not subject to γένεσις and φθορά. They are the stars and planets. Their οὐσία is free from all forms of change except circular movement ; hence each persists as an eternally identical individual ; its cycle is just its cyclical movement, φορά. As against these eternal οὐσίαι, we have such things as Air and Water, men and animals, whose οὐσία is liable to not-being, is " perishable." At first sight, says Aristotle, there seems to be a difference between Air and Water on the one hand and men and animals on the other. The " cycle " in the case of the former is obvious : rain is followed by cloud, cloud by rain, rain by cloud, con-tinually ; but it is not so obvious in the case of men and

APPENDIX A

animals. Although rain entails cloud, and cloud rain, in a continuous cycle, your father's γένεσις does not necessarily entail yours, though yours entails his. But fundamentally the situation is the same in both cases, for (a) γένεσις and φθορά shall never fail (§§ 12 and 14); there must always be a γένος of men, animals and plants (G.A. II), and the race will be continued even if one particular individual does not reproduce itself (this at any rate seems to be implied by Aristotle); (b) in neither case is there persistent identity of the individual: just as you are different ἀριθμῷ from your grandfather, so is the rain which falls to-day different ἀριθμῷ from the rain which fell yesterday or last year.

(17) *De anima* II. 415 a 25 ff. Reproduction is one of the functions of θρεπτικὴ ψυχή (nutritive Soul; see Introd. §§ 41 ff.); and the "most natural" function of all living things is to produce another one like themselves "so that they may partake in the eternal and divine in the way that they can" (ἵνα τοῦ ἀεὶ καὶ τοῦ θείου μετέχωσιν ᾗ δύνανται), since all things strive after this, and for the sake of this they do all that they do κατὰ φύσιν. But they are unable to partake in the eternal and divine by uninterrupted continuance (συνεχείᾳ), because no thing that is φθαρτόν may persist as one and the same ἀριθμῷ; hence they partake in it each in the way in which they can do so, some more, some less; and so the thing persists not as itself but as something like itself (οὐκ αὐτὸ ἀλλ' οἷον αὐτό)—*i.e.*, as one, not ἀριθμῷ, but εἴδει.

[margin: Γένεσις by reproduction a means of attaining eternity.]

(18) Aristotle states more than once that the "matter" for "perishable" things is τὸ δυνατὸν εἶναι καὶ μὴ εἶναι. *E.g.*, (1) in *G. & C.* II. 335 a 24 ff. For things which are εἶναι καὶ μὴ εἶναι δυνατά, the "material cause" (αἴτιον ὡς ὕλη) is τὸ δυνατὸν εἶναι καὶ μὴ εἶναι, which = τὸ γενητὸν καὶ φθαρτόν. (This is twice stated.) Hence, the field in which γένεσις and φθορά take place must be τὸ δυνατὸν εἶναι καὶ μὴ εἶναι: that, then, is their "material" cause. Their "final" cause is their figure or "form"; and there is a third cause or ἀρχή, viz., the "motive" cause. (2) In *Met.* Z 1032 a 15 ff. we read that οὐσίαι *par excellence*, the things which "we consider to have the fullest title to be called οὐσίαι," are animals and plants. And all φύσει γιγνόμενα (as well,

[margin: The "matter" of φθαρτά.]

575

APPENDIX B

of course, as all τέχνη γιγνόμενα) have " matter," for each of them is δυνατὸν εἶναι καὶ μὴ εἶναι, and *this is the " matter* " which is in each of them.

APPENDIX B

Σύμφυτον Πνεῦμα

I. THE FUNCTION OF Σύμφυτον Πνεῦμα IN GIVING PHYSICAL EFFECT TO THE MOVEMENT OF ὀρεκτικὴ ψυχή.

The movement of animals is also caused by an unmoved mover. (1) *M.A.* 700 b 15 ff., *De anima* III. 433 b 11 ff. All the various stimuli (such as intellect, imagination, purpose, wish, appetite, sensation) which " move " animals are reducible to mind and desire (νοῦς and ὄρεξις); hence the πρῶτον κινοῦν of animals is the object of intellect and the object of desire (τὸ ὀρεκτὸν καὶ τὸ διανοητόν). And the πρῶτον κινοῦν κ ι ν ε ῖ οὐ κ ι ν ο ύ μ ε ν ο ν, in virtue of being apprehended in thought or imagination : it is, in fact, τὸ πρακτὸν ἀγαθόν, the good which can be attained in the field of action. We thus have first (1) the *object of desire*, τὸ ὀρεκτόν, which κινεῖ οὐ κινούμενον ; next (2) is *desire* itself, ὄρεξις (or τὸ ὀρεκτικόν, the faculty of desire), and this κινεῖ κινούμενον ; last (3) is *the animal*, which is a κινούμενον οὐ κινοῦν—it gets moved without causing any further movement : it is the last term in the series.

Comparison and contrast of animal movement with that of the universe. (2) *M.A.* 700 b 30. Thus it is evident that in one respect every animal gets set in movement (κινεῖται) in the same manner as that in which the ἀεὶ κινούμενον gets " moved " by the ἀεὶ κινοῦν (which κινεῖ ὡς ἐρώμενον ; see App. A § 3) ; in another respect, however, there is a difference, for it is not " moved " ἀεί, but its every movement has a limit. This limit is τὸ οὗ ἕνεκα, the purpose aimed at by the movement, and when the purpose is achieved the movement ceases.

APPENDIX B

(3) *M.A.* 701 b 34 ff. (ch. 8). Putting the statement in § 1 above in a slightly different form, we can say that the origin of movement in the animal is τὸ ἐν τῷ πρακτῷ διωκτὸν καὶ φευκτόν—the object of pursuit and avoidance in the field of action ; and since τὸ φευκτόν is painful and τὸ διωκτόν is pleasant, and since pain and pleasure are generally accompanied by cooling and heating, therefore the apprehension of these objects in thought or imagination produces *of necessity* (ἐξ ἀνάγκης) cooling and heating. Or again, in other words (ch. 7), desire (ὄρεξις), which as we have just seen (§ 1) is the ultimate, *i.e.*, immediate cause of movement, is effected either through sensation, imagination, or thought, and these bring about ἀλλοίωσις (" alteration," *i.e.*, qualitative change) of various sorts—heating, cooling, expansion, contraction. ¶ 1 Physical accompaniments of desire.

(4) *M.A.*, chh. 8-10. This ὄρεξις, which brings about the animal's movement, must be situated in an ἀρχή (702 a 22) and this ἀρχή is the heart, or the counterpart of the heart in creatures which have no heart (703 a 14) ; besides, we can show independently that the ἀρχή of the κινοῦσα ψυχή must be in a central position (702 b 15) ; and of course ὄρεξις is the ὀρεκτικόν faculty *of the ψυχή*. Thus (701 b 28) when a sensation, or imagination, or thought produces an ἀλλοίωσις in respect of heating or cooling at the region of the heart, a great change or difference is produced in the body—*e.g.*, blushing, blanching, shivering, etc. Seat of desire.

(5) It is important to notice that, according to Aristotle, the movements of the living organism are not *mechanically* caused. In *M.A.*, ch. 7 he compares the small original stimulus (κίνησις) required to set going an automatic puppet (*cf. G.A.*, II. 734 b 8 ff., 741 b 9) with the small change (μεταβολή) that occurs at the ἀρχή (viz., the heart) of a living organism and produces great and numerous changes or " differences " at a distance from the ἀρχή (*cf. G.A.* I. 716 b 3, V. 788 a 11) ; but he takes care to point out that whereas in the automaton there is no ἀλλοίωσις, no qualitative change—the action being entirely mechanical or " clockwork "—in the animal there *is* ἀλλοίωσις ; in an animal one and the same part can become hotter and colder, larger and smaller—it ἀλλοιοῦται. " Alteration " involved in movement of animals.

577

Connate (6) *M.A.*, ch. 10. We have now established that it is ὄρεξις
pneuma the —*i.e.*, ψυχή operating in its faculty of desire—which is the
instrument " formal " cause of movement : it κινεῖ κινούμενον. But
of Soul
acting in ψυχή is not material ; and in living bodies there must be
its faculty some physical substance (σῶμα) too which κινεῖ κινούμενον.
of desire. And this is the ΣΠ. It κινεῖ κινούμενον—κινούμενον by
the ἀρχή which is the ψυχή ; and that is why the ΣΠ is
where it is. In fact, ΣΠ is the " organ " or " instrument " of movement (see also *De anima* III. 433 b 18),
capable of expanding and contracting, and in virtue of
that capability it can exert force and so cause movement.
And it causes movement *by other means than* ἀλλοίωσις
(μὴ ἀλλοιώσει) ; it undergoes no qualitative change itself,
although it brings about changes of that sort in the parts
of the body (and in the embryonic material, as we shall
see).

Summary.   Thus we must insert a fourth term in the series as originally
stated in § 1 :

(1) The object of desire, τὸ ὀρεκτόν, which κινεῖ οὐ
κινούμενον ;
(2) Desire itself, ὄρεξις, which κινεῖ κινούμενον ;
(2a) Σύμφυτον Πνεῦμα, which also κινεῖ κινούμενον ;
(3) The animal, which κινεῖται, but κινεῖ nothing further.

For further references to the action of the heart and the
pneuma, see below, §§ 31, 32.

II. THE FUNCTION OF Σύμφυτον Πνεῦμα IN GIVING
PHYSICAL EFFECT TO THE MOVEMENT OF
θρεπτική (= γεννητική) ψυχή.

Embryo (7) *G.A.* II. 741 b 37 ff. The parts of the embryo get deformed by limited, marked out from each other (διορίζονται), by
means of *pneuma*, but this is neither the *pneuma* of the female
connate
pneuma. parent nor the embryo's own *pneuma*. This is proved
by the case of birds, fishes and insects : some are separate
from the parent, since they get their articulation in the
egg ; some do not breathe at all, being produced out of
larvae or eggs ; and even those which breathe and get
articulated in the womb do not breathe until their lungs

APPENDIX B

are perfected, and both the lungs and the parts which precede them get articulated before the creatures breathe. Further, the fissipede quadrupeds (dogs, etc.) are born blind, and the articulation of the eyelid is effected later. Thus we conclude that the same causes that are responsible for delimiting the young creature qualitatively are also responsible for its quantitative development—for actualizing its latent quantitative potentialities.[a] And of necessity *pneuma* must be present, ὅτι ὑγρὸν καὶ θερμόν, τοῦ μὲν ποιοῦντος, τοῦ δὲ πάσχοντος.

(8) The understanding of this last remark may be helped by a passage in *M.A.*, ch. 8 and other passages. As we saw (§ 3), the ἀρχή of movement in the animal is "the object of pursuit and avoidance in the field of action"; and the thought and imagination of such objects is *of necessity* (ἐξ ἀνάγκης) accompanied by heat and cooling (§ 3). Bodily pleasures and pains are accompanied by heat and cooling either in some part of the body or all over the body. Hence there is good reason in the way the inner regions of the body and the regions around the ἀρχαί of the instrumental parts have been fashioned— these regions change from solid to fluid and from soft to hard and *vice versa*. This being so, and "the passive factor" and "the active factor" (more exactly, "that which is so constituted as to act," and "that which is so constituted as to be acted upon") having the character which they in fact have, when it so happens that the one is active and the other passive, and neither of them lacks any of the ingredients included in its *logos*, then immediately the one acts and the other is acted upon, and we get simultaneously, *e.g.*, the thought "I must walk" and the movement of the limbs in walking—because the imagination produces the desire, the desire produces the affections, and these suitably prepare the instrumental parts.

(9) Now we must remember that the "organ" or "instrument" of movement, that which bridges the gap between the immaterial ὄρεξις on the one hand and the material limbs of the body on the other, is the ΣΠ (§ 6); it is this which gives actual physical effect to the ὄρεξις. ὄρεξις thus, as Aristotle says, stands to the limbs in the relation

Physical accompaniments.

Instrumental function of connate pneuma (a) in desire;

---

[a] This means that the same causes produce both the "uniform parts" (flesh, sinew, etc.) and also the "non-uniform parts" (face, hand, leg, etc.).

579

APPENDIX B

of ποιοῦν to πάσχον, κινοῦν to κινούμενον ; *but so does the*
ΣΠ too (§ 6).   In fact, it is the ΣΠ which brings about the
" preparation of the instrumental parts " by causing in
them the ἀλλοίωσις of which they are capable : it actualizes
their potentialities of changing from soft to hard, etc.

(b) in
develop-
ment of
embryo.
(10) Returning now to the passage of *G.A.*, it would appear
that in the developing embryo also the ΣΠ plays a
similar rôle.   It will be the ΣΠ which gives effect to the
formal cause in the semen so as to produce an embryo
of a particular kind, just as in the other case it gives
effect to the formal cause (viz., ὄρεξις) and produces
movement of the limbs ; here, too, then it will actualize
the latent potentiality of the material, bringing about
in it (741 b 12 ff.) the ἀλλοίωσις of which it is capable—
making it soft, hard, etc.

Connate
pneuma
the instru-
ment of
generative
Soul.
(11) With this in mind we can go on and interpret the rest of
the passage which follows in *G.A.* II. 742-743.   (1) The
heart must be formed first, *because it is the seat of the*
ΣΠ.[a]   (2) The φλέβες extend from the heart all over the
body, and thus can act as channels for the blood (which
is the " matter ") *and for the ΣΠ*[a] (which is the vehicle
of the " form," 729 b 20)—because (*De resp.* 480 a 10)[b]
all the φλέβες pulsate simultaneously with the heart, and
this pulsation is the *pneumatization* of the fluid as it gets
heated in the heart.   (3) Some of the " uniform parts "
(by which term Aristotle means such things as flesh,
nail, horn, sinew, bone) are formed by heat, others by
cold ;   and (740 b 18) the reasons why they are formed
are (*a*) that the female's " residue " is *potentially* what
the fully-formed animal itself is : all the parts are present
potentially in the residue ;   and (*b*) that (*cf.* the very
similar passage referring to ὀρεκτικὴ ψυχή quoted in § 8
above) when " the active factor " and " the passive
factor " come into contact " in that way in which the
one is active and the other passive " (which means in the
right place and at the right time),
then immediately the one acts and the other is acted

[a] These italicized phrases do not actually occur in the passage *G.A.*
742-743, but they are to be supplied from the doctrine of other passages
here examined (see below, § 32) ;   and we must realize that they represent perhaps the chief consideration, though unexpressed, in Aristotle's
mind as he writes the present passage.
[b] See § 31 below.

upon ; the male supplying the ἀρχή of " movement,"
the female supplying the material. It is θρεπτικὴ ψυχή
which is the source of this movement (just as in the
other case it was ὀρεκτικὴ ψυχή which was the source of
the movement)—it brings about both generation and
growth, for θρεπτικὴ ψυχή and γεννητικὴ ψυχή are one
and the same (see 735 a 17, 18). And the " organs "
or " instruments " which it uses are heat and cold : its
movement is " in " them. (This last sentence serves to
emphasize the dual nature of ΣΠ, dealt with in §§ 20 ff.
below ; for of course ΣΠ is the primary " instrument "
of θρεπτικὴ ψυχή.)
[Further important statements on these subjects are
found in *Meteor.* IV. Hot substance and cold substance, says Aristotle, are " active " (because they bring
things together, are συγκριτικά), solid substance and
fluid substance are " passive." Γένεσις, *i.e.,* natural
change, is the work of these *dynameis* ; so is natural
(κατὰ φύσιν) φθορά ; these processes occur in plants,
animals, and their parts, and are brought about by hot
and cold substance, when those ἔχωσι λόγον (*cf. G.A.*
777 b 28), out of the substrate matter underlying each
natural thing, viz., out of the " passive " *dynameis.* If
hot and cold fail to gain the mastery over the matter,
ἀπεψία results. Apart from destruction by force, the
end of all natural objects is putrefaction : it may be
defined as the φθορά of the proper and natural (κατὰ
φύσιν) heat in any fluid thing by the agency of alien
heat (that of the environment), due to lack of proper
heat, *i.e.,* owing to cold ; hence hot and cold are the
causes of putrefaction as they are of γένεσις. Animals
are generated in putrefying substances because the
heat that was secreted in these substances is natural
and is able συνιστάναι (see Introd. § 54). *Cf.* the whole
Book, especially 390 b 2 ff.]

(12) *G.A.* II. 743 a 20. It is not any chance material which
gets made into flesh or bone, nor does it get made in any
chance manner or at any chance time, but only the
material ordained by Nature, and in the manner and at
the time ordained by Nature : that which is *potentially*
X will not be made, actualized, into X by any motive
agent other than one which possesses the *actuality* ; nor

Requisites
for forma-
tion of
embryo.

APPENDIX B

will a motive agent which possesses the *actuality* make an X out of any chance material. Heat is present in the seminal residue, possessing the right movement and actuality (ἐνέργεια) to suit each of the parts ; and in the case of spontaneous generation the heat and movement of the season fulfil this same function.[a]

Connate *pneuma* analogous to *aither*: both are generative.

(13) *G.A.* II. 736 b 30 ff. Every faculty of ψυχή is connected with [b] a physical substance more divine than any of the four " elements " Fire, Air, Water, Earth, and this substance differs according to the degree of value of the ψυχή concerned. There is present in the semen of every animal and in " the foam-like stuff " [c] the so-called " hot substance," which causes the semen to be generative : this is not of course Fire, but it is the *pneuma* which the semen contains, " the substance in the *pneuma*," [d] which is " analogous to the element of the heavenly bodies," viz., the *aither*. That is why the heat of the Sun (*cf.* App. A §§ 9, 10) and the heat of animals (as contained in semen or any other such " residue ") is able to generate, whereas Fire cannot : the Sun, as we know already, consists of *aither*, and here we are told that there is in semen " something analogous " to *aither*.

(14) It is now possible to see what Aristotle means when he says (737 a 17) : " It has now been determined in what way fetations and semen have ψυχή : they have it *potentially*, but not *in actuality*." This *pneuma* or vital heat is not *in actuality* ψυχή ; but semen κινεῖται with a movement that is identical with that which moves the animal's body when the body is growing out of the " ultimate nourishment " (blood), and therefore when the semen gets into the uterus it sets in movement the

---

[a] See further, § 17 and additional note appended there.

[b] ἔοικε κεκοινωνηκέναι, a usefully vague term ; but at any rate it must be intended to denote a close relationship. We might express it perhaps by saying that this substance (viz., the *pneuma*, or more precisely "the substance in the *pneuma* ") with which ψυχή is thus associated is the physical vehicle *par excellence* of ψυχή ; anyway, it is the first physical substance to give expression to the movements of ψυχή ; it is its immediate instrument.

[c] Perhaps intended to include the " frothy bubble " concerned in spontaneous generation ; see §§ 17, 19 below.

[d] *Cf.* the substance which is " in " Air, Water, etc., which is also " in " *aither*, and which makes Air, Water, etc., transparent (§ 26).

APPENDIX B

female's " residue " with the same movement as that by which it κινεῖται itself.

(15) Thus we have an exact parallel with the action of ὀρεκτικὴ ψυχή already examined above, § 6 : ὀρεκτικὴ ψυχή sets in movement the *pneuma*, the *pneuma* sets in movement the limbs ; θρεπτικὴ (= γεννητικὴ) ψυχή sets in movement the *pneuma* in the semen, the *pneuma* in the semen sets in movement the material supplied by the female. There is also a close parallel with the art of the carpenter (730 b 15 ff.) : the carpenter, in whose ψυχή is the " form " of the chair, moves his hands and *instruments* with a movement appropriate to the object that is to be made, and they in turn move the material so as to produce the chair.[a] In all three cases no material part passes from the motive agent to the material on which it is working, but the agent imparts the " form " to the material *by means of the movement which it sets up in the instrument.* Three parallel theories.

(16) We have thus satisfied the requirement that only what is X *in actuality* can produce another X out of material which is *potentially* X : the parent which is X *in actuality* produces another X out of the female's residue which is X *potentially*, but there is an intermediary, viz., the *pneuma* in the semen, which is an instrument possessing the requisite movement, a movement which is identical throughout, in parent, semen, and embryo (see also 734 b). The semen thus is ψυχή *potentially* (735 a 8) ; and the first things which it produces *in actuality* are θρεπτικὴ ψυχή and the physical seat thereof, viz., the heart. Later it produces *in actuality* sensitive ψυχή as well. (Rational ψυχή, having no connexion with any physical substance at all, comes in independently from without ; 736 b). Heart formed first.

(17) A similar situation obtains in the case of spontaneous generation (762 a 18). Animals and plants are formed in earth and in fluid because there is water in earth, and there is *pneuma* in water, and there is Soul-heat (θερμότης ψυχικῆ) in all *pneuma* ; so that " in a way all things are full of ψυχή." Hence plants and animals quickly form once this gets enclosed ; and when this enclosing Spontaneous generation.

[a] For another such reference to *pneuma* as an instrument used by Nature, see *G.A.* 789 b 8 ff.

happens, when the corporeal liquids get heated, a sort of " frothy bubble " is formed. Now the differences between the various creatures which are produced in this way are due to the stuff which makes up the envelope around the Soul-ἀρχή (*cf.* also 738 b 34: foreign seeds produce plants varying according to the soil in which they are sown, for it is the soil that provides them with their material and their body). We can now answer the question, What corresponds in cases of spontaneous generation to the " residue " of the female and the semen of the male in cases of sexual generation ? Just as in sexual generation the female by means of its heat concocts the " residue " (the menstrual fluid) out of the nourishment, so here the heat of the season by a similar process of concoction puts into shape a substance out of the seawater and the earth (762 b 14). That which corresponds here to the male principle in sexual generation is " that portion of the Soul-ἀρχή which is enclosed in the *pneuma* " as described above ; this, just as the semen does, makes a fetation out of the material and implants movement in it.[a]

[*Note.*—It is, however, not clear in what sense there is anything in the case of spontaneous generation which is X *in actuality* (*i.e.*, which possesses the " form " of X) comparable to the parent in ordinary sexual generation. The relationship of agent and material here would appear to resemble rather that of carpenter and timber (for which see § 15) ; but even so, granted that the requisite " movement " is present, it is difficult to see whence its *specific character* is derived ; for the Sun, etc., are " motive," *not* " formal," causes (App. A § 9).

In the case of the carpenter, of course, the " form " is in the carpenter's ψυχή (§ 15). From the passage referred to in § 17 it looks as though Aristotle falls back on the surprising explanation that it is the *material* only that determines what sort of creature is to be formed. If so, then we must assume that, given the agents, or "motive" causes, viz., ψυχή, *pneuma*, and the movement therein contained, *though they are of no specific quality*, the matter is formed by them into whatever creature it happens potentially to be.

[a] *Cf.* § 12 above.

APPENDIX B

But in fact Aristotle himself is prepared to go even further than this. At *Met.* Z 1034 b 5 ff. he actually asserts that in the case of spontaneous generation of natural objects their matter can be set in movement *by itself*: it can supply itself with the same movement as that which the semen supplies (ὅσων ἡ ὕλη δύναται καὶ ὑφ' αὑτῆς κινεῖσθαι ταύτην τὴν κίνησιν ἣν τὸ σπέρμα κινεῖ). That is to say, it can supply itself with everything that in the normal way would have to be supplied by the " form " in the parent creature which is already X *in actuality*, or (in the case of *artefacta*) by the " form " in the ψυχή of the craftsman.

Perhaps Aristotle felt that this startling admission was in some degree justified by the notion that even " that out of which " animals are generated is in a sense φύσις (the ἐξ οὗ as well as the καθ' ὅ and the ὑφ' οὗ of their generation is " φύσις," *Met.* Z 1032 a 24) [a] ; and, as we know, φύσις never acts idly but always has a τέλος in view. Regarded in this way, " matter," the ἐξ οὗ of living things, might be looked upon as considerably more than mere lifeless, inert material ; and in *G.A.* Aristotle does in fact ascribe even the possession of ψυχή to it, as we have seen. Thus, to classify the statements he makes in *G.A.* : (1) The case of Testacea, which arise in sea-water. Water contains *pneuma*, and *pneuma* contains Soul-heat (§ 17). (2) The case of animals and plants spontaneously formed out of putrefying matter. Mistletoe and similar plants are formed when either the soil or certain parts in plants or trees become putrescent (715 b 27 ff.). Now (i) Earth contains Water (§ 17), and, as we saw just now (*ibid.*), Water contains *pneuma*, which contains Soul-heat. And Soul is obviously present already in the plants and trees upon which mistletoe, etc., grow. (ii) As stated in § 13 above (*G.A.* 737 a 3 ff.), the heat of the Sun and of animals can effect generation, and not only the heat of animals which operates through semen, but also *any other natural residue which there may be* has within it a principle of life. This is no doubt intended to cover putrefying animal and vegetable matter (expressly mentioned at *H.A.* 539 a 23 and 551 a 1 ff.), out of which some insects were supposed by Aristotle to arise, and " putrefying soil " as well, which would also qualify under (i) above.

A further palliative might perhaps be found in the con-

---

[a] See also the passages quoted at 741 a 1, n.

585

sideration that in the case of animals it is *sentient* Soul alone which has to be supplied by the male parent, and for plants no sentient Soul is required. Testacea, too, were considered by Aristotle to be plant-like (see 715 b 17, 731 b 8 ff., 761 a 12 ff.).]

III. THE NATURE AND PROPERTIES OF
Σύμφυτον Πνεῦμα

Semen contains *pneuma*.

(18) To repeat first what we have heard so far of the nature of ΣΠ (736 b 30 ff. ; see § 13 above) : There is in the semen of all animals the so-called θερμόν, which causes the semen to be generative. This θερμόν is not Fire, for Fire cannot generate any animal, but the heat of the Sun and of animals (the heat that operates through their semen or some other residue) can do so : for this does contain a vital principle (ζωτικὴ ἀρχή). This substance which is contained in the semen is *pneuma*, and it is "analogous to the element of the stars," viz., *aither*. One obvious way in which it is analogous to *aither* is that it is generative, for the Sun, which is of *aither*, is generative (see App. A §§ 9, 10). We shall find other points of analogy later on (§ 25).

Pneuma contains Soul-heat.

(19) In the passage 735 a 29—736 a 20 we are told that semen when it leaves the body is thick and white, because it has in it much hot *pneuma* owing to the animal's internal heat ; when the heat in the semen has evaporated and the Air has cooled, then it turns liquid and becomes dark in colour. Thus semen is a combination of *pneuma* (here described as " hot Air ") and water (κοινὸν πνεύματος καὶ ὕδατος, τὸ δὲ πνεῦμά ἐστι θερμὸς ἀήρ, 736 a 1) ; in fact, it is a foam, a mass of tiny bubbles. Similarly (762 a 20 ff.) in the case of spontaneous generation we have " a sort of frothy bubble " formed, and this too contains *pneuma*, which contains Soul-heat (see § 17) ; *cf.* too the reference to " the foam-like stuff " (736 b 36) in which, as in the semen, there is enclosed *pneuma*, and in the *pneuma* a substance analogous to the *aither*. Thus *pneuma* is closely associated with heat—a special sort of heat, not the heat of Fire ; and at 762 a 20 we read that " there is Soul-heat in all *pneuma*."

APPENDIX B

(20) Now although in all these passages the heat seems to take the chief place, as it also seems to take the leading part in the formation of embryos, Aristotle says more than once that the embryo is formed by means of cold as well as heat (see § 11 above ; 743 a, 762 b 15, etc.). And it would seem that *pneuma* really has a dual nature. This is true of it when functioning as the instrument of ὀρεκτικὴ ψυχή, and also when it is functioning as the instrument of γεννητικὴ ψυχή (see § 10 above). Thus (*M.A.* 702 a 10) the instrumental parts of the body can change from solid to fluid, soft to hard, and *vice versa*, and it is *pneuma* which brings about these changes. Aristotle tells us (703 a 22) that *pneuma* contracts and expands, and " has heaviness compared with fiery things and lightness compared with the opposite things "; and that this power of contracting and expanding is indispensable to it in view of the functions it has to perform, because the actions of movement are *pushing and pulling*.

Dual character of pneuma.

(21) *De anima* III. 433 b 18 ff. With further reference to pushing and pulling, Aristotle in a brief reference in the *De anima* to the *De motu* states that " the instrument used by ὄρεξις in causing movement " is to be found where a beginning and an end coincide, *e.g.*, at a ball-and-socket joint : one remains at rest and the other is moved : and the two though separable in definition are not separable spatially ; for everything gets moved *by pushing and pulling*. (See also *Phys.* VII. 243 a 12 ff.) Compare too *M.A.* 703 a 12 : The ΣΠ stands in a similar relation to the Soul-ἀρχή as the point in a joint (which κινεῖ κινούμενον) stands to that which is unmoved.

Pneuma effects movement by pushing and pulling.

(22) There is a passage in the *De caelo* (IV. 301 b 20 ff.), where again Aristotle is discussing the way in which movement is brought about, and although he is talking here of Air (ἀήρ) and not specifically of the kind of Air known as *pneuma*, the passage is apposite to our present subject. Now of course according to Aristotle, some of the movement which takes place in the sublunary world can be accounted for by his theory that the " simple natural substances " Fire, Air, Water, Earth have a " natural " movement (see App. A § 2). But movement is also caused *forcibly* ; and force can either

Air as an instrument for effecting movement.

587

accelerate natural movement (*e.g.*, it can make a stone go downwards more quickly than it would do naturally) or it can produce movement contrary to Nature (*e.g.*, it can make a stone go upwards) ; it is in fact the sole source of unnatural movement.   And in either case it *uses Air as its instrument* (ὥσπερ ὀργάνῳ χρῆται τῷ ἀέρι), because *Air is naturally constituted to be light and heavy* (πέφυκε καὶ κοῦφος εἶναι καὶ βαρύς) ; the Air, *qua* light, will cause an object to be carried upwards, for the Air gets pushed and receives the ἀρχή from the force which is exerting itself ; *qua* heavy, it will cause the object to be carried downwards : the force " as it were *hitches the movement on to* (ἐναφάψασα) *the Air* " and so transmits it to the object in either case.   Hence an object which is set moving forcibly (*i.e.*, contrary to Nature) continues travelling although that which set it moving does not follow it up ; and if there were no such physical substance as Air there could be no such thing as enforced movement.[a]   *In the same way*, says Aristotle, *Air gives a fair wind to* (συνεπουρίζει), helps on, *natural movement.*

Dual char- (23) This dual nature of Air is not really so surprising as it
acter of Air.   sounds at first hearing, for (*De caelo* IV. 311 b 5 ff.) *all* the physical substances possess heaviness except Fire, and they all possess lightness except Earth.   *In its own place*, each possesses heaviness, even Air ; thus, except in Water and Earth, Air possesses heaviness.   At 312 a 12 ff. Aristotle lays down that the distinction of " form " and " matter " is to be found in the category of " place " as well as in the categories of " quality " and " quantity " : thus, τὸ ἄνω belongs to the determinate, τὸ κάτω belongs to " matter."   And taking the special instance of the " matter " of " the heavy and light," *qua* potentially X it is the matter of the heavy, *qua* potentially Y it is the matter of the light : it is the same " matter," but its εἶναι is not the same (*cf.* 310 b, 311 a).

(24) For the important rôle of Air as a medium between the objects which give rise to sensations and the sense-organ,

---

[a] It should be remembered that according to Aristotle nothing can exert any effect upon ("move") another thing unless it is in contact with it ; see *Phys.* II. 244 a, b, and *G.A.* II. 734 a 3.   That is why the movement must be "hitched on" to the Air ; *cf. H.A.* VII. 586 a 17 οὐθὲν γὰρ ῥιπτεῖται πόρρω ἄνευ βίας πνευματικῆς.

and for importance of the rôle of *pneuma* in conveying the effects made upon the sense-organ to the heart and so to the ψυχή, see below, §§ 26 ff.

(25) We may now notice two other ways in which *pneuma* is " analogous " to *aither*. (*a*) We noted above (§ 6) that *pneuma* causes " movement " (both ἀλλοίωσις and spatial movement) μὴ ἀλλοιώσει, *i.e.*, without itself undergoing any qualitative change. In this respect it is similar to *aither*, for this too is not liable to any sort of " movement " (except circular φορά) ; Aristotle expressly says that *aither* is not subject to ἀλλοίωσις (*De caelo* I. 270 a 14 ff.), and he even goes so far as to suggest that it is " divine " (270 b 10). (*b*) *Pneuma*, like *aither*, acts as an intermediary between an immaterial mover and material objects. As we have seen, the unmoved mover moves the Heaven and the heavenly bodies which are made of *aither*, and the heavenly bodies in turn " move " sublunary bodies, viz., they bring about the transformation of the elements into one another, and also they bring about γένεσις and φθορά. So too the immaterial ψυχή moves *pneuma*, and *pneuma* in turn causes ἀλλοίωσις, thereby (i) moving the limbs of the body or (ii) causing the " movement " which is the development of the embryo.

Pneuma and aither analogous.

IV. THE FUNCTION OF Σύμφυτον Πνεῦμα IN SENSATION

The following outline of Aristotle's theory of Sensation will indicate the important part played in it by Air and *pneuma*. It will be seen that just as *pneuma* transmits to the parts of the body the movements caused by ψυχή and thereby produces ἀλλοίωσις and movement, so in the reverse direction it apparently transmits to ψυχή the movement of the ἀλλοίωσις caused in the sense-organs by the movements of external stimuli.

It will be convenient to divide this account into two parts :

A. dealing with what goes on outside the sentient body ;
B. dealing with what goes on inside the sentient body.

APPENDIX B

A

Vision. (26) Vision.—Vision is effected in the following way (*De anima* II. 418 a 27 ff.). There are three main factors : Colour, the medium, and the sense-organ.

" Colour " means " that which has the power to set in movement that which is *actually* transparent " (τὸ κατ' ἐνέργειαν διαφανές), and the latter acts as the medium. The medium extends continuously from the object to the sense-organ, and in its turn sets the sense-organ in movement. The medium is indispensable, because colour cannot set the sense-organ in movement direct. According to *G.A.* V. 780 b 34 ff., accuracy in seeing distant objects depends upon the movement of the medium not being dissipated, but " getting a direct passage " (εὐθυπορεῖν) ; indeed, the best results would be obtained if there were a continuous tube between the object and the eye (781 a 9). Compare the case of Hearing, § 27.

Examples of transparent media are Air, Water, and certain solids. Their transparency is due not to themselves, but to the fact that they contain a certain substance which is also found in the " eternal substance of the Upper Cosmos " (ἐν τῷ ἀιδίῳ τῷ ἄνω σώματι), *i.e.*, in the *aither*. Of this substance the *actualization* is Light ; and its actualization is brought about by the agency of Fire or something of a similar kind as the substance of the Upper Cosmos—for this selfsame substance is present in both.[a] Thus Light is essential if vision is to take place, because it is only when the substance in the medium is *actually* (not merely *potentially*) transparent that it can be set in movement by colour.

Hearing. (27) In the case of the other senses too a medium is indispensable ; one example may suffice. In Hearing there are again three main factors : the sounding object, the Air, and the sense-organ.

" A sounding object " (ψοφητικόν) means " an object which can set in movement a continuous volume of Air as far as the ἀκοή " (the organ of hearing), and the movement of the Air constitutes sound only when the

---

[a] The obscurity of this sentence is due to Aristotle's text, not to my presentation of it.

590

APPENDIX B

Air is thus set in movement as one continuous entity and is prevented from being dissipated. (This requirement necessitates that the object struck should have a smooth surface, otherwise the Air cannot be moved as a unity.) Hence here too the medium must be continuous between the sounding object and the sense-organ ; and its movement in turn sets in movement the Air in the ear (*De anima* II. 420).

B

(28) Since (*De sensu* 438 b 7) there must be light within the eye as well as in the external medium, the eye also will have to be transparent ; hence the eye, or rather that part of the eye which sees, viz., the κόρη or pupil, is made of Water (*H.A.* I. 491 b 20, *De sensu* 438 a 13 ff., *P.A.* II. 656 b 1, *G.A.* V. 779 b 23 ff.). Thus the external medium and the internal constituent are both transparent. The substance used for the eye is Water and not Air because Water is more easily kept in a confined space than Air (*De sensu* 438 a 15 ; *P.A.* II. 656 b 2). And it is of course the movement of this part *qua* transparent, not *qua* fluid, that constitutes sight (*G.A.* V. 780 a 4 ; *cf. De sensu* 438 a 13 ff.). If the fluid in the eye is already in violent movement owing to some earlier stimulus, it cannot respond to a fresh movement from without (*G.A.* V. 780 a 8 ff. ; *cf.* a 23).

(29) The sense-organ of Hearing is of Air (*De anima* III. 425 a 4 ; *cf. P.A.* II. 656 b 17 ; *G.A.* V. 781 a 23) ; and the Air in the ear is built into a chamber (ἐγκατῳκοδόμηται) in order to keep it free from disturbance (πρὸς τὸ ἀκίνητος εἶναι), so that it may take up the movements conveyed to it from without, ὅπως ἀκριβῶς αἰσθάνηται πάσας τὰς διαφορὰς τῆς κινήσεως (*De anima* II. 420 a 10; *cf.* the very similar phrase frequently used in *G.A.* V. 779 b—781 b). This Air in the ear is also described as " connate " συμφυής ; *De anima* II. 420 a 12) ; and it is this Air with which we hear. It is itself always in movement with a proper movement of its own (οἰκεία κίνησις) ; sound, however, is of course not this proper movement, but a movement derived from something else (ἀλλότριος).

Sense-organs connected to the φλέβες. (30) Now sensation arises from the heart, the seat of αἰσθητικὴ ψυχή (ἡ αἴσθησις ἀπὸ τῆς καρδίας, P.A. II. 656 b 24 ; cf. 656 a 28, III. 666 a 12, also II. 647 a 25 and G.A. II. 743 b 25), for no bloodless part has the power of sensation, nor has blood itself ; the power resides in " one of the parts that are made out of blood " (P.A. III. 666 a 17, II. 656 b 19). Hence the movement in the sense-organ must somehow be conveyed to the heart. Now it is evident that the senses of touch and taste are connected to the heart (P.A. II. 656 a 29 ; cf. De sensu 439 a 1) ; so are the others, though perhaps not so obviously and directly. Thus, from the eyes " passages " (πόροι) run to the φλέβες around the brain, and similarly from the ears a " passage " connects to the back of the head (P.A. II. 656 b 17). This is confirmed and amplified by G.A. II. 744 a 2, where smell and hearing are said to be " passages " full of ΣΠ, connecting with the external Air, and terminating at the φλέβια which come from the heart and extend around the brain.

Φλέβες connected to the heart, the source of the connate pneuma. (31) In the passage of G.A. V. 781 a 23 ff., which is perhaps out of place and possibly slightly corrupt, some important statements are fortunately clear. We read there that the " passage " of the organ of hearing terminates in the region where the ΣΠ produces the pulsation (deriving, as will be seen, from the heart) ; and we also read of the " movement " which comes through the sense-organ of hearing (presumably to its destination in the heart) being reproduced again through the voice ; at any rate, it is clear that the heart is the ἀρχή of the voice (IV. 776 b 12 ; cf. V. 787 b—788 a). Further details about the pulsation are given in De resp. 479 b 30 ff. Pulsation, says Aristotle, is similar to boiling, which occurs when fluid substance is pneumatized by τὸ θερμόν : the fluid rises up owing to increase of bulk. Pulsation is produced in the heart by the increase of bulk, caused by heat, of the fluid which is continually being supplied to the heart from the nourishment. This action goes on continuously, because the blood is fashioned first of all in the heart, and the inflow of the fluid out of which the blood is produced goes on continuously. And all the φλέβες pulsate too, simultaneously with each other, because they are all connected to the heart. Pulsation is, in fact, "the pneumatization of the fluid as it gets heated."

(32) This seems to give us the key to the theory of sensation as well as the explanation of the upkeep of the ΣΠ. The fluid, as it gets heated and thereby concocted and turned into blood, is "pneumatized." This no doubt implies that the *pneuma* which is already present in the fluid (as it is in any fluid ; see § 17 above), and which contains Soul-heat, acquires some special character or rather "movement" by being brought into contact with the heart, and with the Soul which has its seat there and whose "instrument" the *pneuma* is destined to become ; indeed, we must assume this, because semen contains the *pneuma* which possesses the specific "movement" that is to fashion the embryo (§§ 9, 14 above), and it is from blood that semen is made by further concoction. Hence blood will contain ΣΠ, and we may say that all the φλέβες are instinct with ΣΠ as well as with blood. Hence there is continuity of ΣΠ (or of "the substance similar to *aither*," if this is really to be distinguished from ΣΠ) from the sense-organ, through the "passages" and then the φλέβες, right up to the heart. We have Aristotle's explicit statement that the "passages" of smell and hearing, which are full of ΣΠ, terminate at the φλέβια which come from the heart, and that the "passage" from the eyes does so too. And the φλέβες of course pulsate owing to the "pneumatizing" action set up in the heart.

Continuity of the connate pneuma from sense-organ to heart.

(33) As Beare says on the last page of his book, *Greek Theories of Elementary Cognition* (p. 336), "if we could discover all the properties and functions of ΣΠ, we should have penetrated to the inmost secrets of sense-perception" as envisaged by Aristotle ; for "the ΣΠ was the profoundest cause and the most intimate sustaining agency from the beginning to end of life and sensory power."

Conclusion.

INDEX

The Index is to be regarded as supplementary to the Contents-Summary on pp. lxxi ff. ; see also the Introduction and Appendix.

The method of reference is this :

Roman numerals refer to pages of the Preface.

I denotes paragraphs of the Introduction.

A and B denote paragraphs of Appendix A and B.

The numbers 15a to 89b (standing for 715a to 789b) refer to the pages and columns of the Berlin edition which are printed at the top of each page of the Greek text. The lines are referred to in units of five lines : thus

$$17a1 = 717a1-4$$
$$17b5 = 717b5-9$$

f, ff = following section(s) of five lines, following page(s) etc., as the case may be.

In the text references, each entry is separated from the preceding one by a dash (/), unless they both have the same Berlin page-number.

References throughout include footnotes. (This applies equally to the entries which refer to the Greek text. For example the mention of W. W. Jaeger in the footnote to 719a11 is listed as 19a10.)

INDEX

INDEX

INDEX

597

INDEX

INDEX

INDEX

natural=general or normal
I 8, 12 f / 27b25 / 70b10 ff /
72a35 / 77a15 f
— science 48a10 ; *see also*
physiologers
Nature, natural I 8, 12 ff /
24b20 ff / 41a1 / 44b20 /
52b15 / 53a5 / 65a5 /
70b10 / 76a1 / 77b15 ff /
81b20 / A 12 / *et passim*
—=prime matter 29a30
— and Art 34b20 / 35a1 /
62a15 / 75a20
— compared to a carpenter
30b15 / 43a25
— — clay-modeller 30b25
— — cook 76a1 f
— — craftsman 31a20 /
89b1 ff
— — housekeeper 44b15
— — painter 43b20
— — runner 41b20
— does nothing idly 39b20 /
41b1 / 44a35 / 88b20 f
Necessity, necessary I 5 ff /
17a15 f / 17b35 / 42a15 /
44a10, b10 / 67b10 / A 14
— *versus* Better I 5 ff /
31b20 f / 38a30 ff / 39b25 /
43a35 ff / 55a20 / 76a15
ff, b30 / 78a30 ff, b15 /
82a20 / 89a5 f, b1 ff
Needham, J. x, xi / 33b20 /
63b30
net 34a20
night-blindness 80a15 f
north and south 67a10 /
83a30
Notidanus griseus 16b25
" nourishment " 24b25 ff /
40a25 ff / 76a25, b5 / 77a1 /
86b1

— fluid 53b25 / 67a30
— grades of I 64 / 28a30 /
40b30 f / 44b10 ff / 51b5,
20 / 52b15 / 53b10 /
62b15 f / 66a10 / 70a20 /
77a5
— stages of I 61 ff / 25a10 ff,
b10 / 26b1 ff / 28a15 /
40a20 / 65b25 f / 66a30,
b5 f / 76a25 f / 84a35 /
86b1 / B 31
" number " 78a5 / A 11-13

observation, importance of
viii / 60b30 / 88b10 ff
Octopus 17a5 / 20b30 ff
offspring, mutilated 21b15 /
24a1 f
— same or different in
kind as parents 15a20ff /
23b1 f
oil 35b10 ff / 85a30
oil-flotation 35b15
olive and oleaster 55b10
onion 61b30
opposites (contraries) 24b1 /
66a10
optic nerves 44a10
ore 35b15 / 61b15
organ : *see* instrument
" organizers " I 55
Orpheus 34a15
os uteri 39a30 f / 73a15
ostrich 49b15 / 52b30
ourion 53a20, 30, b5
oviducts : *see* uterus
Ovovivipara 18b30 / 20a15
ovum, mammalian xii / I 77 /
27b30
ox 79a30 / 85b20 f
ox-fish 16b25
oysters 63a30 f

603

INDEX

Poland 83a10
poplar 26a5
Prächter, K. xvi
" preformation " x / 33b25 ff
primum mobile A 3
" principle " I 11, 51 / 16b1, 10 / 40a1 ff / 41b15, 25 / 42b1 ff / 51b5, 20 / 52a10 ff / 62a25, b15 / 63b20 f / 65b10 / 66a15, 25 f / 78a5 / 88a10 f / *et passim*
— of male 57b10
— of movement 15a5 / 34b20 / 42b35 / 44a30
" prior," meaning of 42a20 ff
protoplasm 36a10
pulsation 81a25 / B 31
pupa 58b30
puppets 34b10 / 41b5
purpura 61b30 / 63b5
pus 77a10
pushing and pulling B 20 f
putrefaction 15a20, b25 / 21a5 / 53a30 / 62a10, b25 / 77a10 / 84b5 ff
pygmies 49a1
Pyrrha 63b1

quadrupeds, eggs of oviparous 52b30 ff
qualitative change : see " alteration "
quintessence vi ; *see also aither* ; fifth element

rain A 8
raven 56b15 ff / 85b35
reason I 44 ; *see* Soul, rational
" recapitulation " x

redundance of parts 70b25 ff
regeneration 74b30
" relapsing " 68a15 ff
rennet I 54 / 29a10 / 39b20 f / 72a20
reproduction A 17
resemblance to parents 21b20 ff / 22a15 ff / 26a10 / 38b30 / 64b25 / 66b10 / 67a35 ff / 69a5 ff / 70b5
" residue " I 8, 20, 64 ff / 15a20 / 19b30 / 24b25 ff / 28b15 ff / 37a1, b25 ff / 39b1 / 43a25 / 45b15 f / 49b1 ff / 62a1 f, b1 ff / 76b10 / 83a25 / B 13
respiration 81a25
rhine 46b5
rhinobates 46b5
Rhodes 63a30
Richards, H. xxviii
Rickard, T. A. 35b15
right hotter than left 65b1
Rose, V. 19a10
Ross, W. D. viii
Rudberg, G. xxiii, xxx
Rueff, Jacob 27b30
Russell, E. S. x
Russia 83a10

Saint-Hilaire, J. Barthélemy- xxxii f
Saint-Hilaire family xx
salamander 61b15 ·
Sarmatia 83a10
satyriasis 68b35
Schmidt, J. p. 565
Schneider, J. G. xxviii f
Scot, Michael xxi, xxiii f, xxix ff
Scythia 48a25 / 82b30 / 83a10

605

INDEX